3D Bioprinting for Reconstructive Surgery

Related Titles

Rapid Prototyping of Biomaterials: Principles and Applications
(ISBN 978-0-85709-599-2)

3D Bioprinting and Nanotechnology in Tissue Engineering and Regenerative Medicine (ISBN 978-0-12-800547-7)

Essentials of 3D Fabrication and Translation
(ISBN 978-0-12-800972-7)

Biofabrication: Micro and Nano-fabrication, Printing, Patterning and Assemblies
(ISBN 978-1-4557-2852-7)

3D printing in Medicine
(ISBN 978-0-08-100717-4)

3D Bioprinting for Reconstructive Surgery

Techniques and Applications

Edited by

Daniel J. Thomas

Zita M. Jessop

Iain S. Whitaker

Woodhead Publishing is an imprint of Elsevier
The Officers' Mess Business Centre, Royston Road, Duxford, CB22 4QH, United Kingdom
50 Hampshire Street, 5th Floor, Cambridge, MA 02139, United States
The Boulevard, Langford Lane, Kidlington, OX5 1GB, United Kingdom

Library of Congress Cataloging-in-Publication Data
A catalog record for this book is available from the Library of Congress

British Library Cataloguing-in-Publication Data
A catalogue record for this book is available from the British Library

ISBN: 978-0-08-101103-4 (print)
ISBN: 978-0-08-101216-1 (online)

For information on all Woodhead publications visit our website at https://www.elsevier.com/books-and-journals

Publisher: Matthew Deans
Acquisition Editor: Laura Overend
Editorial Project Manager: Natasha Welford
Senior Production Project Manager: Priya Kumaraguruparan
Cover Designer: Greg Harris

Typeset by SPi Global, India

Contents

List of contributors

E. Abelardo Swansea University Medical School, Swansea, United Kingdom

A. Al-Sabah Reconstructive Surgery and Regenerative Medicine Research Group, Swansea University Medical School, Swansea, United Kingdom

M.P. Chae Monash University, Clayton; Peninsula Health, Frankston, VIC, Australia

C.-W. Chen Institute of Polymer Science and Engineering, National Taiwan University, Taipei, Taiwan, ROC

E. Combellack Reconstructive Surgery and Regenerative Medicine Research Group, Swansea University Medical School; The Welsh Centre for Burns and Plastic Surgery, Morrison Hospital, Swansea, United Kingdom

D. Eggbeer PDR, Cardiff Metropolitan University, Cardiff, United Kingdom

M.W. Findlay Peninsula Health, Frankston; University of Melbourne, Parkville; The Peter MacCallum Cancer Center, Parkville, VIC, Australia

J.-Z. Fu Zhejiang University, Hangzhou, China

X. Fu General Hospital of PLA, Beijing, China

Q. Gao Zhejiang University, Hangzhou, China

N. Gao Edinburgh University Medical School, Edinburgh, United Kingdom

M. Gelinsky Dresden University of Technology (TU Dresden), Dresden, Germany

Y. He Zhejiang University, Hangzhou, China

S.-h. Hsu Institute of Polymer Science and Engineering, National Taiwan University, Taipei, Taiwan, ROC

N. Hu Harvard Medical School, Cambridge, MA, United States

S. Huang General Hospital of PLA, Beijing, China

D.J. Hunter-Smith Monash University, Clayton; Peninsula Health, Frankston, VIC, Australia

A. Ibrahim Institute of Child Health, London, United Kingdom

Z.M. Jessop Reconstructive Surgery and Regenerative Medicine Research Group, Swansea University Medical School; The Welsh Centre for Burns and Plastic Surgery, Morrison Hospital, Swansea, United Kingdom

X. Jin University of British Columbia, Kelowna, BC, Canada

T.H. Jovic Reconstructive Surgery and Regenerative Medicine Research Group, Swansea University Medical School; The Welsh Centre for Burns and Plastic Surgery, Morrison Hospital, Swansea, United Kingdom

K. Kim University of British Columbia, Kelowna, BC, Canada

S. Kyle Reconstructive Surgery and Regenerative Medicine Research Group, Swansea University Medical School, Swansea, United Kingdom

Z. Li General Hospital of PLA, Beijing, China

P. Li University of Sussex, Brighton, United Kingdom

S. Manivannan Cardiff University Medical School, Cardiff, United Kingdom

S.V. Murphy Wake Forest University School of Medicine, Winston-Salem, NC, United States

J. Rosser Welsh Centre for Printing and Coating, Swansea University, Swansea, United Kingdom

K. Sakthivel University of British Columbia, Kelowna, BC, Canada

D. Singh Yale University, New Haven, CT, United States

S.P. Tarassoli Swansea University Medical School, Swansea, United Kingdom

D.J. Thomas 3Dynamic Systems Group, Swansea, United Kingdom

C. Thornton Department of Human Immunology, Swansea University Medical School

Z. Wang University of British Columbia, Kelowna, BC, Canada

I.S. Whitaker Reconstructive Surgery and Regenerative Medicine Research Group, Swansea University Medical School; The Welsh Centre for Burns and Plastic Surgery, Morrison Hospital, Swansea, United Kingdom

F.F. Yang Zhejiang University, Hangzhou, China

Y.S. Zhang Harvard Medical School, Cambridge, MA, United States

H.M. Zhao Zhejiang University, Hangzhou, China

Foreword

Tissue engineering is an emerging multidisciplinary field, one that involves biology, medicine, materials science, and manufacturing in ways that will revolutionize how we improve the health and quality of life for billions of people worldwide. This interface between engineering and medicine opens up exciting new avenues of translational research, including (a) engineered tissues capable of transplantation and regeneration within a patient; (b) diagnostic screens where engineered tissues can be used in place of costly, inaccurate animal studies to test drug metabolism and uptake, as well as toxicity and pathogenicity; and (c) pathology studies, where libraries of diseased tissues that would ordinarily be difficult to obtain could be precisely created and examined.

To actualize the promises of tissue engineering, novel approaches must be taken to accelerate the creation of anatomically correct tissue constructs. Thus, we consider three-dimensional (3D) bioprinting, which uses biomaterials, cells, proteins, and other biological compounds as building blocks to manufacture additively 3D tissue models that can mimic normal and pathological physiology. 3D bioprinting has spawned a collection of emerging technologies, and their synergistic integration can not only redefine the clinical capabilities of regenerative medicine, but also transform the toolsets available for drug discovery and fundamental research in the biological sciences.

This book—*3D Bioprinting for Reconstructive Surgery: Techniques and Applications*—provides a comprehensive discussion on the biomaterials, techniques, and applications crucial to structural regenerative medicine and 3D constructive tissue engineering. Important techniques such as 3D computational design, stem cell preparation, bioreactors, 3D bioprinting, and navigating regulatory issues are reviewed by experts in their respective fields. Biomaterials issues, such as the formulation of bio-inks, hydrogels, and other biopolymers are addressed in detail. Applications to connective tissues such as bone, cartilage, fat, muscle, skin, and blood vessels are discussed by the collaborative groups that use them. I believe that these important topics define an exciting, emerging field of 3D bioprinting, and connective tissue engineering, and they will spur innovation in all fields of precision medicine, from research and development, to education and industrial growth.

Shaochen Chen
University of California, San Diego, CA, United States

Preface

Advances in three-dimensional (3D) bioprinting technologies have created a new frontier in medicine. Bioprinting is evolving as a process, which has the possibility to generate significant breakthroughs, yielding new treatments, and change the foundation of reconstructive medicine. Over the recent years, 3D bioprinting has generated significant research interest among scientists, engineers, and medical practitioners. This rapidly evolving technology is now being harnessed to engineer tissue systems that offer tremendous potential for subsequent transplantation. It is the paramount 3D bioprinting processes as described within this book, which can be used to build transplantable tissues.

3D bioprinting technology is now becoming a powerful application of additive manufacturing technology. It offers the potential to fabricate organized tissue constructs to repair and/or replace damaged or diseased human tissues. This direct bearing for developing safer and more effective healthcare treatments. This is a top-down method for engineering new ways in making treatments more patient centred. It also opens an opportunity for cost-effective patient-specific medicine to evolve. The potential of producing functional tissues on demand, bioprinted in a controlled and safe way for use in humans may one day revolutionize healthcare and have a huge global impact on healthcare and economy.

3D bioprinting technology is currently being progressed as a method to engineer complex tissues and organs. This rapidly growing technology allows for precise placement of multiple types of cells, biomaterials, and biomolecules in spatially predefined locations within 3D structures. Many researchers are focussed on the further development of bioprinting technology and its applications. In this book, we introduce the general principles and limitations of widely used bioprinting systems and applications for tissue and organ regeneration. In addition, the current challenges facing the clinical applications of bioprinting technology are addressed.

Initial methods using the deposition of porous scaffolds biomaterials have been an essential foundation component used for tissue biofabrication. These processes for the production of viable tissue systems have had limited success. This is due to issues with the fabrication of 3D scaffolds with uniformity and microarchitecture deposition, while maintaining interconnection within the structure. This process has therefore been superseded with the use of 3D printing methods. These have been developed as methods to produce complex 3D constructs precisely by deposition biomaterials during fabrication of 3D structure. It is the flexibility of 3D printing technologies that allows for patient-specific structures, which are fabricated from computerized tomography and magnetic resonance imaging data. As a result, we are able to biofabricate structures that have the correct geometrical shape prior to an operative procedure being carried out.

Traditional scaffolds have been used successfully in numerous clinical applications over the past number of years. However, there are strict limitations, which exist in the fabrication of complex and composite tissues. Our inability to control precisely cell migration throughout 3D scaffolds has been one of the key limitations. These compounded by allowing for the deposition of different cell types at critical locations is a further complication. 3D biomanufacturing processes that allow for improved precision and the ability to position cells is critical for future progress toward the generation of complex tissues for surgical transplantation.

These requirements have resulted in the introduction of 3D bioprinting technology as an advanced method that has the possibility to create complex tissues. 3D bioprinting overcomes the limitations in scaffold-based tissue engineering methods by placing multiple cell types and biomaterials in a 3D area at defined locations. Furthermore, tissue constructs can be formed by using methods of generating self-assembling bioprinted cells. Although bioprinting technology represents a new technique in tissue engineering, it is now being used by research teams across the world as a viable method for producing tissue systems. This is because of its ability to spatially position living cells during the biofabrication of a tissue construct.

3D bioprinting technology is now being refined to deposit precisely multiple layers of hydrogels seeded with different cell types to produce tissue structures in 3D. As a result, this technology has emerged as a new tool for the fabrication of 3D tissues. This book discusses the principles of bioprinting technology and the current exciting research being undertaken to generate tissue for use in future reconstructive surgery.

Introduction: Inception, evolution and future of 3D bioprinting

1

Z.M. Jessop, D.J. Thomas†, I.S. Whitaker**
*Reconstructive Surgery and Regenerative Medicine Research Group, Swansea University Medical School, Swansea, United Kingdom, †3Dynamic Systems Group, Swansea, United Kingdom

Regenerative medicine and tissue engineering, underpinned by the core principles of rejuvenation, regeneration, and replacement, are shifting the paradigm in healthcare from symptomatic treatment of the 20th century to curative treatment of the 21st century [1–3]. The current dilemmas for modern day healthcare, such as an aging population and the increasing prevalence of chronic disease require solutions that limit organ dysfunction and tissue degeneration and potentially offer replacement [4]. This was first addressed through transplantation, a field that advanced rapidly in the 1950s through a combination of surgical innovations and fundamental scientific breakthroughs in immunosuppression [5]. In contrast to the allogenic replacement of transplantation, tissue engineering seeks to apply stem cell research with developmental biology principles to regenerate cells, tissues, and organs de novo [6,7]. Regenerative medicine and tissue engineering have been recognized worldwide as "emerging disciplines that hold promise of revolutionizing patient care in the 21st century" [1,2,8].

The principles of tissue engineering are underpinned by the incorporation of cells with biodegradable scaffolds to engineer replacement tissues. Clinically used examples range from individual tissues such as dermis or cartilage [9] to organs such as the trachea and bladder [10,11]. The long-term outcomes have been mixed. Limitations of synthetic polymer scaffolds, such as infection, extrusion, and degradation product toxicity have encouraged interest in decellularized matrices as well biologics for use as scaffolds, as one of the more effective ways of replicating native tissue anisotropy [1,2,12,13]. Decellularized matrices provide durability, enhanced integration, and biocompatibility whilst avoiding allosensitization [11,13]. This may explain why many of the significant breakthroughs and first in man studies have utilized this technique combined with autologous cell-seeding with varying success [10,11,14,15] and even showed promise in vitro for more complex structures such as pulmonary and aortic valves as well as whole organs such as heart and liver [16–18]. Despite early interest and investment in tissue-engineering research, with annual R&D spending estimated at US$580 million [19], initial clinically applicable product release has been slow but steady [20].

The ability to print biological "inks," rather than the plastic and metal inks of traditional 3D printing, has resulted in the birth of the exciting new bioprinting research field [21]. The global 3D bioprinting market was estimated to be $487 million in 2014 and this is predicted to reach $1.82 billion by 2022 [22]. The bioprinter, used to dispense "bioinks," consisting of cells, scaffolds, and biomolecules in a spatially controlled manner, gives multiple advantages over traditional tissue-engineering methods

3D Bioprinting for Reconstructive Surgery. https://doi.org/10.1016/B978-0-08-101103-4.00001-6

of assembly, consisting of nonspecific cell seeding of scaffolds [23]. By controlling the nano-, micro- and macrostructure, 3D bioprinting may replicate complex native-like tissue architecture more faithfully in the laboratory [24,25]. The ability to bioprint physiologically relevant multicellular tissue constructs on demand would obviate the need for autologous tissue harvest and dependency on organ donors as well as transform reconstructive surgery [26,27].

The success of this platform technology ultimately depends not only on the process itself, but answers to the fundamental scientific questions regarding the correct blend of cell source, suitable scaffold, and ideal microenvironment [25]. The potential benefits of bioprinting over other types of tissue assembly include repeatability, customization (personalized medicine), the incorporation of channels for vascularization, high-resolution manufacture, automation, and ability to scale-up production [28,29]. These features may provide the key for successful clinical translation. Given the future potential and synergistic goals of bioprinting and reconstructive surgery in restoring "form and function" [30], we propose that surgeons together with cell biologists, material scientists, computer scientists, and engineers, should be well versed in the principles and intimately involved in the future developments of 3D bioprinting to ensure it maintains clinical applicability [25].

Medical breakthroughs require the convergence of multiple scientific advances for which interdisciplinary collaboration is fundamental. Similar to transplant medicine, tissue engineering through bioprinting requires the convergence of a number of complementary technological advances such as stem cell biology, biomaterial science, reconstructive surgery, biochemistry, computer science, engineering, rheology as well as improved understanding of developmental and molecular biology, before the emergence of a new era in healthcare research.

Despite the significant worldwide laboratory research in the field, there are few reports of successful translation into surgical practice. This text, written by international experts in their respective fields, is a synopsis or current knowledge. The advantages of 3D bioprinting over traditional tissue-engineering techniques in assembling cells, biomaterials and biomolecules in a spatially controlled manner to reproduce native tissue macro-, micro- and nanoarchitectures are discussed, together some examples of ongoing work related to a range of tissue types.

If successful, 3D bioprinting has the potential to manufacture autologous tissue for reconstruction, remove the need for donor tissues, and transform personalized medicine [1,2,31]. We believe this text will be another step towards overcoming the biological, technological, and regulatory challenges ahead by encouraging an integrated approach from a variety of fields and will be a valuable reference book for clinicians, biomaterials scientists, biomedical engineers, and students who wish to broaden their knowledge of 3D bioprinting technology.

References

[1] Jessop ZM, Al-Sabah A, Francis WR, Whitaker IS. Transforming healthcare through regenerative medicine. BMC Med 2016;14:115.

[2] Jessop ZM, Javed M, Otto IA, Combellack EJ, Morgan S, Breugem CC, et al. Combining regenerative medicine strategies to provide durable reconstructive options: auricular cartilage tissue engineering. Stem Cell Res Ther 2016;7:19.

[3] Nelson TJ, Behfar A, Terzic A. Strategies for therapeutic repair: the "R3" regenerative medicine paradigm. Clin Transl Sci 2008;1:168–71.
[4] Trounson A. New perspectives in human stem cell therapeutic research. BMC Med 2009;7:29.
[5] Calne R. The history and development of organ transplantation: biology and rejection. Baillieres Clin Gastroenterol 1994;8:389–97.
[6] Haseltine W. Regenerative medicine 2003: an overview. J Regen Med 2003;4:15–8.
[7] Mironov V, Visconti RP, Markwald RR. What is regenerative medicine? Emergence of applied stem cell and developmental biology. Expert Opin Biol Ther 2004;4:773–81.
[8] O'Dowd A. Peers call for UK to harness "enormous" potential of regenerative medicine. BMJ 2013;347:f4248.
[9] Cao Y, Vacanti JP, Paige KT, Upton J, Vacanti CA. Transplantation of chondrocytes utilizing a polymer-cell construct to produce tissue-engineered cartilage in the shape of a human ear. Plast Reconstr Surg 1997;100:297–304.
[10] Atala A, Bauer SB, Soker S, Yoo JJ, Retic AB. Tissue-engineered autologous bladders for patients needing cystoplasty. Lancet 2006;367:1241–6.
[11] Macchiarini P, Jungebluth P, Go T, Asnaghi MA, Rees LE, Cogan TA, et al. Clinical transplantation of a tissue-engineered airway. Lancet 2008;372:2023–30.
[12] Badylak SF, Gilbert TW. Immune response to biologic scaffold materials. Semin Immunol 2008;20(2):109–16. Epub 2008 Feb 20.
[13] Hoshiba T, Lu H, Kawazoe N, Chen G. Decellularized matrices for tissue engineering. Expert Opin Biol Ther 2010;10(12):1717–28.
[14] Gonfiotti A, Jaus MO, Barale D, Baiguera S, Comin C, Lavorini F, et al. The first tissue-engineered airway transplantation: 5-year follow-up results. Lancet 2014;383(9913):238–44.
[15] Hollander A, Macchiarini P, Gordijn B, Birchall M. The first stem cell-based tissue-engineered organ replacement: implications for regenerative medicine and society. Regen Med 2009;4(2):147–8.
[16] Knight RL, Wilcox HE, Korossis SA, Fisher J, Ingham E. The use of acellular matrices for the tissue engineering of cardiac valves. Proc Inst Mech Eng H 2008 Jan;222(1):129–43.
[17] Ott HC, Matthiesen TS, Goh SK, Black LD, Kren SM, Netoff TI, et al. Perfusion-decellularized matrix: using nature's platform to engineer a bioartificial heart. Nat Med 2008;14(2):213–21.
[18] Soto-Gutierrez A, Zhang L, Medberry C, Fukumitsu K, Faulk D, Jiang H, et al. A whole-organ regenerative medicine approach for liver replacement. Tissue Eng Part C Methods 2011;17(6):677–86.
[19] Kratz G, Huss F. Tissue engineering—body parts from the Petri dish. Scand J Surg 2003;92(4):241–7.
[20] Kemp P. History of regenerative medicine: looking backwards to move forwards. Regen Med 2006;1(5):653–69.
[21] Mironov V, Reis N, Derby B. Review: bioprinting: a beginning. Tissue Eng 2006;12(4):631–4.
[22] Grand View Research, 3D Bioprinting Market Analysis By Technology (Magnetic Levitation, Inkjet Based, Syringe Based, Laser Based), By Application (Medical, Dental, Biosensors, Consumer/Personal Product Testing, Bioinks, Food And Animal Product Bioprinting) and Segment Forecast; San Francisco, 2015.
[23] Vacanti CA. The history of tissue engineering. J Cell Mol Med 2006;10:569–76.
[24] Chia HN, Wu BM. Recent advances in 3D printing of biomaterials. J Biol Eng 2015;9:4.
[25] Jessop ZM, Al-Sabah A, Gardiner M, Combellack E, Hawkins K, Whitaker IS. 3D Bioprinting for Reconstructive Surgery: Principles, Applications and Challenges. J Plast Reconstr Aesthet Surg 2017;70(9):1155–70.

[26] Jessop ZM, Al-Himdani S, Clement M, Whitaker IS. The challenge for reconstructive surgeons in the twenty-first century: Manufacturing tissue-engineered solutions. Front Surg 2015;2:52.

[27] Murphy SV, Atala A. 3D bioprinting of tissues and organs. Nat Biotechnol 2014;32(8):773–85.

[28] Mironov V, Visconti RP, Kasyanov V, Forgacs G, Drake CJ, Markwald RR. Organ printing: tissue spheroids as building blocks. Biomaterials 2009;30(12):2164–74.

[29] Sun W, Darling A, Starly B, Nam J. Computer-aided tissue engineer-ing: overview, scope and challenges. Biotechnol Appl Biochem 2004;39:29–47.

[30] Ingber DE. Mechanical control of tissue growth: function follows form. PNAS 2005;102:11571–2.

[31] Al-Himdani S, Jessop ZM, Combellack E, Naderi N, Zhang Y, Ibrahim A, et al. Tissue Engineered Solutions in Plastic and Reconstructive Surgery: Principles and Practice. Front Surg 2017;4:4.

PART A

3D Bioprinting Techniques

Practical laboratory methods for 3D bioprinting

2

Q. Gao, H.M. Zhao, F.F. Yang, J.-Z. Fu, Y. He
Zhejiang University, Hangzhou, China

2.1 Introduction

Due to its ability of fabricating complex structures and the weak relationship between quantity and cost, three-dimensional (3D) printing is now widely used in many areas such as aerospace, architecture, industrial design, education, medicine, and biology [1–7]. As there are numerous applications that are needed to be customized, 3D bioprinting has experienced explosive growth [8–10]. Bioprinting is focused on how to precisely control the deposition of biological materials (such as ECM proteins, cells, and cell-matrix suspensions) into 3D structures [11]. Many 3D bioprinting methods are developed from extrusion-based bioprinting to photocuring-based bioprinting. Numerous biomaterials can be printed to acquire scaffolds and cell-laden structures such as gelatin, alginate, silk fibroin, fibrin, and collagen [12–20]. Many printed applications are demonstrated, which include bones, ears, vessels, skins, kidneys, and livers [21–28]. Although 3D bioprinting is developing rapidly, it still faces many challenges, for instance how to mimic the cell-cell interface of the real organ structures and how to reconstruct the complex vascular networking in the printed organs.

In the last decade, the number of publications on bioprinting has increased rapidly, as shown in Fig. 2.1, and some important progress has been made [29–37]. As different 3D bioprinting methods have different resolution and fabrication ability, in this chapter, we try to give a practical introduction about 3D bioprinting with a special concern on how to use them in the laboratories. First, the 3D bioprinting methods are briefly reviewed. Then, the biomaterials used in bioprinting are presented, including the discussion on the printability of hydrogels. Next, we discuss the components of a bioprinter and how to assemble a simple one in the laboratory, and a typical extrusion-based bioprinting process was described. At last, we give some future trends of 3D bioprinting.

2.2 Overview of 3D bioprinting methods

3D bioprinting is also known by the names, cell printing and organ printing. Generally speaking, if a biomaterial is used in 3D printing, this process can be called 3D bioprinting, while cell-laden biomaterials are usually used in cell printing or organ printing. In this chapter, we mainly focused on the bioprinting methods that can directly print cell-laden structures. In other words, the three terms can replace one another.

3D Bioprinting for Reconstructive Surgery. https://doi.org/10.1016/B978-0-08-101103-4.00003-X

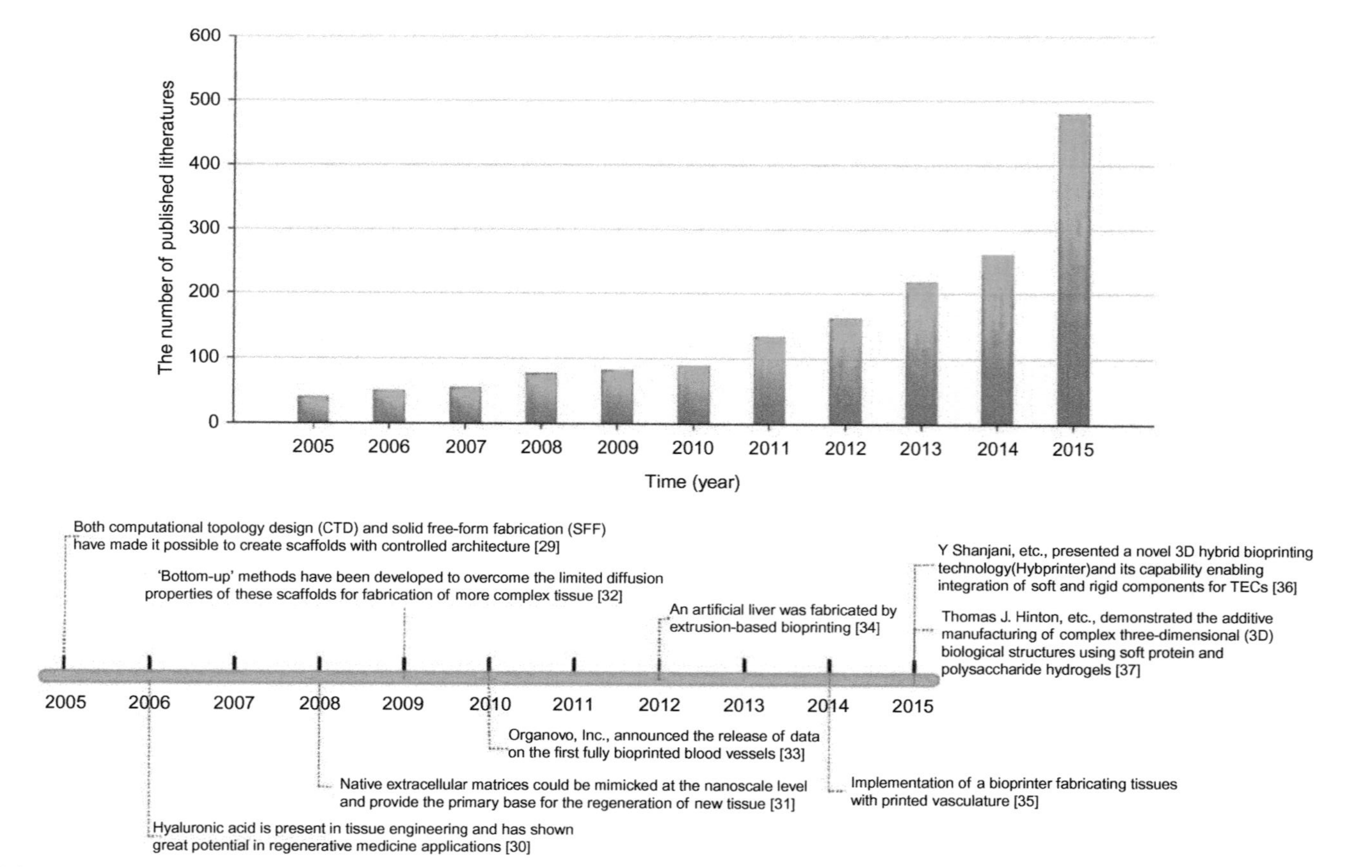

Fig. 2.1 Number of publications related to "3D bioprinting" or "cell printing" or "organ printing" according to ISI Web of Science and Google Scholar (data obtained in September 2016), and some important progress is given.

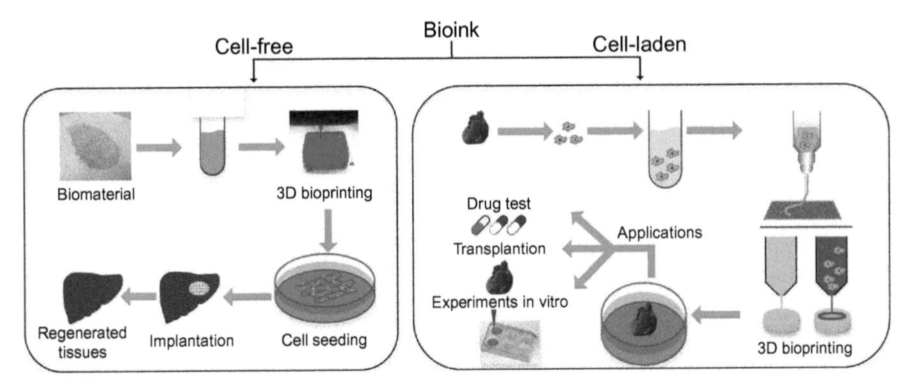

Fig. 2.2 Comparison of cell-free printing and cell-laden printing.

Compared with traditional printing of tissue engineering scaffolds, as shown in Fig. 2.2, in bioprinting, live cells are directly disposed mixed with biomaterials, while cells are seeded onto the scaffold afterwards in cell-free printing [38–40]. According to different material distribution and forming ways, bioprinting could be divided into inkjet-based bioprinting, extrusion-based bioprinting, laser direct writing bioprinting, photocuring-based bioprinting, and cell ball assembling bioprinting.

2.2.1 Inkjet-based bioprinting

Inkjet-based bioprinting, introduced in 2003, can be regarded as the first bioprinting technology [41]. Similar to conventional 2D inkjet printing, bioinks (mixture of cells and precursor gels) are printed as a series of droplets with a piezoelectric- or thermal-driven nozzle, as shown in Fig. 2.3A. After a layer-by-layer printing, a 3D cell-laden structure can be printed. As microdroplets are the basic units to construct the 3D object, it is very important to ensure the gluing or fusing together of droplets.

As the commercial inkjet printer is easily accessible, it is a low-cost bioprinting method. Another merit of this method is that different types of cells can be easily printed, which is suitable for printing cell-cell interfaces. Many researchers have reported studies on cell viability [42], droplet formation with cell-laden bioink [43], and how to print complex structures [44–47].

However, one disadvantage limits its wide use. Due to the small squeezing force of the inkjet nozzle, the viscosity of the bioink is limited and high-density cells cannot be printed. Low viscous ink will limit the structures' strength and the weak strength of the printed structures will be difficult to perfuse or implant. Moreover, owing to the low viscosity of bioinks, the choice of suitable biomaterials is restricted to a narrow range. Another problem that should be addressed is the potential mechanical or thermal cell damages that can occur during inkjetting [48].

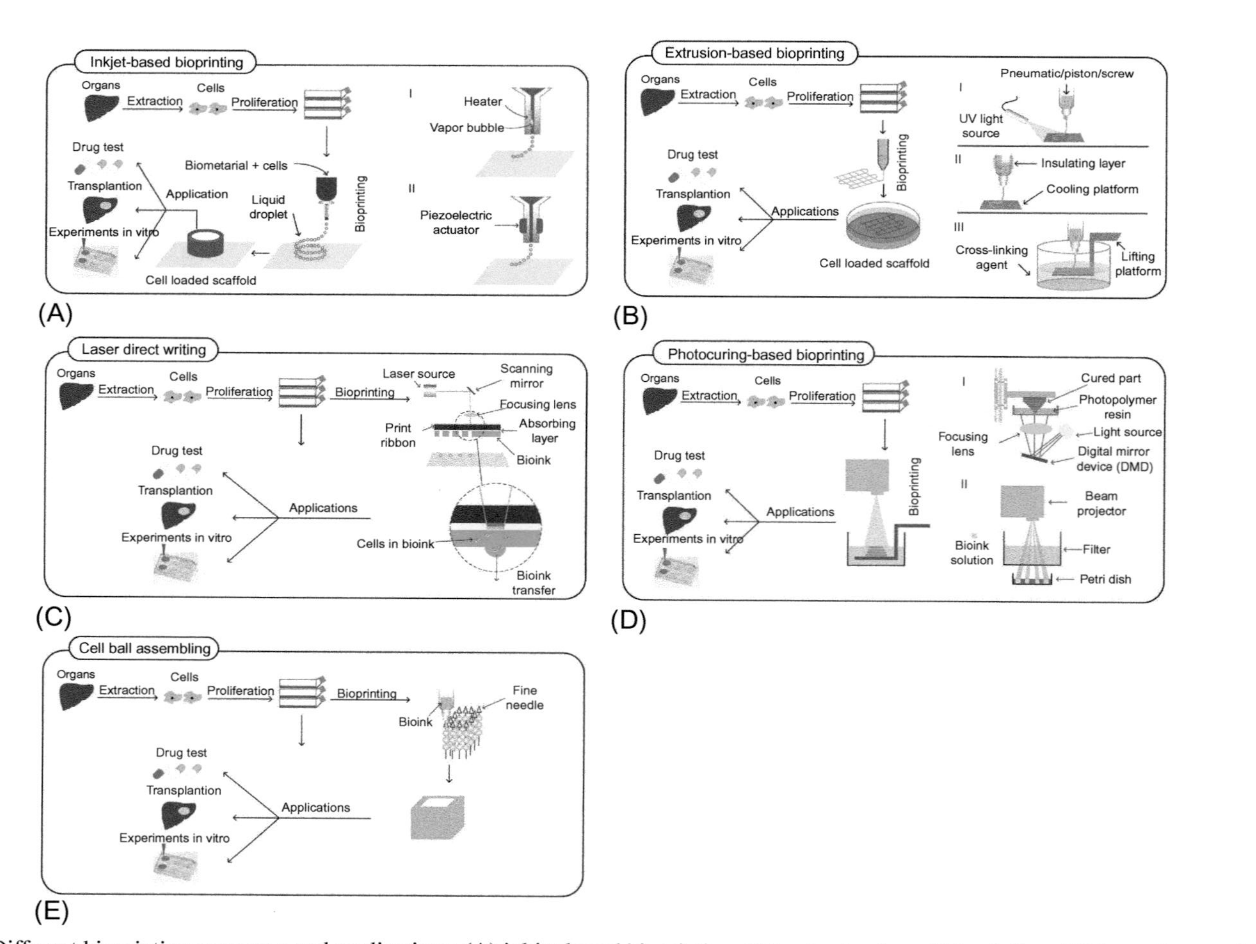

Fig. 2.3 Different bioprinting processes and applications: (A) inkjet-based bioprinting; (B) extrusion-based bioprinting; (C) laser direct writing bioprinting; (D) photocuring-based bioprinting; and (E) cell ball assembling bioprinting.

2.2.2 Extrusion-based bioprinting

As the most widely used bioprinting method, extrusion-based bioprinting utilizes nozzles driven by air pressure or motor to extrude biomaterials in a controlled manner to construct 3D structures. As shown in Fig. 2.3B, filaments are extruded from the micronozzle and deposited on a substrate to form 2D structures. With the movement of the *Z* axis, the layer-by-layer 2D structures will constitute a 3D structure.

A 3D plotter was used to print 3D hydrogel (agar and gelatin solutions) scaffolds in 2002 [49], however, as the melting temperature is 90°C, it cannot be directly used in printing cell-laden structures. The first report on extrusion-based bioprinting of 2003 can be found in Ref. [50], and this method is used in many tissue applications [51–55]. The advantages of extrusion-based bioprinting include: (1) With a pneumatic or a mechanical nozzle, high-density bioink can be extruded, which means that cell-laden structures with enough strength for implantation or perfusion can be printed. Moreover, many biomaterials can be used as printing materials. (2) Simplicity and easy access of this type of bioprinters. A commercialized plotter or a desktop 3D printer (D3DP) can be easily modified to be an extrusion-based bioprinter. Now the D3DP has been totally open sourced and it is very suitable for customization with low cost [7]. In the following chapter, we will discuss how to assemble a typical extrusion-based bioprinter.

2.2.3 Laser direct-write bioprinting

Laser direct writing was first used to write patterns of metals for processing mesoscopic conformal passive electronic devices [56]. In 2000, Odde et al. reported direct writing of living cells using laser [57], and following laser-assisted cell printing method had been developed [26,58–63]. As shown in Fig. 2.3C, a glass substrate coated with a layer of laser-absorbing material is used to generate the microbubbles and protect the cells from direct contact with the high-power laser. First, the mixture of cells and cell encapsulation (usually a kind of hydrogel) is uniformly spread on the laser-absorbing layer. Then, the laser pulses are applied on the substrate. As the high energy is absorbed by the laser-absorbing layer, the cell mixture is vaporized, forming microdroplets. Using an *XYZ* moving platform to drive the glass substrate or the receiver, a 3D object can be printed. In a sense, laser-based bioprinting can be treated as the nozzle-free inkjet-based bioprinting.

The nozzle-free bioprinting method has the ability to create precise user-generated patterns for a variety of cell types and biomaterials [47]. However, most researches on laser-based bioprinting are still focused on the discussion of the process itself. There are several reasons that perhaps limit its current wide application: (1) the high cost of laser-based bioprinters and lack of commercialized printers; (2) the time-consuming step of spreading the cell mixture when printing each layer; and (3) the repeatability of microdroplets requires further research.

2.2.4 Photocuring-based bioprinting

Photocuring-based technique is originally used to fabricate cell-free scaffolds [64–68]. In most studies using this technique, cells are seeded on the scaffolds after fabrication, rather than encapsulated within the scaffold during fabrication. As shown in Fig. 2.3D, light-sensitive materials are selectively solidified layer by layer using UV light. According to the scanning mode, it can be divided into stereolithography technology (SLA) and digital light projection technology (DLP). Bioink is cured point by point in SLA and plane by plane in DLP. Many types of cells including HUVECs and HepG2 were reported to be seeded on the printed scaffold to mimic biological structures. However, it is difficult to use photocuring-based technique in fabricating cell-laden scaffolds because UV light is harmful to cells. To allow concurrent printing cells and scaffold, Tuan et al. [69] adopted a visible light-based system to print human adipose-derived stem cells (hADSCs)-laden PEG/LAP solution and the result shows high viability (>90%). Similarly, Kim et al. [65] presented a custom-built visible light-based stereolithography system in which PEGDA/GelMA was used as the scaffold to load NIH 3T3 cells and 85% cell viability was demonstrated.

The advantages of photocuring-based technique are: (1) Time saving and convenient: No matter how complex the structure of each layer is, printing time is the same, because one layer is solidified at a time. (2) High resolution. However, it has the limitations of material selection and the common light source has damage to cells.

2.2.5 Cell ball assembling bioprinting

"Modular" tissue engineering approaches have been employed to fabricate macroscale tissue architectures by assembling shape and functional controlled microscale tissue building blocks [70], among which cell balls are always used in bioprinting through a self-fusion process [71–73], as shown in Fig. 2.3E. Gabor Forgacs et al. [74] applied a fully biological, scaffold-free tissue engineering technology to fabricate small-diameter multilayered tubular vascular grafts in a two-step process. First, cells were aggregated into multicellular spheroids, and then they were printed layer by layer concomitantly with agarose rods, used as a molding template. Single- and double-layered vascular tubes were created by fusion of the discrete units. As an alternative strategy, Jens M. Kelm et al. [75] presented an approach to fabricate rapidly small-diameter tissue-engineered living blood vessels with living cell reaggregates containing HAFs and HUVECs and the result shows that the maturation/differentiation capacity has been enhanced.

Cell ball assembling bioprinting is a promising method for rapid manufacturing of 3D structures with self-assembly and self-organizing properties. However, fabricating complex structures with sufficient strength is a major challenge.

In the bioprinting process, the basic forming units includes microparticles, microfilaments, and planes, and it is important to glue each unit together and form a real 3D object. There are several methods to achieve this. As hydrogel is the common cell-laden material, in this study hydrogel is taken as an example to illustrate the different ways of fusion. In inkjet-based bioprinting and laser direct writing bioprinting, alginate microparticles are usually used as building blocks and they fuse together based on cross-linking reaction [76,77]. As shown in Fig. 2.4A, two adjacent alginate microparticles are deposited on the $CaCl_2$ liquid level, then Ca^{2+} diffuses into the sodium alginate solution, causing the cross-linking to begin. In this process, due to inadequate

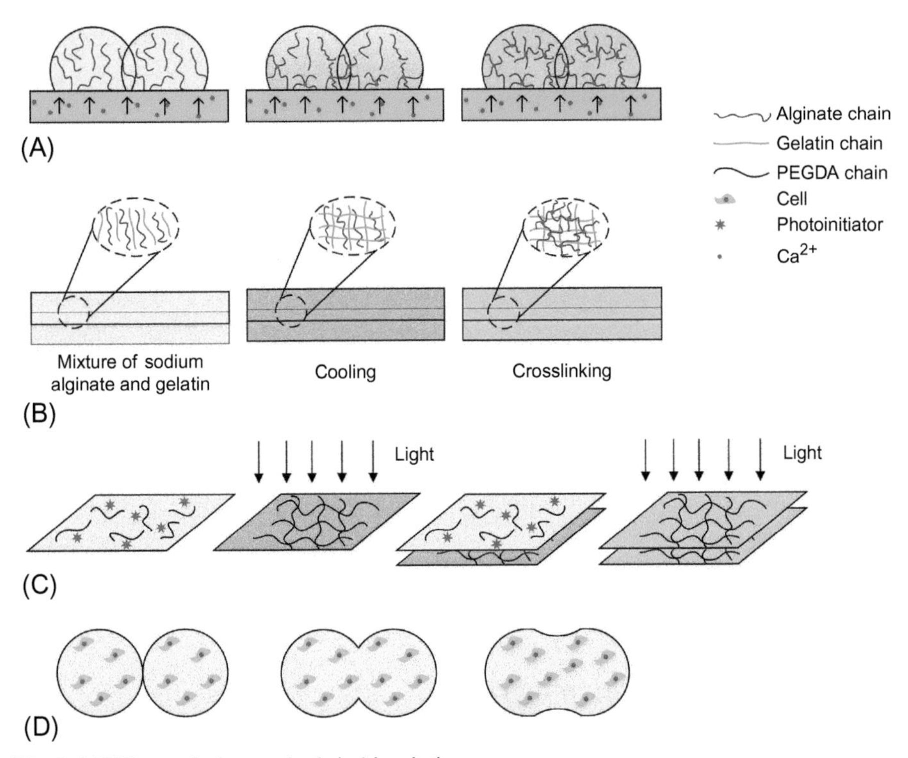

Fig. 2.4 Different fusion methods in bioprinting.

cross-linking, the two microparticles could glue together. In a typical extrusion-based bioprinting method, a 3D structure is formed due to the fusion of microfilaments. As shown in Fig. 2.4B, first, sodium alginate and gelatin are mixed and extruded from the nozzle to form a continuous microfilament. Then adjacent microfilaments glue together spontaneously and can be preformed when gelatin undergoes a temperature change. Finally, the entire structure is immersed into $CaCl_2$ solution for irreversible cross-linking. In the photocuring-based bioprinting method, PEGDA is usually used as the scaffold material. As shown in Fig. 2.4C, under the effect of light and photoinitiator, a layer of PEGDA can rapidly polymerize into the hydrogel and could, at the same time, bond with the surrounding cured materials. Since the cell ball assembling bioprinting is a scaffold-free method, the principle of forming a real 3D object is relied upon in the in-vitro fusion of two cell aggregates, as shown in Fig. 2.4D.

2.3 Biomaterials for printing

The term "biomaterials" in bioprinting mainly refers to soft hydrogels, which could directly load cells and form a designed shape through different bioprinting methods mentioned in the above section [78–81]. In this chapter, we first describe the printability of biomaterials. Then we give the criterion of ideal biomaterials for printing. At last we take several kinds of commonly used hydrogels as examples to discuss the role of biomaterials in different bioprinting processes.

2.3.1 Printability of biomaterials

The term "printability" of a biomaterial means that it can be disposed in a controlled manner and form a designed structure through the corresponding fusion mechanism. In general, a biomaterial will undergo three stages during the entire printing process.

First, the biomaterial is disposed into the basic forming unit including the microparticles, microfilaments, and planes. In this stage, the biomaterial should be in liquid or sol state before printing. An important property of a suitable biomaterial is that it has an adjustable viscosity, such as it changes with temperature and shear thinning as the different printing methods may request different viscosity. Second, the printed units will undergo sol-gel reaction to glue each other. This requires the materials' solidification/gelatinization after printing as soon as possible to keep the shape and ensures fusion at the same time. At last, the materials are stacked into an integrated structure layer by layer. In this process, the biomaterials should have retention capacity for continuous stacking in the vertical direction to achieve good shape fidelity.

2.3.2 Ideal printing biomaterials

In the bioprinting process, biomaterials not only serve as a scaffold to directly load cells, but also provide structure to the printed constructs. In addition to meet the demand of printability, an ideal printing biomaterial should also have other properties. (1) Mechanical properties: To hold its designed 3D shape, the material should maintain adequate mechanical properties, such as strain, stress, shear stress, and so on. (2) Biocompatibility: An ideal printing biomaterial should have no damage to cells and be favorable for cell adhesion and growth. (3) Degradability: Since the cells encapsulated in the biomaterial will produce an extracellular matrix to develop into a functional tissue, the biomaterial should have a degradation kinetic matching the process of cell growth.

2.3.3 Currently used biomaterials

Alginate gel is a biocompatible hydrogel that is often used to encapsulate cells or other biological materials. It forms egg-box-shaped bonding structures upon contact with calcium ion, and offers an aqueous environment for cells [82–84]. Based on its good gelation ability, alginate is often used in bioprinting. The viscosity of alginate could be adjusted by preparing it in different concentrations, which allows it to apply to different bioprinting methods. Low viscous alginate could be used in inkjet-based bioprinting and high viscous alginate is often used in laser direct-write bioprinting or extrusion-based bioprinting. In the bioprinting process, alginate solution is firstly processed into microparticles or microfilaments, then the adjacent units undergoe a progressive cross-linking reaction to glue together, and at last the printed structure is immersed in $CaCl_2$ to form a completely cross-linked structure with adequate mechanical strength. Although alginate is widely used in tissue engineering, it is not so good for cell adhesion without chemical modification, and the cross-linking reagent has a bad effect on cell viability [85]. Therefore, modification of alginate is essential for bioprinting.

Gelatin is a thermosensitive hydrogel. It is in liquid form when the temperature is more than 35°C which is below physiological temperature ranges [78]. In bioprinting it cannot be directly used as a cell scaffold and always requires combination with other biomaterials. Gelatin is often mixed with alginate in an extrusion-based bioprinting method [81]. Here, gelatin is used for shape stabilization before gelatinization of the alginate. Due to its thermosensitive property, gelatin is often used as a sacrificial bioink to build channels inside thick hydrogel scaffolds [86] or as supporting material to produce complex 3D biological structures [37]. By chemical modification with methacrylic anhydride, photosensitive hydrogel (GelMA) could be synthesized and could be used in extrusion-based bioprinting and photocuring-based bioprinting [87,88].

2.4 Owning a 3D bioprinter

In brief, the two main components of a bioprinter system are the spatial motion control unit and the dispense unit [22]. To maintain high-resolution structures, the positioning of the control unit should be very precise and operates under the CNC software. According to each printing technology, the dispense unit can be classified into two forms: nozzle based and nozzle free.

2.4.1 Commercialized 3D bioprinters

In this section, some currently commercialized 3D bioprinters will be introduced including inkjet-based bioprinters [50], photocuring-based bioprinters [89–91], extrusion-based bioprinters, and cell ball assembling bioprinters [71,92,93]. What's more, we collected some other types of bioprinters on the market.

Inkjet-based bioprinting is a very common type of bioprinting. This technology is based on the rapid creation and release of liquid droplets, followed by their precise deposition on a substrate [94]. When combined with a high-throughput deposition capability it could be employed for in-vitro tissue-engineered models, and can even be applied in tissue regeneration and organ bioprinting.

The 3D printer manufacturer Pokit recently developed an inkjet bioprinter Edison Invivo [95]. As shown in Fig. 2.5, Edison Invivo uses a bioink to produce cell structures in the form of organic tissues, with the goal of making it possible to create personalized and transplantable tissues from the cells collected from a specific patient. This in-situ bioprinting system will be applied in skin regeneration according to the company's project [96].

As another familiar bioprinting method, 3D bioprinting by photocuring allows fabrication of internal pores and defined macroscopic shapes. Most commercial photocuring printers can be remolded into a bioprinter by applying hybrid biocompatible resin. Fig. 2.6 shows a commercial SLA printer developed by Formlabs

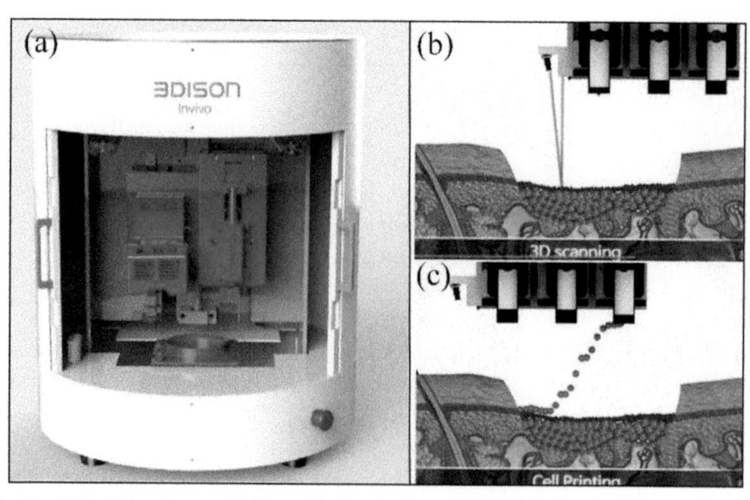

Fig. 2.5 The inkjet bioprinter of Edison Invivo (released by Pokit, Korean): (A) the structure of Edison Invivo; (B) 3D scanning process to acquire the injured skin surface; and (C) skin regeneration by printing the bioink onto the injured surface.

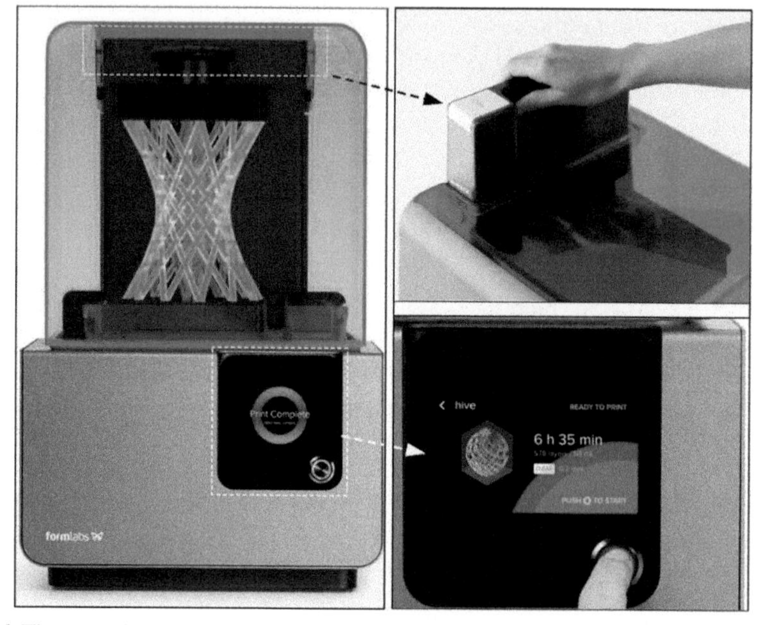

Fig. 2.6 The stereolithography bioprinter of Form 2 (Formlabs, Somerville, MA, United States).

(United States). In this bioprinter, a self-heating resin tank is designed for stable temperature control, and the sliding peel process with a wiper will guarantee the fine dimensional accuracy of the printed structure. Besides, a removable and stackable resin tank and a touch panel are added into the design. However, the total price of this bioprinter is affordable for ordinary customers.

Fig. 2.7 The bioprinter of NovoGen MMX BioprinterTM (left) and Envision TEC 3D bioprinter (right)

Extrusion-based 3D bioprinter is a very common member of the bioprinter family. Bioink is extruded by air pressure or an injection pump when plotting a 3D biological structure [97]. NovoGen MMX Bioprinter (Organnovo, United States) is the world's first commercial extrusion-based 3D bioprinter [98]. As shown in Fig. 2.7, the bioprinter includes two robotically controlled precision print nozzles: one for placing human cells and the other for placing a hydrogel, scaffold, or a support matrix. A computer-controlled, laser-based calibration system is developed to achieve the required repeatability in consistently positioning the cell dispensing capillary tip attached to the print head to within microns.

Envision TEC launched a bioprinter with a built-in camera, temperature-controlled build platform and sensor ports, five cartridge slots, and two print heads with independent temperatures.

To match various bioprinting demands, the machine is equipped with heating and cooling capabilities on a building platform, an automatic z-height controlling system, and automated nozzle cleaning structures.

Some enterprises and institutions have delivered their extrusion-based bioprinters in addition to the two already introduced famous commercial bioprinters as show in Fig. 2.8. Fig. 2.8A demonstrated a bioprinter produced from 3D Bioprinting Solution,

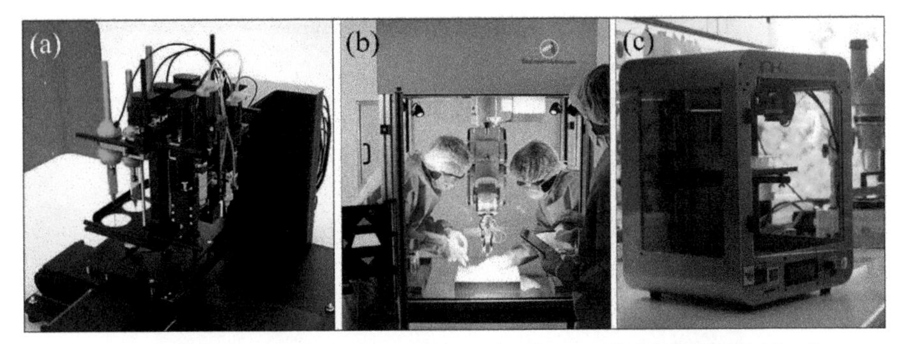

Fig. 2.8 Some typical commercial extruder-based bioprinters: (A) FABION 3D bioprinter (made by 3D Bioprinting Solution, Russia) [99]; (B) BioAssembely 3D bioprinter (Advanced Solutions, United States); and (C) a desktop bioprinter from CELLINK (Sweden).

a Russia company. The bioprinter shown in Fig. 2.8B is developed by an US company, owning a polyarticular mechanical arm as its actuator. Due to its flexible mechanical degree of freedom, bioink can be fabricated onto the 3D curve. Moreover, a desktop bioprinter seems to be a good choice for the customers with affordable consumption power [54]. CELLNK form Sweden seized such opportunity and developed their desktop bioprinter as shown in Fig. 2.8C.

One of those bioprinting companies was Cyfuse Biomedical, a Japanese corporation that has developed a novel form of bioprinting known as the Kenzan method [100]. By placing cellular spheroids in fine needle arrays according to a predesigned 3D data, the cells will fuse together into a solid in the next few days; the tissues can be made firm by placing them in a bioreactor (Fig. 2.9).

Except for the 3D bioprinters introduced above, there are some special 3D bioprinters on the market as shown in Fig. 2.10. Instead of the extrusion-based nozzle, the printAlive bioprinter adopted a microfluidic print head to extrude cell-compatible hydrogel into heterogeneous structures. With the rolling axis (Fig. 2.10A and C), the hydrogel containing cells can be continually fabricated into bandage-like structure simulating human skin. By replacing the modularized print head, many more 3D structures can be printed by this printer.

Nano three-dimensional (n3D) Biosciences developed a new type of 3D bioprinter Nano 3D Biosciences bioprinter. In this design, the cells are rapidly printed using magnetic 3D bioprinting (Fig. 2.10B and D). In contrast to magnetic levitation using the Bio-Assembler, in magnetic 3D bioprinting, cells incubated with NanoShuttle (NS) overnight are printed into spheroids by placing a 96-well plate full of magnetized cells atop a drive of magnets. The magnets below the well aggregate the cells using mild magnetic forces to form a spheroid at the bottom of the well. After only 15 min to a few hours, the plate of spheroids can be removed from the magnets and cultured long-term. The reason why the spheroid can be assayed in any number of ways is that NS does not interfere with fluorescence or other biochemical assays.

Fig. 2.9 Ragenova 3D bioprinter (Cyfuse Biomedical, Japan) [101].

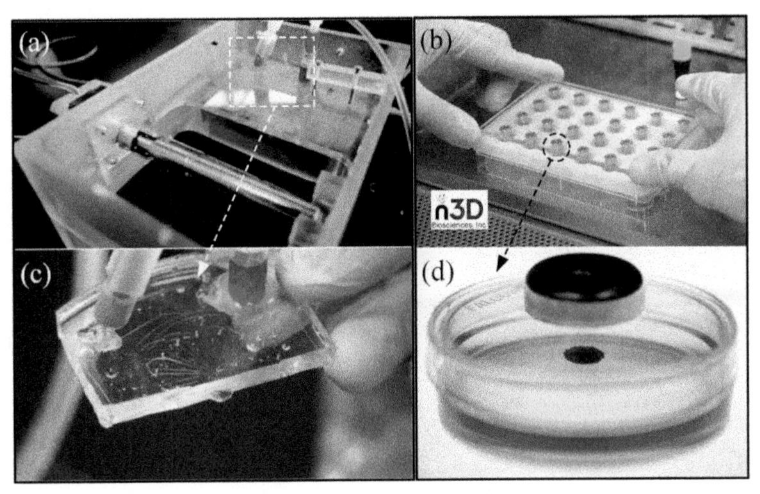

Fig. 2.10 Some other special 3D bioprinters on the market: (A) PrintAlive 3D bioprinter(); (B) Nano 3D Biosciences magnetic bioprinter (n3D Biosciences, Inc, United States); (C) Microfluidic print head of the printAlive 3D bioprinter; and (D) magnetic assembly of spheroids mimicking native tissue environments.

2.4.2 Assembling a bioprinter in laboratories

3D bioprinters for research have been extensively described in many literatures, and the most adopted sort is the extrusion-based bioprinter [102–106]. This class of 3D bioprinter has several advantages over the others: (1) It can be set up easily in a personal laboratory with less efforts. (2) A diversity of materials can be used with a wide range of viscosity and cell density. (3) The capability of fabricating 3D gradient and heterogeneous artificial tissue/organ is viable by applying multiple extruders [107,108].

In this part, we will introduce the main process of assembling an extrusion-based bioprinter in laboratories.

As shown in the system diagram in Fig. 2.11, the extrusion-based bioprinter consists of one *XYZ* orthometric motion platform, a printer controller, a fluidic dispense module, a pressure controller, a monitor module, and a computer to run the printer controller. This configuration of a bioprinter is very typical which can satisfy different applications.

The printer controller is integrated with a motion controller and a temperature controller. Both extrusion control and platform movement are conducted by the motion control which means that at least five axis channels should be supported for cooperative work. When dealing with a complex mission like fabricating a gradient structure, the motion controller should be capable of four-axis simultaneous control. For this purpose, the frame of the bioprinter is designed as an identical three-axis linear motion system to release the calculation burden of the printer controller; meanwhile, the kinematic accuracy will not be affected.

The pressure-based heads are managed by the pressure controller, which are very suitable in extruding low viscosity fluid or materials with a biggish dispensing nozzle.

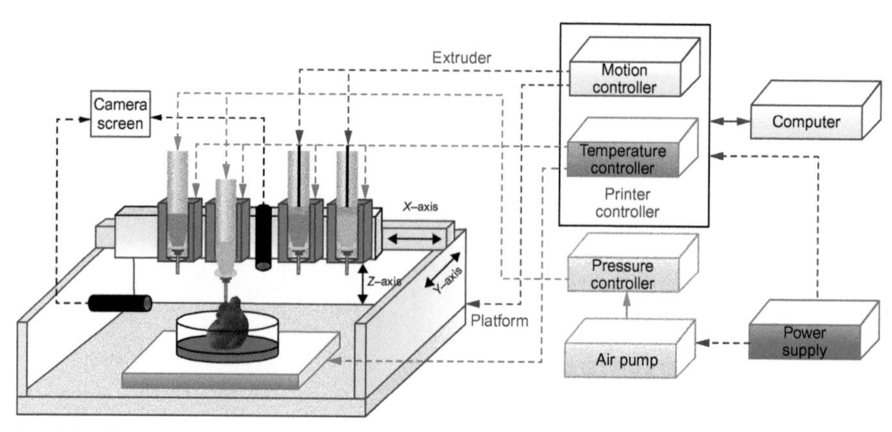

Fig. 2.11 The main system of an extrusion-based bioprinter with two types of dispense head.

When applying pressure-based nozzles, the ejection process corresponds to the pressure's high response, which makes the ejection sluggishly small.

As for the nozzles driven by the plunger pumps, they will be admired for their precision of ejection position and large extruding force. In other words, this kind of nozzle is more appropriate when fluid with high viscosity is needed to be ejected from a smaller micronozzle. In addition, as the biomaterial is loaded by the injection syringe, the risk of cell contamination reduces owing to the relatively closed space [109]. The separation between the ejection syringe and the plunger pump makes it easy for replacing the biomaterials.

It is to be insured that the cell environment is always suitable during printing and a stable temperature control is of great significance to guarantee cell density and printing performance. We demonstrated two general kinds of hardware devices applying temperature control: extruder and platform.

To monitor the extrusion process, a UBS microscope is installed on the platform moving along with the nozzle. In some occasions, the monitor system also acts as a visual positioning equipment for zero adjustment of the nozzles or other attachments.

The prototype of a bioprinter emphasized the use of standard extrusion-based bioprinting with full function of motion control, visualization, temperature control, and extrusion nozzle as shown in Fig. 2.12. The framework of the bioprinter is built by aviation aluminum profile. To breed bacteria, the Z-plate and baseboard are anodized which are easy to clean and can be disinfected using alcohol (95%) or ultraviolet light. The three-axis kinematic pairs consist of a three linear module driven by a stepper motor.

The printer is equipped with two different nozzles, which are fixed on the perforated plate, and switched by the lifting motor. When the perforated plate moves along the CNC path, the nozzles are firmly fixed. To support more extrusion channels, the biomaterials are injected by long-distance extruders for the extrusion-based nozzle. For the pressure-based nozzle, materials are prepared in the metal syringe.

A computer is necessary for communicating with the printer controller and showing the online image of the nozzle visual photographed by the USB microscope.

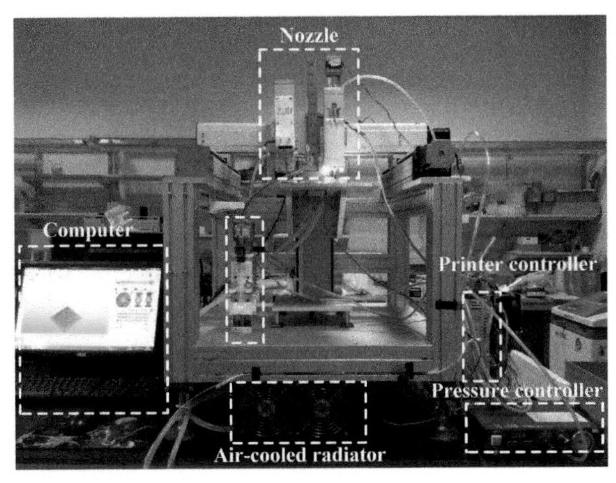

Fig. 2.12 The prototype of a bioprinter.

To handle the whole printing procedure, the printer controller is facing a great task in motion control, temperature control, injection control, manual operation, and communication with the computer. Under the comprehensive consideration, the CPU of the printer controller is selected as ATmega2560 with 16 MHz operating frequency and up to 86 IO pins.

The modularized nozzle system makes it easy to allocate nozzles flexibly, combing different types of printing methods as shown in Fig. 2.13. Two nozzles are mounted on the printer in this design. Fig. 2.13B shows two long-distance extruders, which are reconstructed by the plunger pump. With the use of the specific fixtures made by fused deposition modeling, injectors can be quickly replaced online. The air pressure

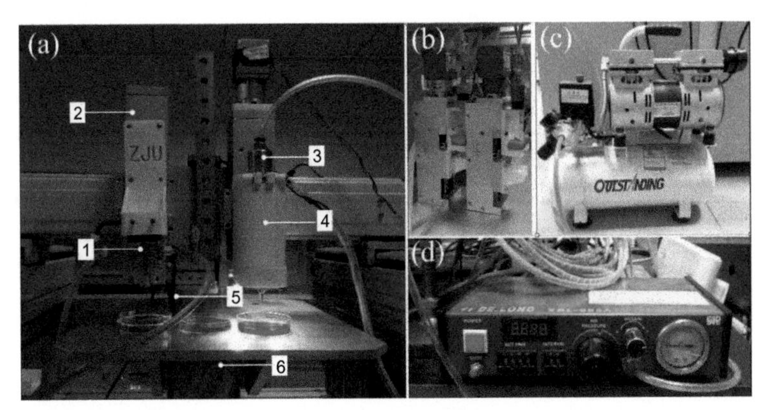

Fig. 2.13 Details of the bioprinter: (A) nozzle part of the bioprinter: (1) extrusion nozzle, (2) lifting linear guid2way, (3) metal syringe, (4) heating shield, (5) UBS microscope; and (6) semiconductor refrigeration system; (B) long-distance extruders; (C) air pump; and (D) pressure controller.

Table 2.1 **The main parameters of extrusion-based bioprinting**

Item	Parameters
Range of motion	220×140 mm (height <140 mm)
Material	Biological hydrogel
Forming platform	Petri dish, tissue culture plates, etc.
Disinfection method	UV light, alcohol (95%)
Forming precision	>100 μm
Temperature	−25°C to 37°C (platform), 25°C to 37°C (nozzle)
Forming speed	<1500 mm/min (nozzle speed)
Single loading capacity	>100 mL
Cell load	Support

is supplied by the air pump (Fig. 2.13C), and it was controlled by the pressure controller (Fig. 2.13D). Air pressure is switched by the relay which was lined with the printer controller making it realizable following with the intermittent CNC path.

Heating of syringe is achieved by the flexible healing film when ejecting with cell-laden bioink [110], which is inserted onto the inner side of the healing shield as shown in Fig. 2.13A. On the Z-platform, three semiconductor refrigeration chips are fixed at the bottom to cool the fabricated bioink for better shape. The main parameters of the bioprinter are listed as shown in Table 2.1. Fabrication of tissues is a tough and charming task; this bioprinter based on extrusion is suitable for simple applications such as fabrication of blood vessels, tissue units, organoids, and so on [55,111,112].

2.5 Typical 3D bioprinting processes

In this section, a typical extrusion-based bioprinting is given [113], as shown in Fig. 2.14. In this process, the printing hydrogel is a mixture of sodium alginate (SA) and gelatin in an appropriate proportion. And L929 mouse fibroblasts are chosen as a representative cell line. The mixture was extruded on a cool substrate for the solidification of gelatin and for fixing the biostructures. After printing, the structures were immersed in calcium chloride ($CaCl_2$) solution for the cross-linking of SA to obtain the cell-laden hydrogel structures. The bioprinter used in this process is described in Section 2.4.2, which includes an *XYZ* moving platform with a cooling substrate (Peltier cooler) to contain the printed hydrogels and a nozzle could be heated to extrude the hydrogel easily under air pressure. The detailed process is as follows.

2.5.1 Data acquiring

First, by scanning a real organ, a CT or an MRI dataset is received. Next the dataset of the structure needs to print is extracted from the redundancy data by Mimics and meshed into STL files. Since the model received from CT or MRI has several defects such as lack of local information, surface triangles redundancy, and topological

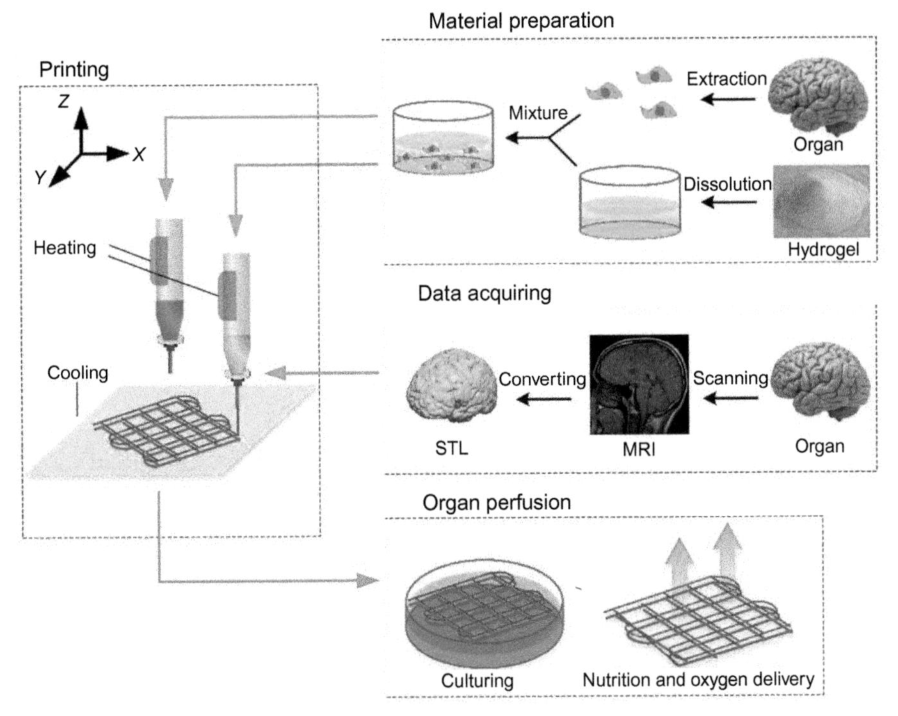

Fig. 2.14 A typical extrusion-based bioprinting process.

structure errors, the STL file should be remeshed according to the required accuracy by Magics and then can be directly read by a 3D bioprinter.

2.5.2 Material preparation

The alginate/gelatin hydrogel was prepared in advance. Sodium alginate powder and gelatin powder were sterilized under UV light for half an hour; hydrogel solution was prepared using deionized water and placed in a magnetic stirrer for 24h at 120rpm at 37°C to make the final hydrogel solution with a concentration of 5% alginate and 16% gelatin (w/v). $CaCl_2$ solution, which is the cross-linking agent, was prepared by dissolving $CaCl_2$ power into deionized water to obtain a final concentration of 2% (w/v).

For cell-laden hydrogel solution preparation, culture flasks with 90% L929 cells confluency were washed with phosphate buffer saline (PBS), and incubated with 0.25% Trypsin-EDTA for 3min at 37°C in 5% of CO_2 to detach the cells from the culture flasks. The cell suspension was centrifuged at 1000rpm for 5min at room temperature; the supernatant was discarded; and the cells were resuspended in MEM cell culture medium to a concentration of 2×106 cells/mL. Then the cell suspension was mixed with the hydrogel solution at a volume ratio of 1:1 and placed in a magnetic stirrer for 5min at 100rpm at 37°C in 5% of CO_2, resulting in a cell density of 1×106 cells/mL and hydrogel solution with a concentration of 2.5% alginate and 8% gelatin (w/v).

2.5.3 Printing

A 3D cell-laden structure is printed layer by layer based on the fusion of microfilaments. Hydrogel is extruded by air pressure from the nozzle to form a continuous microfilament, and then the adjacent microfilaments glue together with the help of a computer-aided moving platform. In this process, the quality and accuracy of the microfilaments are influenced by the printing parameters including air pressure, feed rate, and printing distance. The relationship between them is shown in Fig. 2.15A.

The air pressure, P, is the driving force that extrudes the hydrogel out from the nozzle against surface tension. P is mainly determined by the viscosity of the hydrogel and will affect the printing stability and the printed line width. After defining the parameter, Ds, the distance from the separating location to the nozzle, Fig. 2.15B gives an accessible pressure range combined with hydrogel viscosity to obtain high printing quality. The width of microfilament is effected by the nozzle feed rate, F, and the printing distance, H. As shown in Fig. 2.15C, the filament width increased with the nozzle feed rate and the printing distance.

There are some unavoidable errors in printing special structures including the overlap problem in sharp angle printing and diffusion phenomenon in lattice scaffold printing as shown in Fig. 2.15D and E, and several strategies can be used to decrease the errors by controlling the extrusion rate. In a word, the printing parameters should be harmonized to obtain high shape fidelity.

2.5.4 Organ perfusion

To get a functional organ, cells are not simply stacked together, but rather require adequate perfusion in vitro prior to implantation to allow delivery of growth factors, oxygen, and other nutrients. Thus, higher request to bioprinting needs to be put forward and appropriate structures should be designed and printed. An indirect way is to print porous tissue structures by mixing porogens into printing materials [114]. Another way is to print grid structures with interconnected channels [55]. However, when printing a complex structure, a vascular network should be integrated into thick tissues. There are many approaches to fabricate a vessel-like structure, either printing using sacrificial bioinks to build perfused microchannels [115–117] or direct printing a vessel-like structure [118–120].

2.6 Trends of bioprinting

2.6.1 Limitations of bioprinting

Although bioprinting has developed rapidly, there are still some limitations hindering its extensive use in tissue engineering. First of all, printable biomaterials are very limited. The existing biomaterials used in bioprinting are all commonly used materials in other fields, rather than new materials specially designed for existing bioprinting methods. Even for a commonly used biomaterial, the printing method is single. Another limitation is that there is a great difference between the

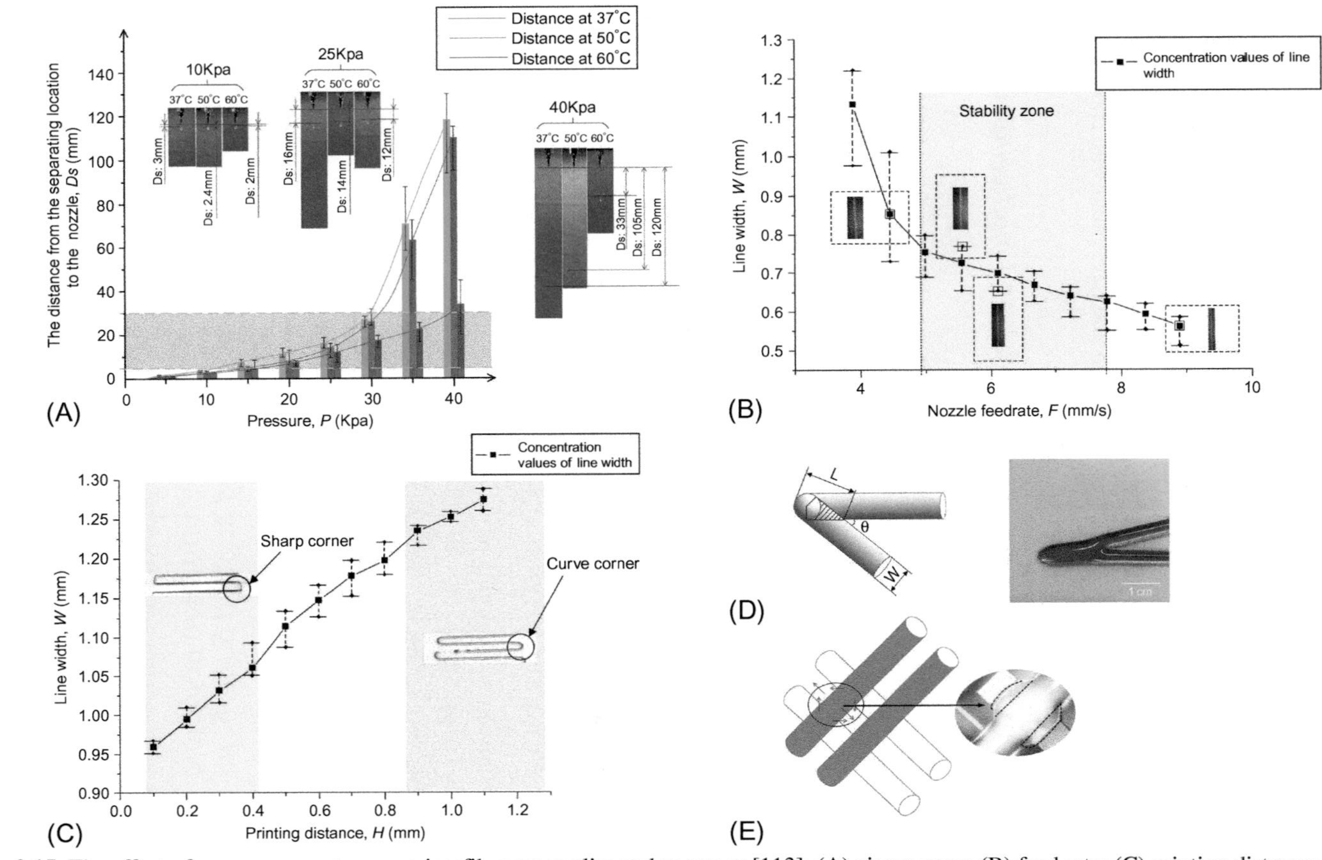

Fig. 2.15 The effect of process prameters on microfilament quality and accuracy [113]: (A) air pressure; (B) feed rate; (C) printing distance; (D) overlap; and (E) diffusion.

current printed structures and real tissues or organs. Now most of the laboratory 3D Bioprinting methods are still in the stage of shape control, rather than the function control. As bioprinting is being developed so rapidly, research on the cellular function, for example, cell migration, cell differentiation, cell-cell communication, is being reported.

2.6.2 Future trends

Due to its individuation and controllable properties, in the future, bioprinting may develop into a potential tool for organ regeneration and have promising applications in tissue engineering.

2.6.2.1 Printing multiscale, multimaterial, and multicellular structures

An organ is composed of several tissues which have different sizes, diverse hardness and various kinds of cells. Taking for example printing a vessel-like structure. Real blood vessels have diameters ranging from micrometer scale to centimeter scale (~20 μm to ~2.5 cm for capillaries to aorta [44,45]), and contains three layers of cells: fibroblasts, smooth muscle cells, and endothelial cells with different functions. This would require developing a more appropriate method to printing multiscale, multimaterial, and multicellular vascular structures. Due to its advantage in fabricating complicated tissues with controllable size and shape, patterning multiple cell types accurately, and building a gradient environment of extracellular matrix, bioprinting would better achieve this goal by combining different bioprinting methods and using more suitable materials.

2.6.2.2 Exploring cell damage mechanism during printing process

In the currently used bioprinting process, cells are more or less damaged due to different reasons. For example, cells will suffer shear stress during extrusion through a form nozzle in extrusion-based bioprinting. Cell deformation will happen and cell viability will decrease. In photocuring-based bioprinting, cells will expose to light even if there is visible light, this may affect the cell function. There are few researches on cell damage mechanism, and in the near future, the mechanism could be explained better and further guide the development of bioprinting.

2.6.2.3 Integrating printing process, organ perfusion, and biological detection

Compared with other abiotic 3D printing, bioprinting is a technology in which living cells are deposited and assembled to a tissue structure. This means the cell-laden structure will continue to be cultured in vitro after printing and be detected to obtain the required biodata. Currently, organ perfusion and biological detection are independently conducted. If these two processes are integrated into the printing process, with a timely feedback, bioprinting could become more convenient in laboratory use.

2.6.2.4 Combining bioprinting with organ-on-a-chip technology

As a novel analytical tool, organ-on-a-chip technology has attracted increased interest and attention [121–123], since Donald E. Ingber introduced the concept in 2010 [124]. And some organ-on-a-chip models were built based on this idea to simulate the microenvironment of real organs in vitro, including vessel-on-a-chip, liver-on-a-chip, heart-on-a-chip, and tumor-on-a-chip. However, only several tissue cells were cultured in these chips, and if a bioprinted tissue that has more structure and function similarity to real organ can be integrated into these organ-on-a-chip devices, the method based on organ-on-a-chip could be more popular. Some researches have been reported to integrate bioprinting into organ-on-a-chip platform for continuous perfusion and we are looking forward to some deep applications.

2.7 Conclusions

In this chapter, we have presented the current practical bioprinting methods used in laboratories. In addition, the components of a bioprinter and how to assemble a simple one in the laboratory is discussed in detail, and a typical extrusion-based bioprinting process is described. However, most of the bioprinting methods and their applications are still under laboratory research. We believe that through the efforts of researchers with different discipline backgrounds, bioprinting will mature and find more practical applications in clinical medicine.

References

[1] Kroll E, Artzi D. Enhancing aerospace engineering students' learning with 3D printing wind-tunnel models. Rapid Prototyp J 2011;17(5):393–402.
[2] Gibson I, Rosen D, Stucker B. Additive manufacturing technologies: 3D printing, rapid prototyping, and direct digital manufacturing. Springer; 2014.
[3] Schubert C, van Langeveld MC, Donoso LA. Innovations in 3D printing: a 3D overview from optics to organs. Br J Ophthalmol 2013;98:159–61.
[4] Gross BC, et al. Evaluation of 3D printing and its potential impact on biotechnology and the chemical sciences. Anal Chem 2014;86(7):3240–53.
[5] Wong KV, Hernandez A. A review of additive manufacturing. ISRN Mech Eng 2012;2012.
[6] Lipson H, Kurman M. Fabricated: the new world of 3D printing. John Wiley & Sons; 2013.
[7] He Y, Xue G-H, Fu J-Z. Fabrication of low cost soft tissue prostheses with the desktop 3D printer. Sci Rep 2014;4.
[8] Mironov V, Reis N, Derby B. Review: bioprinting: a beginning. Tissue Eng 2006;12(4):631–4.
[9] Guillemot F, Mironov V, Nakamura M. Bioprinting is coming of age: report from the international conference on bioprinting and biofabrication in Bordeaux (3B'09). Biofabrication 2010;2(1). 010201.
[10] Dababneh AB, Ozbolat IT. Bioprinting technology: a current state-of-the-art review. J Manuf Sci Eng 2014;136(6). 061016.

[11] Chang CC, et al. Direct-write bioprinting three-dimensional biohybrid systems for future regenerative therapies. J Biomed Mater Res B Appl Biomater 2011;98(1):160–70.
[12] Nakamura M, et al. Biomatrices and biomaterials for future developments of bioprinting and biofabrication. Biofabrication 2010;2(1). 014110.
[13] Leukers B, et al. Hydroxyapatite scaffolds for bone tissue engineering made by 3D printing. J Mater Sci Mater Med 2005;16(12):1121–4.
[14] Dimas LS, et al. Tough composites inspired by mineralized natural materials: computation, 3D printing, and testing. Adv Funct Mater 2013;23(36):4629–38.
[15] Billiet T, et al. The 3D printing of gelatin methacrylamide cell-laden tissue-engineered constructs with high cell viability. Biomaterials 2014;35(1):49–62.
[16] Das S, et al. Bioprintable, cell-laden silk fibroin–gelatin hydrogel supporting multilineage differentiation of stem cells for fabrication of three-dimensional tissue constructs. Acta Biomater 2015;11:233–46.
[17] Chia HN, Benjamin MW. Recent advances in 3D printing of biomaterials. J Biol Eng 2015;9(1):1.
[18] Lee Y-B, et al. Bio-printing of collagen and VEGF-releasing fibrin gel scaffolds for neural stem cell culture. Exp Neurol 2010;223(2):645–52.
[19] Inzana JA, et al. 3D printing of composite calcium phosphate and collagen scaffolds for bone regeneration. Biomaterials 2014;35(13):4026–34.
[20] Duan B, et al. 3D bioprinting of heterogeneous aortic valve conduits with alginate/gelatin hydrogels. J Biomed Mater Res 2013;101(5):1255–64.
[21] Bose S, Vahabzadeh S, Bandyopadhyay A. Bone tissue engineering using 3D printing. Mater Today 2013;16(12):496–504.
[22] Lee J-S, et al. 3D printing of composite tissue with complex shape applied to ear regeneration. Biofabrication 2014;6(2). 024103.
[23] Hockaday LA, et al. Rapid 3D printing of anatomically accurate and mechanically heterogeneous aortic valve hydrogel scaffolds. Biofabrication 2012;4(3). 035005.
[24] Ikegami T, Maehara Y. Transplantation: 3D printing of the liver in living donor liver transplantation. Nat Rev Gastroenterol Hepatol 2013;10(12):697–8.
[25] Sochol RD, Gupta NR, Bonventre JV. A role for 3D printing in kidney-on-a-chip platforms. Curr Transplant Rep 2016;3(1):82–92.
[26] Koch L, et al. Skin tissue generation by laser cell printing. Biotechnol Bioeng 2012;109(7):1855–63.
[27] Miller JS, et al. Rapid casting of patterned vascular networks for perfusable engineered three-dimensional tissues. Nat Mater 2012;11(9):768–74.
[28] Lee VK, et al. Creating perfused functional vascular channels using 3D bio-printing technology. Biomaterials 2014;35(28):8092–102.
[29] Hollister SJ. Porous scaffold design for tissue engineering. Nat Mater 2005;4(7):518–24.
[30] Fan J, Xiong D-S, Huang Y-S. Application of hyaluronic acid in tissue engineering. Chin Med Equip J 2006;5. 011.
[31] Engel E, et al. Nanotechnology in regenerative medicine: the materials side. Trends Biotechnol 2008;26(1):39–47.
[32] Lena S, et al. Guanosine hydrogen-bonded scaffolds: a new way to control the bottom-up realisation of well-defined nanoarchitectures. Chem Eur J 2009;15(32):7792–806.
[33] Sears NA, et al. A review of three-dimensional printing in tissue engineering. Tissue Eng B Rev 2016;22(4):298–310.
[34] Murphy SV, Atala A. 3D bioprinting of tissues and organs. Nat Biotechnol 2014;32(8):773–85.
[35] Hammer J, et al. A facile method to fabricate hydrogels with microchannel-like porosity for tissue engineering. Tissue Eng Part C Methods 2013;20(2):169–76.

[36] Shanjani Y, et al. A novel bioprinting method and system for forming hybrid tissue engineering constructs. Biofabrication 2015;7(4). 045008.
[37] Hinton TJ, et al. Three-dimensional printing of complex biological structures by freeform reversible embedding of suspended hydrogels. Sci Adv 2015;1(9). e1500758.
[38] Sun M, et al. Systematical evaluation of mechanically strong 3D printed diluted magnesium doping wollastonite scaffolds on osteogenic capacity in rabbit calvarial defects. Sci Rep 2016;6. 34029.
[39] Liu A, et al. The outstanding mechanical response and bone regeneration capacity of robocast dilute magnesium-doped wollastonite scaffolds in critical size bone defects. J Mater Chem B 2016;4:3945–58.
[40] Liu A, et al. 3D printing surgical implants at the clinic: a experimental study on anterior cruciate ligament reconstruction. Sci Rep 2016;6. 21704.
[41] Tuan RS, Boland G, Tuli R. Adult mesenchymal stem cells and cell-based tissue engineering. Arthritis Res Ther 2002;5(1):1.
[42] Xu T, et al. Viability and electrophysiology of neural cell structures generated by the inkjet printing method. Biomaterials 2006;27(19):3580–8.
[43] Xu C, et al. Scaffold-free inkjet printing of three-dimensional zigzag cellular tubes. Biotechnol Bioeng 2012;109(12):3152–60.
[44] Xu C, et al. Study of droplet formation process during drop-on-demand inkjetting of living cell-laden bioink. Langmuir 2014;30(30):9130–8.
[45] Christensen K, et al. Freeform inkjet printing of cellular structures with bifurcations. Biotechnol Bioeng 2015;112(5):1047–55.
[46] Kim JD, et al. Piezoelectric inkjet printing of polymers: stem cell patterning on polymer substrates. Polymer 2010;51(10):2147–54.
[47] Cui X, Boland T. Human microvasculature fabrication using thermal inkjet printing technology. Biomaterials 2009;30(31):6221–7.
[48] Cui X, et al. Cell damage evaluation of thermal inkjet printed Chinese hamster ovary cells. Biotechnol Bioeng 2010;106(6):963–9.
[49] Landers R, et al. Rapid prototyping of scaffolds derived from thermoreversible hydrogels and tailored for applications in tissue engineering. Biomaterials 2002;23(23):4437–47.
[50] Mironov V, et al. Organ printing: computer-aided jet-based 3D tissue engineering. Trends Biotechnol 2003;21(4):157–61.
[51] Ozbolat IT, Hospodiuk M. Current advances and future perspectives in extrusion-based bioprinting. Biomaterials 2016;76:321–43.
[52] Colosi C, et al. Microfluidic bioprinting of heterogeneous 3D tissue constructs using low-viscosity bioink. Adv Mater 2016;28(4):677–84.
[53] Trachtenberg JE, et al. Extrusion-based 3D printing of poly (propylene fumarate) in a full-factorial design. ACS Biomater Sci Eng 2016;2:1771–80.
[54] Faulkner-Jones A, et al. Bioprinting of human pluripotent stem cells and their directed differentiation into hepatocyte-like cells for the generation of mini-livers in 3D. Biofabrication 2015;7(4). 044102.
[55] Yu Z, et al. Three-dimensional printing of Hela cells for cervical tumor model in vitro. Biofabrication 2014;6(3). 035001.
[56] Piqué A, et al. A novel laser transfer process for direct writing of electronic and sensor materials. Appl Phys 1999;69(1):S279–84.
[57] Odde DJ, Renn MJ. Laser-guided direct writing of living cells. Biotechnol Bioeng 2000;67(3):312–8.
[58] Ringeisen BR, et al. Laser printing of pluripotent embryonal carcinoma cells. Tissue Eng 2004;10(3–4):483–91.

[59] Schiele NR, et al. Laser-based direct-write techniques for cell printing. Biofabrication 2010;2(3). 032001.
[60] Barron JA, et al. Biological laser printing: a novel technique for creating heterogeneous 3-dimensional cell patterns. Biomed Microdevices 2004;6(2):139–47.
[61] Barron JA, Krizman DB, Ringeisen BR. Laser printing of single cells: statistical analysis, cell viability, and stress. Ann Biomed Eng 2005;33(2):121–30.
[62] Guillemot F, et al. High-throughput laser printing of cells and biomaterials for tissue engineering. Acta Biomater 2010;6(7):2494–500.
[63] Gruene M, et al. Laser printing of three-dimensional multicellular arrays for studies of cell–cell and cell–environment interactions. Tissue Eng Part C Methods 2011;17(10):973–82.
[64] Ho C-T, et al. Liver-cell patterning lab chip: mimicking the morphology of liver lobule tissue. Lab Chip 2013;13(18):3578–87.
[65] Wang Z, et al. A simple and high-resolution stereolithography-based 3D bioprinting system using visible light crosslinkable bioinks. Biofabrication 2015;7(4). 045009.
[66] Gauvin R, et al. Microfabrication of complex porous tissue engineering scaffolds using 3D projection stereolithography. Biomaterials 2012;33(15):3824–34.
[67] Zhang AP, et al. Rapid fabrication of complex 3D extracellular microenvironments by dynamic optical projection stereolithography. Adv Mater 2012;24(31):4266–70.
[68] Soman P, et al. Spatial tuning of negative and positive Poisson's ratio in a multi-layer scaffold. Acta Biomater 2012;8(7):2587–94.
[69] Lin H, et al. Application of visible light-based projection stereolithography for live cell-scaffold fabrication with designed architecture. Biomaterials 2013;34(2):331–9.
[70] Nichol JW, Khademhosseini A. Modular tissue engineering: engineering biological tissues from the bottom up. Soft Matter 2009;5(7):1312–9.
[71] Mironov V, et al. Organ printing: tissue spheroids as building blocks. Biomaterials 2009;30(12):2164–74.
[72] Jakab K, et al. Tissue engineering by self-assembly and bio-printing of living cells. Biofabrication 2010;2(2). 022001.
[73] Marga F, et al. Toward engineering functional organ modules by additive manufacturing. Biofabrication 2012;4(2). 022001.
[74] Norotte C, et al. Scaffold-free vascular tissue engineering using bioprinting. Biomaterials 2009;30(30):5910–7.
[75] Kelm JM, et al. A novel concept for scaffold-free vessel tissue engineering: self-assembly of microtissue building blocks. J Biotechnol 2010;148(1):46–55.
[76] Boland T, et al. Application of inkjet printing to tissue engineering. Biotechnol J 2006;1(9):910–7.
[77] Pataky K, et al. Microdrop printing of hydrogel bioinks into 3D tissue-like geometries. Adv Mater 2012;24(3):391–6.
[78] Skardal A, Atala A. Biomaterials for integration with 3-D bioprinting. Ann Biomed Eng 2015;43(3):730–46.
[79] Munaz A, et al. Three-dimensional printing of biological matters. J Sci Adv Mater Dev 2016;1(1):1–17.
[80] Malda J, et al. 25th anniversary article: engineering hydrogels for biofabrication. Adv Mater 2013;25(36):5011–28.
[81] Wüst S, et al. Tunable hydrogel composite with two-step processing in combination with innovative hardware upgrade for cell-based three-dimensional bioprinting. Acta Biomater 2014;10(2):630–40.
[82] Wong TW. Alginate graft copolymers and alginate–co-excipient physical mixture in oral drug delivery. J Pharm Pharmacol 2011;63(12):1497–512.

[83] Lee KY, Mooney DJ. Alginate: properties and biomedical applications. Prog Polym Sci 2012;37(1):106–26.
[84] Pawar SN, Edgar KJ. Alginate derivatization: a review of chemistry, properties and applications. Biomaterials 2012;33(11):3279–305.
[85] Cohen J, et al. Survival of porcine mesenchymal stem cells over the alginate recovered cellular method. J Biomed Mater Res A 2011;96(1):93–9.
[86] Zhao L, et al. The integration of 3-D cell printing and mesoscopic fluorescence molecular tomography of vascular constructs within thick hydrogel scaffolds. Biomaterials 2012;33(21):5325–32.
[87] Bertassoni LE, et al. Direct-write bioprinting of cell-laden methacrylated gelatin hydrogels. Biofabrication 2014;6(2). 024105.
[88] Soman P, et al. Digital microfabrication of user-defined 3D microstructures in cell-laden hydrogels. Biotechnol Bioeng 2013;110(11):3038–47.
[89] Arai K, et al. Three-dimensional inkjet biofabrication based on designed images. Biofabrication 2011;3(3). 034113.
[90] Pereira RF, Bártolo PJ. 3D bioprinting of photocrosslinkable hydrogel constructs. J Appl Polym Sci 2015;132(48). 42458.
[91] Morris VB, et al. Mechanical properties, cytocompatibility and manufacturability of chitosan: PEGDA hybrid-gel scaffolds by stereolithography. Ann Biomed Eng 2016;1–11.
[92] Mehesz AN, et al. Scalable robotic biofabrication of tissue spheroids. Biofabrication 2011;3(2). 025002.
[93] Jakab K, et al. Tissue engineering by self-assembly of cells printed into topologically defined structures. Tissue Eng A 2008;14(3):413–21.
[94] Tse C, et al. Inkjet printing Schwann cells and neuronal analogue NG108-15 cells. Biofabrication 2016;8(1). 015017.
[95] Leng L, Ba Q, Amini-Nik S, et al. Skin printer: microfluidic approach for skin regeneration and wound dressing. In: 17th international conference on miniaturized systems for chemistry and life sciences, MicroTAS 2013, 3; 2013.
[96] Vijayavenkataraman S, Lu WF, Fuh JYH. 3D bioprinting of skin: a state-of-the-art review on modelling, materials, and processes. Biofabrication 2016;8(3). 032001.
[97] Köpf M, et al. A tailored three-dimensionally printable agarose–collagen blend allows encapsulation, spreading, and attachment of human umbilical artery smooth muscle cells. Biofabrication 2016;8(2). 025011.
[98] Robbins JB, et al. A novel in vitro three-dimensional bioprinted liver tissue system for drug development. FASEB J 2013;27(872):812.
[99] 3D bioprinting solutions. Laboratory for biotechnological research. Moscow, Russia. http://www.bioprinting.ru [Accessed May 2015].
[100] Faulkner-Jones A, et al. Development of a valve-based cell printer for the formation of human embryonic stem cell spheroid aggregates. Biofabrication 2013;5(1). 015013.
[101] Cyfuse Biomedical K.K. University of Tokyo Entrepreneur Plaza. Tokyo, Japan. http://www.cyfusebio.com [Accessed May 2015].
[102] Reid JA, et al. Accessible bioprinting: adaptation of a low-cost 3D-printer for precise cell placement and stem cell differentiation. Biofabrication 2015;8(2). 025017.
[103] Shim J-H, et al. Bioprinting of a mechanically enhanced three-dimensional dual cell-laden construct for osteochondral tissue engineering using a multi-head tissue/organ building system. J Micromech Microeng 2012;22(8) 085014.
[104] Ragaert K, et al. Machine design and processing considerations for the 3D plotting of thermoplastic scaffolds. Biofabrication 2010;2(1) 014107.
[105] Hiller J, Hod L. Methods of parallel voxel manipulation for 3D digital printin. In: Proceedings of the 18th solid freeform fabrication symposium, vol. 200; 2007.

[106] Ozbolat IT, Chen H, Yin Y. Development of 'Multi-arm Bioprinter' for hybrid biofabrication of tissue engineering constructs. Robot Comput Integr Manuf 2014;30(3): 295–304.
[107] Zhao Y, et al. The influence of printing parameters on cell survival rate and printability in microextrusion-based 3D cell printing technology. Biofabrication 2015;7(4). 045002.
[108] Snyder J, et al. Hetero-cellular prototyping by synchronized multi-material bioprinting for rotary cell culture system. Biofabrication 2016;8(1). 015002.
[109] Merceron TK, et al. A 3D bioprinted complex structure for engineering the muscle–tendon unit. Biofabrication 2015;7(3). 035003.
[110] Kim BS, et al. Three-dimensional bioprinting of cell-laden constructs with polycaprolactone protective layers for using various thermoplastic polymers. Biofabrication 2016;8(3). 035013.
[111] Fedorovich NE, et al. Organ printing: the future of bone regeneration? Trends Biotechnol 2011;29(12):601–6.
[112] Mandrycky C, et al. 3D bioprinting for engineering complex tissues. Biotechnol Adv 2016;34(4):422–34.
[113] He Y, et al. Research on the printability of hydrogels in 3D bioprinting. Sci Rep 2016;6. 29977.
[114] Huang H, et al. Avidin–biotin binding-based cell seeding and perfusion culture of liver-derived cells in a porous scaffold with a three-dimensional interconnected flow-channel network. Biomaterials 2007;28(26):3815–23.
[115] Kolesky DB, et al. Three-dimensional bioprinting of thick vascularized tissues. Proc Natl Acad Sci 2016;113(12):3179–84.
[116] Kolesky DB, et al. 3D bioprinting of vascularized, heterogeneous cell-laden tissue constructs. Adv Mater 2014;26(19):3124–30.
[117] Gao Q, et al. Coaxial nozzle-assisted 3D bioprinting with built-in microchannels for nutrients delivery. Biomaterials 2015;61:203–15.
[118] Zhang Y, et al. Characterization of printable cellular micro-fluidic channels for tissue engineering. Biofabrication 2013;5(2). 025004.
[119] Luo Y, Lode A, Gelinsky M. Direct plotting of three-dimensional hollow fiber scaffolds based on concentrated alginate pastes for tissue engineering. Adv Healthc Mater 2013;2(6):777–83.
[120] Onoe H, et al. Metre-long cell-laden microfibres exhibit tissue morphologies and functions. Nat Mater 2013;12(6):584–90.
[121] Wang Z, et al. Organ-on-a-chip platforms for drug delivery and cell characterization: a review. Sensor Mater 2015;27(6):487–506.
[122] Maschmeyer I, et al. A four-organ-chip for interconnected long-term co-culture of human intestine, liver, skin and kidney equivalents. Lab Chip 2015;15(12):2688–99.
[123] Lee H, Cho D-W. One-step fabrication of an organ-on-a-chip with spatial heterogeneity using a 3D bioprinting technology. Lab Chip 2016;16(14):2618–25.
[124] Huh D, et al. Reconstituting organ-level lung functions on a chip. Science 2010;328(5986):1662–8.

Further Reading

[1] Hoch E, Tovar GE, Borchers K. Bioprinting of artificial blood vessels: current approaches towards a demanding goal. Eur J Cardiothorac Surg 2014;46(5):767–78.
[2] Kinstlinger IS, Miller JS. 3D-printed fluidic networks as vasculature for engineered tissue. Lab Chip 2016;16(11):2025–43.

Computational design of biostructures

3

D. *Eggbeer*
PDR, Cardiff Metropolitan University, Cardiff, United Kingdom

3.1 Introduction

The application of medical imaging, CAD, and additive manufacturing (AM) in the field of biostructures is evolving at a rapid pace. On its initial conception in the 1980s through to the 1990s, CAD and AM were largely developed for the aerospace, automotive, and other engineering sectors. As technologies became more commercialized and affordable, bureau services began adopting them, making them more widely available in product development. In parallel to other industrial sectors, interest in the application of advanced computer-based technologies to medical applications has also grown dramatically, particularly in the last two decades. This has been driven by advances in medical imaging modalities such as computer tomography (CT), improvements in specifically developed CAD software, and increased accessibility to high-performance computer technologies. By 1993, methods of using AM to produce anatomical models were becoming increasingly refined, driven by research at the University of Leuven [1], yet the processes were still not in widespread use. Materialise (Leuven, Belgium) were early pioneers in the development of specialist software to enable processing of medical scan data for translation into computer models. Since 1994, they have developed a range of software and services dedicated to processing CT data into CAD models, designing custom devices, and preparing data for AM fabrication. As early as 1995, applied CAD/machining technologies were utilized in industrial engineering practices to assist the design of implant prostheses [2]. However, a major impediment to the application of CAD in medical applications such as implant design and prosthetic design was the fact that it requires the integration of existing anatomical forms with the creation of complex, naturally occurring free form shapes.

Developments in both CAD software and AM processes have resulted in much more widespread use in biostructure fabrication since the early 2000s. Researchers began to focus on how CAD could be used to plan complex surgical procedures and design custom-fitting devices such as guides (physical jigs used in the surgical procedure to drill through, cut along, or position anatomy) [3]. AM became a tool to produce end-use devices that allowed complex surgical plans to be translated into the operating theater using custom physical devices.

As AM technologies have reached a point of functional and regulatory readiness, their benefits can be widely exploited. There is also increasing evidence that products such as patient-specific guides and implants are associated with reductions in expensive surgery time, reduced infection risks, and recovery periods compared with off-the-shelf

3D Bioprinting for Reconstructive Surgery. https://doi.org/10.1016/B978-0-08-101103-4.00004-1

alternatives [4]. Research illustrates that they can deliver a more accurate fit, help reduce theater time, reduce the likelihood of surgical revisions [5], decrease stress shielding and bone resorption [6], incorporate tailored mechanical properties [7], and improve osseointegration [8]. Publications demonstrate this in AM case studies ranging from cranioplasty plates [9], to orbital floor reconstruction implants [10], and complex osteotomies [11].

In more recent years, lower-cost medical image processing software and AM technologies (now commonly termed three-dimensional (3D) printing), have begun filtering into hospitals, introducing the possibilities to an entirely new audience of nonengineering specialist users. The scope of application is also increasing from alloplastic polymer and metal devices to include research into the fabrication of autologous biostructures. This hybrid of biology, design, and engineering is beginning to challenge the way medicine is practiced and providing exciting new opportunities in the way disease, trauma, and congenital conditions are treated.

This chapter describes the application of medical imaging, CAD, and AM in the field of alloplastic biostructures, particularly focusing on cranio-maxillofacial, orthopedic and facial prosthetic applications. It covers how these techniques are used to create computer models of patient anatomy, plan surgical procedures, design, and produce custom devices. The range of technologies and processes used in each stage are covered in general terms for those not familiar with them or new to the field, and case studies are used to illustrate how these techniques have been used in practice in recent years. These case studies will illustrate where CAD/AM have been combined to enable otherwise difficult complex procedures to be undertaken efficiently. A discussion on the barriers that currently prevent more widespread adoption of CAD/AM and the research that is being undertaken to address them are also presented. Implications for more efficient and widespread adoption of CAD/AM are concluded.

3.2 Essential steps in the computational design of biostructures

Five essential steps are required when designing alloplastic biostructures:

1. Medical imaging (to acquire the anatomical structures of interest);
2. Data processing (to segment and prepare the data, then create a CAD model);
3. Design (to create a CAD model of the final structure or a mold tool to create the final part);
4. Fabrication (to create the final physical part); and
5. Postprocessing (to complete the part prior to sterilization and use).

These steps are discussed in detail.

3.3 Medical imaging

Acquiring good-quality images of anatomy is a prerequisite to producing virtual or physical models using CAD and AM. There are a number of applicable scanning modalities that can be used to capture biostructures, some of which require substantial hospital-based infrastructure and others which can be deployed in small labs or clinics.

There are two broad categories of scanning modality for biostructures: those designed to capture both internal and external anatomical details and those that capture only external topography. Most hospital-based scanners capture both internal and external anatomical details and are typically large, sophisticated pieces of equipment also used for diagnosis purposes. The two most commonly used modalities, which have already been introduced briefly are CT and magnetic resonance (MR) imaging. Positron emission tomography (PET) is also used in some situations to identify specific disease locations within the body. Both CT & MR use different physical methods to generate cross-sectional images through the human body. The patient is typically scanned lying down on a table that is fed through the center of the scanner while the images are acquired. The result is a stacked set of two-dimensional (2D) images.

Medical imaging professions have adopted an internationally agreed standard for the exchange of patient data, DICOM. DICOM stands for digital imaging and communications. The standard was initiated in the 1970s by a joint committee from the American College of Radiology (ACR) and the National Electrical Manufacturers Association (NEMA). They first published an ACR-NEMA standard in 1985 and updated it in 1988. The name was changed to DICOM in in 1993. The standard now covers all kinds of medical images but also includes other data such as patient name, reference number, study number, dates, and reports. Since then, most manufacturers adhere to the standard, and data transfer problems are much less likely to occur than was previously the case.

The DICOM standard (ISO 12052) enables the transfer of medical images to and from software and scanners from different manufacturers and aids the development of picture archiving and communication systems (PACS), which can be incorporated with larger medical information or records system. DICOM images are the usual start point from which most medical imaging to CAD to AM production methods are undertaken.

More information on DICOM can be found at http://medical.nema.org/.

3.3.1 Computer tomography (CT) scanning

CT scanning was developed by Godfrey Hounsfield of EMI Laboratories in 1972 and was first clinically available by around 1974–76. This allowed medical specialists to investigate internal body structures and create a 3D visualization of bony anatomy.

CT works by passing focused ionizing X-rays from an emitter through the body and measuring the amount of the X-ray energy absorbed on the detector. The person being scanned lays on a table, which travels along the longer axis of the body through a rotating emitter and detector array. The amount of X-ray energy absorbed by a known slice thickness is proportional to the density of the body tissue. Cross-sections perpendicular to the longer axis of the patient are acquired. This is therefore usually termed axial or transverse images. Taking measurements from many angles allows the tissue densities to be composed as a gray scale cross-sectional image. Modern scanners perform a continuous spiral around the longer axis of the patient, which enables scans to be performed rapidly (a few seconds or less). 3D CT scans are therefore often referred to as helical CT. Many modern CT scanners also use multiple arrays to increase the rate of data capture and improve volume acquisition. Tissue density is indicated by shades within the gray scale. A quantitative scale for describing radiodensity, termed the Hounsfield scale (named after Sir Godfrey Newbold Hounsfield), provides an accurate density for the type of tissue. Air is represented by a value of −1000 (black on the gray scale) and bone/enamel between +700 (cancellous bone) and +3000 (dense bone) (white on the gray scale). An example CT slice image through the head is shown in Fig. 3.1. Images are much like computer bitmaps with a pixel resolution of typically 512x512 or 1024x1024.

As Fig. 3.1 illustrates, CT shows bones more clearly than surrounding soft tissues, making it an important imaging modality when investigating skeletal biostructures. The density difference between soft tissues and air is also high, which allows structures such as sinuses to be easily visualized. The relatively narrow Hounsfield range between soft tissue structures means that they are more difficult to be visualized on CT image slices; adjacent soft tissue structures are difficult to differentiate between. Contrast agents that absorb x-ray energy can be used to help boost the contrast of some structures and make them more visible when reviewing CT slice images. This is useful, for example, when attempting to identify vascular systems in relation to bony anatomy when planning autogenous reconstruction (Fig. 3.2).

The quality of CT images and subsequent reconstructions can vary greatly depending on the slice distances, image resolution, computer reconstruction algorithms, and other factors. Some factors can be controlled to help ensure the data captured can be used for subsequent visualization of biostructures, whereas others cannot. Some of the

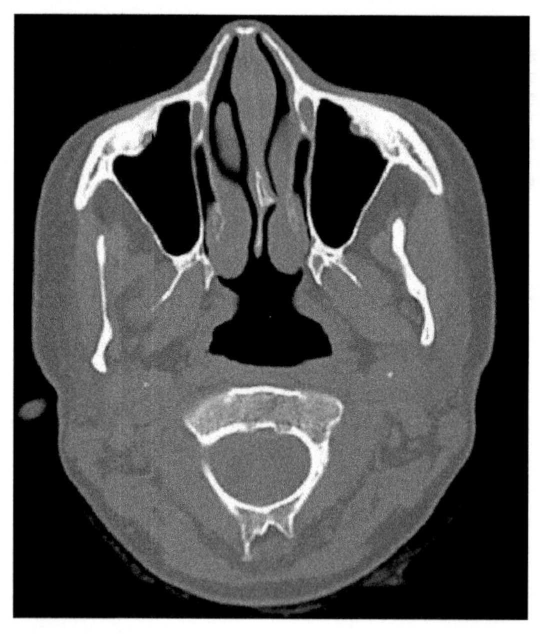

Fig. 3.1 A typical Axial CT slice image through the midface with cortical (dense) bone showing as white, air as black, and soft tissues as shades of gray.

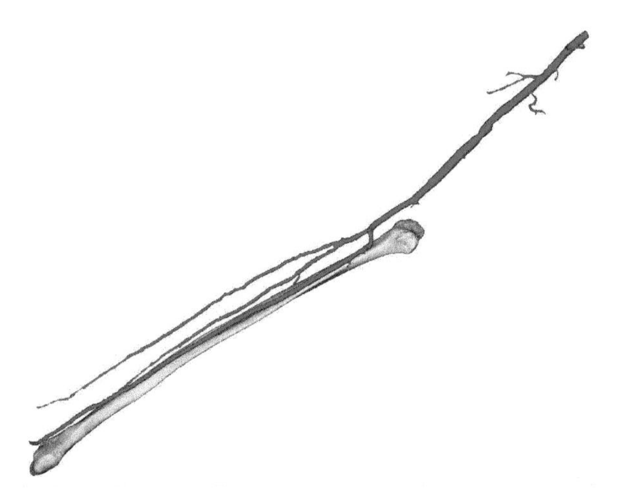

Fig. 3.2 Blood vessels highlighted in relation to the fibula bone to aid in fibula freeflap reconstruction.

primary factors that influence image quality include slice distance, partial pixel effect, gantry tilt, noise, scatter, and movement.

A higher number of slices (decreased slice distance) will provide better detail in areas of high anatomical detail and thin bones. However, increasing the number of slices will also increase the level of ionizing radiation a patient is exposed to. Fig. 3.3 illustrates how a low number of CT slices results in a poor 3D reconstruction. Therefore, a balance must be struck between the level of detail required and acceptable radiation dose. The balance between obtaining sufficient data resolution and minimizing radiation dose is particularly important around sensitive areas of anatomy such as the orbits. The bones underneath the contents of the eye are extremely thin (in the order of 0.3 mm), so can fall between the slice thickness of CT scanning. Imaging the orbits accurately is also made more difficult due to the partial pixel effect. Where areas of high

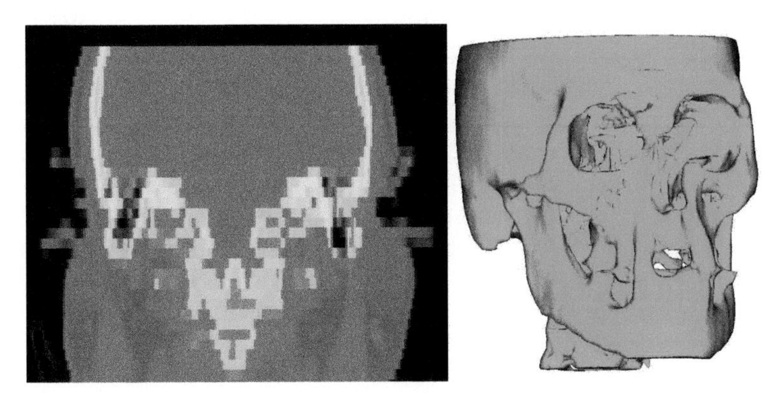

Fig. 3.3 The effect of a low number of CT slices on a 3D reconstruction of the skull. Left: coronal view of the CT data. Right: 3D reconstruction of the CT data.

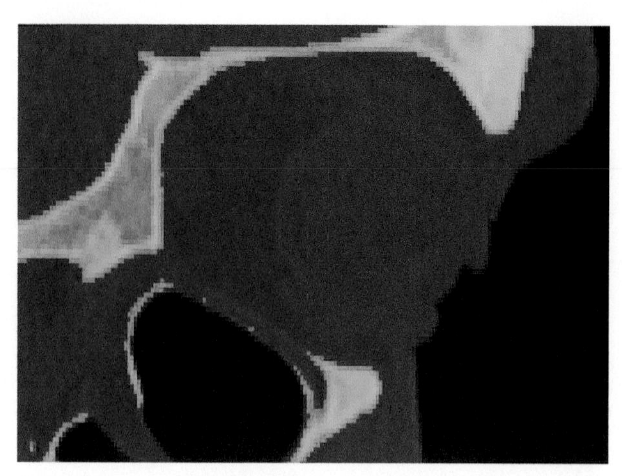

Fig. 3.4 Shown in the coronal view—Thin bones of the orbital floor that lie next to air in the maxillary sinus can appear as gray rather than clear white in CT data due to the limited resolution of the image and partial pixel effect.

tissue density such as the thin bones of the orbital floor are adjacent to air, the limited pixel resolution of the image slice means that the boundaries cannot be defined accurately. Hard tissues typically appear blurred and are an approximated gray between black air and white hard tissue. The effect is missing portions data where there should be hard tissue anatomy. This is illustrated in Fig. 3.4. The partial pixel effect should be considered when reconstructing computer and physical models of anatomy, especially involving the midface. Case study 1 illustrates an example where the orbital floors of a patient were reconstructed in CAD to enable the design of a custom implant.

Gantry tilt is sometimes used to optimize the orientation of CT slices to improve the capture of anatomy. The bed on which the patient lies down is moved through the scanner at an angle. Generally, gantry tilt should be avoided when data are intended for subsequent CAD reconstruction as although software is capable of picking up the degree of gantry tilt, whether it is positive or negative is not specified. Where small gantry tilt values are used, it is therefore possible to create a CAD reconstruction with the anatomy 'skewed' and therefore inaccurate (Fig. 3.5).

Noise in CT images makes it difficult to distinguish clear tissue boundaries and has the effect of creating a lumpy and porous surface for CAD reconstructions. Noise is typically worse in deep areas of anatomy. Fig. 3.6 illustrates the effect of noise. To a degree, noise can be controlled. Generally, smooth CT reconstruction algorithms result in data that are better for CAD and physical model reconstruction. Liaising with radiographers to optimize the image processing algorithms and kernels to generate smooth images can assist in accurate CAD surface model generation.

Dense objects such as dental fillings and implants can cause x-rays to scatter, causing streaks and shadows in CT slice images (Fig. 3.7). This is a particular problem in the facial region, where details such as teeth or thin midface bone structures are commonly affected if the person being scanned has fillings or previous reconstruction

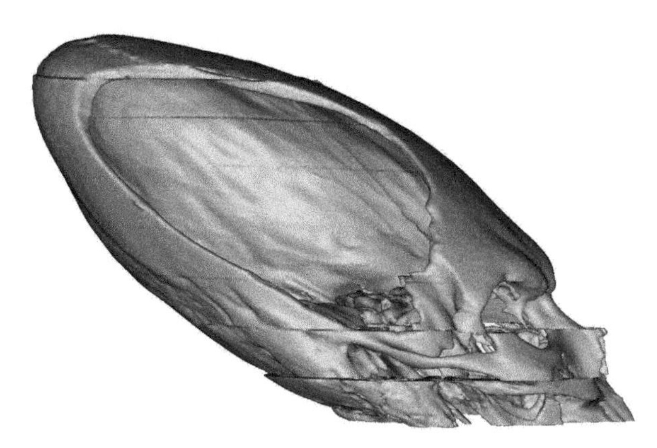

Fig. 3.5 Illustrated in this image of the skull, extreme gantry tilt becomes obvious (in this case it is also shown with steps in the data) in 3D reconstructions.

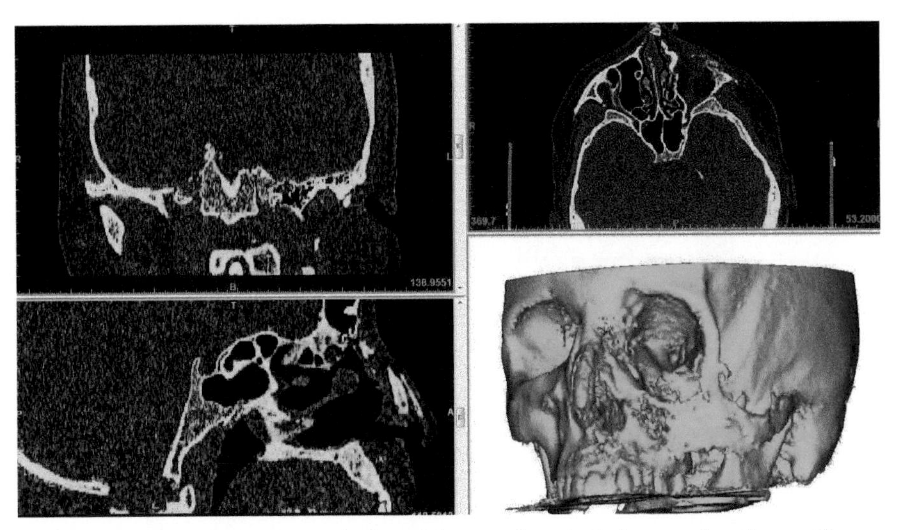

Fig. 3.6 The effect of noise shown in the axial, coronal, sagittal views, and on the resulting 3D reconstruction.

implants placed. Little can be done to overcome the effect of scatter, so manual editing of data is required in later stages to clean up the effects. This is discussed in detail in the Image processing and 3D reconstruction section.

Patient movement can also cause problems when CT scanning. Like in photography, if the object being captured moves, a ghosting or step effect is seen in the images. Even with capture times of only a few seconds, movement results in an inaccurate 3D reconstruction of the CT slice images when they are stacked together. Fig. 3.8 illustrates this. It is not usually possible to correct for movement, so care must be taken to ensure the patient remains still during image acquisition.

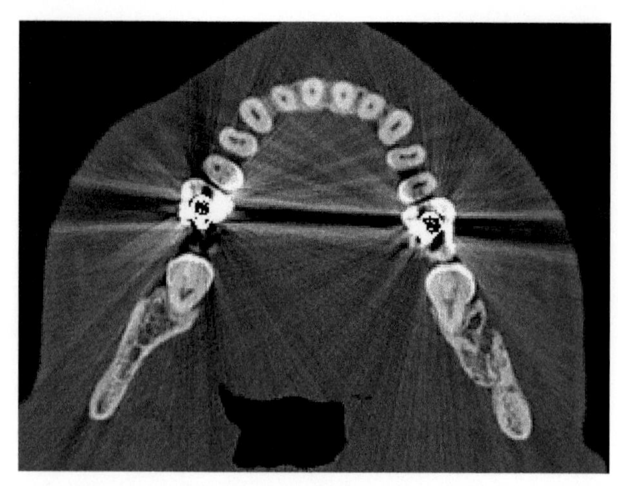

Fig. 3.7 Illustrated in an axial slice of the mandible region, scatter causes streaks and dark shadows in CT images. In this case, scatter is caused by teeth fillings.

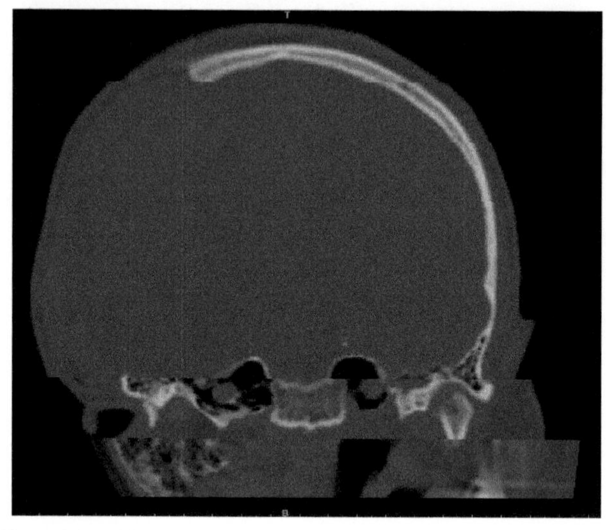

Fig. 3.8 The effect of a patient moving during a CT scan illustrated in a coronal reconstruction of a head CT scan.

3.3.2 Cone beam CT

Cone beam CT (CBCT) has become popular in orthodontic, dental, and maxillofacial applications. It is a more recent development that utilizes the same basic principle as conventional CT scanning, but in a way that results in lower patient radiation doses. Unlike conventional CT, the patient sits upright in a CBCT scanner. Unlike

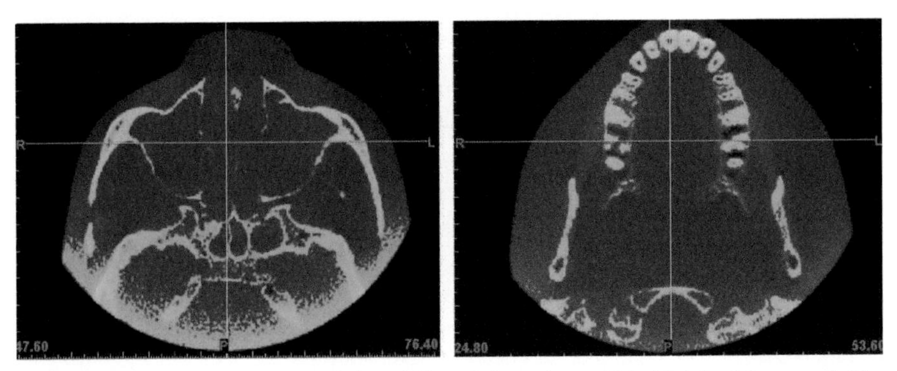

Fig. 3.9 CBCT axial image through the midface (left) and mandible (right) with an overlaid mask of green selecting the Hounsfield range of bone.

conventional CT (which utilizes a narrow fan of x-rays and a narrow detector), the x-rays in CBCT are emitted in a cone shape and detected by a flat panel detector. Hundreds of images are captured from different angles in a single rotation of the emitter and detector around the head (termed isocenter). Software algorithms are used to calculate volumetric data. CBCT is a much lower-cost and smaller-footprint imaging modality that is designed to focus on a smaller field of view in greater detail. CBCT produces isotropic (equal in the x-y-z dimensions) voxels as low as 0.076 mm, allowing fine details of structures such as individual teeth to be imaged. There are a number of notable practical and technical limitations. The field of view is usually limited to anterior portions of the head (the technology has been most widely developed for dental specialties). Image quality is also compromised compared with conventional CT. Grayscale/Hounsfield values typically alter in each image and throughout the volume, making it much more difficult to segment tissue in subsequent processing stages. This is a particular problem at the extremities of the field of view, where noise increases and contrast decreases. This is illustrated in Fig. 3.9. Contrast between soft tissues is also poor, making CBCT less useful for the evaluation of pathologies. These issues can make subsequent data segmentation in postprocessing software using automatic algorithms challenging.

3.3.3 *Magnetic resonance imaging*

MR imaging works by using powerful magnet fields to alter the alignment of randomly magnetic aligned protons (hydrogen ions) within the human body temporarily. Radio frequency pulses push the aligned protons to a higher energy level. When the magnetic field is turned off, the ions return to their natural alignment and release the energy they absorbed as radio waves. To construct an MR image, the strength of the radio waves emitted by the atoms is measured at precise locations. The repeated alignment, radio wave excitement, and relaxing of protons is necessary to detect the very small amount of emitted energy. By collecting signals from many locations, a cross-sectional image can be created. Different radio coils and sequences are used

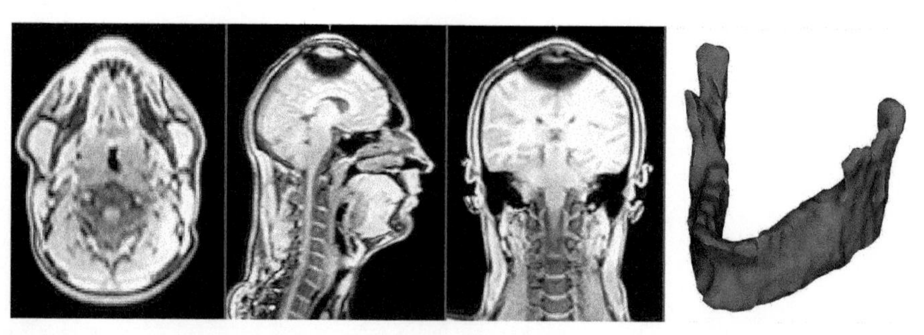

Fig. 3.10 Illustrated in an axial, sagittal, coronal, and 3D reconstruction of the mandible (left to right), MR data show good soft tissue contrast, but is not well-adapted for bony reconstruction.

depending on the type of image required and tissue being imaged. As in CT scanning, the resulting cross-sectional image is a gray scale pixel image, the shade of gray being proportional to the strength of the signal.

MR imaging targets the hydrogen nuclei present in water molecules. Soft tissues that have a high water content show up in lighter shades of gray and areas containing little or no water show up darker. Air, cortical bone, and teeth show as black, making MR less suitable for imaging bony details. An example of MR image slice data and a resulting 3D reconstruction of the mandible is shown in Fig. 3.10.

MR images also suffer from similar shadowing of details caused by implants, fillings, etc.

As MR imaging does not expose the body to ionizing radiation, research has attempted to develop it as an alternative to CT for visualization [12]. However, longer acquisition times exacerbate the issues of patient movement during scans and leads to higher operating costs.

3.3.4 3D Surface scanning techniques

When capturing the surface topography of human anatomy, light-based surface scanning methods are typically more feasible to use than dedicated medical imaging modalities. Initially developed for industrial applications, surface scanning methods are increasingly being adopted for medical applications to replace the need for physical impression taking. Particularly prominent applications include prosthetic and orthotics [13,14]. Unlike CT, CBCT, or MR, surface scanning only captures external details, but is nonionizing and can be extremely fast. The noncontact nature also prevents the distortion of tissues.

There are an increasing range of surface scanning/capture methods including structured white light, fringe project, laser, photogrammetry, infrared depths sensing, and hybrids. Each has benefits and limitations, but there are general guidelines which, if followed, help ensure that the data captured is of good quality. As in photography and medical imaging modalities, subject movement should be avoided to prevent the data being blurred or inaccurate. Faster data acquisition methods such as photogrammetry have capture times of tenths of a second, but subject stillness helps improve accuracy.

Also like photography, surface scanners can only capture what is visible, so areas hidden from view result in gaps. All light-based scanners are also not able to capture highly reflective, transparent, or noncoherent surfaces such as eyes or hair well. These surfaces usually result in missing data and gaps in the reconstructed surfaces.

The raw data captured are usually in the form of a point cloud, comprising thousands of x-y-z plotted points in space. Postprocessing is required to filter outlaying points. Further processing is undertaken to create surfaces from the point cloud, smooth those surfaces, and fill holes. Many modern scanners and proprietary software undertake this processing with minimal manual intervention, thereby making it easier for the user.

3.4 File formats

Alongside the numerous proprietary file formats associated with software programs, there are a small number of common cross-platform ones that are used to exchange CAD data and in preparation for AM/3D printing production. The most widely used, the STereoLithography (STL) file, describes surfaces of a CAD object by a series of triangles, each triangle being specified by its normal (vector perpendicular to the surface) and the three points of its vertices. The result is a faceted model; the number of triangles describing an object is directly proportional to the appearance of facets and how accurate the STL surface is to the original CAD model (Fig. 3.11). The STL file was developed originally by 3D-Systems for the Stereolithography AM production method described later in this chapter. Other similar file formats, such as OBJ (developed by Wavefront Technologies, USA) or PLY (Also known as the Stanford Triangle Format) also use triangles to describe the surface of a CAD model, but also include the ability to describe color.

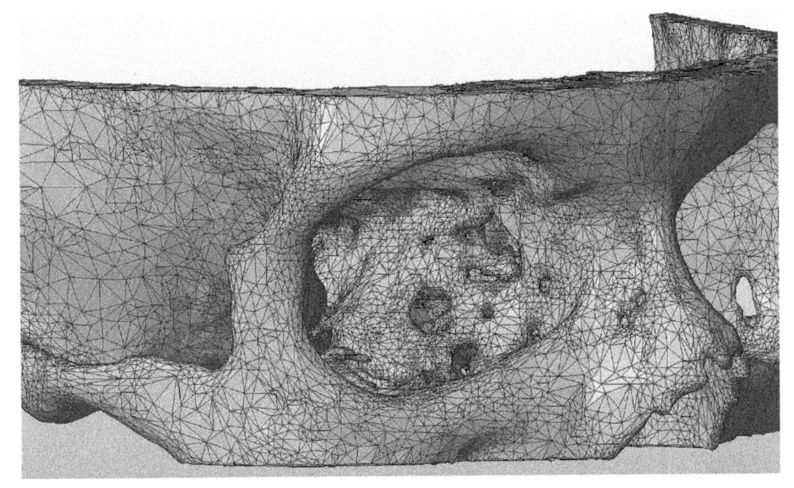

Fig. 3.11 The triangular mesh structure of an STL file shown on a 3D reconstruction of the orbits.

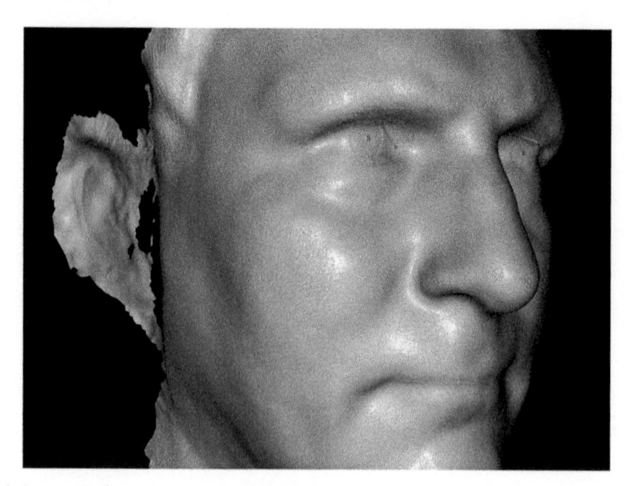

Fig. 3.12 A high-resolution, reconstructed surface scan of the face.

STL, OBJ, and PLY only describe the outer surface of a CAD geometry. Fig. 3.12 illustrates the result of a 3D surface scan of a face. This surface has been reconstructed automatically by the proprietary software by filtering the point cloud data and creating an STL surface based on defining triangle edges between the points. Where no point cloud was created, there is no surface. Further processing is required to give the surface an enclosed volume. This is a necessary step prior to AM fabrication.

3.5 Image processing and 3D reconstruction

This section covers the processing of medical imaging data for the creation of computer-based biostructure models. Details of the common software approaches and associated methods of creating useable data are covered.

There is a range of commercial software programs available to convert DICOM data to a CAD model—even freeware—and the general processes are the same in each. It is important however, to ensure that the software used in clinical cases meets regulatory and quality requirements. The essential steps are: import the slice data into the software (typically relying on automated wizards that ensure it is correctly orientated and scaled); save the project; select the desired range of tissue density of interest based on the Hounsfield units (typically bone); create a 'mask' that represents the anatomy of interest (multiple masks can typically be created); manually edit 'mask(s)' to improve the accuracy (including the clean-up of scatter, noise, and re-processing of information lost due to the partial pixel effect); turn the slices into a CAD surface (3D volumetric rendering); and output the volume for subsequent planning and device design.

One of the longest-established and common software packages used to process DICOM data is Mimics (Materialise NV, Belgium) and another common alternative is Simpleware (Synopsys, Inc., UK). Fig. 3.13 illustrates the basic process of DICOM to

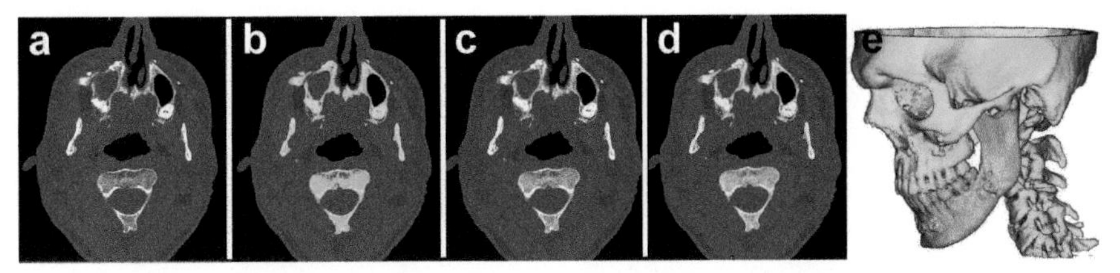

Fig. 3.13 Steps to process CT data to a CAD model in Mimics. DICOM format CT slice image (A), an overlaid 'mask' selecting tissue within a given Hounsfield value range (B), 'Region Growing' to remove unwanted parts of the 'mask' (C), separating the mandible and maxilla (D), and calculating a 3D rendering (E).

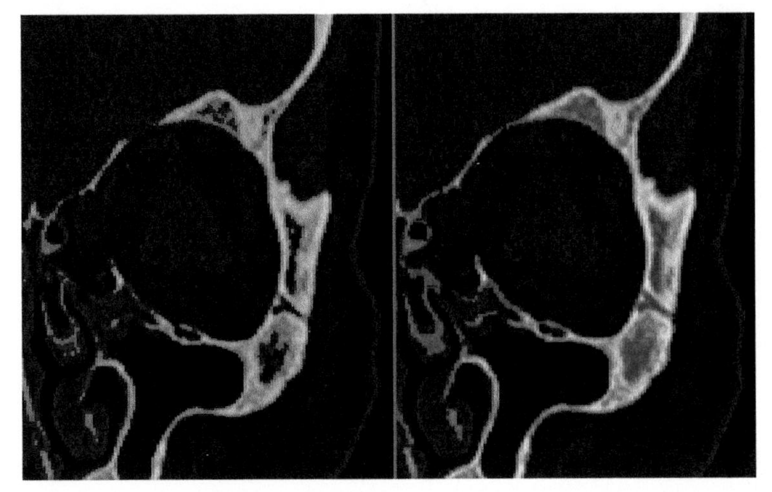

Fig. 3.14 Left: coronal view of the original mask (pink) of an orbital floor showing some gaps. Right: the same orbit having had the orbital floor manually edited back in using localized thresholding.

CAD. Fig. 3.14 illustrates how Mimics can be used to ensure details such as the thin bones of the orbital floor can be manually edited back into the data if they are missed due to the partial pixel effect. Fig. 3.15 illustrates the removal of scatter.

3.6 CAD

There are a large number of dedicated software programs for surgical planning and biostructure modelling. Some are dedicated to the planning and execution of specific surgical procedures, some dictate the use of a specific service provider to supply the end use devices, whereas others allow a higher degree of design and production freedom. Many of the medical procedure-specific software tools have the benefit of allowing a non-CAD specialist to access planning and production tools. Nonmedical-specific software is inherently more versatile, but is not tailored to non-CAD or non-engineering users. There is an inherent barrier to their use by nondesign experts.

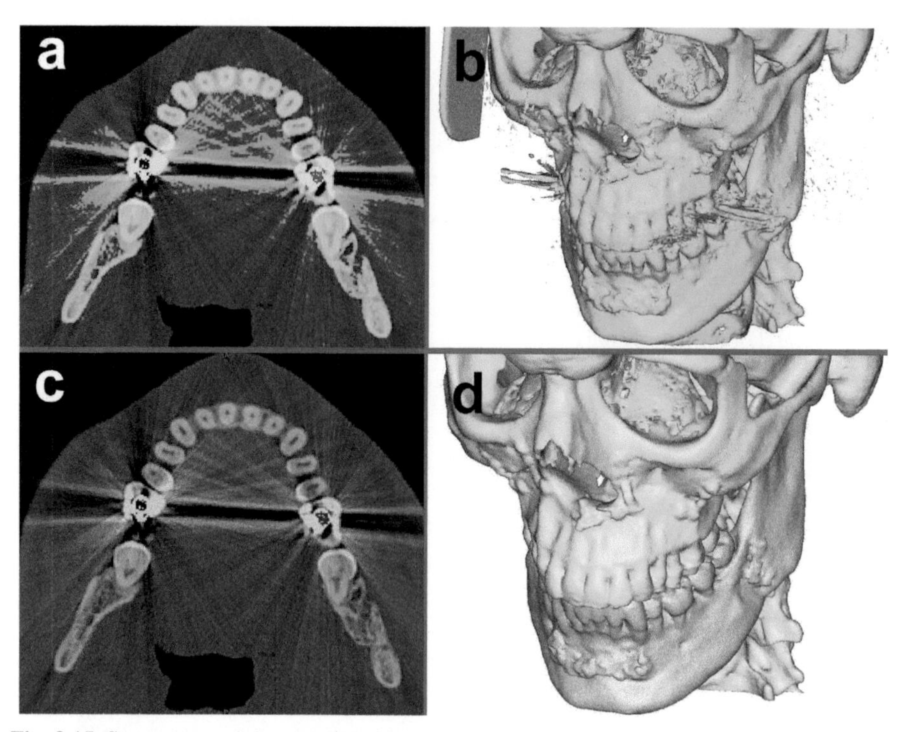

Fig. 3.15 Scatter around the mandible in an axial view (A), resulting 3D reconstruction (B), manual clean-up of the mask (C), resulting 3D reconstruction once cleaned, and the mandible and maxilla have been separated (D).

FreeForm Plus (Geomagic 3D-Systems, USA) represents one of the most widely used and published on design tools for medical applications [15–17]. FreeForm uses voxel modelling (3D pixels) combined with a haptic (sense of touch) device, which is akin to hand carving, but in a digital environment (Fig. 3.16). It is therefore appropriate for designing complex, unconstrained 3D shapes and forms.

Materialise offers an alternative, 3-Matic, which relies on the native STL file format for geometry design. There are also an increasing number of engineering CAD software packages and 'freeware' capable of relatively complex form of design, but these tend to be more limited than both FreeForm and 3-Matic.

3.7 An overview of additive manufacture and 3D printing

According to Swaelens and Kruth [1], and Santler et al. [18], Alberti was the first to publish the idea of making 3D physical models from CT data in 1980. Early techniques relied on cutting out and stacking sheets of aluminum according to the individual CT slice image. Toth et al. later published some of the earliest attempts at using computer technologies to manufacture craniofacial reconstructions [19]. Some initial

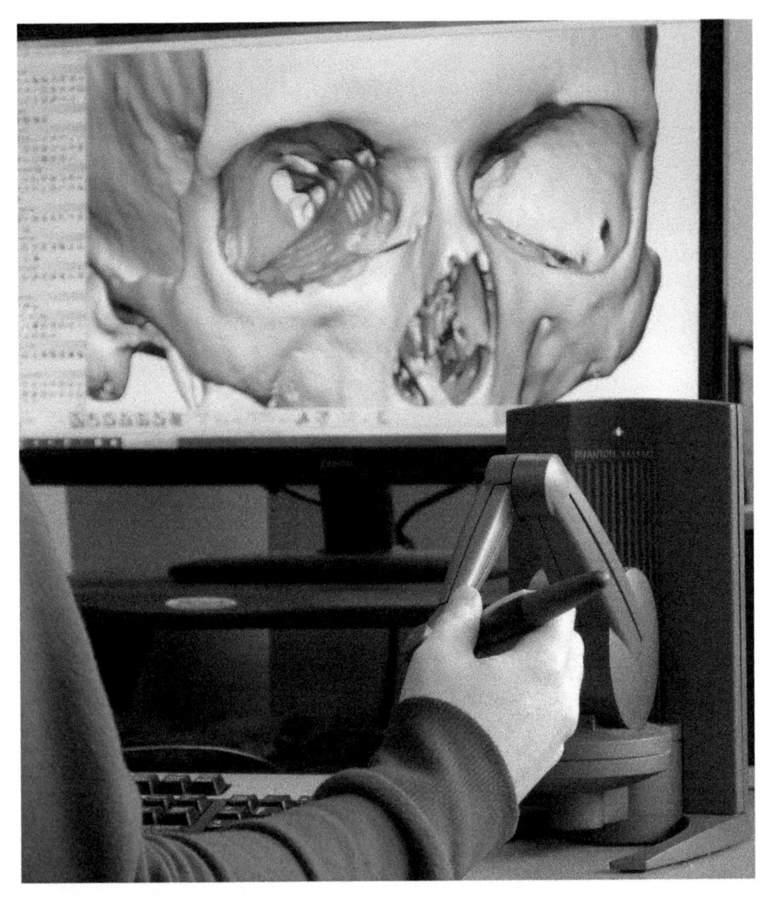

Fig. 3.16 The haptic interface design used with FreeForm.

attempts to create physical anatomical models from 3D images generated from medical image data and by CAD involved cutting down a block of foam or polyurethane by computer-controlled milling machines [20].

In 1986, it was Chuck Hull who developed the first patent for a manufacturing system that fabricated physical parts in layers. The process was termed Stereolithography (commonly referred to as SLA). This revolutionary process fabricated physical replicas of 3D CAD models by curing photopolymer resin, cross-linked by an ultraviolet spectrum laser layer-by-layer. The term Rapid Prototyping (RP) was born since the process was initially used to produce prototypes prior to the final production. Hull formed the company 3D-Systems (Rock Hill, USA) with R. Freed, which is still in operation today [21]. RP is still a widely used term in the industry. The term AM is commonly used in industry to describe the layer-additive production of end use parts as opposed to prototypes.

In recent years, what was an expensive technology has evolved from being accessible by only large organizations to something that even small organizations or individuals can use in their offices or homes at a basic level. The popular term has become 3D printing.

AM/3d printing processes do not require tooling and so are relatively inexpensive for one-off or batch production. However, AM machines are comparatively slow on a per-part basis. The layer-by-layer production method offers a platform to overcome many of the constraints that previously limited the ability to fabricate complex anatomical-based forms. For example, internal detail and undercut features are relatively simple for AM as it is not limited by tool path access, draft angle considerations, and other similar traditional constraints. However, as the case studies later in this chapter illustrate, design for AM still requires careful consideration to ensure parts are optimiezd for accuracy and production efficiency.

Common principles are shared by the increasing range of AM technologies. 3D CAD geometry is automatically sliced into layers which are then sequentially recreated by a machine in (typically) polymer, metal, or a composite. Depending on the process, layer thickness is usually in the order of 0.1–0.3 mm. Thin layers result in the ability to create finer details, but at the compromise of speed.

There are four main polymer AM material classes: StereoLithography Apparatus (SLA) in which a laser solidifies liquid layers of photo-sensitive monomer in a vat; Multi-Jet Printing (MJP) in which a print-head jets droplets of material onto a build plate; fused deposition modelling (FDM) in which a nozzle extrudes a bead of semi-molten material onto a build plate; and selective laser sintering (SLS) where powdered material is partially melted onto to the layer below by a laser beam. Variations of these overarching methods are frequently being added to the market, making it challenging to present all possible permutations in a book. This chapter will concentrate on the SLA process, which remains one of the most widely used for biostructure fabrication. The SLA process requires the fabricated part to be supported by a lattice-like structure, which supports overhangs and attaches the part to the bed beneath the resin. Once a layer is cured, the machine moves the part being built down into the liquid bath of monomer resin by one layer thickness, the surface is recoated, then the next layer is drawn. Fig. 3.17 is a long exposure photograph that illustrates a layer of resin being

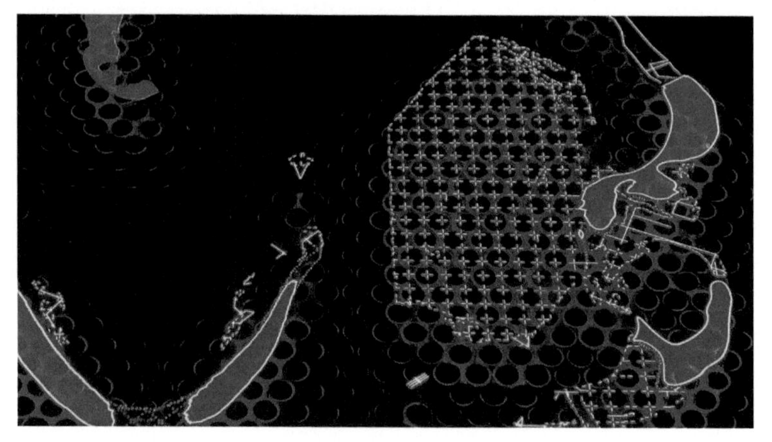

Fig. 3.17 A twenty-second exposure image of an SLA layer being drawn by the ultraviolet spectrum laser.

cross-link polymerized by the ultraviolet spectrum laser. The outline of the part layer is shown in bright blue, the hatched inner area of the part cross-section is also shaded blue, and the support structures are small crosses in blue.

Once complete, the part raises above the resin for draining, it is removed from the bed, before being cleaned, supports removed and postcured. A completed SLA model of a midface still in the SLA machine build chamber is shown in Fig. 3.18. The model is anatomically upside down, with support structures attaching it to the perforated metal bed and visible inside the orbits.

Fig. 3.19 illustrates the layer step effect of the SLA process at 35x magnification.

Low-cost, FDM-based polymer AM (most typically referred to as 3D printing) technologies are increasingly being used for biostructure fabrication. Extremely low machine, material, and software prices, combined with an increasing range of material

Fig. 3.18 A completed SLA build of a midface model upside-down still on the perforated build platform.

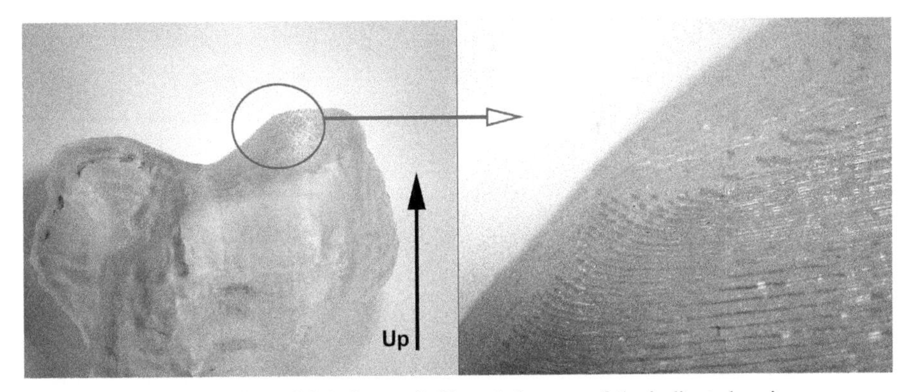

Fig. 3.19 SLA model of the tibial plateau (left) and close up of the indicated region illustrating the layer steps of the process (right).

Fig. 3.20 A few second exposure of the laser melting process (image courtesy of Renishaw PLC).

options makes them a potentially useful method, where parts only need to serve a temporary reference or limited functional purpose. There is a vast range of FDM-based technologies, and details about their application are readily available via internet searches.

The last decade has also seen the development of metal materials AM machines under two major umbrella categories—selective laser melting (SLM) and electron beam melting (EBM). SLM uses a laser beam to melt powdered metal materials (shown in Fig. 3.20). SLM has the ability to produce highly detailed metal components. EBM uses a beam of electrons in the same way—but is quicker and has fewer issues with heat-stress as a result of preheating each new layer of powder. When utilized within a robust production process and regulatory framework, metal AM is an ideal technology for producing patient-specific medical devices including guides and implants. Both SLM and EBM were used for the examples in this chapter.

AM/3D printing is also complemented by computer-aided machining (CAM). For example, CAM is commonly used to produce custom alloplastic implants in materials such as Polyether Ether Ketone (PEEK). Although research is being undertaken to refine the AM of PEEK, it is currently not in widespread use.

3.8 Postprocessing

Postprocessing requirements vary significantly between AM processes, but all require some level of cleaning and treatment to ensure the parts are suitable for biostructure application. The layer-additive nature of the processes means that is possible to trap residual material within a structure, which in biostructure applications, could lead to undesirable contamination. The ability to clean and postprocess should be considered from the outset of part design. Polymer-based processes typically require the removal of residual support material or powder. This can require aggressive chemicals, compressed air, and other specialist clean up equipment. FDM 3D printing technologies require fewer specialist clean-up facilities, but instead rely on manual support structure removal. Part orientation and the effect this has on the quality of surfaces are therefore paramount when designing and building a biostructure using AM/3D printing. Laser-based metal AM processes also require further postprocessing beyond cleaning and support removal, including heat treatment to remove residual stresses, feature finishing, and surface finishing. Most of these processes are manual or semi-automated and can take around two days. Electron beam melted parts do not require heat treatment, but still require cleaning, feature and surface finishing prior to use.

3.9 Application case studies

CAD and AM of biostructures have an important role in allowing surgeons to carry out procedures more confidently, safely, and accurately. These benefits are possible through: being able to design devices which fit the anatomy perfectly; being able to rehearse, trial, and optimize different surgical approaches before getting to theater; being able to translate digital planning into reality by using guides; and by removing or minimizing the artisanal or 'by-eye' stages of complex procedures.

A highly competent suite of image capture, design, and manufacturing technologies currently exist for use in cranio-maxillofacial and orthopedic surgery. The techniques are now well into the mainstream, but development is still ongoing to refine the application and ensure new AM methods are adopted efficiently. Application of these technologies has evolved from the production of simple anatomical replica models toward the design and production of end-use surgical devices. CAD methods and AM technologies and materials allow efficient one-off production of patient-specific devices. There is a wide range of published literature describing the use of these technologies in patient-specific drilling, cutting, and positioning guides [3,22–28] and implants [29–31,10], both of which can be used to assist in posttraumatic reconstruction, disease management, and congenital case planning. Cutting, drilling, and repositioning guide materials vary between AM polymers and titanium, which have been approved for transient use.

CAD and AM workflows are favorable to more traditional laboratory or in-theater-based approaches, such as stock-plate bending, or pressing custom implants in sheets of titanium, as they offer a higher degree of design freedom. In addition to application in more routine operations, previously impossible or extremely risky procedures have been made viable using CAD/AM. This section illustrates examples where CAD/AM techniques have been used in trauma and disease surgery to improve the safety and accuracy of procedures. The case studies illustrate the use of Mimics and FreeForm for planning procedures and designing patient-specific devices.

3.9.1 Case study 1: Orbital reconstruction using a patient-specific implant

This case was undertaken at the Royal United Hospital, Bath by surgeon Andrew Felstead and colleagues. Consent has been obtained to present the technical approach in this book.

This case study illustrates the application of CAD and AM in the design and production of a patient-specific orbital floor and wall implant. Orbital floor/wall implants are often required to reconstruct the fragile bones beneath the eye, which are easily damaged in blunt midface trauma (as in this case study). Considerations for orbital implant design include: ensuring they are sufficiently small (acknowledging the limited access enabled by typical transconjunctival or subciliary incision approaches); if possible, ensuring the posterior part of the implant can rest on solid bone, but not extending the implant more than 35-mm posteriorly from mid-infra orbital rim (to avoid critical structures such as the optic nerve); trying to locate the implant in a secure position on the orbital rim (either through the use of positioning guides or ensuring the implant locates on a well-contoured section of bone); and minimizing the size of the fixation tabs and ensuring screws are countersunk to avoid palpability beneath the thin skin; adding holes that prevent blood pooling on the implant.

The initial stages of CT data to CAD model and additional step of manually drawing the missing details of the orbital floor outlined in previous sections were carried out first. The resulting STL files were imported into FreeForm using the 'Buck' setting and fill holes option (Fig. 3.21). This setting prevents material being removed from the CAD model, thus preserving the accuracy of the original anatomy.

CAD provides the ability to base reconstructions on accurately mirroring unaffected anatomy or the morphology of the underlying maxillary sinus [32]. Fig. 3.22 illustrates the essential steps of selecting the unaffected side of the midface and mirroring to achieve an initial reconstruction of the defective side. Various smoothing, carving, and modelling tools in FreeForm were used to refine the newly reconstructed orbit CAD model, and flat planes were used to assess the accuracy of the reconstruction throughout (bottom of Fig. 3.22).

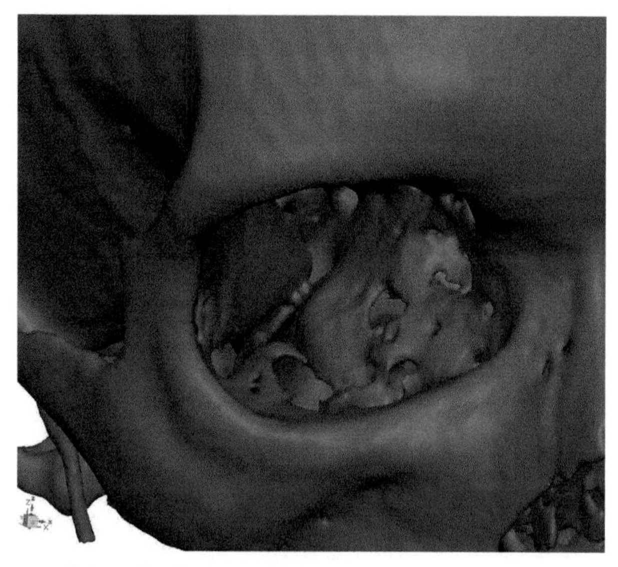

Fig. 3.21 Close-up of the patient's right orbit illustrating the orbital floor defect.

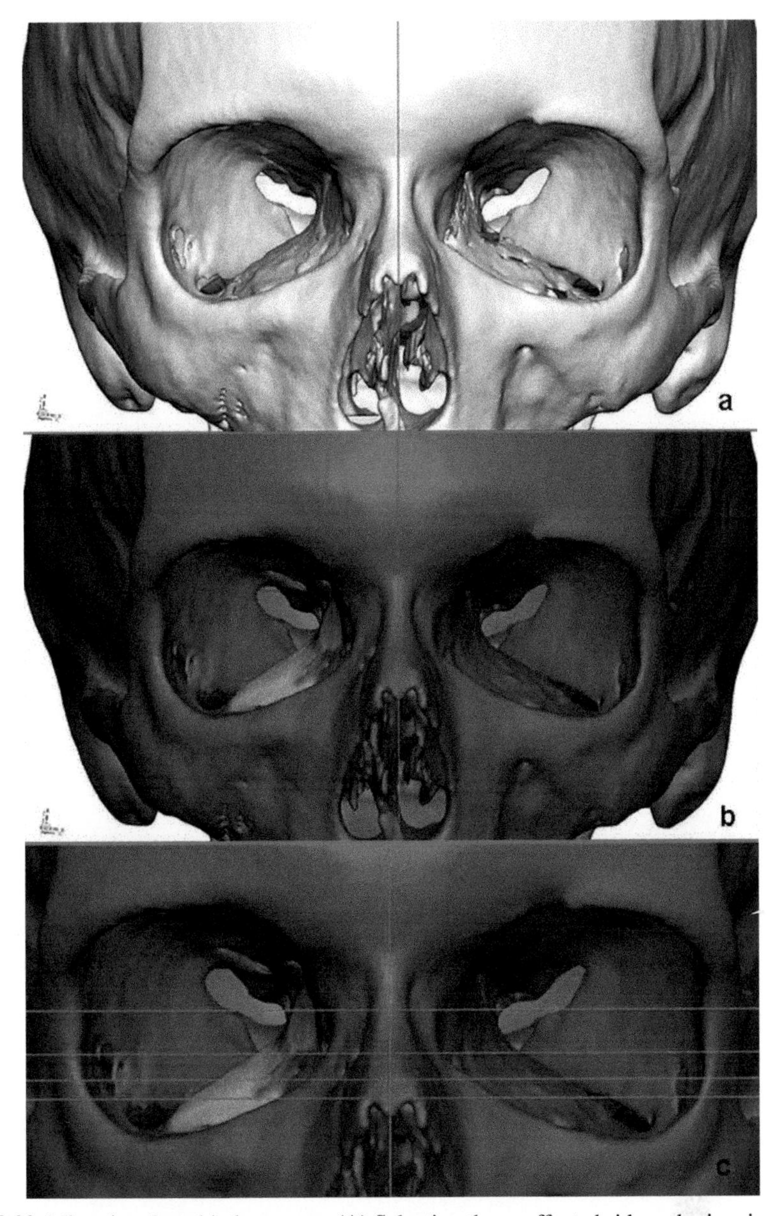

Fig. 3.22 Mirroring the orbital anatomy. (A) Selecting the unaffected side and mirroring it across the midsagittal plane, (B) blending the mirrored portion into the protected 'Buck' anatomy, and (C) assessing the reconstruction using planes.

FreeForm provides many ways of achieving a suitable implant or device design. One approach is illustrated here. An outline representing the extent of the proposed implant was sketched on to the surface of the reconstructed right orbital floor and wall (Fig. 3.23). The 'emboss with curve' tool was used to create a raised thickness of

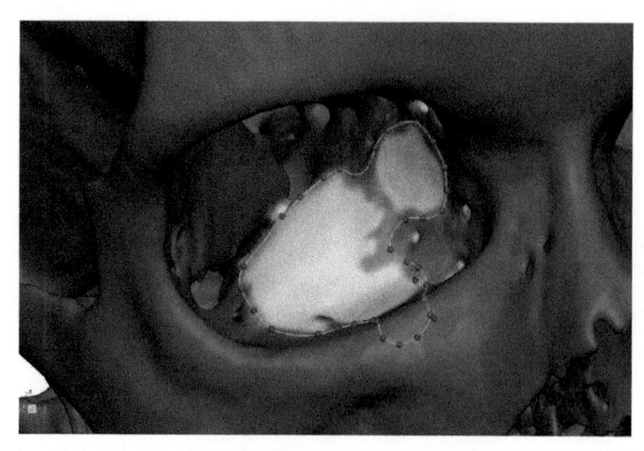

Fig. 3.23 Drawing the extent of the intended reconstruction directly on to the CAD model surface.

0.5 mm. CAD versions of the intended screws were then imported and aligned according to the best bone quality and drill access. The area of the implant around the screw heads was locally thickened. This ensured that the screws were fully flush with the implant surface and would not cause damage or be felt through the thin soft tissue. Cut out fluid drain holes were then 'extruded' through design. The initial design was then confirmed with the prescribing surgeon. The completed implant design is illustrated in Fig. 3.24. An additional positioning guide, designed to ensure the correct placement of the implant was then developed. This engaged a larger area of the infraorbital rim to prevent the implant moving into the wrong position during the procedure.

With the design agreed by the prescribing surgeon, the 'Buck' anatomy was removed, the devices smoothed, and any unattached pieces of virtual 'Clay' removed. The implant and guide were then converted into a 'Mesh' structure, and Boolean subtraction methods were used to remove the screws from the devices. The implant, guide and screws were then exported as STL files and imported back into the Mimics file

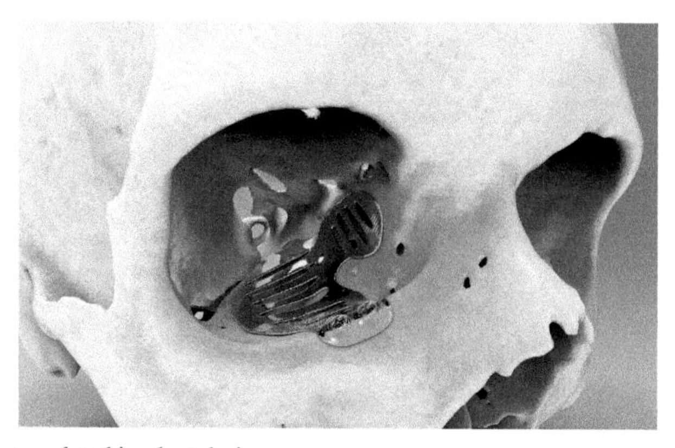

Fig. 3.24 The completed implant design.

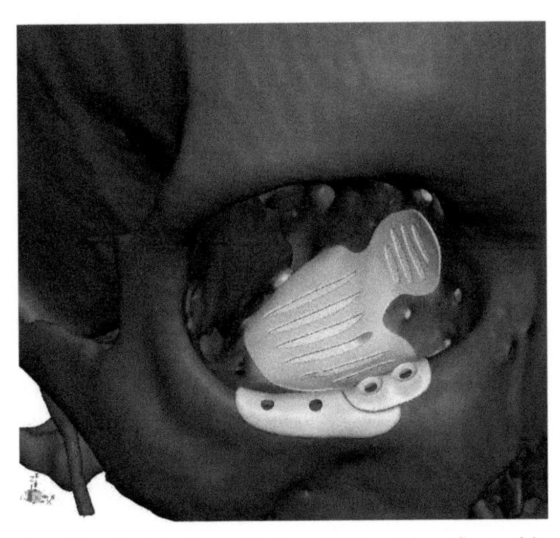

Fig. 3.25 The completed implant design with the additional location guide.

containing the original CT slice data. Device fit was analyzed to ensure sensitive anatomical structures were avoided. The completed implant and guide design are shown in Fig. 3.25.

As with any custom implantable device, the design was shared with the prescribing surgeon to ensure their satisfaction prior to fabrication. The STL files and instruction for critical features including screw hole dimensions and desired surface finishes were sent for fabrication in titanium using laser melting (Renishaw PLC, UK) in an appropriate regulatory compliant environment. Once the fabrication process was complete, a heat treatment cycle was undertaken to relieve residual part stress (ensuring accuracy), automated and manual finishing techniques were used to achieve the desired surface qualities and feature accuracy, and a cleaning cycle undertaken to remove contamination. The parts were supplied in a clean, nonsterile state along with operational techniques to the hospital.

A subciliary incision was used to access the path for plate insertion. As with most orbital floor/wall reconstructions, one of the main issues is ensuring that the bony margin of the defect are delineated and that all herniated tissue is retrieved with as little trauma as possible. In this case, one of the biggest advantages of the custom plates was that the surgeon had an appreciation of how they needed to carry out dissection and they could try it in when they thought the limit of dissection was reached. The plate then confirmed that no extra or excessive dissection was required.

The other major advantage was that no implant bending or trimming was required to achieve the planned correct location and restoration. In this case, the guide provided was not required as the plate sat in the correct place first time and did not side along the infraorbital rim.

Postoperative reviews confirmed good healing and signs of good functional outcome.

3.9.2 Case study 2: Design of mandibular re-contouring guides

This case was undertaken at the North Manchester General Hospital by Mr Richard Graham and his colleagues. Consent has been obtained to present the technical approach, use images taken during the procedure and of the patient for educational purposes, including in this book.

This case study illustrates the application of CAD and AM in the production of mandibular re-contouring guides. The patient presented with a bony defect due to a central giant cell granuloma that had previously been treated with intralesional steroids—these stabilized the lesion and calcified it internally. The preoperative facial contours are illustrated in Fig. 3.26.

Techniques described in the earlier sections and the previous case study were used to segment the bony anatomy, export the STL files of the upper and lower jaws and import into FreeForm CAD software. An additional step was also undertaken in Mimics to highlight the inferior dental/alveolar nerve (IDN), which was considered a critical structure to avoid during the surgical procedure (Fig. 3.27).

An extra-oral approach via neck skin incision that would provide sufficient access was agreed prior to surgical planning and device design. To establish a target 'ideal' mandible contour, the left side was mirrored across the mid-sagittal plane, making it easier to interpret where cutting planes needed to remove the calcified lesion (Fig. 3.28).

The direction of cuts that would help achieve the 'ideal' contour was visualized prior to sketching profiles that would eventually guide the surgeon's saw. The sketch profiles were used to construct three guides using various modelling tools available in FreeForm. Care was taken to ensure that the guides could be used in turn and still fit

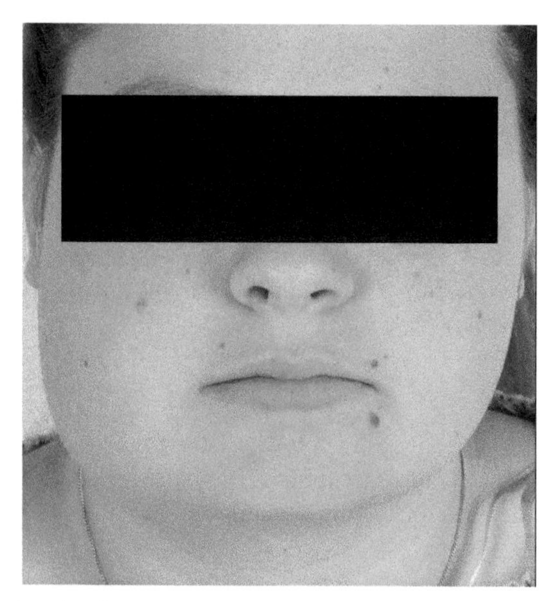

Fig. 3.26 Preoperative facial asymmetry.

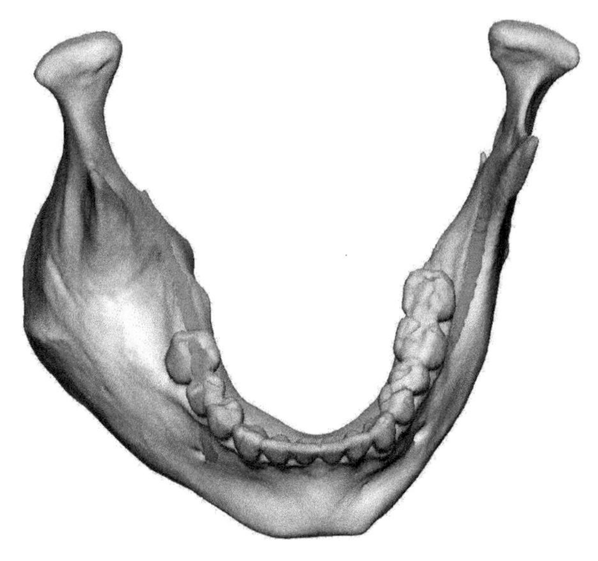

Fig. 3.27 Highlighting the dental/alveolar nerve.

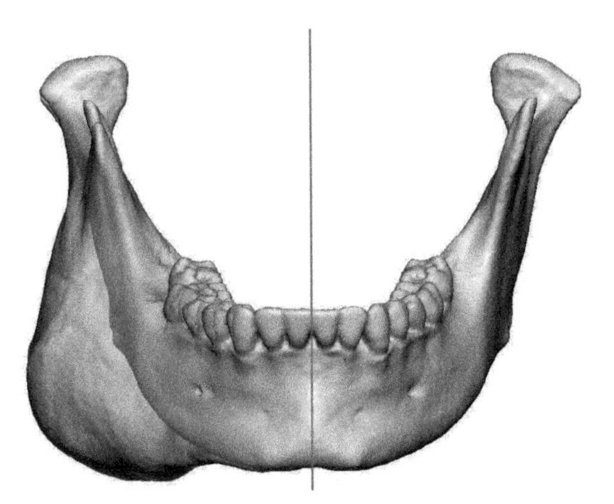

Fig. 3.28 Mirror-based reconstruction to illustrate the 'ideal' mandible contours.

securely to the remaining anatomy while being used. Tabs and holes for temporarily screwing the guides in position were also added (using Synthes standard 2.0x6 mm screws). The three completed guide designs are shown in Fig. 3.29. The guide colored green trimmed the inferior margin of the right side of the mandible, the guide colored orange trimmed a more natural angle on the lateral side of the mandible and the guide colored purple was used to trim the lateral aspect further to match the 'ideal' contour. The guides were fabricated in titanium using the same laser melting AM techniques described previously.

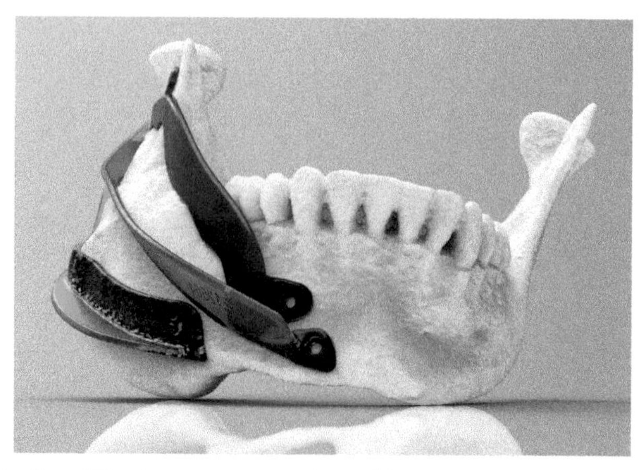

Fig. 3.29 Illustration of the three separate cutting guides.

Fig. 3.30 shows the first, inferior margin-cutting guide in use during the operation. It is shown secured in place with two Synthes screws. The surgeon cut along the lower edge using the face to guide the saw to achieve the correct contour. Fig. 3.31 shows the completed cut with the excess lesion portion of bone being removed.

The other two guides were then used to assist the other cuts. The procedure went smoothly and all guides functioned as intended. The postoperative radiographs illustrated that good aesthetic recontouring had been achieved (Fig. 3.32). Follow-up consultation indicated positive patient satisfaction and healing. No postoperative patient pictures were available at the time of writing.

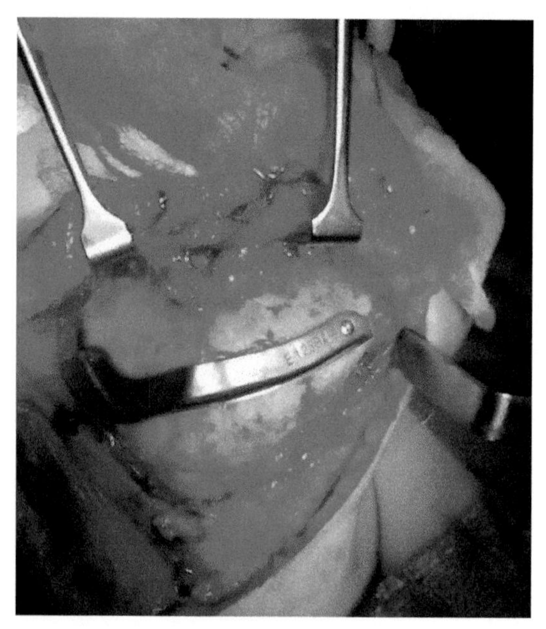

Fig. 3.30 The first stage, inferior cutting guide in place.

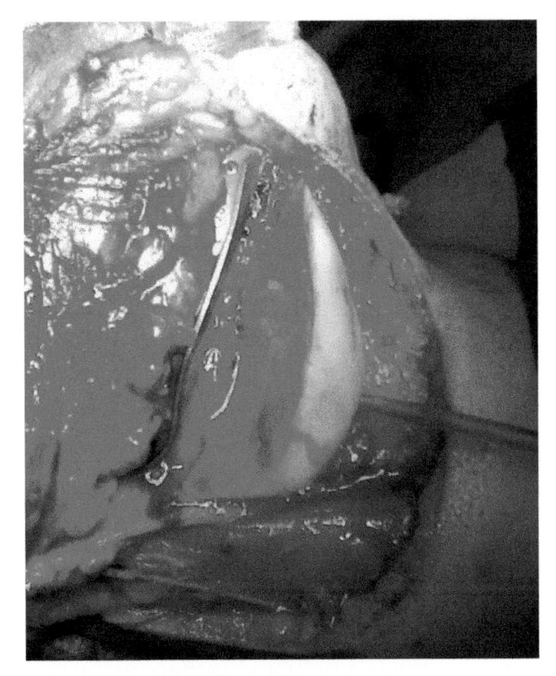

Fig. 3.31 The first cut and removed portion of bone.

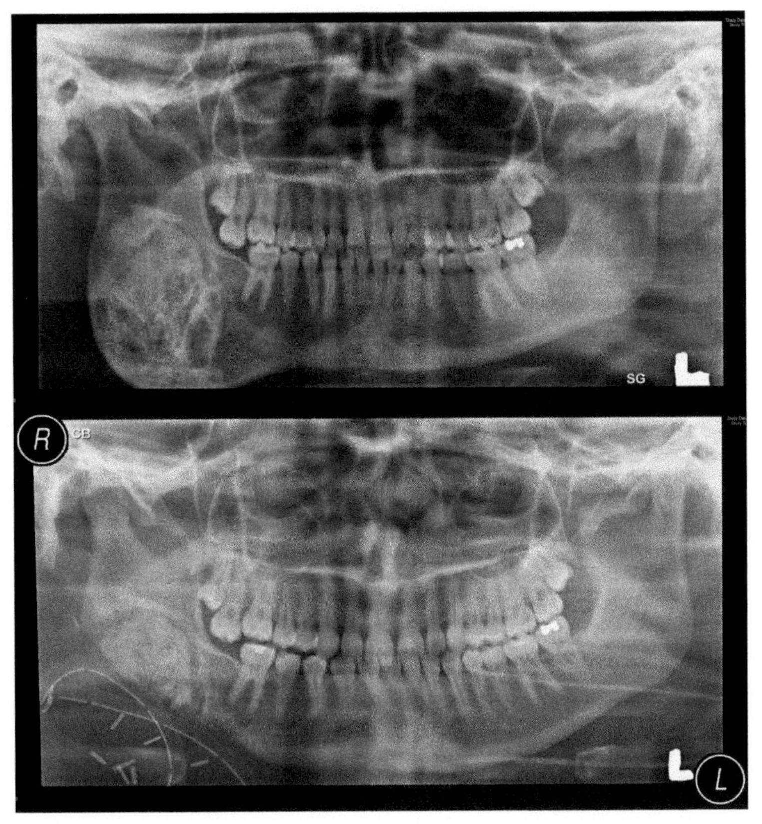

Fig. 3.32 The preoperative mandible contours (top) and postoperative contours (bottom).

3.10 Case study 3: Tumor removal and custom craniofacial implant

This case was led by Prof. Dr. Aleksandar Kiralj and his team at the Faculty of Medicine Novi Sad, Hajduk Veljkova 3, Novi Sad, Republic of Serbia. Device design assistance was also provided by Dr Igor Budak (Associate Professor), Faculty of Technical Sciences, University of Novi Sad. Consent has been obtained to present the technical approach in this book.

A 20-year-old patient presented with a congenital deformity and a benign tumor that has spread across the cheekbone, which resulted in significant bony and soft tissue facial asymmetry. A CT scan was undertaken and techniques described previously were used to create a CAD model of the bony anatomy and the tumor separately (Fig. 3.33). This allowed the cut margins to be determined.

Planes and surfaces were used to define the intended cut margins accurately that would assist the surgeon to remove the tumor. Various modelling tools in FreeForm were used to translate the cut margins into guide designs that located on two key parts of the facial anatomy, the supraorbital rim and posterior area of the zygomatic arch. The guides indicated the desired cutting vectors and included triangular holes designed to predrill the screw holes for the final implant (Fig. 3.34). Triangular holes were used as they allow better irrigation during drilling and provide a high degree of angular accuracy. The anterior cut was intended to be undertaken without a guide as this area is typically difficult to access.

With the tumor margin agreed, reconstruction of the defect was undertaken using a mirror image of the patient's unaffected right facial anatomy and various modelling tools within FreeForm (Fig. 3.35).

Implant design was undertaken based on the contours of the reconstructed anatomy. Details including screw locations were defined, the implant design was shelled to a thickness of 1 mm and the inside surfaces removed to leave an open structure intended

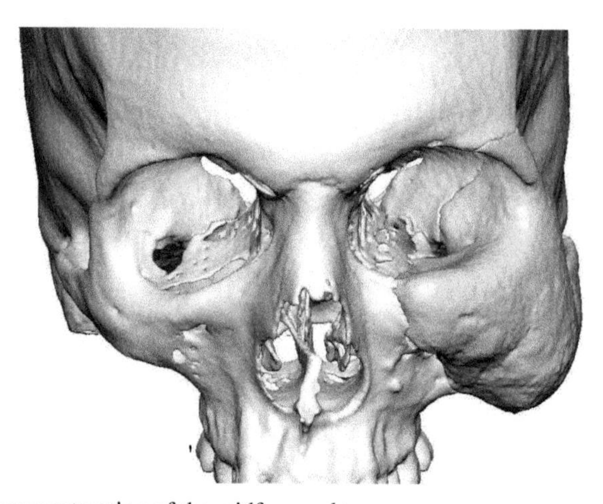

Fig. 3.33 CAD reconstruction of the midface and tumor.

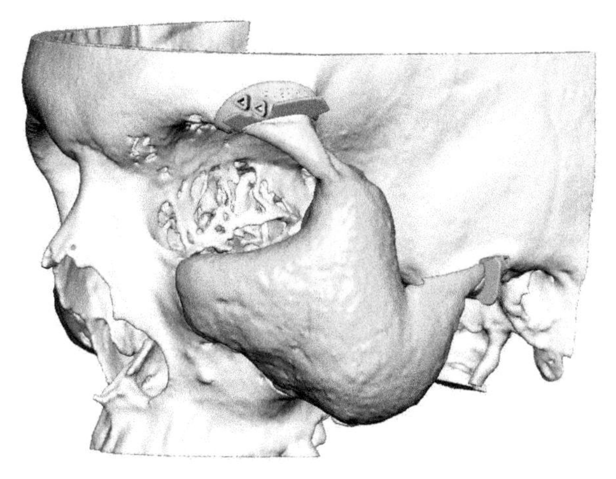

Fig. 3.34 Cutting guides designed to assist at the zygomaticofrontal suture and zygomatic arch.

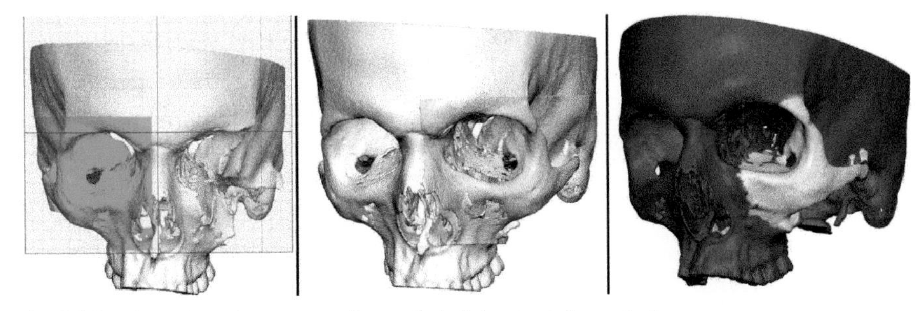

Fig. 3.35 Mirror-based reconstruction and draft implant shape design.

to reduce the potential for infection. As with most custom implant designs, the goal was also to minimize the amount of alloplastic material introduced into the body. The completed design is shown in Fig. 3.36. The surgical approach and the implant located during surgery are illustrated in Fig. 3.37.

The operation lasted about 3 hours. Reviews have confirmed a positive outcome and the patient was very happy with the results, especially considering that this was a very visible part of the face.

3.10.1 Case study 3: Custom surgical guides and implant for pelvis reconstruction

This case was led by Prof Ian Pallister, Trauma and Orthopaedic Surgeon at Morriston Hospital, Swansea. Early surgical planning was undertaken with the Maxillofacial Unit at Morriston Hospital. Consent has been obtained to present the technical approach in this book.

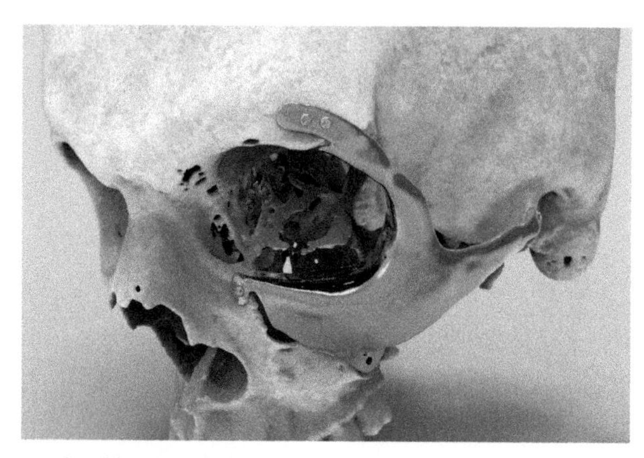

Fig. 3.36 The completed implant design.

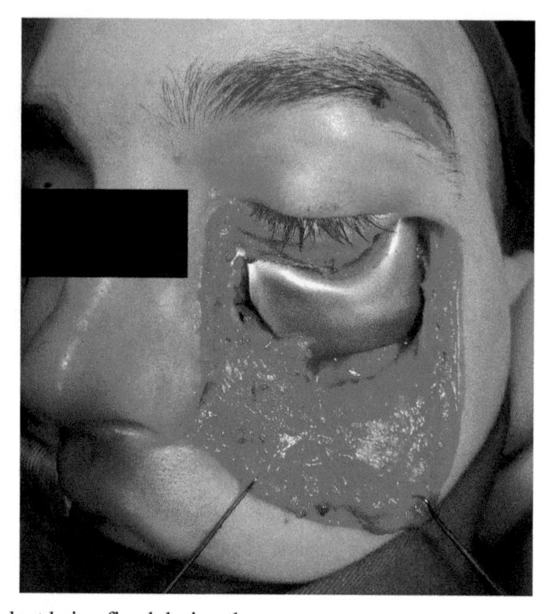

Fig. 3.37 The implant being fixed during the surgery.

Surgery to treat pelvic fractures is a highly specialized field due to the complex geometry of the pelvis and the important major nerves and blood vessels that pass from the pelvis to the lower limbs. When a pelvis has deformed and begun to heal, corrective osteotomy and surgical fixation of this is even more difficult. In this case, due to the patient's severe osteoporosis, multiple fractures of the pelvis had occurred, but displacement resulted in reduced pelvic diameter and severe shortening of the right leg compounded by extraosseous bone growth along the fracture lines. The prescribing surgeon concluded that standard plating systems would not be able to secure the bones reliably after they were realigned following osteotomy. Computer-aided

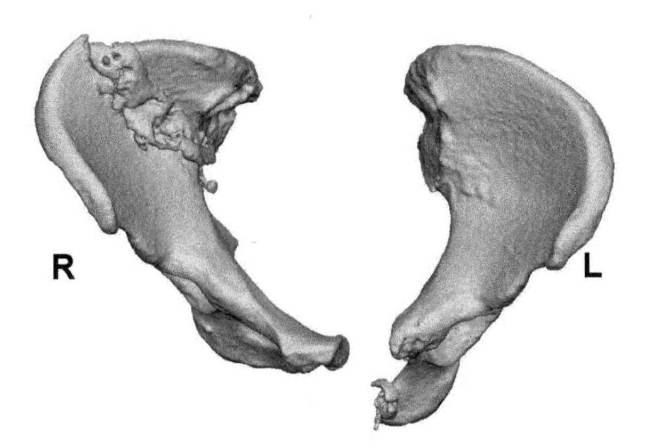

Fig. 3.38 Preoperative 3D reconstruction of the left and right pelvis illustrating the major fracture site on the patient's right side.

planning combined with the design and production of patient-specific guides/implants was the only option to allow the necessary surgical precision.

The patient's right hemi-pelvis had fractured in the posterior region of the iliac crest vertically. This caused the anterior region of the right pelvis to swing medially, closing the pubic arch and displacing the position of the acetabulum resulted in shortening of the right leg. Additional fractures of the left pubic crest also compounded the complexity (illustrated in Fig. 3.38).

Methods described in previous case studies were used to generate a CAD model of the left and right hemi-pelvis. FreeForm software was used to plan the osteotomy cut on the right side and reposition the portion of bone based on a mirror of the left side across the midsagittal plane.

With the osteotomy agreed, the implant design was drafted over multiple iterations. CAD screw geometries were carefully positioned to avoid critical structures and engage portions of bone less affected by the osteoporosis. Additional 'wing' features were added to the posterior aspects of the implant to provide supplementary stability, bracing the medially tilted right hemi-pelvis in a corrected position. The freedom to locate screws in exactly the desired position and create a single implant that spanned the entre pelvis was considered to provide greater stability over the use of prebent stock plates. The completed design is shown in Fig. 3.39.

Bone-mounted guides were then designed to ensure that the screw locations and vectors could be accurately predrilled prior to the implant being placed. The guides were designed to be thin and locate on highly contoured areas of the pelvis to provide stability. Fig. 3.40 illustrates the guide designs.

Low-cost polymer 3D printed replicas of the proposed implant, guides, and pelvis were fabricated for the surgeon to rehearse the procedure. Once the designs were approved, the implant was fabricated using Electron Beam Melting (EBM, Arcam AB, Sweden) and the guides using Laser Melting (Renishaw PLC, UK). Appropriate heat treatment, postprocessing, and cleaning procedures were used to complete the fabrication processes prior to sterilization and use.

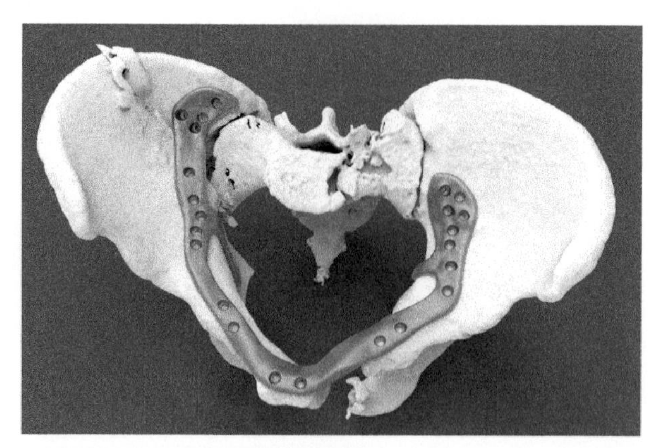

Fig. 3.39 The completed, one-part implant design.

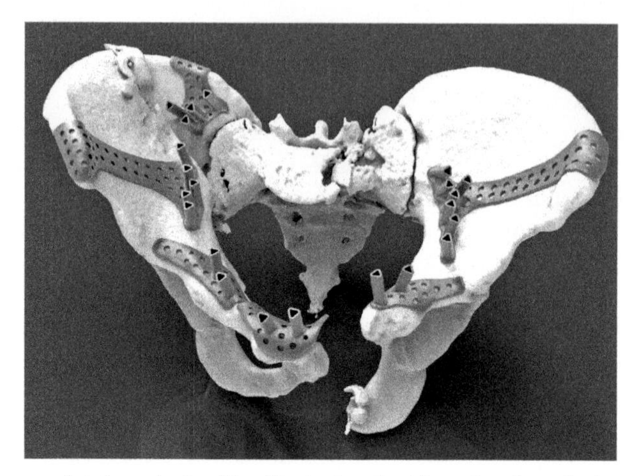

Fig. 3.40 The completed surgical guides illustrating the triangular-shaped tubes used to ensure guide the drill in the correct vector.

The operation proceeded as planned and the predrilling of the screw holes allowed the procedure to progress smoothly once the plate was maneuvered into position. The patient recovered rapidly and resumed almost pain-free full weight bearing within ten weeks of the surgery.

3.10.2 Case study 4: Extraoral facial prosthetics

The facial prosthesis cases illustrated here were undertaken in collaboration with Peter Evans at the Maxillofacial Unit at Morriston Hospital, Swansea. The cases have been presented in various journal papers, conferences, and PhD work. Appropriate consent has been received for use in this book.

Facial prosthetics is a discipline closely related to maxillofacial surgery and addresses the needs of patients who have a missing facial anatomy as a result of either disease, trauma, or congenital reasons. Increasing patient numbers, the need to improve process efficiency, the desire to add value to the profession, and the lack of access to maxillofacial prostheses provision in some areas of the world has led researchers to investigate the potential benefits of computer-aided technologies in facial prosthetics. Best practice is for prosthetists (or anaplastologists in some parts of the world) to work closely with maxillofacial surgeons in the planning of facial prostheses. The current state-of-the-art in prosthesis retention relies on placing implants into the patient's bone, exposing the implants through the skin and mounting the prosthesis using a bar and clips or magnets. Computer-aided planning and CAD are commonly used to assist in accurately placing implants, design surgical guides and elements of the definitive prosthesis. An overview of these methods is provided.

3.10.2.1 Implant placement planning

Early work on the application of AM technologies in the manufacture of surgical guides concentrated on the production of drilling guides for oral osseointegrated implants [33–35]. The essential steps are similar to implant design—CT data of the patient is translated into a 3D CAD model, bone quality and desired implant location are assessed, a guide is designed, and then prepared using AM. Developments have now led to dental implant planning software and drilling guides being offered commercially (SimPlant, SurgiGuides, Materialise N.V.), but FreeForm and other nonmedical-specific CAD tools can also be used. Published research carried out in the use of surgical guides for implant placement in facial prosthetic rehabilitation and orthopedic surgery is growing [11,36–39]. To be appropriate for the manufacture of surgical guides, the RP processes have to be accurate, robust, rigid, and able to withstand sterilization [3]. Both polymer and metal guides can be used, but there is a need to reinforce polymer components with metal inserts or 'keys' that prevent the surgeons' drill accidentally damaging the material. Guides designed to locate implants for extraoral prosthesis retention also need to avoid placement on/near sensitive anatomical structures, locate with stability on the bone, and be sufficiently small yet easy to handle. This is especially critical for long zygomatic implants, which are drilled from the frontal process region, through the maxillary sinus (under the eye), and into the zygomatic bone (illustrated in Fig. 3.41). These implants are typically used to retain nasal prostheses. A surgical guide for placing a zygomatic implant is illustrated being used in combination with a metal 'key' in Fig. 3.42.

3.10.2.2 Prosthesis design

CAD of prosthesis is still relatively under-developed in clinical application due to technical challenges and access to appropriate technologies. Research has illustrated how—through a combination of 3D scanning, CAD, and AM—prosthesis design can be carried out using a hybrid of digital approaches and laboratory methods [40–45]. Incorporation of the implant retention mechanisms is still one of the most technically challenging elements of prosthesis CAD [46].

Prosthesis design typically begins with 3D scanning the face to capture the defect and surrounding regions. This replaces the process of physical impression taking

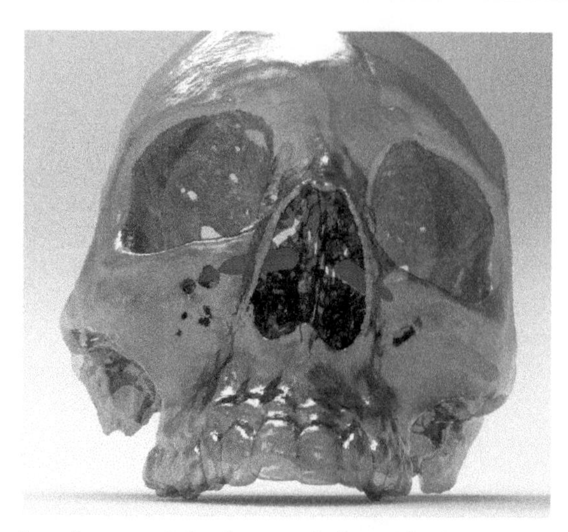

Fig. 3.41 The location of zygomatic implants are indicated in pink.

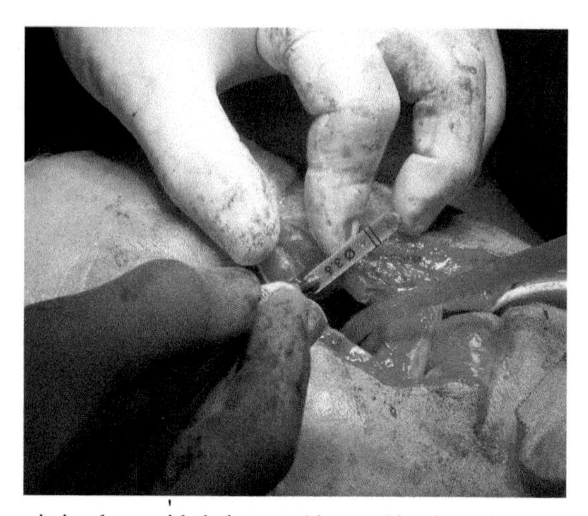

Fig. 3.42 A zygomatic implant guide being used in combination with a metal surgical key.

(illustrated in Fig. 3.43). However, 3D scanning of anatomy such as ears and the globes of the eyes is particularly challenging. Capturing intricate, sharp edge, engineered features that are typically situated in skin folds/crevices such as the implant abutments (which protrude through the skin) is also challenging.

The resulting 3D surface can be imported into software such as Freeform for prosthesis design. Prosthesis shapes can be detailed in FreeFrom using donor anatomical geometry, or the patient's own presurgical data. Additional skin and wrinkle details can also be added to help make the prosthesis more lifelike. Retention component design remains a challenge, primarily due to the difficulty in accurately capturing the location of the implants and abutment structures in relation to the gross anatomy using 3D scanning. Some research has attempted to address this, but the outcomes are not

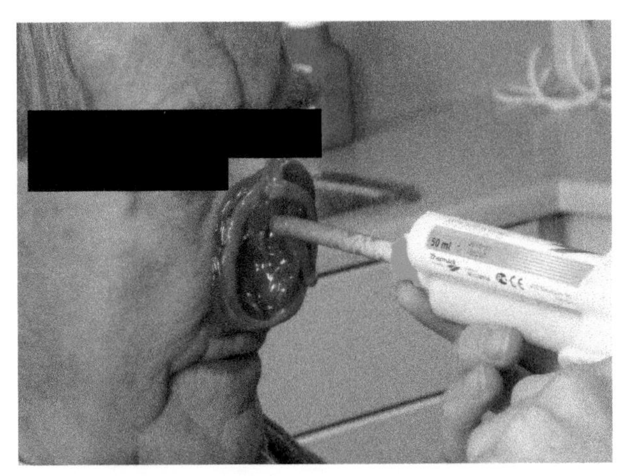

Fig. 3.43 Taking an impression of a patient's nasal defect using a silicone material. Noncontact 3D scanning can replace this stage in facial prosthesis design.

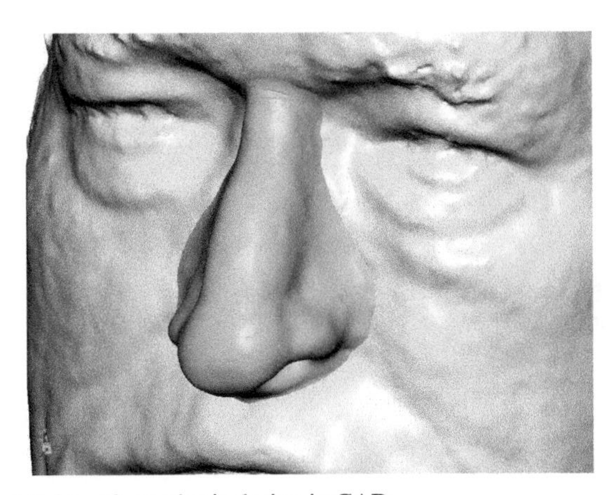

Fig. 3.44 A completed nasal prosthesis design in CAD.

in widespread clinical application [46]. The current work-around involves combining conventional laboratory methods of producing retention mechanisms with CAD of the prosthesis form. A nasal prosthesis design completed in Freeform is illustrated in Fig. 3.44.

AM technologies are currently unable to produce a realistic prosthesis directly from CAD, so it is necessary to design a mold tool into which the definitive prosthesis can be fabricated in an appropriate color-matched material. 3D printing has become a viable method for producing mold tools. This process has been described [45] and an example mold tool design for a nasal prosthesis is shown in Fig. 3.45. Silicone is currently the favored material for prosthesis fabrication as it can be easily processed and colored to match the patients' skin tones. The molding process is illustrated in Fig. 3.46.

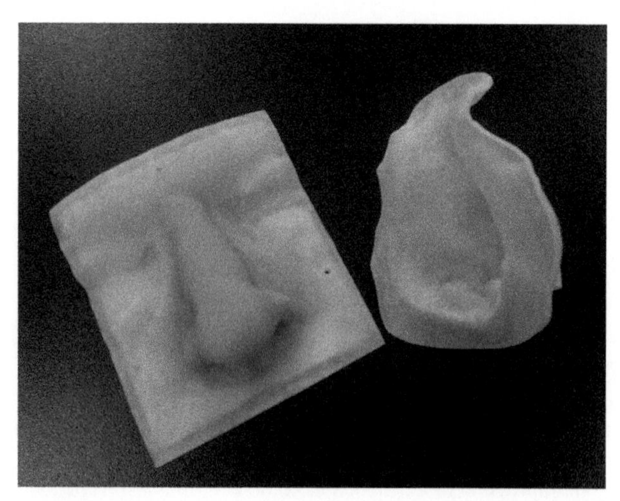

Fig. 3.45 A two-part 3D printed/AM-produced mold tool used to create a nasal prosthesis.

Fig. 3.46 Color matching silicone to create an orbital prosthesis.

Fig. 3.47 shows a completed nasal prosthesis designed using a combination of CAD/AM techniques and laboratory methods.

3.11 Discussion

The application of medical imaging, CAD, and AM in the field of biostructures can help reduce the risk associated with complex surgical procedures, improve the likelihood of an accurate outcome, and reduce the duration. In prosthetics, these approaches

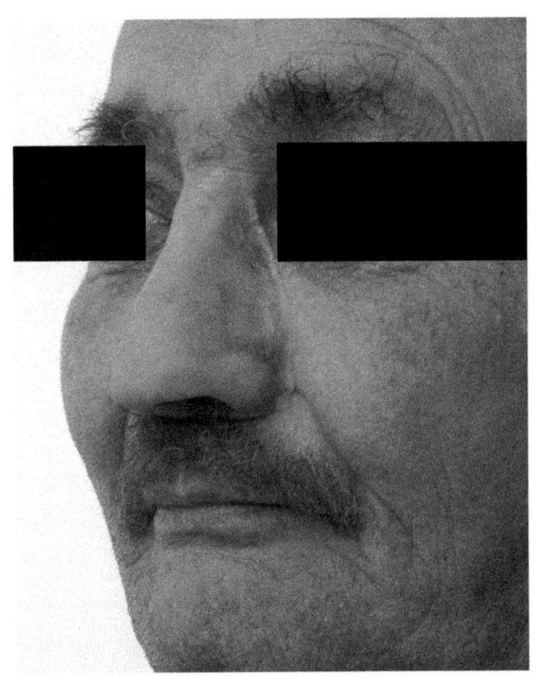

Fig. 3.47 A completed nasal prosthesis designed and produced using CAD and laboratory methods.

offer the potential to reduce the duration of appointments and improve process repeatability. However, despite these benefits, consideration should still be given to when a CAD/AM approach can be used most effectively.

Complex surgical reconstructions require careful planning that benefits from a multidisciplinary approach—ideally involving the prescribing surgeon and a well-qualified design engineer. Clinical knowledge should be complemented with a comprehensive understanding of software design freedoms and limitations associated and with AM manufacturing processes. Each of the surgical case studies illustrated required a high degree of collaborative up-front planning and design time. Although this is typically offset during the procedure, it can be a barrier to using these approaches more routinely. The economics of using AM over alternative methods is another area not widely addressed in academic literature. CAD/AM costs are perceived, as being significantly higher upfront than alternative lab-based methods, but this does not necessarily reflect overall treatment costs. Time saved in theater, improved patient outcomes, improved treatment compliance, and the ability to redeploy medical specialists away from manufacturing tasks can all positively influence overall treatment costs and offset the extra costs of using CAD/AM. However, it can be a challenge to convey this to a finance department when trying to justify an initially higher upfront cost.

AM of metal devices in particular is also a relatively slow process, with current best-practice being around four days from a build starting to part completed and cleaned. In lifesaving trauma cases, it is usually not possible to produce a custom implant/metal guide in time, so stock plates, bent on the patient or on a rapidly produced replica model is still the only viable solution.

For the reasons stated above, despite the benefits, CAD/AM is still only used in a relatively small percentage of cases where it could be.

Some companies already in the patient-specific device market have taken the approach of creating highly specialist software for planning specific surgical procedures. Many of these solutions are wizard driven. This helps deliver a robust solution, but offers limited flexibility to the end user. At the other end of the production chain, AM technologies are still, in the case of metal, in the early and very complex stage of development; they require highly specialist knowledge to operate effectively. The accelerated rate of technology innovation is likely to lead to much wider applications in the near future.

The mass popularization of AM/3D printing also presents opportunities and threats to the way technology is used in biostructure fabrication. Remarkable as these technologies are, they can also be inefficient, expensive, and potentially dangerous if used outside a robust design process. There is a danger that these approaches are being used in situations where they offer no clinical benefit or cost saving over alternative, more traditional techniques. Accurate reporting on the healthcare economic and technical benefits of CAD/AM in biostructure fabrication will only be possible when sufficiently high numbers of cases have been critically reviewed.

AM or 3D bioprinting of autologous and active biostructures is another promising field of research where the application of design methodologies will become crucial to ensure clinically viable approaches are developed. There is still a significant amount of research required to develop 3D bioprinting to the clinical readiness level of AM/3d printing of alloplastic biostructures.

References

[1] Swaelens B, Kruth JP. Medical applications of rapid prototyping techniques. In: Proceedings of the 4th international conference of rapid prototyping; 1993. p. 107–17.

[2] Wehmoller M, Eufinger H, Kruse D, Massberg W. CAD by processing of computed tomography data and CAM of individually designed prostheses. Int J Oral Maxillofac Surg 1995;24(1 Pt 2):90–7.

[3] Bibb R, Eggbeer D, Bocca A, Evans P, Sugar A. Design and manufacture of drilling guides for osseointegrated implants using rapid prototyping techniques. In: Proceedings of the 4th National Conference on Rapid & Virtual. Prototyping and Applications. Professional Engineering Publishing, London; 2003. p. 3–12.

[4] Peel S, Bhatia S, Eggbeer D, Morris D, Hayhurst C. Evolution of design considerations in complex craniofacial reconstruction using patient-specific implants. In: Proceedings of the Institution of Mechanical Engineers, Part H: Journal of Engineering in Medicine [preprint]; 2017. Available at, http://journals.sagepub.com/doi/abs/10.1177/0954411916681346?ai=1gvoi&mi=3ricys&af=R [accessed 14th March, 2017].

[5] Singare S, Lian Q, Wang WP, Wang J, Liu Y, Li D, et al. Rapid prototyping assisted surgery planning and custom implant design. Rapid Prototyping J 2009;15(1):19–23.

[6] Harrysson O, Cansizogly O, Marcellin-Little DJ, Cormier DR, West Ii HA. Direct metal fabrication of titanium implants with tailored materials and mechanical properties using electron

beam melting technology. In: Advanced Processing of Biomaterials Symposium, Materials Science and Technology Conference and Exhibition. Cincinnati, OH; 2008. p. 366–73.
[7] Parthasarathy J, Starly B, Raman S. A design for the additive manufacture of functionally graded porous structures with tailored mechanical properties for biomedical applications. J Manufact Process 2011;13(2):160–70.
[8] Palmquist A, Snis A, Emanuelsson L, Browne M, Thomsen P. Long-term biocompatibility and osseointegration of electron beam melted, free-form–fabricated solid and porous titanium alloy: Experimental studies in sheep. J Biomater Appl 2011;27(8):1003–16.
[9] Poukens J, Laeven P, Beerens M, Nijenhuis G, Sloten JV, Stoelinga P, et al. A classification of cranial implants based on the degree of difficulty in computer design and manufacture. Int J Med Robot Comput Assist Surg 2008;4(1):46–50.
[10] Salmi M, Tuomi J, Paloheimo K, Björkstrand R, Paloheimo M, Salo J, et al. Patient-specific reconstruction with 3D modeling and DMLS additive manufacturing. Rapid Prototyping J 2012;18(3):209–14.
[11] Peel S, Eggbeer D, Sugar A, Evans P. Post traumatic zygomatic osteotomy and orbital floor reconstruction using digital tools and 3D-printing – a case report. Rapid Prototyping J 2016;22(6):878–86.
[12] Eley KA, Watt-Smith SR, Sheerin F, Golding SJ. Black Bone" MRI: a potential alternative to CT with three-dimensional reconstruction of the craniofacial skeleton in the diagnosis of craniosynostosis. Eur Radiol 2014;24(10):2417–26. https://doi.org/10.1007/s00330-014-3286-7. Epub 2014 Jul 20.
[13] Pathak VK, Nayak C, Singh AK, Chaudhary H. A novel approach for customized prosthetic socket design. Biomed Eng: Appl Basis Commun 2016;28(5):https://doi.org/10.4015/S1016237216500228. 1650037 (9 pages).
[14] Palousek D, Rosicky J, Koutny D, Stoklásek P, Navrat T. Pilot study of the wrist orthosis design process. Rapid Prototyping J 2014;20(1):27–32. https://doi.org/10.1108/RPJ-03-2012-0027.
[15] Evans P, Eggbeer D, Bibb R. Orbital Prosthesis wax Pattern Production using Computer Aided Design and Rapid Prototyping Techniques. J Maxillofac Prosthet Technol 2004;7:11–5. ISSN: 1366–4697.
[16] Eggbeer D, Evans P. Computer-aided methods in bespoke breast prosthesis design and fabrication. Proc Inst Mech Eng H, J Eng Med 2011;225(1):94–9.
[17] Binder WJ, Dhir K. Internet access to advanced 3-dimensional software for the prototyping and design of complex and precise custom mandibular implants. Am J Cosmet Surg 2016;33(2):83–90. https://doi.org/10.1177/0748806816648677.
[18] Santler G, Karcher H, Geggl A, Kern R. Stereolithography versus milled three-dimensional models: comparison of production method, indication and accuracy. Comput Aided Surg: Off J Int Soc Comput Aided Surg 1998;3(5):248–56.
[19] Toth BA, Ellis DS, Stewart WB. Computer-designed prostheses for Orbitocranial Reconstruction. Plastic Reconstruct Surg 1986;81(3):315–22.
[20] Eufinger H, Wehmöller M, Harders A, Heuser L. Prefabricated prostheses for the reconstruction of skull defects. Int J Oral Maxillofac Surg 1995;24(1):104–10.
[21] Jacobs PF. Rapid Prototyping & Manufacturing: Fundamentals of Stereolithography. 1st ed. Dearborn, USA: Society of Manufacturing Engineers; 1992.
[22] Abdel-Moniem Barakat A, Abou-ElFetouh A, Hakam MM, El-Hawary H, Abdel-Ghany KM. Clinical and radiographic evaluation of a computer-generated guiding device in bilateral sagittal split osteotomies. J Cranio-Maxillofac Surg 2013;42:e195–203. Available at, http://www.sciencedirect.com/science/article/pii/S1010518213002503 [last accessed 14th March, 2017].

[23] Murray DJ, Edwards G, Mainprize JG, Antonyshyn O. Advanced technology in the management of fibrous dysplasia. J Plast Reconstr Aesthet Surg 2008;61(8):906–16.
[24] Li B, Zhang L, Sun H, Yuan J, Shen SGF, Wang X. A novel method of computer aided orthognathic surgery using individual CAD/CAM templates: a combination of osteotomy and repositioning guides. Br J Oral Maxillofac Surg 2013;51(8):e239–44. Available at, http://www.sciencedirect.com/science/article/pii/S026643561300096X [last accessed 14th March, 2017].
[25] Modabber A, Legros C, Rana M, Gerressen M, Riediger D, Ghassemi A. Evaluation of computer-assisted jaw reconstruction with free vascularized fibular flap compared to conventional surgery: A clinical pilot study. Int J Med Robot Comput Assist Surg 2012;8(2):215–20.
[26] Bibb R, Eggbeer D, Evans P, Bocca A, Sugar A. Rapid manufacture of custom-fitting surgical guides. Rapid Prototyping J 2009;15(5):346–54.
[27] Ciocca L, Fantini M, De Crescenzio F, Persiani F, Scotti R. Computer-aided design and manufacturing construction of a surgical template for craniofacial implant positioning to support a definitive nasal prosthesis. Clin Oral Implants Res 2011;22(8):850–6.
[28] Darwood A, Collier J, Joshi N, Grant WE, Sauret-Jackson V, Richards R, et al. Re-thinking 3D printing: A novel approach to guided facial contouring. J Cranio-Maxillofac Surg 2015;43(7):1256–60.
[29] Guevara-Rojas G, Figl M, Schicho K, Seemann R, Traxler H, Vacariu A, et al. Patient-specific polyetheretherketone facial implants in a computer-aided planning workflow. J Oral Maxillofac Surg 2014;72(9):1801–12.
[30] Bibb R, Eggbeer D, Paterson A. Medical modelling: the application of advanced design and rapid prototyping techniques in medicine. 2nd ed. Cambridge, UK: Woodhead Publishing; 2015.
[31] Lethaus B, Bloebaum M, Koper D, Poort-Ter Laak M, Kessler P. Interval cranioplasty with patient-specific implants and autogenous bone grafts—success and cost analysis. J Craniomaxillofac Surg 2014;42(8):1948–51.
[32] Mustafa SF, Key SJ, Evans PL, Sugar AW. Virtual reconstruction of defects of the orbital floor using the morphometry of the opposite maxillary sinus. Br J Oral Maxillofac Surg 2010;48(5):392–3.
[33] Sarment DP, Al-Shammari K, Kazor CE. Stereolithographic surgical templates for placement of dental implants in complex cases. Int J Periodont Restorat Dent 2003;23(3):287–95.
[34] Sarment DP, Sukovic P, Clinthorne N. Accuracy of implant placement with a stereolithographic surgical guide. Int J Oral Maxillofac Implants 2003;18(4):571–7.
[35] Di Giacomo GAP, Cury PR, de Araujo NS, Sendyk WR, Sendyk CL. Clinical application of stereolithographic surgical guides for implant placement: preliminary results. J Periodontol 2005;76(4):503–7.
[36] Goffin J, van Brussel K, van der Sloten J, van Audekercke R, Smet MH, Marchal G, et al. 3D-CT based, personalized drill guide for posterior transarticular screw fixation at C1-C2: technical note. Neuro-Orthopedics 1999;25(1/2):47–56.
[37] Goffin J, van Brussel K, van der Sloten J, van Audekercke R, Smet MH. Three-dimensional computed tomography-based, personalized drill guide for posterior cervical stabilization at C1-C2. Spine 2001;26(12):1343–7.
[38] Verdonck HWD, Poukens J, Overveld HV, Riediger D. Computer-assisted maxillofacial prosthodontics: a new treatment protocol. Int J Prosthodont 2003;16(3):326–8.

[39] van Brussel K, Haex B, van der Sloten J, van Audekercke R, Goffin J, Lauweryns P, et al. Personalised drill guides in orthopaedic surgery with knife-edge support technique. In: Proceedings of the 6th International Symposium on Computer Methods in Biomechanics and Biomedical Engineering. First Numerics, Cardiff, UK (CD-ROM). 2004; 2004.
[40] Chen LH, Tsutsumi S, Iizuka T. A CAD/CAM technique for fabricating facial prostheses: a preliminary report. Int J Prosthodont 1997;10(5):467–72.
[41] Coward TJ, Watson RM, Wilkinson IC. Fabrication of a wax ear by rapid-process modelling using stereolithography. Int J Prosthodont 1999;12(1):20–7.
[42] Chua CK, Chou SM, Lin SC, Lee ST, Saw CA. Facial prosthetic model fabrication using rapid prototyping tools. Integrated Manufac Syst 2000;11(1):42–53.
[43] Runte C, Dirksen D, Delere H, Thomas C, Runte B, Meyer U, et al. Optical data acquisition for computer-assisted design of facial prostheses. Int J Prosthodont 2002;15(2):129–32.
[44] Cheah CM, Chua CK, Tan KH, Teo CK. Integration of Laser Surface Digitizing with CAD/CAM Techniques for Developing Facial Prostheses. Part 1: Design and Fabrication of Prosthesis Replicas. Int J Prosthodont 2003;16(4):435–41.
[45] Eggbeer D, Bibb R, Evans P, Ji L. Evaluation of direct and indirect additive manufacture of maxillofacial prostheses. Proc Inst Mech Eng H J Eng Med 2012;226(9):718–28.
[46] Daniel S, Eggbeer D. A CAD & AM process for maxillofacial prostheses bar-clip retention. Rapid Prototyping J 2016;22(1):170–7.

Cell preparation for 3D bioprinting

4

A. Al-Sabah, Z.M. Jessop*, I.S. Whitaker*, C. Thornton†*
*Reconstructive Surgery and Regenerative Medicine Research Group, Swansea University Medical School, Swansea, United Kingdom, †Department of Human Immunology, Swansea University Medical School

4.1 Introduction

Tissue damage associated with trauma, pathological, or congenital disorders often lead to the need to reconstruct or replace the affected areas [1]. Depending on the extent of damage, in most cases tissue or organ replacement is necessary; however, it is sometimes not feasible due to the limited availability of donors and tissue compatibility issues. Autologous tissue is the preferred choice for tissue transplants, as it bypasses the need for immunosuppressive treatments, but due to the limited availability of the specific tissue type and morbidity associated it is not a feasible option particularly if a significant amount of tissue or a whole organ is required to restore homeostasis. As a result, tissue engineering provides a novel alternative means to utilize the benefits of using autologous tissue to achieve similar functionality as endogenous tissue restoration of function.

One of emerging technologies that hold great promise in the field of tissue engineering is 3D bioprinting. 3D bioprinting allows the creation of tissue engineered constructs that allow precise depositioning of cells to create a microenvironment similar to native tissue. Regardless of the method of 3D bioprinting, whether the cells are combined with a biomaterial or printed directly on a scaffold the rate-limiting step of this process is cell number. The number of cells required depending on the type of 3D bioprinter utilized can range from 1×10^6 to $1 \times 10^8\,\mathrm{mL}^{-1}$ [2–5]. The large number of cells required for this process is necessary to synthesize extracellular matrix needed to provide support and functionalization of the construct. As a result, expansion of cells is an essential step in preparation for 3D bioprinting in order to have sufficient number of cells. As some cells have limited proliferation capabilities, the optimal cell population commonly used in 3D bioprinting is progenitor or stem cells. This encompasses the isolation of stem cells from the host as well as in vitro expansion of this population and once the number is sufficient for the tissue construct, stem cells are gradually differentiated to the specific cell type required. However, the in vitro expansion process requires the cells to be out of their natural microenvironment for extended periods of time which introduces a myriad of potential complications.

Prolonged expansion of primary cells in culture has been associated with numerous complications including chromosomal abnormalities, dedifferentiation, and tumor formation as well as genetic-based complications, adverse reactions to chemical contaminants which are introduced during cell culture processing. As a result, it is essential to assess cell health prior to implantation of cells in tissue engineering constructs. This review will provide an overview of different methodologies available for cellular expansion and assess cell health.

3D Bioprinting for Reconstructive Surgery. https://doi.org/10.1016/B978-0-08-101103-4.00006-5

4.2 Cell expansion

4.2.1 Bioreactors

Providing the optimal growth condition for the specific cell type to be expanded requires understanding of the native microenvironment of the cell within the tissue and controlling these conditions. Bioreactor systems provide a controlled environment to maintain and monitor several external factors such as temperature, nutrient delivery, gas exchange, and removal of metabolites (Fig. 4.1) [6]. These devices recapitulate the mass transfer that occurs in vivo. In most tissues, cells are situated in very close proximity to blood capillaries which enables required mass transfer needed for the survival of cells [7]. There are several types of bioreactor designs that are used for tissue engineering applications which includes perfusion systems, rotating wall, and spinner flasks. A perfusion system was shown to be an ideal system that allows media

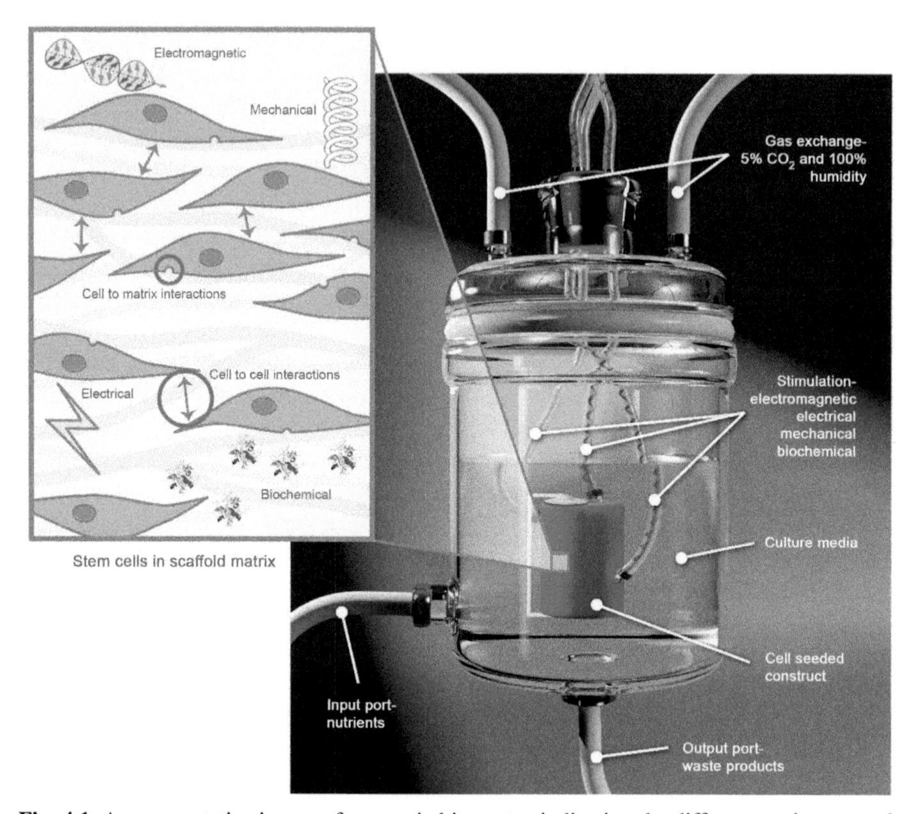

Fig. 4.1 A representative image of a generic bioreactor indicating the different environmental factors controlled within this system.
Courtesy of Steve Atherton.

perfusion throughout the extracellular matrix of a monolayer culture or tissue constructs [8,9]. Rotating wall and spinner flask systems were demonstrated to ensure homogeneity of media; however, it is not effective to ensure media perfusion through a tissue construct [10–12].

The use of bioreactors has been reported in the literature to enhance cell proliferation and differentiation with cell types which are susceptible to some form of mechanical forces (shear, tensile, compressive forces) in vivo [13,14]. As most tissues are vascularized, i.e., there is constant exchange of metabolites and gases through interstitial fluid or blood flow, thus cells become exposed to shear forces. To represent a physiologically relevant environment, bioreactors must be dynamic in nature by exposing cells to mechanical forces as a result of media movement exerting shear forces on cells. Fluid shear forces particularly with bone cells such as osteoblasts and osteocytes have shown to respond through increasing mineralization and promoting osteogenic differentiation [15,16].

4.2.2 Microcarriers

The most common issue with prolonged expansion of some cell types is the dedifferentiation of cells. To overcome this issue, studies have used microcarriers to expand cells. Microcarriers are microspheres in which adherent cells are seeded upon to encourage cell expansion. The surface area provided by these microcarriers is the equivalent of fifteen 75 cm^2 culture flasks [17]. The use of microcarriers has been shown to increase cell expansion by 17-fold with respect to conventional monolayer culture [18]. Microcarriers are commercially available composed of either synthetic or natural polymers. Some of the composition of microcarriers includes materials which include dextran, plastic, glass, cellulose, gelatin, and collagen [19,20]. The options available are based on whether the aim of the cell expansion is for testing and optimization of an experiment protocol or the cells will be implanted in vivo and require the microcarriers to biodegrade at an optimal rate at the intended site [17]. Furthermore, the surface of microcarriers can be modified to enhance cell attachment by coating with matrix macromolecules such as collagens or manipulation of the surface charge. Manipulation of microcarrier surface is dependent on the cell type. A study has shown that the addition of positively charged groups on the microcarrier surface has increased stem cell adherence compared to ECM molecule coating [21]. In contrast, culturing human articular chondrocytes with collagen-coated microcarriers demonstrated a 20-fold increase in cell proliferation within 2 weeks [22]. Studies culturing bone and cartilage cells on microcarriers in a bioreactor system have identified the newly synthesized matrix from these cells is representative of native tissue [19,23]. In addition to providing a large surface area for cell expansion, microcarriers have been used as slow release growth factor delivery systems. The delivery of growth factors using microcarriers provides a reliable method of mimicking the cell microenvironment thus enhancing the cell phenotype and inhibiting dedifferentiation [24]. The use of microcarriers presents an ideal method for cost-effective cell expansion as well as maintenance of cell phenotype during the process.

4.2.3 Viability

a. Trypan blue exclusion method
Whether cells are expanded for experimental processes such as testing a new scaffold or prepared to be implanted in patients, viability tests are necessary to determine that the growth conditions or prolonged culture of cells is not resulting in cell death [25]. There are multiple viability tests available to determine cell viability. Trypan blue exclusion method is one of the earliest and simplest viability assays. Trypan blue is a negatively charged dye which only stains cells with a compromised cell membrane, hence indicating cell death [26]. In contrast, viable cells are absent of trypan blue due to both the cell membrane and dye being negatively charged.

b. MTT assay
In addition to dye exclusion, some viability assays use metabolic activity as an indicator of cell viability. One of these assays is the MTT assay which is a commonly used colorimetric viability assay based on detecting cellular metabolic activity. The assay utilizes 3-(4,5-dimethylthiazol-2-yl)-2,5-diphenyltetrazolium bromide (MTT), a yellow tetrazolium salt, to evaluate the efficiency of mitochondrial enzymes. MTT is reduced via mitochondrial dehydrogenases to an insoluble purple formazan product [27]. As the formazan product is insoluble it is entrapped within live cells. The amount of formazan produced would be proportional to the number of viable cells. A spectrophotometer is used to measure the absorbance values of formazan to determine cell number. Similar in principle to the MTT assay, Alamar Blue assesses cell growth and viability via cell metabolism [28,29]. Alamar Blue is a nontoxic permeable dye which is reduced by cells to a pink fluorescence product which is then measured by a spectrophotometer. The nontoxic nature of Alamar Blue allows the monitoring of cells in culture over time, hence providing information regarding proliferation rates as well as viability [30].

c. Live/dead assay
The live/dead assay is the most informative viability assay available which presents information regarding the number of viable and nonviable cells, cell morphology, and the distribution of cells. The live/dead assay is a dual color fluorescence assay which discriminates viable from dead cells based on intracellular esterase activity [31]. The two dyes used in this assay are calcein acetoxymethyl ester (AM) and ethidium homodimer. Viable cells have an intact cell membrane where calcein AM, a nonfluorescent dye, is able to permeabilize and be hydrolyzed by intracellular esterases into green fluorescence. In contrast, ethidium homodimer, a nonpermeable red fluorescent dye, detects nonviable cells by penetrating the damaged cell membrane and binding to DNA [32].

4.3 Sterility assessment

During culturing of cells, sterility conditions can be easily comprised by introduction of fungal or bacterial contamination. In normal experimental culture conditions, antibiotics are used to prevent growth of potential contaminants, in contrast the use of antibiotics is not permitted in human cell cultures processed for implantation. The presence of exogenous compounds such as antibiotics may elicit an immune response, and may mask the presence of contaminants as their growth is temporarily stunted and may be inadvertently introduced into the patient [33]. Cells that will be processed as potential treatment for patients must be processed according to the current good manufacturing practices (CGMP) guidelines regarding cell therapies. One of the essential requirements to ensure lowering risk of contamination is to conduct all cell cultures in clean room with high efficiency particular absorbing (HEPA) filter to prevent airborne contaminants as mentioned in the guidelines. Confirmation of culture sterility is commonly confirmed by testing small amounts from each cell batch during the expansion process [34,35]. For a smaller scale testing methods, commercially available kits for detection of common types of contaminants (bacterial or fungal) could be used to verify sterility of cultures. These tests utilize fluorescent stains to distinguish the type of contamination present in cultures by testing a small volume of media taken from each cell batch [36].

Scaling up cell production for commercial purposes requires additional thorough sterility tests as set by regulatory bodies such as the Food and Drug Administration (FDA) and European Pharmacopoeia (EP) [37]. The two common methods of sterility are either the inoculation or filtration methods. For the inoculation method, samples must be incubated in aerobic and anaerobic conditions for up to 2 weeks to confirm sterility of sample [38]. Samples are considered sterile based on the turbidity of the mediums used. In contrast, the filtration method encompasses the use of a membrane filter with $<0.45\,\mu m$ in pore diameter, to prevent microbes from passing through, which media or other tissue culture reagents could be filtered through. The filtration process is commonly facilitated by a vacuum apparatus or by the use of a syringe [39].

As well as potential microbial and viral contamination, testing of process-related impurities must be conducted particularly if these cells were prepared to be implanted in patients. Due to the expansion and passaging of cells over a prolonged period may result in accumulation of process-related impurities derived either from the culture medium or released from cells [40]. Some of the impurities identified in culture include DNA and host cell proteins (HCP) which are released in culture by secretion or disruption in the cell membrane [41]. These impurities are detected by using an Enzyme-linked immunosorbent assay (ELISA).

4.3.1 Enzyme-linked immunosorbent assay

The ELISA method involves the usage of specific antibodies to detect target proteins in a colorimetric assay allowing the quantification of antigens in a sample. There are two ELISA methods, direct and indirect which refers to whether the primary antibody used to detect the antigen is conjugated (direct) or unconjugated (indirect) and thus requires a second antibody for visualization [42]. A sample is first incubated to a plastic well to allow adherence of the antigen. Once antigen is attached to plastic, an antibody specific to target protein which is conjugated to an enzyme binds to the antigen [43]. A substrate is the added to the well in which the enzyme catalyzes to form a colored product. The amount of colored product is proportional to antigen present [44]. Levels of impurities in culture that may be harmful or initiate an immunological reaction postimplantation must be assessed in final cell culture prior to implantation [45].

4.4 Cell phenotype characterization

4.4.1 Immunophenotyping and immunohistochemical analysis

As prolonged in vitro expansion is known to cause cellular dedifferentiation, it is essential to confirm cellular phenotype throughout the scaling up process. Immunophenotyping and immunohistochemical analysis are the two commonly used techniques to confirm cell phenotype. Immunophenotyping is one of the main clinical applications of flow cytometry. Flow cytometry is based on the screening of a stream of cells labeled with a cell surface marker antibody conjugated with a fluorescence probe [46]. The fluidics systems focus the cell population into a stream of individual cells. As individual cells pass through the narrow channel, a light is illuminated on each cell passing hence exciting the fluorescent probes [47]. The fluorescent emission is detected by sensors which provide information about the phenotypic properties of each individual cell. Flow cytometric data outputs detect changes in DNA content, cell surface expression, and morphology [48]. Analyzing cells based on their phenotypic properties allows clear distinction of healthy cells from abnormal cells. In contrast, immunohistochemistry staining provides an alternative method to visualize a wide range of specific cell markers and their distribution to confirm cell type, phosphorylation states of proteins, as well as their localization. Immunohistochemistry is commonly used clinically for identification of tumor markers and determination of malignancy status of tumors. A panel of antibodies which are specific to proteins commonly found at target cell are used for the characterization of a cell type [49]. Due to the specificity of antibodies-antigen reaction, immunohistochemistry is one of the best methods in detecting protein expression. However, a reporter is needed in order to visualize the antibody-antigen reaction. There are numerous reporters that can be used to stain this reaction and they fall under two categories, chromogenic or fluorescent reporters.

a. Chromogenic reporters
Chromogenic reporters generate a color from a colorless substrate as a result of an enzymatic reaction. An enzyme-coupled secondary antibody, which is raised against the species used to generate the primary antibody, will bind to the primary antibody, however the reaction is colorless [50]. A substrate, specific to the enzyme linked to the secondary antibody, is added to the sample resulting in an enzymatic reaction forming an insoluble colored compound. The commonly used enzyme labels are horseradish peroxidase and calf intestinal alkaline phosphatase, which results in catalyzing their respective substrates 3,3′-diaminobenzidine (DAB) and a combination of nitro blue tetrazolium chloride (NBT) and 5-bromo-4-chloro-3-indolyl phosphate (BCIP) for ALP into a colored compound to be visualized under a light microscope [51].

b. Fluorescent reporters
The alternative method of protein labeling is via fluorescent reporters. In this case secondary antibodies have a fluorophore, a fluorescent molecule, conjugated. The benefit of fluorescent labeling is it enables multiplex staining, thus having thorough high resolution phenotype analysis in a relatively short amount of time [52,53]. Fundamentally, it is ideal to confirm as many cell markers at once to assess cell phenotype; however, it is not cost effective if applied at a larger scale such as a clinical setting. A more

cost-effective method would be using ELISA assays which are additionally used for protein expression profiling to determine key protein markers to confirm cell identity. Commercially available multiplexed arrays enable measurements of up to 16 proteins per well which provides an overall protein profile of a given sample and is very cost effective rather than running individual ELISA assays particularly for a large-scale screening [54,55].

4.4.2 Genomic assessment

a. Quantitative PCR
Quantitative PCR (qPCR) is one of the earliest genetic screens that is conducted to identify cell phenotype. Each cell type expresses a set of markers creating a unique phenotypic profile. Using qPCR, the levels of mRNA of these specific markers could be detected to establish if cultured cells have retained their phenotype. Primers specific to the markers are used to conduct this test. PCR arrays, a systematic screen, of numerous genes up to 384 genes could be conducted if more genomic screening is required [56]. Each array is comprised of either 96- or 384-well-plate format in which each well contains a specific gene primer [57]. As a result, the cell type in question could be screened for multiple genes simultaneously. The disadvantage of this setup is that there are no replicates for each gene to account for errors.

b. Karyotyping
In addition to loss of cellular phenotype, prolonged passaging of cells is also known to result in chromosomal instability. Karyotyping techniques are used to assess chromosomal arrangement and stability in cells [58,59]. Traditional karyotyping of chromosome is achieved by using a chemical stain. The staining of chromosomes highlights the compacted (heterochromatin) and accessible (euchromatin) areas of each chromosome resulting in a banding pattern which is visualized using a light microscope [60]. Chemical stains also provide information on the overall size and morphological structure [61]. As classical karyotyping methods detect subtler alterations in chromosomes by the banding method, modern karyotyping techniques permit the detection of chromosomal aberrations of a million base pairs [62].

c. Fluorescent In Situ Hybridization (FISH)
One such technique is Fluorescent In Situ Hybridization (FISH). FISH is based on the use of fluorescent probes which are complementary to specific segments in a sample of DNA. Chromosomes in the metaphase stage of the cell cycle are fixed and double-stranded DNA is separated to allow annealing of the respective FISH probe onto the DNA segment [63,64]. As the probes target specific sequences, it is highly accurate in determination of any chromosomal mutations that may have occurred. Traditional karyotyping is cheaper as a screen for abnormalities; however, for a clinical setting FISH is commonly used due to its high accuracy and resolution.

d. Telomere length assay
An important parameter to investigate in prolonged cell cultures is assessment of cellular senescence. Each cell has a limited proliferative capacity and ultimately reaches senescence. Cellular senescence is the state in which a cell is no longer able to undergo cellular division [65]. Telomere length assay is one of the common methods of cellular senescence assessment. Telomeres are repetitive DNA sequence present at the end of chromosomes. Studies have associated the shortening of telomere length in cells to cellular senescence [66–68]. Telomere length is measured in samples using commercially available kits. Briefly, genomic DNA is digested and is separated by gel electrophoresis and transferring fragments into a

membrane. Fragments on membrane are then hybridized to a labeled probe specific to telomere sequence and visualized by chemiluminescence reaction [69].

e. Quantification of β-galactosidase
An alternative method for detection of senescence is quantification of β-galactosidase levels. B-galactosidase was suggested by studies as a biomarker of cell senescence. Studies have demonstrated that expression of β-galactosidase is predominantly detectable in senescent cells and not in immortalized cell lines [70,71]. This method involves staining the cells with a chemical, X-gal, which is cleaved by β-galactosidase into a blue precipitate [72]. β-galactosidase is a good overall screen to determine cellular senescence; however, a thorough and more robust determinant would be looking at telomere length as it has been thoroughly studied and established as a marker of cell senescent.

4.5 Metabolic assessment

Cell metabolism provides a reflection of the health status of the cell. The mitochondrion is the main powerhouse of the cell in which bioenergetic processes occur by the uptake of fuel sources such as glucose and fatty acids and converts them into energy in a series of enzymatic reactions [73,74]. Bioenergetics processes are tightly regulated as they are essential for cell viability. Deregulation of cellular bioenergetics has been previously associated with numerous pathologies such as cancer, diabetes, and neurodegenerative diseases [75].

Measurement of cellular bioenergetics will present a clearer picture for understanding alterations in cell phenotype and physiology. Bioenergetics of cells is quantified via measurement of extracellular flux and the alteration in oxygen and protons concentrations in the media. The rate of oxygen consumption rate (OCR)—an indicator of mitochondrial respiration and rate of acid efflux (ECAR)—the amount of lactic acid formation during glycolytic metabolism are parameters which determine cell physiology and their metabolic state [76–78]. Measurement of OCR and ECAR simultaneously provides unique bioenergetic profiles that occur due to culture conditions or specific treatments.

It is well established that the environmental growth conditions influence cellular bioenergetics. Studies examining the culture of stem cells in nonphysiological levels of oxygen have shown that it is detrimental to cell viability and genetic stability. With higher oxygen levels, the bioenergetic mechanisms of stem cells have shifted from glycolytic to mitochondrial respiration (oxidative phosphorylation) [79]. Consequently, the increased consumption of oxygen results in the generation of reactive oxidative species (ROS) which is implicated in promoting cellular senescence, cell death, and genetic instability [80,81]. Physiological oxygen levels were shown to enhance cell stemness in prolonged culture of stem cells thus emphasizing the necessity to recapitulate the microenvironment of the cell [82]. In addition to increasing stemness, low oxygen levels have resulted in increased stem cell proliferation and extension of cellular life span [83]. In vivo cell bioenergetics is dependent on the microenvironment of the cell such as available vascularization and oxygen tension of the tissues. Most stem cells reside in a hypoxic environment which have enabled them to utilize glycolysis as their preferred means of energy production [84]. As a result, it is relatively simple to confirm whether the cells of interest have retained their metabolic phenotype by assessing the extracellular flux after prolonged expansion.

Culture media composition and, in particular, glucose concentration is additionally a potential source which can alter cellular bioenergetics. As mentioned previously, stem cells primarily utilize glycolysis for energy production, whereas it is suggested that differentiated cells shift toward oxidative phosphorylation [85]. This phenomenon was observed with embryonic stem cells cultured in high glucose concentrations which have shown to result in formation of ROS promoting cell differentiation [86]. A study by Rogers et al. have identified that culturing endothelial cells in either low or high glucose accelerated cell senescence [87]. These findings must be considered when selecting the optimal concentration of glucose supplementation for cell culture media. Using the glucose concentration that is not physiologically relevant for the cell of interest could lead to dedifferentiation of cells and premature cell aging.

4.6 Conclusion

Expansion of cells for 3D bioprinting is a long process which can be made efficient by using nonconventional methods of tissue culture such as the use of bioreactors and microcarriers. Having the ability to modify these tools allow for further customization of what is required to create the optimal protocol for cell expansion of a specific cell type. As mentioned previously, the variability in measures taken to initiate efficient cell expansion based on cell type adds an additional obstacle to standardize the expansion process based on cell type. However, the combination of bioreactors and microcarriers in cell culture has provided a cost-effective method in expanding cells to generate the large quantities needed for 3D bioprinting and other tissue engineering applications.

Cells prepared for 3D bioprinted constructs with the intent to be implanted in humans require additional parameters to be evaluated to confirm the safety of these cells in vivo. Contamination of cell culture reagents which can elicit an immune response presents a significant risk as a result of prolonged cell expansion, thus cultured cells must be adequately assessed for presence of any potential contaminants (Fig. 4.2). The prolonged expansion and passaging of cells has been known to result in dedifferentiation of cells as well as affecting genomic stability and cell physiology. Phenotypic and genomic alterations as a consequence of prolonged cell expansion has been well documented in the literature to lead to loss of cell identity and functionality [88]. Confirmation of cell identity and functionality plays an integral role to the success and maintenance of a 3D bioprinted construct as the matrix synthesized by cells will provide robustness and integrity to the construct. Therefore it is essential to confirm cell identity and functionality by examining parameters such as cell viability, genomic assessment, phenotype characterization, and cellular metabolism.

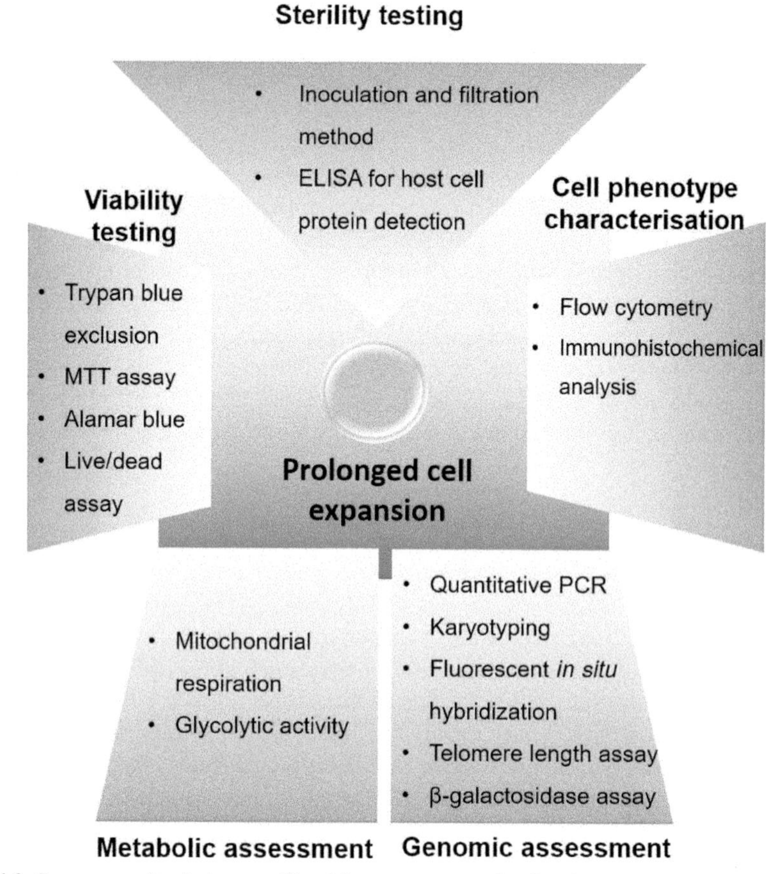

Fig. 4.2 Summary of techniques utilized for assessment of cells after prolonged cell culture.

References

[1] Eberli D, Atala A. Tissue engineering using adult stem cells. Methods in enzymology. Cambridge, MA: Academic Press; 2006.

[2] Guillotin B, Souquet A, Catros S, Duocastella M, Pippenger B, Bellance S, et al. Laser assisted bioprinting of engineered tissue with high cell density and microscale organization. Biomaterials 2010;31:7250–6.

[3] Marga F, Jakab K, Khatiwala C, Shepherd B, Dorfman S, Hubbard B, et al. Toward engineering functional organ modules by additive manufacturing. Biofabrication 2012;4:022001.

[4] Mironov V, Kasyanov V, Markwald RR. Organ printing: from bioprinter to organ biofabrication line. Curr Opin Biotechnol 2011;22:667–73.

[5] Xu T, Jin J, Gregory C, Hickman JJ, Boland T. Inkjet printing of viable mammalian cells. Biomaterials 2005;26:93–9.

[6] Martin I, Wendt D, Heberer M. The role of bioreactors in tissue engineering. Trends Biotechnol 2004;22:80–6.

[7] Datta AK. Biological and bioenvironmental heat and mass transfer. New York: Taylor & Francis; 2002.

[8] Bancroft GN, Sikavitsas VI, Mikos AG. Design of a flow perfusion bioreactor system for bone tissue-engineering applications. Tissue Eng 2003;9:549–54.
[9] Grayson WL, Frohlich M, Yeager K, Bhumiratana S, Chan ME, Cannizzaro C, et al. Engineering anatomically shaped human bone grafts. Proc Natl Acad Sci U S A 2010;107:3299–304.
[10] Meinel L, Karageorgiou V, Fajardo R, Snyder B, Shinde-Patil V, Zichner L, et al. Bone tissue engineering using human mesenchymal stem cells: effects of scaffold material and medium flow. Ann Biomed Eng 2004;32:112–22.
[11] Stiehler M, Bünger C, Baatrup A, Lind M, Kassem M, Mygind T. Effect of dynamic 3-D culture on proliferation, distribution, and osteogenic differentiation of human mesenchymal stem cells. J Biomed Mater Res A 2009;89A:96–107.
[12] Zhang Z-Y, Teoh SH, Chong W-S, Foo T-T, Chng Y-C, Choolani M, et al. A biaxial rotating bioreactor for the culture of fetal mesenchymal stem cells for bone tissue engineering. Biomaterials 2009;30:2694–704.
[13] Fung YC. Biomechanics: mechanical properties of living tissues. New York, NY: Springer; 2013.
[14] Hay E. Cell biology of extracellular matrix. 2nd ed. Trends in cell biology. vol. 2. Cambridge, MA: Cell Press; 1992.
[15] Bancroft GN, Sikavitsas VI, Van Den Dolder J, Sheffield TL, Ambrose CG, Jansen JA, et al. Fluid flow increases mineralized matrix deposition in 3D perfusion culture of marrow stromal osteoblasts in a dose-dependent manner. Proc Natl Acad Sci U S A 2002;99:12600–5.
[16] Bilodeau K, Mantovani D. Bioreactors for tissue engineering: focus on mechanical constraints. A comparative review. Tissue Eng 2006;12:2367–83.
[17] Malda J, Frondoza CG. Microcarriers in the engineering of cartilage and bone. Trends Biotechnol 2006;24:299–304.
[18] Melero-Martin JM, Dowling MA, Smith M, Al-Rubeai M. Expansion of chondroprogenitor cells on macroporous microcarriers as an alternative to conventional monolayer systems. Biomaterials 2006;27:2970–9.
[19] Malda J, Kreijveld E, Temenoff JS, Van Blitterswijk CA, Riesle J. Expansion of human nasal chondrocytes on macroporous microcarriers enhances redifferentiation. Biomaterials 2003;24:5153–61.
[20] Malda J, Van Blitterswijk CA, Grojec M, Martens DE, Tramper J, Riesle J. Expansion of bovine chondrocytes on microcarriers enhances redifferentiation. Tissue Eng 2003;9:939–48.
[21] Frauenschuh S, Reichmann E, Ibold Y, Goetz PM, Sittinger M, Ringe J. A microcarrier-based cultivation system for expansion of primary mesenchymal stem cells. Biotechnol Prog 2007;23:187–93.
[22] Frondoza C, Sohrabi A, Hungerford D. Human chondrocytes proliferate and produce matrix components in microcarrier suspension culture. Biomaterials 1996;17:879–88.
[23] Shikani AH, Fink DJ, Sohrabi A, Phan P, Polotsky A, Hungerford DS, et al. Propagation of human nasal chondrocytes in microcarrier spinner culture. Am J Rhinol 2004;18:105–12.
[24] Perez RA, El-Fiqi A, Park JH, Kim TH, Kim JH, Kim HW. Therapeutic bioactive microcarriers: co-delivery of growth factors and stem cells for bone tissue engineering. Acta Biomater 2014;10:520–30.
[25] Freshney RI, Stacey GN, Auerbach JM. Culture of human stem cells. Hoboken, NJ: Wiley; 2007.
[26] Tran SL, Puhar A, Ngo-Camus M, Ramarao N. Trypan blue dye enters viable cells incubated with the pore-forming toxin HlyII of Bacillus cereus. PLoS One 2011;6: e22876.

[27] Liu Y, Peterson DA, Kimura H, Schubert D. Mechanism of cellular 3-(4,5-dimethylthiazol-2-yl)-2,5-diphenyltetrazolium bromide (MTT) reduction. J Neurochem 1997;69:581–93.
[28] Hamid R, Rotshteyn Y, Rabadi L, Parikh R, Bullock P. Comparison of alamar blue and MTT assays for high through-put screening. Toxicol In Vitro 2004;18:703–10.
[29] Hughes D, Mehmet H. Cell proliferation and apoptosis. Oxfordshire: Taylor & Francis; 2003.
[30] O'Brien J, Wilson I, Orton T, Pognan F. Investigation of the Alamar Blue (resazurin) fluorescent dye for the assessment of mammalian cell cytotoxicity. Eur J Biochem 2000;267:5421–6.
[31] Periasamy A. Methods in cellular imaging. New York, NY: Springer; 2013.
[32] Maltaris T, Kaya H, Hoffmann I, Mueller A, Beckmann M, Dittrich R. Comparison of xenografting in SCID mice and LIVE/DEAD assay as a predictor of the developmental potential of cryopreserved ovarian tissue. In vivo 2006;20:11–6.
[33] Mcgarrity GJ. Detection of contamination. Methods in enzymology. Cambridge, MA: Academic Press; 1979.
[34] Giancola R, Bonfini T, Iacone A. Cell therapy: cGMP facilities and manufacturing. Muscles Ligaments Tendons J 2012;2:243–7.
[35] Stormer M, Radojska S, Hos NJ, Gathof BS. Protocol for the validation of microbiological control of cellular products according to German regulators recommendations--Boon and Bane for the manufacturer. Vox Sang 2015;108:314–7.
[36] Veal DA, Deere D, Ferrari B, Piper J, Attfield PV. Fluorescence staining and flow cytometry for monitoring microbial cells. J Immunol Methods 2000;243:191–210.
[37] Hourd P, Ginty P, Chandra A, Williams DJ. Manufacturing models permitting roll out/ scale out of clinically led autologous cell therapies: regulatory and scientific challenges for comparability. Cytotherapy 2014;16:1033–47.
[38] Stormer M, Wood EM, Schurig U, Karo O, Spreitzer I, McDonald CP, et al. Bacterial safety of cell-based therapeutic preparations, focusing on haematopoietic progenitor cells. Vox Sang 2014;106:285–96.
[39] Council of Europe. Biological tests. European Pharmacopoeia 5 2005.
[40] Rathore AS, Sofer G. Process validation in manufacturing of biopharmaceuticals. 3rd ed. Oxfordshire: Taylor & Francis; 2012.
[41] Zhou W, Kantardjieff A. Mammalian cell cultures for biologics manufacturing. Heidelberg, New York: Springer; 2014.
[42] Van Emon JM. Immunoassay and other bioanalytical techniques. Boca Raton, FL: CRC Press; 2016.
[43] Hnasko R. ELISA: methods and protocols. New York, NY: Springer; 2015.
[44] Crowther JR. The ELISA guidebook. 2nd ed. New York: Humana Press; 2010.
[45] Yano K, Watanabe N, Tsuyuki K, Ikawa T, Kasanuki H, Yamato M. Regulatory approval for autologous human cells and tissue products in the United States, the European Union, and Japan. Regen Ther 2015;1:45–56.
[46] Shapiro HM. Practical flow cytometry. Hoboken, NJ: Wiley; 2005.
[47] Rahman M. Introduction to flow cytometry. AbD serotec 2006.
[48] Givan AL. Flow cytometry: first principles. Hoboken, NJ: Wiley; 2013.
[49] Ramos-Vara JA. Principles and methods of immunohistochemistry. In: Gautier J-C, editor. Drug safety evaluation: methods and protocols. Totowa, NJ: Humana Press; 2011.
[50] Buchwalow IB, Böcker W. Immunohistochemistry: basics and methods. Berlin, Heidelberg: Springer; 2010.
[51] Petersen KH. Novel horseradish peroxidase substrates for use in immunohistochemistry. J Immunol Methods 2009;340:86–9.

[52] Robertson D, Savage K, Reis-Filho JS, Isacke CM. Multiple immunofluorescence labelling of formalin-fixed paraffin-embedded (FFPE) tissue. BMC Cell Biol 2008;9:13.

[53] Suzuki T, Fujikura K, Higashiyama T, Takata K. DNA staining for fluorescence and laser confocal microscopy. J Histochem Cytochem 1997;45:49–53.

[54] Jin P, Han TH, Ren J, Saunders S, Wang E, Marincola FM, et al. Molecular signatures of maturing dendritic cells: implications for testing the quality of dendritic cell therapies. J Transl Med 2010;8:4.

[55] Panelli MC, White R, Foster M, Martin B, Wang E, Smith K, et al. Forecasting the cytokine storm following systemic interleukin (IL)-2 administration. J Transl Med 2004;2:17.

[56] Arikawa E, Quellhorst G, Han Y, Pan H, Yang J. RT2 Profiler™ PCR arrays: pathway-focused gene expression profiling with qRT-PCR. SA biosciences technical article, 11 2011.

[57] Boone DR, Micci M-A, Taglialatela IG, Hellmich JL, Weisz HA, Bi M, et al. Pathway-focused PCR array profiling of enriched populations of laser capture microdissected hippocampal cells after traumatic brain injury. PLoS One 2015;10: e0127287.

[58] Fazeli A, Liew CG, Matin MM, Elliott S, Jeanmeure LF, Wright PC, et al. Altered patterns of differentiation in karyotypically abnormal human embryonic stem cells. Int J Dev Biol 2011;55:175–80.

[59] Lund RJ, Nikula T, Rahkonen N, Narva E, Baker D, Harrison N, et al. High-throughput karyotyping of human pluripotent stem cells. Stem Cell Res 2012;9:192–5.

[60] Shemilt L, Verbanis E, Schwenke J, Estandarte AK, Xiong G, Harder R, et al. Karyotyping human chromosomes by optical and X-ray ptychography methods. Biophys J 2015;108:706–13.

[61] Yunis JJ, Sanchez O. G-banding and chromosome structure. Chromosoma 1973;44:15–23.

[62] Korf BR, Irons MB. Human genetics and genomics. Chichester, West Sussex: John Wiley & Sons; 2013.

[63] Kearney L. Molecular cytogenetics. Best Pract Res Clin Haematol 2001;14:645–69.

[64] Pinkel D, Landegent J, Collins C, Fuscoe J, Segraves R, Lucas J, et al. Fluorescence in situ hybridization with human chromosome-specific libraries: detection of trisomy 21 and translocations of chromosome 4. Proc Natl Acad Sci U S A 1988;85:9138–42.

[65] Shay JW, Wright WE. Senescence and immortalization: role of telomeres and telomerase. Carcinogenesis 2005;26:867–74.

[66] Allsopp RC, Vaziri H, Patterson C, Goldstein S, Younglai EV, Futcher AB, et al. Telomere length predicts replicative capacity of human fibroblasts. Proc Natl Acad Sci U S A 1992;89:10114–8.

[67] Chang E, Harley CB. Telomere length and replicative aging in human vascular tissues. Proc Natl Acad Sci U S A 1995;92:11190–4.

[68] Weng NP, Levine BL, June CH, Hodes RJ. Human naive and memory T lymphocytes differ in telomeric length and replicative potential. Proc Natl Acad Sci U S A 1995;92:11091–4.

[69] Gan Y, Engelke KJ, Brown CA, Au JL. Telomere amount and length assay. Pharm Res 2001;18:1655–9.

[70] Debacq-Chainiaux F, Erusalimsky JD, Campisi J, Toussaint O. Protocols to detect senescence-associated beta-galactosidase (SA-[beta]gal) activity, a biomarker of senescent cells in culture and in vivo. Nat Protoc 2009;4:1798–806.

[71] Gary RK, Kindell SM. Quantitative assay of senescence-associated β-galactosidase activity in mammalian cell extracts. Anal Biochem 2005;343:329–34.

[72] Dimri GP, Lee X, Basile G, Acosta M, Scott G, Roskelley C, et al. A biomarker that identifies senescent human cells in culture and in aging skin in vivo. Proc Natl Acad Sci U S A 1995;92:9363–7.

[73] Rolfe DF, Brown GC. Cellular energy utilization and molecular origin of standard metabolic rate in mammals. Physiol Rev 1997;77:731–58.
[74] Sinclair R. Response of mammalian cells to controlled growth rates in steady-state continuous culture. In Vitro 1974;10:295–305.
[75] Ferrick DA, Neilson A, Beeson C. Advances in measuring cellular bioenergetics using extracellular flux. Drug Discov Today 2008;13:268–74.
[76] Eklund SE, Kozlov E, Taylor DE, Baudenbacher F, Cliffel DE. Real-time cell dynamics with a multianalyte physiometer. In: Rosenthal SJ, Wright DW, editors. Nanobiotechnology protocols. Totowa, NJ: Humana Press; 2005.
[77] Fleischaker RJ, Weaver JC, Sinskey AJ. Instrumentation for process control in cell culture. In: Perlman D, Allen IL, editors. Advances in applied microbiology. Academic Press; 1981.
[78] Wiley C, Beeson C. Continuous measurement of glucose utilization in heart myoblasts. Anal Biochem 2002;304:139–46.
[79] Estrada JC, Albo C, Benguria A, Dopazo A, Lopez-Romero P, Carrera-Quintanar L, et al. Culture of human mesenchymal stem cells at low oxygen tension improves growth and genetic stability by activating glycolysis. Cell Death Differ 2012;19:743–55.
[80] Droge W. Free radicals in the physiological control of cell function. Physiol Rev 2002;82:47–95.
[81] Parrinello S, Samper E, Krtolica A, Goldstein J, Melov S, Campisi J. Oxygen sensitivity severely limits the replicative lifespan of murine fibroblasts. Nat Cell Biol 2003;5:741–7.
[82] Lennon DP, Edmison JM, Caplan AI. Cultivation of rat marrow-derived mesenchymal stem cells in reduced oxygen tension: effects on in vitro and in vivo osteochondrogenesis. J Cell Physiol 2001;187:345–55.
[83] Fehrer C, Brunauer R, Laschober G, Unterluggauer H, Reitinger S, Kloss F, et al. Reduced oxygen tension attenuates differentiation capacity of human mesenchymal stem cells and prolongs their lifespan. Aging Cell 2007;6:745–57.
[84] Ito K, Suda T. Metabolic requirements for the maintenance of self-renewing stem cells. Nat Rev Mol Cell Biol 2014;15:243–56.
[85] Sharifpanah F, Wartenberg M, Hannig M, Piper H-M, Sauer H. Peroxisome proliferator-activated receptor α agonists enhance cardiomyogenesis of mouse ES cells by utilization of a reactive oxygen species-dependent mechanism. Stem Cells 2008;26:64–71.
[86] Crespo FL, Sobrado VR, Gomez L, Cervera AM, Mccreath KJ. Mitochondrial reactive oxygen species mediate cardiomyocyte formation from embryonic stem cells in high glucose. Stem Cells 2010;28:1132–42.
[87] Rogers SC, Zhang X, Azhar G, Luo S, Wei JY. Exposure to high or low glucose levels accelerates the appearance of markers of endothelial cell senescence and induces dysregulation of nitric oxide synthase. J Gerontol A Biol Sci Med Sci 2013;68:1469–81.
[88] Wang Y, Zhang Z, Chi Y, Zhang Q, Xu F, Yang Z, et al. Long-term cultured mesenchymal stem cells frequently develop genomic mutations but do not undergo malignant transformation. Cell Death Dis 2013;4: e950.

3D bioprinting for scaffold fabrication

5

D. Singh
Yale University, New Haven, CT, United States

Abbreviations

μm	micrometer
2D	two-dimensional
3DP	3-dimensional printing
ALP	alkaline phosphatase
AM	additive manufacturing
BMP	bone morphogenetic protein
BSM	bladder submucosa
$CaCl_2$	calcium chloride
CAD	computer-aided design
CT scan	computed tomography scan
CMV	cytomegalovirus
Dia	diameter
DNA	deoxyribonucleic acid
ESC	endothelial stem cells
E. coli	*Escherichia coli*
ECM	extracellular matrix
G-blocks	α-L-guluronic acid blocks
ECs	endothelial cells
GAG	glycosaminoglycan
GFP	green fluorescent protein
GF	growth factor
HUVECs	human umbilical vein endothelial cells
hMSCs	human mesenchymal stem cells
IEC 6 cells	small intestine epithelial cells
LBL	layer by layer
MPa	megapascal
NaCl	sodium chloride
NSC	neural stem cells
NHLF	Normal human lung fibroblasts
M-blocks	β-D-mannuronic acid monomers
PAE	porcine aortic endothelial
p-DNA	plasmid DNA
PBS	phosphate buffer saline
PCL	polycaprolactone
PEG	poly (ethylene glycol)
pHEMA	polyhydroxyethylmethacrylate

3D Bioprinting for Reconstructive Surgery. https://doi.org/10.1016/B978-0-08-101103-4.00007-7

PLGA	poly (lactic-co-glycolic acid)
PU	polyurethane
P(LA/CL)	poly(l-lactide/epsilon-caprolactone)
PGA	poly(glycolic acid)
RP	rapid prototyping
RGD	Arg-gly-asp
SFF	solid free-form fabrication
SIS	small intestinal submucosa
UV	ultraviolet
VEGF	vascular endothelial growth factor

5.1 Introduction

Human body has limited ability to regenerate a diseased organ. In case of major trauma, tissue and organs whose integrity is compromised, results in devastating deficits and tissue dysfunction [1]. Using materials to augment or repair damaged tissue dates to ancient times, when natural materials such as wood for repairing bone, or using reeds or tubes for nose reconstruction to structurally replace lost tissue due to trauma or disease [2]. As the engineering and medical technologies advanced the criteria of material selection shifted from availability to suitability. In early 20th century naturally available materials were quickly replaced with synthetic polymers such as metal alloys and ceramics that resulted in increased tissue restoration, better functionality, and higher reproducibility [2–5]. Today, clinicians face an ever increasing burden of congenital abnormalities, trauma, and various degenerative diseases, and with the advent of tissue engineering, it is possible to engineer new biological therapeutics that can repair variety of tissues that were impossible previously (Fig. 5.1) [3]. As a result of advances made by tissue engineers, thousands of lives have been either saved or have given a better quality of life; for example, in case of burnt patient, the scars left behind due to trauma resulted in patient dissatisfaction and poor life quality; however, with new tissue-engineered bioprinted skin, these scars formation could be stopped [6]. Another example such as vascular stents and artificial hips have not only helped patients live pain free but has also allowed them to perform day-to-day functions normally giving better quality of life than before [7,8]. Based on its application biomaterials can be defined as type of materials used as medical device and usually involved multidisciplinary team from materials science to engineering to biologist [3,9,10]. Since these materials have to perform major biofunctions without causing immune rejection there are certain properties that need to be addressed while engineering these matrices [11–14]. The need and importance of these medical devices could be gauzed by the fact that in last two decades biomaterials have become an intellectually distinct discipline.

5.1.1 Tissue engineering history and evolution

Nearly three decades ago Langer and Vacanti introduced a concept that has now grown into independent stream in both academic and industrial arena and the concept was "Tissue engineering." In the paper published in pediatric surgery titled "Selective cell

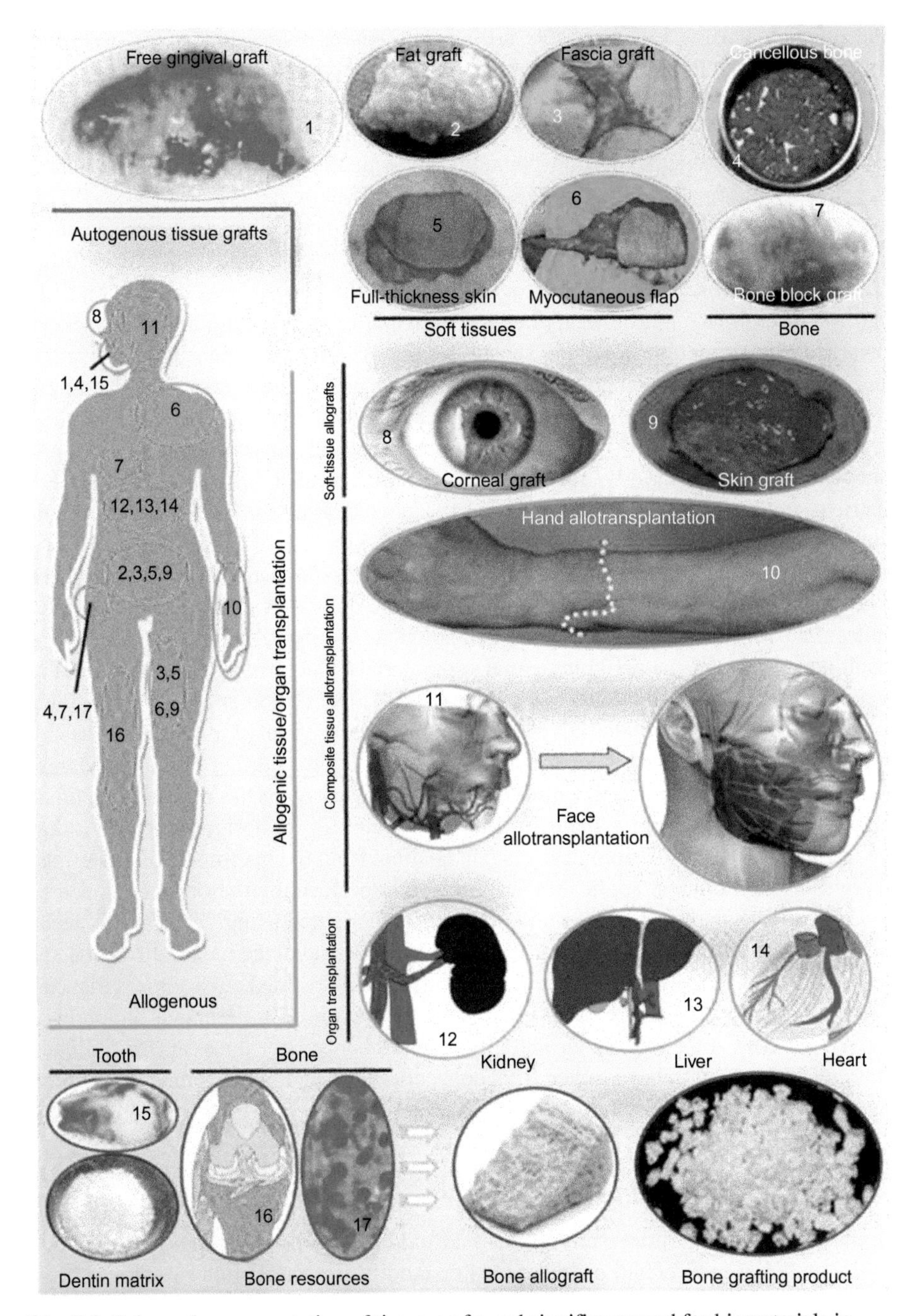

Fig. 5.1 Schematic representation of tissue grafts and significant need for biomaterials in clinical application.
Reproduced with permission from Chen FM, Liu XH. Advancing biomaterials of human origin for tissue engineering. Prog Polym Sci 2016; 53: 86–168.

transplantation using bioabsorbable artificial polymers as matrices" [15] described culturing single cells and clusters of fetal and adult hepatocytes, small intestine cells, and pancreatic islet cells which were seeded onto different biodegradable polymeric matrices for 4 days before implantation in host animal. Some of these implantation resulted in viable hepatocytes transplantation showing mitotic figures and vascularization of cell clusters. This concept was termed as "chimeric neomorphogenesis." This work over the period of time formed the basis of what is now known as "tissue engineering." Same authors in 1993 published a groundbreaking article in science titled "Tissue Engineering" in which a clear definition of the concept was provided stating "an interdisciplinary field that applies the principle of engineering and life sciences towards development of biological substitutes that restores, maintains or improve tissue function"[16]. Ever since this paper came out tissue engineering has become a thriving area with ~100 companies and every academic institution worldwide working in some concept related to it [17–23]. Revolution in molecular biology, proteomics, and genomics further the advanced biomaterial engineering and changes the way these matrices are designed and used, for example, understanding the role of specific molecule such as bone morphogenetic proteins (BMPs) in case of osteogenesis allowed bioengineers to incorporate this into synthetic matrix as bioactive component. These types of combination that could directly influence cells and tissue through precise molecular pathways allowed engineers to manipulate biological response to the matrix resulting in state-of-the-art commercial biomaterials like controlled drug-releasing vascular stents and BMP-loaded bone grafts [22,24]. The skill to engineer the biological activity into artificial materials greatly increases the amount of its potential uses by improving the performance in many traditional applications. Furthermore, the increasing focus on functionality and complexity of tissues and organs has resulted in biomaterials engineers to consider and evolve newer techniques for designing matrixes that are inspired from human biological system. Distinct from the man-made biomaterials, matrices used in clinical application are frequently performing multitask and need to be stable in a dynamic condition and are usually fabricated using "bottom-up" techniques [25]. Both the materials and biophysical process involved in its fabrication are inspiring the design and fabrication of new materials in artificial environment that can potentially be used for wider range of clinical applications. This demands an intellectual shift in how a material is viewed and defined, hence over the years "techniques for scaffold fabrication has evolved from conventional particulate-leaching to complex three-dimensional (3D) organ printing" [26].

5.1.2 Emergence of 3D bioprinting

Limitation with most of the conventional scaffold fabrication techniques is that it does not allow the user to create precise pores or modify its distribution or achieve high interconnectivity with detail geometry [27–29]. Three-dimensional printing (3DP) has been developed to overcome most of these limitations that holds the progress of tissue engineering. 3DP came into light first in 1986, by Charles, W. Hull in the method that he named as stereolithography [30]—a thin layer could be cured with UV lights and sequentially printed into layers to form a solid 3D structure. This process was further applied for creating sacrificial resin molds to fabricate 3D scaffold from biomaterials. This field further progressed with development of solvent-free, aqueous-based systems that allowed direct printing of biomaterials into 3D matrices that could be transplanted with or without cells [31–42]. With this, 3D technique has emerged as a promising tool for fabricating scaffold with highly precise and accurate detailing of intrinsic structure of any tissue or organ. This next step of 3DP as a form of tissue engineering is made possible due to recent advances in this field, cell biology along with material science. An interdisciplinary effort has allowed 3D printed materials to be used for engineering medical devices like stents and bone grafts (Fig. 5.2) [28,43–47]. It is currently being used for achieving 3DP of materials by layer-by-layer process which includes fused deposition, selective laser sintering and stereolithography, rapid prototyping to name a few. All these techniques have been used to engineer scaffold ranging from millimeter to nanoscale size. In last two decades, bioprinting has advanced from 2D printing to additive method in which successive layers are aligned to form 3D structure [48–50]. The printing of complex 3D shapes with intrinsic geometry is applied to enable rapid prototyping and manufacturing industries to produce various products ranging from implantable graft to cosmetic enhancement. In this methodology, layer-by-layer detailed positioning of biopolymers, bioactive agents, and living cells with spatial control of placing a functional component come together to produce a functional 3D tissue. There are many approaches by which this can be achieved such as biomimicry, mini-tissue building blocks, and autonomous self-assembly [51–53]. Researchers across the world are developing these approaches to engineer 3D functional living constructs that have mechanical and biological properties appropriate for clinical restoration of affected tissue and organs. Advantages of using this technique are that it allows fabricating versatile scaffolds with complex geometry that are capable of homogenous distribution of cells and mimics the local extracellular matrix (ECM). However, types of polymer that could be used for this is a major limitation

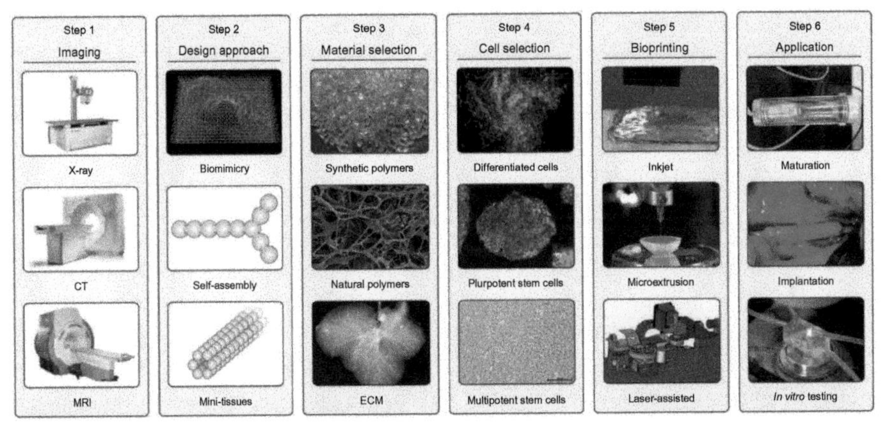

Fig. 5.2 Imaging techniques could be used as guide for bioprinting. Various scaffolds printed can be source for tissue or organ transplant.
Reproduced with permission from Murphy SV, Atala A. 3D bioprinting of tissues and organs. Nat Biotechnol 2014; 32: 773–785.

since not all the polymers provide the final desirable properties required in clinical scenario [54,55]. Evolving from conventional methodology, 3DP allows tissue engineers creative freedom to design scaffolds that are capable of imitating the ECM of any complex organs which, in turn, provides a microenvironment or niche for cell attachment, proliferation, and distribution within this complex structure with potential to develop into functional tissue in the in vitro conditions.

5.1.3 Natural polymers and 3DP

There are many factors to consider while engineering scaffolds for clinical use and crucial ones are biocompatibility, degradation rate, porosity, interconnected pores, and mechanical stability and strength. Among these biocompatibility and degradation are important criteria that biomaterials are required to fulfill, hence it is important to ensure scaffold degrades leaving desirable living tissue without causing any toxic by-products [33,56,57]. Similarly, materials should only induce minimal inflammatory response thereby reducing the chances of immune rejection by the host body. Equally vital is biocompatibility of the scaffold as it is desirable for matrix to support high cell attachment, proliferation, and aid in differentiation. Furthermore, as the seeded cells proliferate and differentiate into specific linage, matrix should be able to withstand the force exerted by these cells without collapsing since it would be to poor rate of diffusion of gases, nutrients, and metabolic waste products resulting in poor neo-tissue formation [58]. There has to be a fine balance between the biocompatibility and mechanical strength as these matrices are ultimately aimed for human use and should withstand the daily wear and tear. It is easier to fabricate matrix with high mechanical strength however that does not always mean better biocompatibility and could also lead to slower rate of degradation [59]. This balance can be achieved by choosing the right combination of polymers used for scaffold fabrication. Naturally derived polymers such as chitosan, alginate, fibronectin, hyaluronic acid, and collagen have innate biological functions and have been successfully employed in different areas of tissue engineering. Since these are derived from the natural ECM they tend to retain its functional surface receptor that makes it attractive for cellular attachment and can regulate its proliferation better than the synthetic materials.

Materials inspired by ECM is optimized milieu that have been developed to maintain homeostasis and enhance tissue formation. Therefore major efforts have been made to use it as a source material for scaffold fabrication to guide morphogenesis in repairing and engineering damaged tissue. It is for this reason many researchers have proposed on isolating natural ECM and use them as base material for scaffold fabrication. As ECM plays an instructive role for many cellular activities, the hypothesis is that the surface motif can be retained during purification process thereby providing space for new tissue formation after cell seeding. This also helps overcoming the major drawback faced by synthetic materials that it lacks the cell recognition signals. For 3DP, collagen [60], alginate [61], hyaluronic acid [62], and gelatin [62] are commonly used; however, some of the materials when used alone does not provide the mechanical strength required for cell seeding, hence the combination of polymers is chosen to avoid failures. Collagen/gelatin is known to contain large glycine, proline, and hydroxyproline residues along with smaller amounts of different aromatic and sulfur-based amino side in the polymer chains[63]. Pyrrolidine that is combination of proline and hydroxyproline is responsible for stabilizing the tertiary superhelix structure via steric hindrance. It is also due to these amino acid collagen is an attractive polymer for it can modulate cellular behavior however, for 3DP it is not suitable to use alone due to poor mechanical integrity upon cross-linking. On contrary, the hydrolyzed product of collagen gelatin does not present this bottleneck which could be due to the fact that helical arrangement is lost in the processing. Gelatin is coiled above 40°C and reverses to alpha helix below 30°C. At lower concentration, chain mobility occurs producing intramolecular bonds [64]. The extent of reversible fold of helix structure in gelatin is directly proportional to its concentration in solution, type of solvent used, and temperature of reaction. Gelatin is one of the most used polymers in tissue engineering and nearer to collagen in biological functionality as it holds the RGD sequence of collagen which promotes cell attachment by integrin receptors on the polymer surface. Besides these, gelatin is also biodegradable, since its denature product of natural ECM but without the concern over, the immunogenicity and pathogen transmission associated with collagen. Thermoresponsiveness of gelatin could be used in bioprinting as it can help in structure maintenance upon printing. However, gelatin as a single polymer cannot be used for bioprinting as it is difficult to optimize temperature and viscosity during printing hence it is usually used with unsaturated methacrylamide side groups to form a photosensitive gelatin derivative. When gelatin methacrylamide is exposed to the UV radiation it causes stable double bond cross-linking between the gelatin polymer chain and methacrylamide thereby resulting in irreversible yet temperature-sensitive copolymer that can be further used for bioprinting[65].

Another commonly used natural polymer is alginate that is an anionic polysaccharide obtained from seaweed. It is linear chain composed of β-D-mannuronic acid monomers (M-blocks) alternating with α-L-guluronic acid blocks (G-blocks) that are intermixed M and G domains. Alginate readily forms a gel with divalent cations due to the ionic bridge formation within the G block and without use of harsh cross-linking agent it is possible to obtain a low strength 3D matrix [66]. As like other natural polymers it has an excellent biocompatibility and since gelatin can be obtained at room temperature it is an attractive material for bioprinting. It is easier to encapsulate cells with alginate solution using calcium chloride as a buffer; a homogenous bead has been made that further fuses together to form 3D gel construct containing cell of choice. This printed cell construct was found to have stable mechanical integrity however, since alginate does not have any cell attractive receptors it could induce apoptosis (cell death due to lack of adhesion molecules) resulting in anoikis. Therefore it is usually used in combination with gelatin or fibronectin to improve its biological function [67].

5.1.4 Clinical need for scaffold-based tissue engineering

Patient suffering from different ailment and diseases may be treated with transplanted tissue or organs. However, ever increasing shortage of donor organs get worse with every year especially in developed countries where more of aging population resides

[68,69]. The real challenge in tissue engineering lies in the fact that it has to imitate what occurs in nature. Various attempts have been made to engineer in vitro practically every tissue and organs in the body. Research has been proceeding in creating tissue-engineered cartilage [70], nerve [71], intestine [72], kidney [73], pancreas [74], liver [75] and heart muscle, values and blood vessels [76]. In field of connective tissues, research has been ongoing across the world for engineered bone [77], cartilage, ligaments, and tendon [78]. So far the highest success has been obtained in skin [42], bladder [79], and larynx [80–83]; with success it means the tissue-engineered grafts have been used in patient yield success regeneration. While major breakthroughs have happened and increase in economic activity within the tissue engineering arena has grown exponentially with ever increasing number of product getting into clinical trials or manufacturing units for mass production. The revenue generated by the products related to biomaterial-based regenerative medicine is approximately US$240 million per annum [84].

One example of tissue that need new treatment approach like engineering-based regenerative medicine is urethra. Different scaffolds fabricated using PGA or acellular collagen-based matrix obtained from the small intestine and bladder have been extensively used for regenerating urethral tissue. Since the structures of this tissue could result in congenital defects, infections, and injuries. One of the most critical elements needed for successful treating of this condition is biomaterial scaffold and different types of materials have been engineered for urethral reconstruction. Mainly, 3 types of scaffold are used:

1) Living autologous tissues such as buccal mucosa [85], skin [86], ureter [87,88], appendix [89], blood vessels [90], and radial forearm free flap [91] all have been used as urethral grafts
2) Synthetic biomaterials including type 1 collagen sponge, co-poly(l-lactide/epsilon-caprolactone) [P(LA/CL)] [92] have been tested by different groups but these materials are generally known to lack appropriate biological active molecules that enables neo-tissue formation across the grafts to proceed quickly.
3) The natural collagen-based biopolymers that are allogeneic or xenogenic sources which include small intestinal submucosa (SIS) [93–96], bladder submucosa (BSM) [97], urethral ECM [98], amniotic membranes [99] to few examples. Acellular matrix from Atala et al. [93] used from experiments and clinical uses. In the in vivo studies performed by this group, segment of the urethra was resected and substituted with acellular scaffold grafts in on lay fashion. Histology testing of these tissues showed complete epithelialization and smaller vessels and muscle infiltration. These grafts have given favorable results and using collagen-based materials provides advantage such as early wound closure, faster tissue formation, and improved functionality of neo-generated tissue. In more recent studies same group engineered a 3D scaffold using small intestinal submucosa (SIS) (Fig. 5.3) which promotes cell-matrix penetration and allows multidimensional structure formation in the in vitro conditions that allows the research to screen for the best possible combination before moving forward to clinical trials and application. Using 3D porous SIS having interconnected macro and micro porous structures which allows cells infiltration into the scaffold is engineered. Importantly, heterogeneous cellular protein was removed to mitigate inflammatory reactions, calcifications, fibrosis, and graft shrinkage that are observed with other types of scaffolds. This study clearly shows how invent of 3D scaffold could be used for treating a condition that could otherwise prove to be fatal for patient while waiting for suitable match.

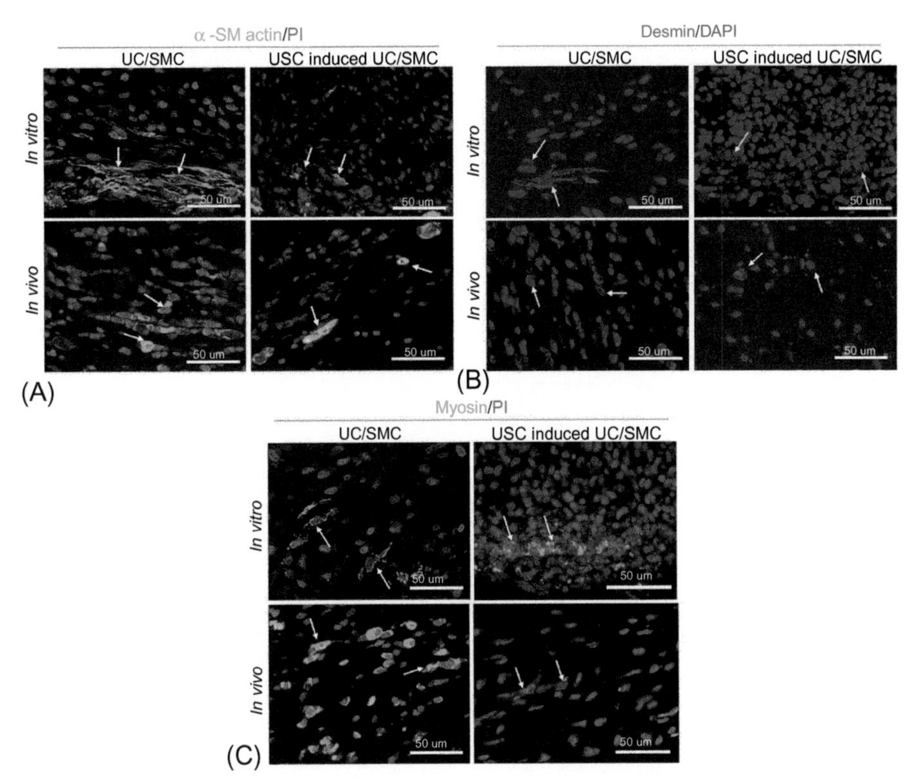

Fig. 5.3 Immunofluorescent staining of myogenic-differentiated USC in vitro and in vivo. Tissue sections of in vitro (top panel) and in vivo (bottom panel) cultured cells that were immune stained (green or red) using (A) α-SM actin, (B) desmin, and (C) myosin antibody. Reproduced with permission from Wu SF, Liu Y, Bharadwaj S, Atala A, Zhang YY. Human urine-derived stem cells seeded in a modified 3D porous small intestinal submucosa scaffold for urethral tissue engineering. Biomaterials 2011; 32: 1317–1326.

5.2 3DP for engineering scaffold

Scaffolds designed previously were focused on macroscale features like interconnected pores for gessoes exchange and nutrient transport and remodeling tissues. One of the strategies that augment the function of tissue-engineered constructs is to imitate tissue and cells microarchitecture and environment [42,100]. Tissues with the body are divided into repeat functional units, for example, kidney-nephron-islet, and 3D architecture coordinates the process within the multiple types of specialized cells. Further, the local cellular environment presents the biochemical and physical cues which can specifically directs both cellular functions such as metabolism and biosynthesis along with cellular fate processes like proliferation, migration, differentiation, and apoptosis. Hence, fabricating a functional 3D tissue construct that brings together both microscale characteristics for suitable cell functions and macroscale architecture and mechanics

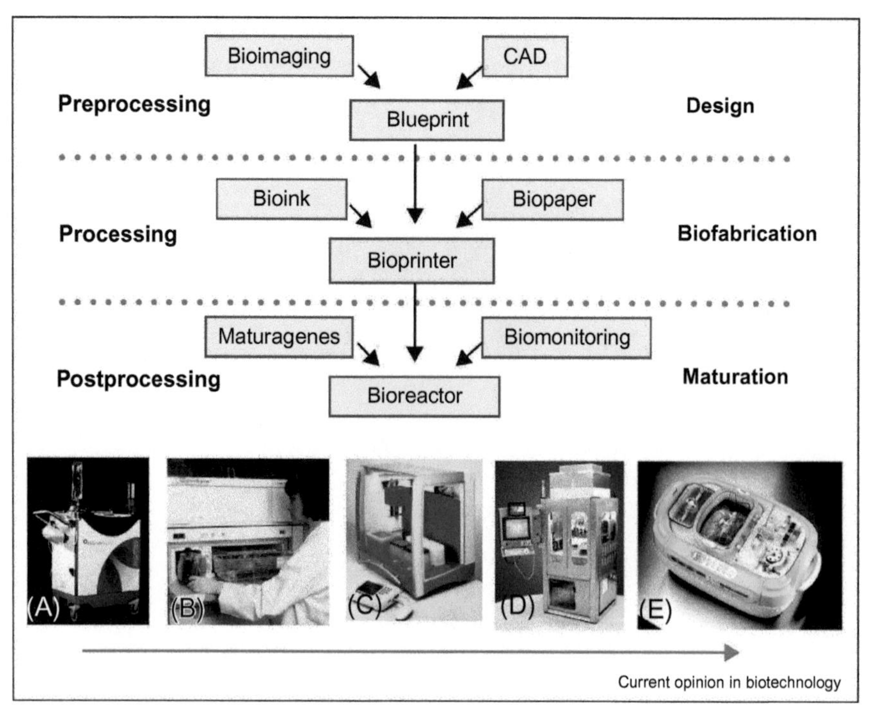

Fig. 5.4 Essential steps in organ printing technology and an organ biofabrication line. (A) Celution—clinical sell sorter for the scalable isolation of autologous adipose tissue-derived adult stem cells (Cytori Therapeutics, United States). (B) Bioreactor for the scalable propagation of stem cells (Aastrom Bioscience, United States). (C) EpMotion-5070—robot for the automated scalable biofabrication of uniformly sized tissue spheroids (Eppendorf, Germany). (D) Bioasembly Tool—robotic bioprinter (Sciperio/nScript, United States). (E) LifePort—Perfusion apparatus (Organ Recovery System, United States). Reproduced with permission from Mironov V, Kasyanov V, Markwald RR. Organ printing: from bioprinter to organ biofabrication line. Curr Opin Biotechnol 2011; 22:667–673.

that allow control over chemistry and topography over multiple length scales. Tissue engineering scaffold that can mimic complex architecture of native tissues is always difficult to engineer than conventional porous polymeric scaffold that supports undirected cell attachment and spreading within homogenous and relatively larger constructs. A step forward in this was made when computer-controlled rapid prototyping technique was employed for fabricating a 3D scaffold with precise architecture with control at macro and microscale (Fig. 5.4) [101]. These 3D fabrication techniques offer numerous prospects to improve the great potential for tissue engineering. For example, the functionality of complex tissue units is expected to depend upon the independent control of macro and microscale features. The incorporation of vascular beds could allow the larger construct to support nutrient diffusion alone. Furthermore, the combination of clinical imaging data along with computer-aided design (CAD)-based free-form techniques allows the engineering of replacement tissues that can be customized to the shape of any particular defect. Finally, the upscale production of identical functional

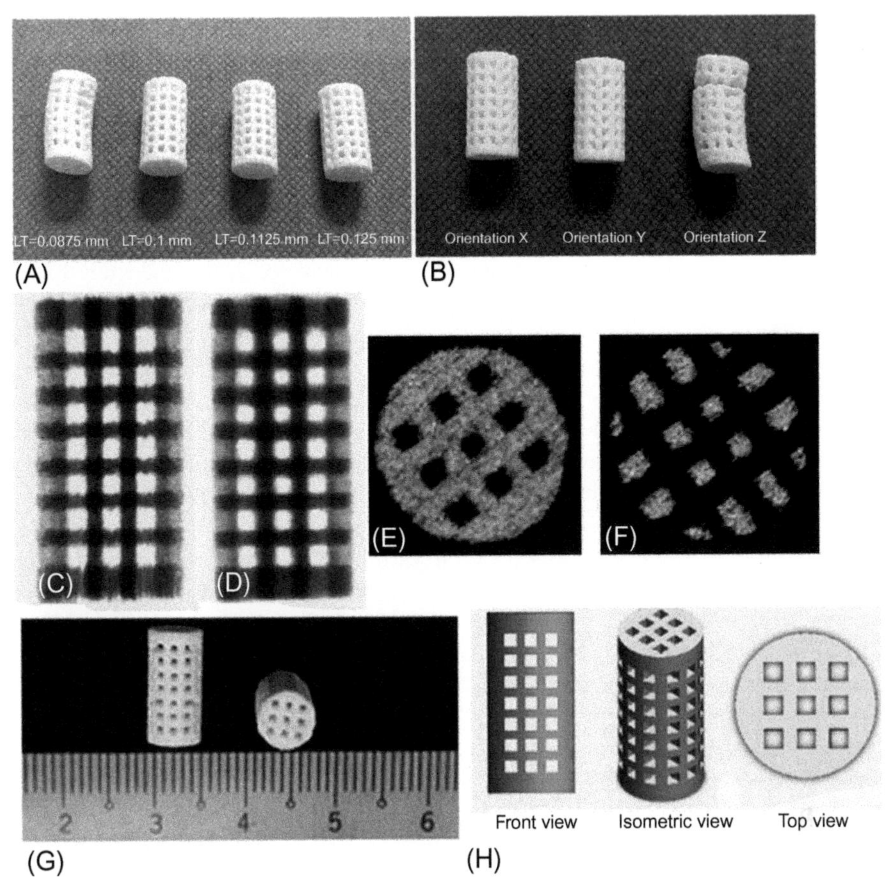

Fig. 5.5 Representative image of printed scaffold of different layer thickness (A) different orientation (B) CT results of lateral view of 90 (C) 180 (D), and a middle cross-sectional view including layer of powders (E), pores and struts (F), 3D printed specimen (G), scaffold designed by Solidworks (H).
Reproduced under open access from Farzadi A, Solati-Hashjin M, Asadi-Eydivand M, Abu Osman NA. Effect of layer thickness and printing orientation on mechanical properties and dimensional accuracy of 3D printed porous samples for bone tissue engineering. PLoS One 2014;9.

tissue units may be used in cell-based assay for drug discovery or to study basic fundamental biological pathways. The solid free-form fabrication (SFF) also known as rapid prototyping (RP) or additive manufacturing (AM) is considered as flexible alternative technique for fabricating highly accurate complex-shaped scaffolds that are difficult to engineer by traditional technique. The combination of AM with optimized computational architecture is highly desirable by surgeons working on developing artificial tissue with well-defined architecture within complex geometry model with precise control and reproducibility (Fig. 5.5) [102]. Whichever technique is used for engineering scaffold there are few critical properties that scaffold must possess in order to achieve successful tissue regeneration.

5.2.1 Mechanical properties of 3D scaffold

Different tissues have different requirement; however, the mechanical strength of scaffold is one of the most critical characteristic and it varies for each tissue and organ. For example, in case of bone, variety of scaffold have been engineered in past three decades [103]. These materials are including bioactive glasses [104,105], bioceramics [106] such as calcium phosphate[107,108], synthetic polymer like polycaprolactone (PCL) [109] and natural materials such as collagen type 1 and hydroxyapatite. However, none of these have the strength that is required to withstand the cyclic and static load in the in vivo environment while maintaining adequately high porosity to allow bone ingrowth, transportation of nutrients, and vascularization. For the scaffold to meet these needs it has to have a porosity of 60%–90% with average pores size >50 μm along with compressive strength in par of cortical bone which lies around 100–150 MPa range along the long axis. Achieving high porosity is possible with traditional scaffold manufacturing techniques; however, higher porosity compromises the mechanical strength directly, hence there is continuous demand for high strength bone scaffold for treating various bone defects especially segmental bone defects [110]. Recently, work done by Eqtesadi et al. [111] has been a step toward meeting this demand. In this work, robocasting technique was used to engineer 13–93 bioactive glass scaffold that had average strut thickness of 274 μm and porosity of 51% along with compressive strength of 86 MPa (brittle strength) and bending strength of 15 MPa. To further improve the mechanical property PCL was infiltrated into the scaffold which significantly improves the mechanical strength of the scaffold. The impregnation technique dramatically enhanced the toughness and damage tolerance capacity of the glass structure. In fact, when if the ceramic skeleton failed, the polymer infiltrate grips the composite preventing it from immediate catastrophic failure of entire structure (Fig. 5.6). However, the toughening agent seems to mask the bioactive site of the scaffold making it difficult for cell interaction and also reduced compression and bending strength.

In step toward improving this Esfahani et al. [112] designed and fabricated 3D scaffold with hexagonal pore geometry to obtain a high contact area using printing technique. Printed scaffold produced a highly anisotropic architecture resulting in improved load transfer compared to various types of conventional scaffold. In this

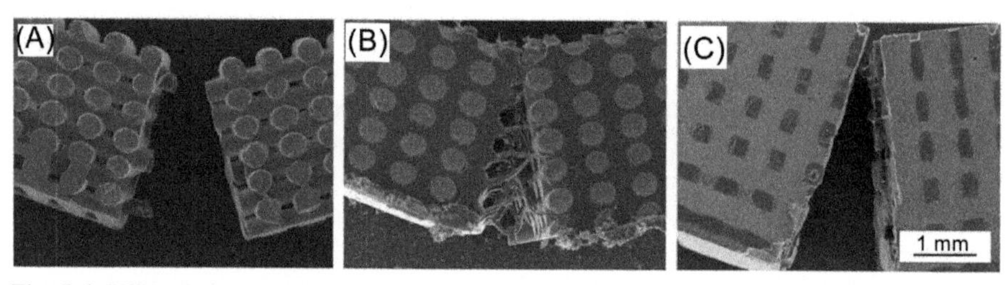

Fig. 5.6 SEM micrographs of samples after bending tests: (A) bare scaffolds, (B) 13–93/PCL, and (C) 13–93/PLA composites.
Reproduced with permission from Eqtesadi S, Motealleh A, Pajares A, Guiberteau F, Miranda P. Improving mechanical properties of 13-93 bioactive glass robocast scaffold by poly (lactic acid) and poly (epsilon-caprolactone) melt infiltration. J Non-Cryst Solids 2016;432:111–119.

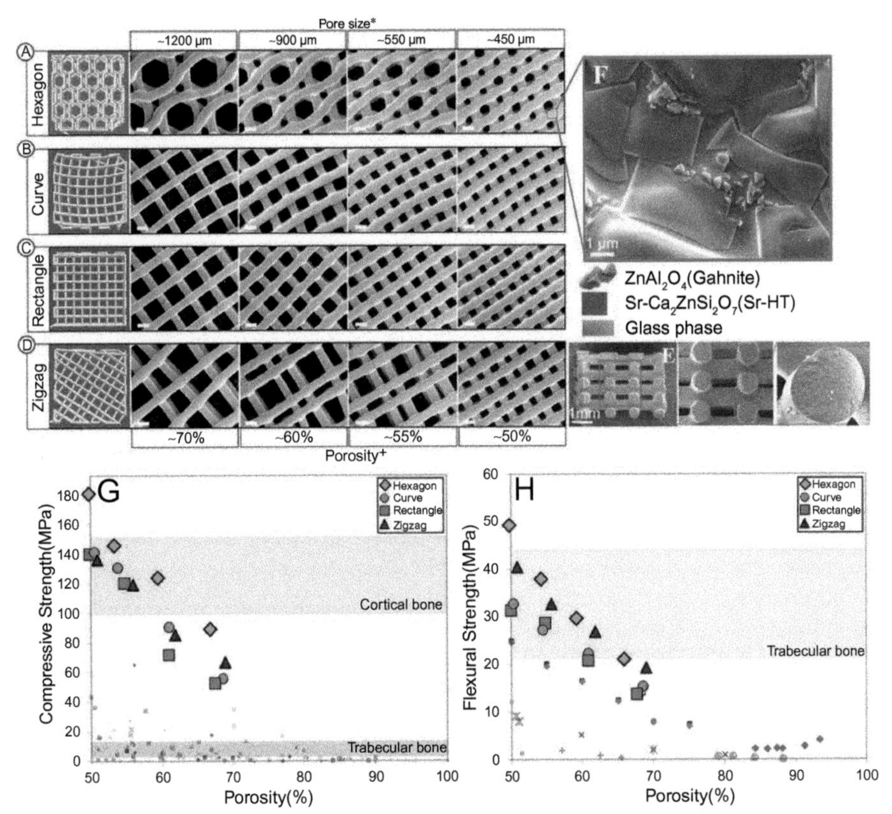

Fig. 5.7 (A) Hexagonal, (B) Curved, (C) Rectangular and (D) Zigzag design. (E) SEM images of fracture surface of a Sr-HT-Gahnite scaffold prepared by direct ink writing (perpendicular to the deposition plane, z direction), revealing the solid struts without any microporosity in the microstructure. (F) The microstructure of Sr-HT-Gahnite scaffolds consisting of three phases of (1) Sr-HT grains, (2) ZnA12O4 crystals, and (3) a glass phase between the grains. (G) Compressive strength of Sr-HT-Gahnite scaffolds with distinct pore geometries vs porosity. Comparison with compiled data from literature studies for polymer, composite, bioactive ceramic and glass scaffolds at porosities between 50% and 95%. (H) Flexural strength of Sr-HT-Gahnite scaffolds with hydroxyapatite and bioactive glass scaffolds. reproduced from Roohani-Esfahani SI, Newman P, Zreiqat H. Design and fabrication of 3D printed scaffolds with a mechanical strength comparable to cortical bone to repair large bone defects. Sci Rep 2016;6.

study, author fabricated four scaffolds with very distinctive pore geometries such as hexagonal, rectangular, zigzag, and curvy of variety of porosities ranging from 50% to 70% and the diameter of deposited struts was ~540 μm obtained by direct ink writing technique (Fig. 5.7) [112]. Sr-HT-Gahnite (Stronntium (Sr) doped Hardystonite ($Ca_2ZnSi_2O_7$, HT grains and cluster of Gahnite ($ZnAl_2O_4$) scaffolds were fabricated via robotic deposition device (Hyrel 3D) printing the inks through 600-μm custom-made nozzle. The ink is loaded onto syringe, mounted on robotic arm, and printed on an oil-coated glass substrate of ~4 mm thickness. This allows the printed scaffold to easily detach from the underlying substrate upon drying. A controlled-heat treatment technique was used for decomposing organics and for sintering the particles into the dense struts. Using different testing conditions it was found that hexagonal architecture was highly porous having compressive strength of 100 MPa. The 3D printed scaffold demonstrated the strength 150 times greater than scaffold fabricated using synthetic polymeric composite and 5 times higher than ceramic and glass scaffold with similar porosity. The authors used crucial technique to test the scaffold strength that mostly is neglected but can serve as important characteristic especially for bone regeneration that is the resistance to cycling loading also known as fatigue resistance. Most of the materials usually falls much below the nominal strength when tested using cyclic loading and unloading technique. Failure of the scaffold by fatigue is associated with adverse body response, whereas cyclic loading of scaffold is reported in some cases to result in reduced healing times. The scaffold intended to be implanted into femoral bone cross-section of 6 cm^2 with body weight of 70 kg should have cyclic strength of 5–14 MPa. The high fatigue resistance tested using compressive cyclic load (1,000,000 cycles at 1–10 MPa) showed the failure reliability and flexural strength of scaffold to be around 30 MPa and all the Sr-Ht-Gahinte scaffolds were close to the native bone strength. This study clearly showed that using recent advances in the printing techniques it is possible to engineer scaffold which can match cortical bone properties. Altering the pore geometry could have effect on the scaffold porosity and mechanical strength and should be used as a technique for modulating scaffold properties. The scaffold with anisotropic architecture especially hexagonal patterns has better compressive strength, flexural strength, and fatigue resistance with reliability in compression. However, every scaffold had sufficient porosity that could support cellular growth, migration, and vascularization which, in turn, can aid in neo-tissue generation. These scaffolds could be ultimately used in treating the load-bearing bone defects which might not be possible using conventional scaffold manufacturing systems.

5.2.2 Techniques for fabricating 3D matrix

A central challenge in fabricating complex 3D printed scaffold on which cellular morphology and behavior could be analyzed is the in-depth characterization of materials' physicochemical properties. Even with well-defined pore alignment and distribution cellular function can be affected if there are no well-controlled 3D geometric, spatial, and physicochemical cues which directly influences cell migration, cell-cell contact, elongation, and alignment [113,114]. Direct ink writing technique which is extrusion-based method for printing 3D scaffold is a promising platform that can manufacture a programmable micro-periodic scaffold. In this technique each of the parameter such as extruded filament diameter, microporosity, pitch, and material combination can be all controlled independently giving a bigger window to manipulate the geometry of the scaffold [115]. Most common technique used for scaffold fabrication is the solid-free-form method in which layer-by-layer polymer deposition is done for obtaining 3D matrix. It involves creating a 3D computer-based model generated by imaging the tissue by X-ray or CT scans and slicing the images to build 2D images using the given software and finally fabricating computer-controlled process of layer-by-layer assembling it into a 3D structure that can be further modified using surface modification to obtain matrix with nano-architecture that is exactly like the natural tissue or organ [116]. Complex 3D detailing such as undercuts, internal voids, cantilevers as simplified as circles and lines thereby achieving shape complexity in each level. This technique gives freedom in form and combined with appropriate material deposition methodology it offers control over tissue engineering trio by simultaneously guiding the spatial distribution of cells, scaffolding substrate during fabrication, and signaling molecules [117]. There are five popular SFF technologies that have been utilized for tissue engineering application and for 3DP; however, 3DP is one of the most recent advances in the field.

5.2.2.1 Three-dimensional printing

Invented at MIT and developed further by other groups, 3DP fabricates matrix onto a powder bed by inkjet printing liquid binder solution [118–128]. The process starts with spreading a layer of fine powered materials evenly across the piston and X-Y positioning system along with print head are coordinated to print the 2D pattern by selective depositing binder droplets onto the powder layer (Fig. 5.8) [129].

The piston, powder bed, and part are lowered and same step is followed again for next layer repeatedly till the entire part is completed. At postfabrication process, powder is removed to reveal the underlying engineered structure. The local composition and architecture can be manipulated by adjusting the print head to deposit at a predetermined volume or by altering the parameters during fabrication. For creating micro-channels within the scaffold, additional distribution of seeding surface within the interior of the device, increasing the seeding density and

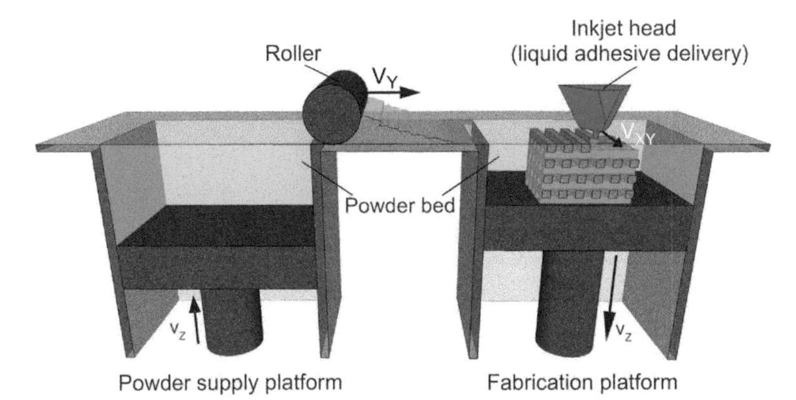

Fig. 5.8 Representative sketch of 3DP setup and functioning. The roller spreads a thin layer of polymeric powder over the previously processed layers which are then solidified by spatially controlled delivery of liquid binder.
Reproduced with permission from Billiet T, Vandenhaute M, Schelfhout J, Van Vlierberghe S, Dubruel P. A review of trends and limitations in hydrogel-rapid prototyping for tissue engineering. Biomaterials 2012; 33: 6020–6041.

uniformity can result in patterned surface chemistry with spatial controlled cell distribution. Using this technique Kim et al. engineered highly porous scaffold that demonstrated cell ingrowth in 3D printed scaffold [130]. Also, since the fabrication is at room temperature it provides window for using thermoresponsive polymers that allows incorporation of drug and other pharmaceutical bioactive agents into the scaffold. Major limitations with this technique are constrains involved with pore size since it relies on the porogen particles sizes achieving the exact pore size is difficult. Second limitation includes achieving precise shape and complexity especially, since powder-based materials require organic solvents as liquid binder which affects the drop-on-demand print head system.

As a step further, Lee et al. [131] used this technique to fabricate indirect 3DP, as in molds were printed and final materials casted into mold cavity, allowed negating limitation associated with direct printing. Scaffold with small villi structure constructed using CAD software and mold with 150 μm struts that were spaced 850 μm apart to have well-defined intervillus spacing and diameter, respectively. To demonstrate the ability to build complex architecture with large pores using indirect 3DP technique directly from medical imaging data zygoma-shaped matrix was directly created from CT data. Images were rendered and sliced into 2D layers that were further processed into each individual sliced layers that were further built sequentially on a 3DP machine (Fig. 5.9). IEC 6 cells were seeded onto these villi shaped scaffold to demonstrate potentiality of these materials to support cell attachment and growth. To exploit the compatibility, the same zygoma-shaped scaffold was fabricated by solvent casting PLGA into plaster mold, followed by porogen and plaster molds removal. Zygoma scaffold was replicated using the negative shape of reconstructed model. The optimal pore size is a critical requirement for various tissue regeneration yet there is no constant value, for example bone growth requires pores to be in the range of 200–400 μm however, achieving this by direct printing is not yet reported. Lee et al. reported in this study the internal pore morphology with well-interconnected pores of range 300–500 μm. Using indirect 3DP it was possible to fabricate scaffold with large pore size with fine features that can use the commercially available 3D systems. Furthermore, these scaffolds were capable of supporting cellular attachment and proliferation and can be used for engineering complex scaffold for other tissue regeneration application.

5.2.3 Bioinks for 3D scaffold

This is by far, the most challenging advances in 3D bioprinting process. Usually, the ink has to fulfill the biological, mechanical, and physical requirements of the printing techniques [132,133]. First, from biological point of view the ink has to be biocompatible while allowing cell attachment and growth. Mechanically, it has to provide sufficient strength and stiffness to be able to maintain structural integrity of the ink even after printing and finally, the physical aspect requires the ink to have precise viscosity to be able to dispense from print head [134,135]. Bioinks are basically made up of living cells usually between 10,000 and 30,000 cells/10–20 μL droplets suspended in pregel solution of polymers that can be cross-linked by photo or thermal stimulus [136] ($CaCl_2$, gelatin, thrombin, NaCl, and fibrinogen) [137] or can be cell medium.

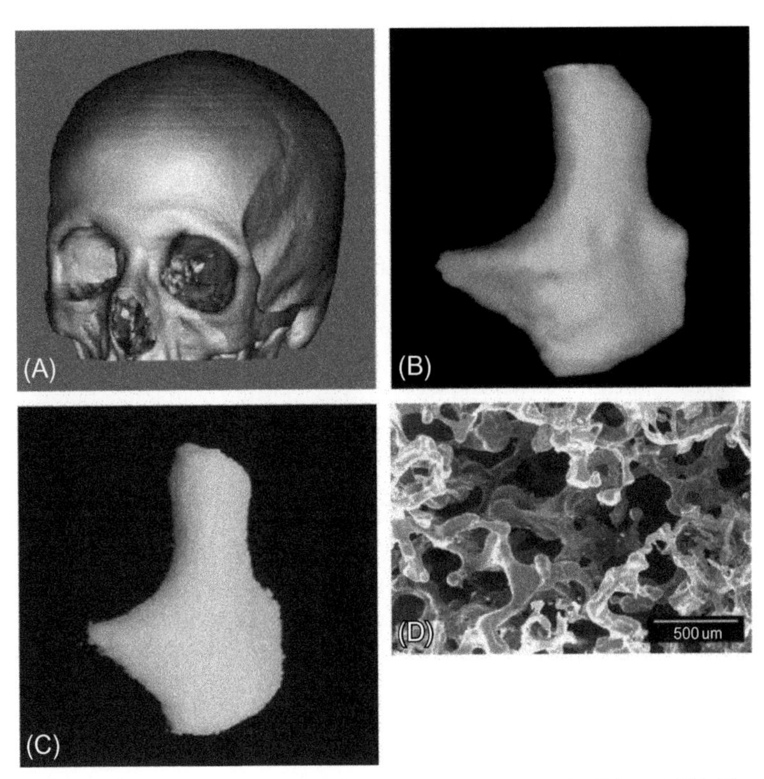

Fig. 5.9 3D reconstruction of zygoma (A, B) and zygoma-shaped PLGA scaffold (C). Zygoma was generated from CT 2D images and zygoma-shaped scaffold was produced from the mold based on reconstructed 3D model. SEM image of the cross-section of scaffold showed high interconnected pores ranging from 300 to 500 μm (D).
Reproduced with permission Lee M, Dunn JCY, Wu BM. Scaffold fabrication by indirect three-dimensional printing. Biomaterials 2005; 26: 4281–4289.

In practice, this process involves same layer-by-layer deposition of cell-loaded bioink that results in AM of patterned architecture with variety of cell types, growth factors (GFs), or mechanical cues that can be positioned with far more precision than can be achieved by conventional scaffold fabrication methodology [138]. These bioinks have to withstand extrusion, maintain the structural integrity for long period all along keep cyto-compatible pH, and allow adequate nutrient diffusion. Due to its intrinsic porosity and high nutrient-bearing capacity, hydrogels are the most promising candidate for bioink design. For example, poly (L) and poly-(D) lactic acid can be dissolved in dioxane along with BMP grounded as particles and suspended into water and this acts as bioink for making bone scaffold materials.

Bioinks without cells are usually used for engineering scaffold that is later used for supporting cells hence using biocompatible materials becomes important. Typically, polymers used for gels are alginate, agarose, fibrin, chitosan, poly(ethylene glycol)-PEG, pHEMA [115,139,140]. Most of these materials can be cross-linked using simple buffers or polar/nonpolar solvents to form a transparent gel at room temperature. Due this ability to form gel rapidly, most of these materials are ideal to be used as bioink for both hard and soft biomaterials and can be later used for supporting cells growth and proliferation.

5.2.4 Bioinks for functional 3DP

The emergence of hybrid multicomponent gels which could be designed with desirable physical properties from each component of the ensemble is exciting. For example, biodegradable polymers are usually strengthened with osteo-inductive ceramic materials such as calcium phosphate or nano-fibrous cellulose to increase the shear thinning of alginate gels, while a mixture of Pluronic and acrylated Pluronic has been used to fabricate a synthetic gel that can be cross-linked using both temperature and light. Point is these hybrid systems does seem to support the cell viability and have precise properties for printing. However, the case of long-term use of 3DP was first reported by Armstrong et al., [141] in which they designed Pluronic acid-alginate multicomponent bioink with a complex phase behavior that was used in two-step 3DP techniques to engineer cartilage and bone architectures. Specifically, 3D structures containing human mesenchymal stem cells (hMSCs) printed by extrusion of shear-thinning cell-loaded gel onto a heated stage, resulting in spontaneous solidification by sol-gel transition of Pluronic. This structure was further stabilized through alginate cross-linking with $CaCl_2$. In this, Pluronic polymer part acts as sacrificial template that gets completely expelled during the cross-linking resulting in the formation of micron-sized pores or micro-channels patterns. Furthermore, the Pluronic-templated alginate gel has shown favorable material properties such as increased shear thinning, shear modulus, and compressive modulus. This provides a stage for printing macroscopic structure such as cartilage, trachea, ear and nose, as well as fine fibers and meshes for vascular and cardiac system. In the study by Armstrong, there was no significant cell loss for over 10 d when cultures on 3D architecture printed using bioink. Furthermore, cells encapsulated within these materials could be differentiated into osteoblast and chondrocytes over 5 weeks to result in printed tissue-engineered construct, including a full-sized tracheal cartilage ring (Fig. 5.10).

The cellularized bioinks were articulated using alginate (sodium salts) and Pluronic F127; different conditions were optimized to overcome the conflicting conditions that were required to solubilize each of the polymer. Briefly, F127 was autoclaved and UV-sterilized alginate (10 wt.%) was mixed at 4°C and temperature was slowly allowed to reach 25°C to obtain homogenous solution that could be loaded with hMSCs. A MendelMax 3D printer was retrofitted with syringe pump that was used to extrude the pregelled polymer-cell fluid onto a stage set at 37°C. The high temperature instigated an instantaneous sol-gel transition led by F127 component of printed layers; this allows the generation of self-supporting 3D architecture which is further cross-linked with $CaCl_2$ wash to obtain a stable 3D scaffold. Hollow square-based rectangular prisms of 10 mm × 10 mm × 2.4 mm outer dimension and wall thickness of 1.6 mm were printed from 6 thick layers (400 μm), and this template was used to allow high throughput printing and superficial assessment of print quality and structural integrity. Bone and cartilage tissue engineering was done for 3 and 5 weeks and it was found that printed

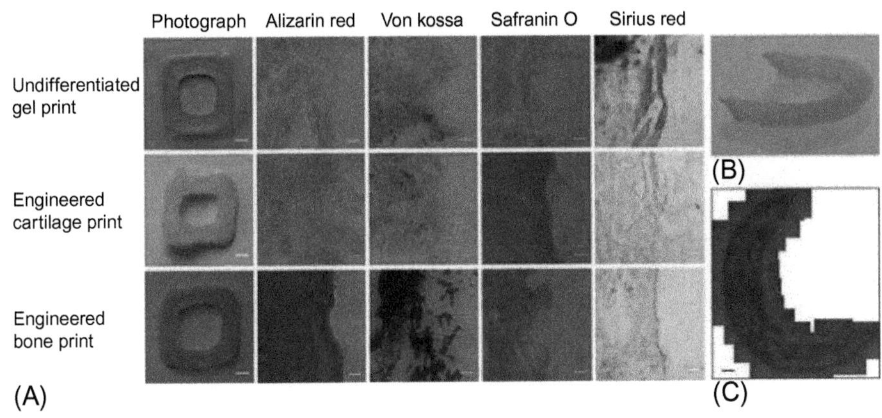

Fig. 5.10 Tissue engineering using 3D printed scaffold loaded with hMSCs laden 13%wt F127-6 wt.% alginate hybrid gel. (A) Printed squares used for cartilage tissue engineering (35 d course) and bone (21 d course) and compared with standard printed scaffold (0 d). A full size trachea cartilage ring printed by this technique (B) that produced ECM rich in glycosaminoglycan (GAG) (C).

material was excellent for supporting cell differentiation assessed by ability of differentiating cells making its own ECM (Fig. 5.10A). This construct exhibited hemispherical geometry which is critical for cartilage tissue and ECM was rich in glycosaminoglycan (GAG) fibers (Fig. 5.10B and C) giving strong evidence that 3D printed scaffold could be used for achieving functional repair or restoration of the tissue. This study provided a platform for cell-laden prints which could make its own ECM over period of cell culture within the confined geometry, opening a new arena to print tissue constructs, and repair tissue and organs even with complex geometry and intrinsic architecture.

5.3 Transduction and bioprinting scaffold for tissue regeneration

Powerful new treatment techniques are being developed in the interdisciplinary area of tissue engineering and regenerative medicine, to address the various structural and functional disorders of human bodies by using a combination of cells, scaffold, and engineering materials. The main aim of these therapies is to repair or restore the normal functions of the affected organs. In cases where the tissue functionality can be restored by normal engineering techniques, alternative approaches such as supplementing the culture with GFs or gene modification are attempted. Moreover, in tissue engineering delivering of functional genes or transfecting into the target cells to aid the formation of functional tissue is becoming increasing vital especially as gene modification has shown to improve cell and tissue function in different areas of cell research. Although, the protocol to transfect is well established and usually is performed by microinjection, electroporation, liposome-mediated or viral assisted. However, following these techniques can be difficult for tissue engineering since it results in significant viral toxicity due to the vector, low rate of transfection in both microinjection and liposome-mediated methods and mostly results in low cell viability when electroporation is performed. Furthermore, most of these protocols require processing of cells before building new tissue, hence there is an urgent need to develop protocol that can effectively and efficiently transfect cells with target genes during the course of tissue building process without compromising on cell viability. The possibility of achieving successful transfection along with cell delivery could only be using one platform for tissue engineering that is 3DP.

In one of the first studies performed as proof of concept simultaneously combined gene transfection and cell delivery on inkjet printed scaffold [142]. HP 692C and 550C printers were used and biopaper substrate was prepared from collagen type I from rat tail. Collagen was diluted at 1.0 mg/mL concentration in chilled phosphate buffer saline (PBS) by adjusting pH 7.5 and dispensed on coverslips. Following, this was curing step that allowed gel to set and matrix was used for further use. Second step was to get the cells ready, plasmid pmaxGFP and pIRES-VEGF-GFP driven by CMV early immediate promoters that were amplified in DH5α strain of E. *coli*. PAE (porcine aortic endothelial) cells were trypsinized and suspended in buffer provided with Nucleofectin. The plasmid was added to this cell suspension and it was loaded directly into ink cartridge for printing. Different test performed showed there was no damage or change in the plasmid structural integrity after printing. These genetically modified cells were further delivered within the 3D fibrin gel and were found to express GFP when implanted into mice. This study clearly indicated the possibility to combine gene augmentation with printing could be a feasible and effective technique for cell-based therapy.

5.3.1 Scaffold-assisted nonviral transduction for musculoskeletal system

Gene delivery by using nonviral vector is very new in the field of bone tissue engineering in comparison to viral gene therapy [143,144]. Even though the viral transduction results in high transfection efficiencies there is also concern for the safety and immunogenicity associated with this technique. Furthermore, virus-dependent recombination technique presents the risk of prolonged expression of protein that exceeds the time needed for healing of bone defects, hence nonviral methodology is usually considered safer and gives a transient expression of target genes [145]. In the field of bone tissue regeneration, three different studies have shown effective way for gene delivery using nonviral transfection method using cationic polysaccharide complexes and acetylate polyethylenimine system to transfect the cells with plasmid DNA-containing gene of interest to provide osteogenic stimulus. One of the most effective stimulants for bone formation is BMP-2 that has been extensively studied and used for clinical applications; however, the effective delivery vehicle is still to be established since most commonly used techniques have caused concern regarding the dose release and its possible side effects. To overcome these issues Loozen et al. [146] developed gene-activated scaffold into larger constructs; they created porous hydrogel using bioprinting methodology which allowed better diffusion of nutrients and ingrowth of blood vessel. This group was able to engineer 3D constructs with precise and reproducible size, shape, and pore architecture. This construct was made up on alginate loaded with multipotent stromal cells along with $CaCl_2$ particles that were printed onto either porous or nonporous/solid fashion (Fig. 5.11). Furthermore, plasmid DNA encoding BMP-2 was included within the construct. Cells were transfected by plasmid DNA and were found to differentiate toward the osteogenic linages which could be noted from increased level of BMP-2 and ALP production and porous scaffold was found to be better than the solid

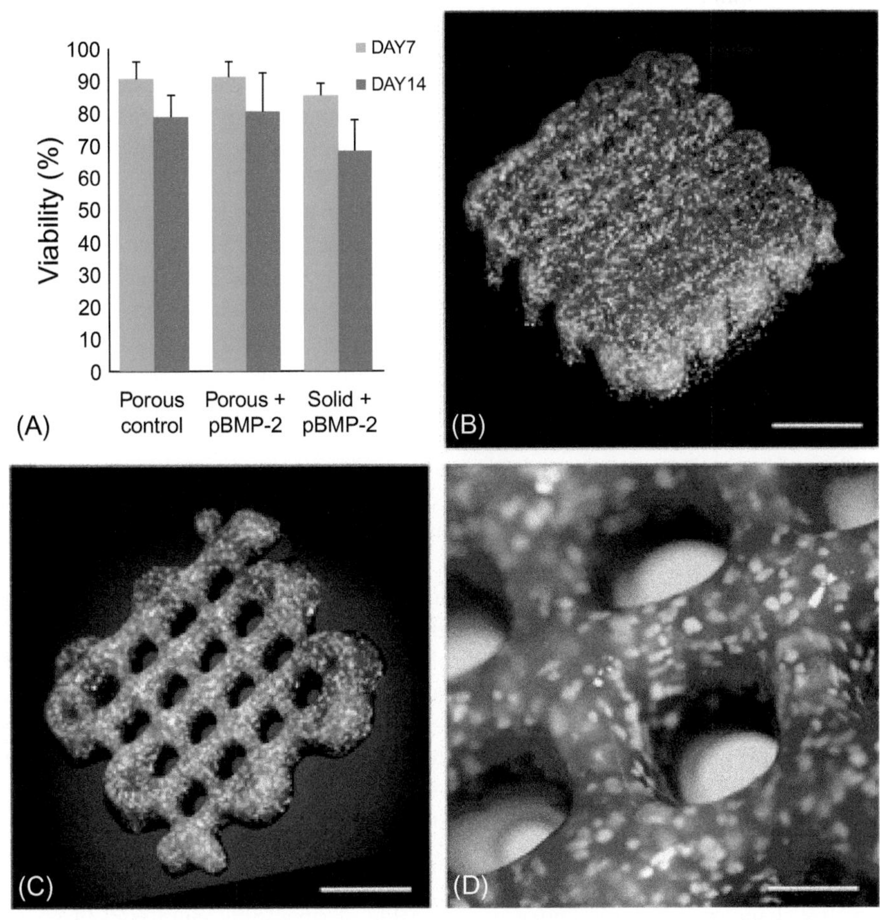

Fig. 5.11 Bioprinted scaffold loaded with MSCs, ceramic particles with and without the plasmid DNA coding for BMP-2. Cells were found to be viable up to 2 weeks (A). Two different types of scaffold were fabricated such as solid (B) and porous (C). Final shaped and architecture of the scaffold directly after the printing shows homogenous pores (D). Reproduced with permission from Loozen LD, Wegman F, Oner FC, Dhert WJA, Alblas J. Porous bioprinted constructs in BMP-2 non-viral gene therapy for bone tissue engineering. J Mater Chem B 2013; 1: 6619–6626.

constructs. This was first time nonviral gene therapy was combined with 3DP and successful transfection was achieved; however, more optimization is needed toward making these constructs for clinical application.

5.3.2 Bioprinting of vascularized tissue and organs

Engineering functional tissue constructs has always been challenging; however, if the tissues are thicker than few 100 μm then tissue faces difficulties surviving and proliferating due to diffusion limit especially for vascular system it is critical to vitalize thick tissue [44,147–149]. Different techniques have been tried to solve this problem and one of the approach was seeding cells (ECs) in scaffold that has predefined macro and micro-channels engineered in free-form manners or micro-fabrication techniques [150]. However, seeding ECs into a prefabricated thick scaffold and allowing it to migrate and populate the matrix is challenging. In additional, the conventional technique for engineering 3D free-form fabrication usually involves UV cross-linking or high temperature or abnormal pH which makes it difficult for viable cells to be deposited during the fabrication step and has to be seeded later thereby, limiting the ability to design complex tissue construct with multiple cell types [151,152]. Another approach is to engineer prevascularized tissue by mixing mural cells and ECSs with hydrogel. Once it is implanted into the animal, the remodeling of vascular system occurs and eventually connects with the host circulation. However, when cells are simply mixed and incorporated into hydrogel, there is no control in spatial organization of cell-gel structure. Furthermore, vascular formation is limited with this technique and flow through vasculature cannot be achieved.

Recently, with 3DP techniques it is possible to develop on-demand control of cells and biomaterials. This allows engineering a complex 3D hydrogel structure and depositing living cells within this construct at same time of the bioprinting step. However, construction of complete capillary network of ~10 μm in dia, using single cells even with bioprinting process is nearly impossible due to limitation in spatial resolution and time. Even in the protocols where single cell level of patterning is possible it requires considerable amount of time ranging from hours to days even for small piece of tissue measuring few millimeters in size, hence incompatible to fabricate live tissue. Therefore it is more practical to process a lower resolution (~100 μm) bioprinting than single cell level (1–10 μm). In a step forward, Lee et al. [153] developed a method by which large (lumen size 1 mm) fluidic vascular channels construct could be developed and for creating adjacent capillaries network, a maturation process was added, thereby, leading to formation of connected networks of capillaries to the large perfused vascular channels. Authors, first developed a methodology in which the fluidic vascular channels with large lumen were capable of supporting mural and ECs and could be easily introduced around the channels for remodeling. Microvasculature network was further formed by embedding ECs and fibroblast cells within the fibrin gel within the two vascular channels and blood vessels sprouting out were derived from larger printed channels to create capillary vascular channel connection (Fig. 5.12). They developed a bioprinting platform which was based on solid free-form fabrication techniques. The printing system consists of a dispenser array that had individual

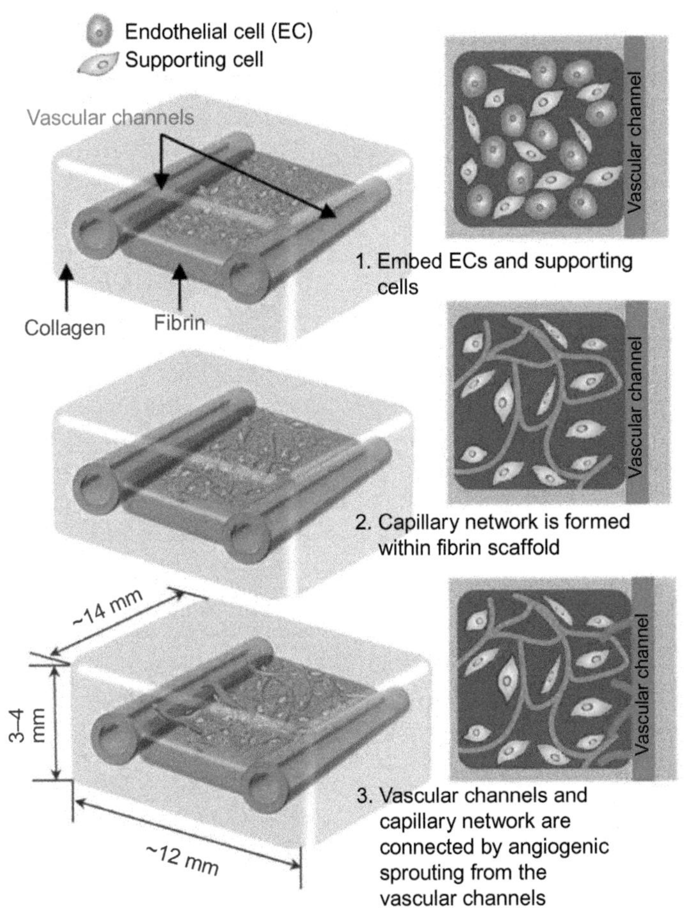

Fig. 5.12 Schematics of growth and maturation process multiscale vascular system fabricated by using 3D bioprinting technology.
Reproduced with permission from Lee VK, Lanzi AM, Ngo H, Yoo SS, Vincent PA, Dai GH. Generation of multi-scale vascular network system within 3D hydrogel using 3D bio-printing technology. Cell Mol Bioeng 2014; 7: 460–472.

controlled value, 3-axis robotic stage, materials loading units, and an attachable temperature control. The liquid-based materials that included GFs, cells, and polymers could be dispensed by air pressure during the opening of gate of micro-value. Volume dispense can be adjusted and so was the air pressure. Under normal operating condition dispensing was done at 1-kHz actuation frequency, providing a high throughput printing capacity. Second, the cells that were used were transfected separately to overexpress the tags. Human umbilical vein endothelial cell (HUVECS) were transfected with lentivirus overexpressing mCherry red and ECs sprouting along with capillary formation were observed due to the transfection with lentivirus overexpressing EGFP. HUVECs were seeded in the inner surface of fluidic channels and GFP-HUVECs that were used to create microvascular bed were seeded in between the fluidic channels.

Collagen hydrogel precursor was used as a material for main scaffold printing and porcine gelatin was used as a sacrificial material for creating fluidic channels. For creating capillary bed, fibrinogen, thrombin, HUVECs, and NHLFs were used. Using layer-by-layer approach two fluidic vascular channels within 3D coll I matrix were printed. The collagen was printed on flow chamber and polymerized and this step was repeated for 5–6 times to completely fill the bottom flow chamber. Next step, gelatin was printed in straight pattern following solidification of gelatin at 4°C, fibrinogen, HUVECs (GFP transfected), and NHLF was deposited in between the two channels during fabrication. Once this layer gelation was complete, several layers of collagen were printed on top of entire structure. After this step the entire system was incubated at higher temperature for 30 min to liquefy gelatin layers thereby, leading to the formation of fluidic channels. Then, HUVECs (mCherry-transfected) were injected into the channels as suspension to seed in the inner channel surface and create a cell lining and since the entire construct was printed on flow chamber it allowed long-term perfusion.

This method enabled the construction of fluidic channels similar to endothelial system, formation of capillary networks, and generation of multiscale vascular system that followed by connecting millimeter scale of vessels with microvasculature. The larger channels created using 3DP technique and angiogenic sprouting of transfected ECs were derived from channel edge. This work showed it was possible to achieve capillary networks and the entire process resembled vasculogenesis. 3DP gave a platform by which control vascular structure could be perfused and biological self-assembling ECs could be dispensed without affecting the viability or the genotype of the cells. The integrated method allowed introduction of several critical factors such as porous ECM that could promote angiogenesis, fibrin that induced EC tube formation, and maturation of capillaries network was possible due to surrounded seeding of multiple polarized ECs to form cell lining. This system could also be used to study various disorders that might be possible to study in the in vivo condition hence proving the potentiality of 3DP in vascular tissue engineering.

5.3.3 Bioprinting for neural tissue engineering

The purpose of tissue engineering is to develop a biological substitute that can improve the functions of neural tissues and unfortunately there are no effective methodologies till date that can reverse or repair neurodegenerative diseases [154,155]. As a critical module in nerve regeneration, 3D scaffolds provide the much needed physical support that facilitates the functioning of cells which, in turn, results in better host tissue engraftments followed by neo-tissue formation. Most of the present techniques used for scaffold fabrication do not offer the step in which intrinsic features could be added; hence 3DP technology acts as a powerful tool for building such tissues. Lee et al. showed that rat embryonic neural cells could be embedded in collagen hydrogel precursor by printing and could be used for neural regeneration [156]. In one of the first studies published by Hsieh et al., used polyurethane dispersions (PU1 and PU2) without any cross-linking agent and it could be easily fine-tuned to obtained solid content of this dispersion. NSCs were embedded into this before gelation and printed at 37°C [157]. The cell printing ink used in this methodology was novel as the aqueous dispersion of PU nanoparticles was able to form hydrogel when temperature

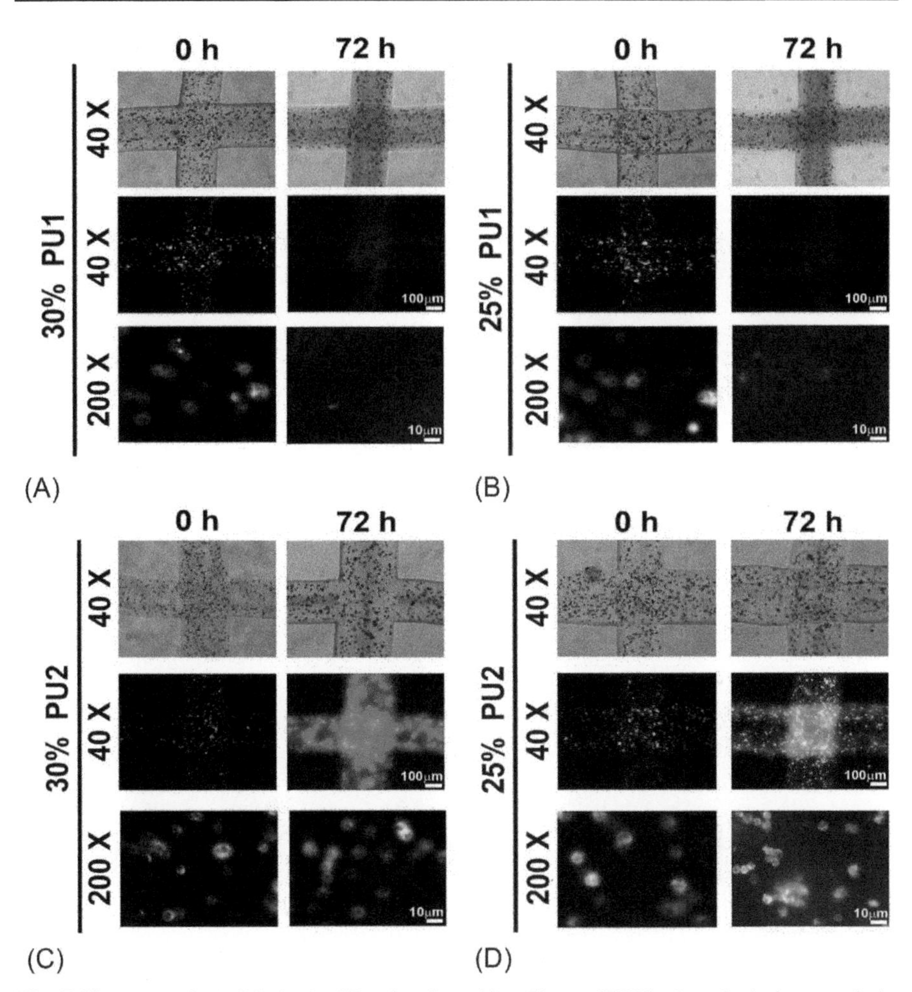

Fig. 5.13 Images for NSCs in the 3D printed stacking fibers of PU hydrogels during a period of 72 h. (A) NSCs embedded in 30% PU1, (B) NSCs embedded in 25% PU1, (C) NSCs embedded in 30% PU2, and (D) NSCs embedded in 25% PU2. Cells were labeled with PKH26 (red fluorescence).
Reproduced with permission from Hsieh FY, Lin HH, Hsu SH. 3D bioprinting of neural stem cell-laden thermoresponsive biodegradable polyurethane hydrogel and potential in central nervous system repair. Biomaterials 2015; 71:48–57.

was increased. The 3D bioprinting of NSC-laden construct on the thermoresponsive PU ink was performed using fused deposition manufacturing equipment that included injection system that was integrated with personal computer and xyz motion platform with heating system. Computer was used to design the structure, plan the manufacturing paths, and control the motion of the platform. Optimized concentration of NSCs and PU was filled in a barrel and printed by stacking fibers into petri dish that was placed at 37°C. Eight layers of fibers were stacked and printed dimension of NSC-laden hydrogel construct was 1.5 cm × 1.5 cm × 1.5 mm (W × D × H) (Fig. 5.13). NSCs

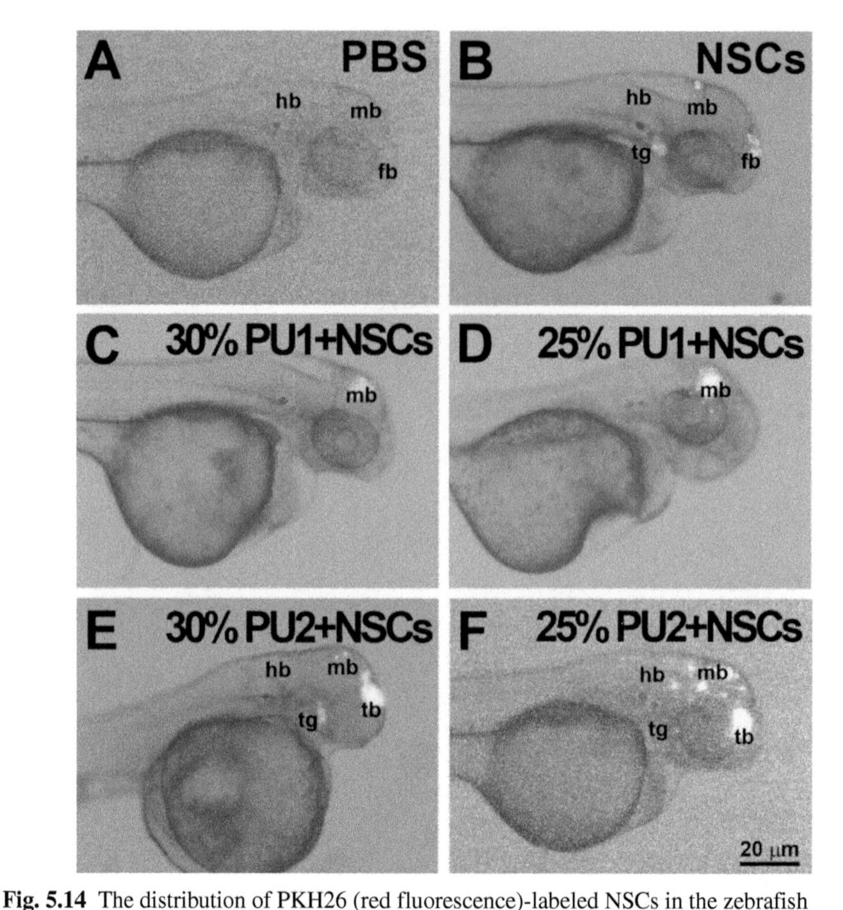

Fig. 5.14 The distribution of PKH26 (red fluorescence)-labeled NSCs in the zebrafish embryos at 48 hpf. (A) Embryos injected with PBS, (B) embryos injected with NSCs, (C) embryos injected with NSC-laden 30% PU1 hydrogel, (D) embryos injected with NSC-laden 25% PU1 hydrogel, (E) embryos injected with NSC-laden 30% PU2 hydrogel, and (F) embryos injected with NSC-laden 25% PU2 hydrogel. Regions marked with fb, mb, hb, and tg represent forebrain, midbrain, hindbrain, and trigeminal ganglion.
Reproduced with permission from Hsieh FY, Lin HH, Hsu SH. 3D bioprinting of neural stem cell-laden thermoresponsive biodegradable polyurethane hydrogel and potential in central nervous system repair. Biomaterials 2015; 71:48–57.

hydrogels suggested PU2 was better material for cell proliferation than the PU1. The NSCs cultured on 3D printed hydrogel construct was implanted into the zebrafish embryos. The NSCs were transfected and labeled with PKH26 (red fluorescence) and path was monitored. NSCs injected directly into the embryo showed lower cell survival and were found to be randomly distributed (Fig. 5.14) whereas NSCs on PU2 were found not only proliferating but migrating and dispersed in all brain area.

This study was a step forward toward optimizing and showing proof of concept that it was possible to deliver the cells into the areas of brain and when performed in combination with tissue engineering concept gives better results than single cell injections. This study further tested on the brain injury model and found the 3D printed NSCs on PU2 could potentially rescue the functional properties of impaired nervous system and could be a new therapeutic strategy needed for neurodegenerative diseases.

5.4 Future directions and concluding remarks

There are many challenges in the field of 3DP especially with the materials that could be used and cannot be used. Currently, biomaterials used for 3DP are selected because of favorable extrusion characteristics or cross-linking feasibility and biocompatibility. An ideal material should be biocompatible that could be easily manipulated into complex 3D structures allowing addition of intrinsic tissue-like features without losing the ability to direct cells into desired linages. The mechanical and rheological properties of the materials also determine the printability of the hydrogel, hence there are very limited range of polymers that can fit into this window such as collagen, hyaluronic acid, photo-cured acrylates, and alginate-modified copolymers. There is a need to develop a system that allows more polymers to be printable and could open the possibility to achieving better success especially for cell viability and functionality in the in vitro condition. The 3D bioprinting for neural or vascular or bone tissue regeneration ultimately replies on the bioink that is used for printing these constructs. Although, various attempts have been made and many improvements have suggested the feasibility of using this system in vivo, the biomimetic properties required from the 3D printed scaffold need more advancement. However, the advantages of using 3DP are more than the limitation and it is the most promising next generation of treatment for tissue and organ repair.

Acknowledgment

I would like thank all the authors whose work has been cited and all the researchers in working in companies for their contribution toward the regenerative medicine and tissue engineering.

References

[1] Gurtner GC, Callaghan MJ, Longaker MT. Progress and potential for regenerative medicine. Annu Rev Med 2007;58:299–312.

[2] Feinberg AW. Engineered tissue grafts: opportunities and challenges in regenerative medicine. Wiley Interdiscip Rev Syst Biol Med 2012;4:207–20.

[3] Chen FM, Liu XH. Advancing biomaterials of human origin for tissue engineering. Prog Polym Sci 2016;53:86–168.

[4] Alghazali KM, Nima ZA, Hamzah RN, Dhar MS, Anderson DE, Biris AS. Bone-tissue engineering: complex tunable structural and biological responses to injury, drug delivery, and cell-based therapies. Drug Metab Rev 2015;47:431–54.

[5] Guermani E, Shaki H, Mohanty S, Mehrali M, Arpanaei A, Gaharwar AK, et al. Engineering complex tissue-like microgel arrays for evaluating stem cell differentiation. Sci Rep 2016;6.
[6] Guvendiren M, Molde J, Soares RMD, Kohn J. Designing biomaterials for 3D printing. ACS Biomater Sci Eng 2016;2:1679–93.
[7] Yoshikawa M, Sato R, Higashihara T, Ogasawara T, Kawashima N. Rehand: realistic electric prosthetic hand created with a 3D printer. IEEE Eng Med Biol Soc 2015;2470–3.
[8] Ripley B, Kelil T, Cheezum MK, Goncalves A, Di Carli MF, Rybicki FJ, et al. 3D printing based on cardiac CT assists anatomic visualization prior to transcatheter aortic valve replacement. J Cardiovasc Comput Tomogr 2016;10:28–36.
[9] Hanker JS, Giammara BL. Biomaterials and biomedical devices. Science 1988;242:885–92.
[10] Wandrey C, Hasegawa U, van der Vlies AJ, O'Neil C, Angelova N, Hubbell JA. Analytical ultracentrifugation to support the development of biomaterials and biomedical devices. Methods 2011;54:92–100.
[11] Harrison RH, St-Pierre JP, Stevens MM. Tissue engineering and regenerative medicine: a year in review. Tissue Eng Part B Rev 2014;20:1–16.
[12] Huang J, Lin XX, Shi Y, Liu W. Tissue engineering and regenerative medicine in basic research: a year in review of 2014. Tissue Eng Part B Rev 2015;21:167–76.
[13] Lin XX, Huang J, Shi Y, Liu W. Tissue engineering and regenerative medicine in applied research: a year in review of 2014. Tissue Eng Part B Rev 2015;21:177–86.
[14] Wobma H, Vunjak-Novakovic G. Tissue engineering and regenerative medicine 2015: a year in review. Tissue Eng Part B Rev 2016;22:101–13.
[15] Vacanti JP, Morse MA, Saltzman WM, Domb AJ, Perezatayde A, Langer R. Selective cell transplantation using bioabsorbable artificial polymers as matrices. J Pediatr Surg 1988;23:3–9.
[16] Langer R, Vacanti JP. Tissue engineering. Science 1993;260:920–6.
[17] Engelmayr Jr. GC, Cheng M, Bettinger CJ, Borenstein JT, Langer R, Freed LE. Accordion-like honeycombs for tissue engineering of cardiac anisotropy. Nat Mater 2008;7:1003–10.
[18] Fischer KM, Morgan KY, Hearon K, Sklaviadis D, Tochka ZL, Fenton OS, et al. Poly(limonene thioether) scaffold for tissue engineering. Adv Healthc Mater 2016;5:813–21.
[19] Radisic M, Park H, Chen F, Salazar-Lazzaro JE, Wang Y, Dennis R, et al. Biomimetic approach to cardiac tissue engineering: oxygen carriers and channeled scaffolds. Tissue Eng 2006;12:2077–91.
[20] Radisic M, Park H, Gerecht S, Cannizzaro C, Langer R, Vunjak-Novakovic G. Biomimetic approach to cardiac tissue engineering. Philos Trans R Soc Lond B Biol Sci 2007;362:1357–68.
[21] Wang J, Bettinger CJ, Langer RS, Borenstein JT. Biodegradable microfluidic scaffolds for tissue engineering from amino alcohol-based poly(ester amide) elastomers. Organogenesis 2010;6:212–6.
[22] Webber MJ, Khan OF, Sydlik SA, Tang BC, Langer R. A perspective on the clinical translation of scaffolds for tissue engineering. Ann Biomed Eng 2015;43:641–56.
[23] Singh D, Singh D, Zo S, Han SS. Nano-biomimetics for nano/micro tissue regeneration. J Biomed Nanotechnol 2014;10:3141–61.
[24] van der Stok J, Koolen MKE, de Maat MPM, Yavari SA, Alblas J, Patka P, et al. Full regeneration of segmental bone defects using porous titanium implants loaded with BMP-2 containing fibrin gels. Eur Cell Mater 2015;29:141–54.

[25] Stevens MM, Mayer M, Anderson DG, Weibel DB, Whitesides GM, Langer R. Direct patterning of mammalian cells onto porous tissue engineering substrates using agarose stamps. Biomaterials 2005;26:7636–41.
[26] Urciuolo F, Imparato G, Totaro A, Netti PA. Building a tissue in vitro from the bottom up: implications in regenerative medicine. Methodist Debakey Cardiovasc J 2013;9:213–7.
[27] Gatenholm P, Martinez H, Sundberg J. Bioprinting of 3D porous nanocellulose scaffolds for tissue engineering and organ regeneration. Abstr Pap Am Chem Soc 2012;243.
[28] Murphy SV, Atala A. 3D bioprinting of tissues and organs. Nat Biotechnol 2014;32:773–85.
[29] Pati F, Gantelius J, Svahn HA. 3D bioprinting of tissue/organ models. Angew Chem Int Ed 2016;55:4650–65.
[30] Hull CW. Apparatus for production of three-dimensional objects by stereolithography. Google Patents, 1986.
[31] Acar P, Hadeed K, Dulac Y. Advances in 3D echocardiography: from foetus to printing. Arch Cardiovasc Dis 2016;109:84–6.
[32] Beaman J, Lopez F. Emerging nexus of cyber, modeling, and estimation in advanced manufacturing: vacuum arc remelting to 3D printing. Mech Eng 2014;136:63–70.
[33] Chia HN, Wu BM. Recent advances in 3D printing of biomaterials. J Biol Eng 2015;9:.
[34] Choi JY, Das S, Theodore ND, Kim I, Honsberg C, Choi HW, et al. Advances in 2D/3D printing of functional nanomaterials and their applications. ECS J Solid State Sci Technol 2015;4:P3001–9.
[35] Daly D. Disruptive potential for 3D printing in advanced manufacturing. Abstr Pap Am Chem Soc 2014;248:.
[36] Lee M, Wu BM. Recent advances in 3D printing of tissue engineering scaffolds. Methods Mol Biol 2012;868:257–67.
[37] Lindegaard JC, Madsen ML, Traberg A, Meisner B, Nielsen SK, Tanderup K, et al. Individualised 3D printed vaginal template for MRI guided brachytherapy in locally advanced cervical cancer. Radiother Oncol 2016;118:173–5.
[38] Michalek P, Richtera L, Krejcova L, Nejdl L, Kensova R, Zitka J, et al. Bioconjugation of peptides using advanced nanomaterials to examine their interactions in 3D printed flow-through device. Electrophoresis 2016;37:444–54.
[39] Neumaier K, Haller D, Khan SH, Grattan KTV. 3D solid modeling of multilayer printed circuit boards. VDI Berichte 1614;2001:269–83.
[40] O'Hara RP, Chand A, Vidiyala S, Arechavala SM, Mitsouras D, Rudin S, et al. Advanced 3D mesh manipulation in stereolithographic files and post-print processing for the manufacturing of patient-specific vascular flow phantoms. Proc SPIE 2016;9789.
[41] Pondrom S. 3D printing in transplantation While 3D printing has made significant advances in recent years, the reality of whole organ generation remains a long way off. Am J Transplant 2016;16:1339–40.
[42] Singh D, Singh D, Han SS. 3D printing of scaffold for cells delivery: advances in skin tissue engineering. Polymers (Basel) 2016;8.
[43] Mironov V, Visconti RP, Kasyanov V, Forgacs G, Drake CJ, Markwald RR. Organ printing: tissue spheroids as building blocks. Biomaterials 2009;30:2164–74.
[44] Norotte C, Marga FS, Niklason LE, Forgacs G. Scaffold-free vascular tissue engineering using bioprinting. Biomaterials 2009;30:5910–7.
[45] Baptista PM, Orlando G, Mirmalek-Sani SH, Siddiqui M, Atala A, Soker S. Whole organ decellularization—a tool for bioscaffold fabrication and organ bioengineering. In: 2009 annual international conference of the IEEE engineering in medicine and biology society, Vols 1-20; 2009. p. 6526–9.

[46] Kajstura J, Rota M, Hall SR, Hosoda T, D'Amario D, Sanada F, et al. Evidence for human lung stem cells. N Engl J Med 2011;364:1795–806.
[47] Guillemot F, Souquet A, Catros S, Guillotin B, Lopez J, Faucon M, et al. High-throughput laser printing of cells and biomaterials for tissue engineering. Acta Biomater 2010;6:2494–500.
[48] Klebe RJ. Cytoscribing—a method for micropositioning cells and the construction of two-dimensional and 3-dimensional synthetic tissues. Exp Cell Res 1988;179:362–73.
[49] Xu T, Zhao WX, Zhu JM, Albanna MZ, Yoo JJ, Atala A. Complex heterogeneous tissue constructs containing multiple cell types prepared by inkjet printing technology. Biomaterials 2013;34:130–9.
[50] Iwami K, Noda T, Ishida K, Morishima K, Nakamura M, Umeda N. Bio rapid prototyping by extruding/aspirating/refilling thermoreversible hydrogel. Biofabrication 2010;2.
[51] Zopf DA, Hollister SJ, Nelson ME, Ohye RG, Green GE. Bioresorbable airway splint created with a three-dimensional printer. N Engl J Med 2013;368:2043–5.
[52] Huh D. Microengineered physiological biomimicry: human organs-on-chips. FASEB J 2015;29.
[53] Huh D, Torisawa YS, Hamilton GA, Kim HJ, Ingber DE. Microengineered physiological biomimicry: organs-on-chips. Lab Chip 2012;12:2156–64.
[54] Ingber DE, Mow VC, Butler D, Niklason L, Huard J, Mao J, et al. Tissue engineering and developmental biology: going biomimetic. Tissue Eng 2006;12:3265–83.
[55] Derby B. Printing and prototyping of tissues and scaffolds. Science 2012;338:921–6.
[56] Li XM, Cui RR, Sun LW, Aifantis KE, Fan YB, Feng QL, et al. 3D-printed biopolymers for tissue engineering application. Int J Polym Sci 2014;2014.
[57] Li XM, Feng QL, Liu XH, Dong W, Cui FH. Collagen-based implants reinforced by chitin fibres in a goat shank bone defect model. Biomaterials 2006;27:1917–23.
[58] Serrano MC, Nardecchia S, Garcia-Rama C, Ferrer ML, Collazos-Castro JE, del Monte F, et al. Chondroitin sulphate-based 3D scaffolds containing MWCNTs for nervous tissue repair. Biomaterials 2014;35:1543–51.
[59] Jakab K, Norotte C, Marga F, Murphy K, Vunjak-Novakovic G, Forgacs G. Tissue engineering by self-assembly and bio-printing of living cells. Biofabrication 2010;2.
[60] Haberstroh K, Ritter K, Kuschnierz J, Bormann KH, Kaps C, Carvalho C, et al. Bone repair by cell-seeded 3D-bioplotted composite scaffolds made of collagen treated tricalciumphosphate or tricalciumphosphate-chitosan-collagen hydrogel or PLGA in ovine critical-sized calvarial defects. J Biomed Mater Res B 2010;93b:520–30.
[61] Lee H, Kim Y, Kim S, Kim G. Mineralized biomimetic collagen/alginate/silica composite scaffolds fabricated by a low-temperature bio-plotting process for hard tissue regeneration: fabrication, characterisation and in vitro cellular activities. J Mater Chem B 2014;2:5785–98.
[62] Ouyang LL, Highley CB, Rodell CB, Sun W, Burdick JA. 3D printing of shear-thinning hyaluronic acid hydrogels with secondary cross-linking. ACS Biomater Sci Eng 2016;2:1743–51.
[63] Wu ZJ, Su X, Xu YY, Kong B, Sun W, Mi SL. Bioprinting three-dimensional cell-laden tissue constructs with controllable degradation. Sci Rep 2016;6.
[64] Lee JB, Wang XT, Faley S, Baer B, Balikov DA, Sung HJ, et al. Development of 3D microvascular networks within gelatin hydrogels using thermoresponsive sacrificial microfibers. Adv Healthc Mater 2016;5:781–5.

[65] Billiet T, Gevaert E, De Schryver T, Cornelissen M, Dubruel P. The 3D printing of gelatin methacrylamide cell-laden tissue-engineered constructs with high cell viability. Biomaterials 2014;35:49–62.
[66] Tumbleston JR, Shirvanyants D, Ermoshkin N, Januszewicz R, Johnson AR, Kelly D, et al. Continuous liquid interface production of 3D objects. Science 2015;347:1349–52.
[67] Tabriz AG, Hermida MA, Leslie NR, Shu WM. Three-dimensional bioprinting of complex cell laden alginate hydrogel structures. Biofabrication 2015;7.
[68] Stocum DL, Zupanc GKH. Stretching the limits: stem cells in regeneration science. Dev Dyn 2008;237:3648–71.
[69] Muschler GE, Nakamoto C, Griffith LG. Engineering principles of clinical cell-based tissue engineering. J Bone Joint Surg Am 2004;86a:1541–58.
[70] Hung KC, Tseng CS, Dai LG, Hsu SH. Water-based polyurethane 3D printed scaffolds with controlled release function for customized cartilage tissue engineering. Biomaterials 2016;83:156–68.
[71] Gu Q, Tomaskovic-Crook E, Lozano R, Chen Y, Kapsa RM, Zhou Q, et al. Functional 3D neural mini-tissues from printed gel-based bioink and human neural stem cells. Adv Healthc Mater 2016;5:1429–38.
[72] Wengerter BC, Emre G, Park JY, Geibel J. Three-dimensional printing in the intestine. Clin Gastroenterol Hepatol 2016;14:1081–5.
[73] Sullivan DC, Mirmalek-Sani SH, Deegan DB, Baptista PM, Aboushwareb T, Atala A, et al. Decellularization methods of porcine kidneys for whole organ engineering using a high-throughput system. Biomaterials 2012;33:7756–64.
[74] Goh SK, Bertera S, Olsen P, Candiello JE, Halfter W, Uechi G, et al. Perfusion-decellularized pancreas as a natural 3D scaffold for pancreatic tissue and whole organ engineering. Biomaterials 2013;34:6760–72.
[75] Uygun BE, Soto-Gutierrez A, Yagi H, Izamis ML, Guzzardi MA, Shulman C, et al. Organ reengineering through development of a transplantable recellularized liver graft using decellularized liver matrix. Nat Med 2010;16:814–20.
[76] Amiel GE, Komura M, Shapira O, Yoo JJ, Yazdani S, Berry J, et al. Engineering of blood vessels from acellular collagen matrices coated with human endothelial cells. Tissue Eng 2006;12:2355–65.
[77] Brunello G, Sivolella S, Meneghello R, Ferroni L, Gardin C, Piattelli A, et al. Powder-based 3D printing for bone tissue engineering. Biotechnol Adv 2016;34:740–53.
[78] Pilipchuk SP, Monje A, Jiao YZ, Hao J, Kruger L, Flanagan CL, et al. Integration of 3D printed and micropatterned polycaprolactone scaffolds for guidance of oriented collagenous tissue formation in vivo. Adv Healthc Mater 2016;5:676–87.
[79] Atala A. Tissue engineering of human bladder. Br Med Bull 2011;97:81–104.
[80] Omori K, Nakamura T, Kanemaru S, Kojima H, Magrufov A, Hiratsuka Y, et al. Cricoid regeneration using in situ tissue engineering in canine larynx for the treatment of subglottic stenosis. Ann Otol Rhinol Laryngol 2004;113:623–7.
[81] Omori K, Tada Y, Suzuki T, Nomoto Y, Matsuzuka T, Kobayashi K, et al. Clinical application of in situ tissue engineering using a scaffolding technique for reconstruction of the larynx and trachea. Ann Otol Rhinol Laryngol 2008;117:673–8.
[82] Ringel RL, Kahane JC, Hillsamer PJ, Lee AS, Badylak SF. The application of tissue engineering procedures to repair the larynx. J Speech Lang Hear Res 2006;49:194–208.
[83] Zheng WX, Shi TC, Yue XY, An X. The research of laryngeal reconstruction with personalized artificial larynx using tissue engineering. Adv Mater Res 2013;655-657:1939–44.

[84] Carlson R. Estimating the biotech sector's contribution to the US economy. Nat Biotechnol 2016;34:247–55.
[85] Mungadi IA, Ugboko VI. Oral mucosa grafts for urethral reconstruction. Ann Afr Med 2009;8:203–9.
[86] Djordjevic ML, Majstorovic M, Stanojevic D, Bizic M, Ducic S, Kojovic V, et al. One-stage repair of severe hypospadias using combined buccal mucosa graft and longitudinal dorsal skin flap. Eur J Pediatr Surg 2008;18:427–30.
[87] Mitchell ME, Adams MC, Rink RC. Urethral replacement with ureter. J Urol 1988;139:1282–5.
[88] Mitchell ME, Adams MC, Saint RB. Urethral replacement with ureter. J Urol 1987;137:A105-A.
[89] Koshima I, Inagawa K, Okuyama N, Moriguchi T. Free vascularized appendix transfer for reconstruction of penile urethras with severe fibrosis. Plast Reconstr Surg 1999;103:964–9.
[90] Foroutan HR, Khalili A, Geramizadeh B, Rasekhi AR, Tanideh N. Urethral reconstruction using autologous and everted vein graft: an experimental study. Pediatr Surg Int 2006;22:259–62.
[91] Dabernig J, Shelley OP, Cuccia G, Schaff J. Urethral reconstruction using the radial forearm free flap: experience in oncologic cases and gender reassignment. Eur Urol 2007;52:547–54.
[92] Kanatani I, Kanematsu A, Inatsugu Y, Imamura M, Negoro H, Ito N, et al. Fabrication of an optimal urethral graft using collagen-sponge tubes reinforced with copoly(L-lactide/epsilon-caprolactone) fabric. Tissue Eng 2007;13:2933–40.
[93] Wu SF, Liu Y, Bharadwaj S, Atala A, Zhang YY. Human urine-derived stem cells seeded in a modified 3D porous small intestinal submucosa scaffold for urethral tissue engineering. Biomaterials 2011;32:1317–26.
[94] Knopp BP, Eppley BL, Prevel CD, Rippy MK, Harruff RC, Badylak SF, et al. Experimental assessment of small-intestinal submucosa as a bladder wall substitute. Urology 1995;46:396–400.
[95] Zhang YY, Kropp BP, Moore P, Cowan R, Furness PD, Kolligian ME, et al. Coculture of bladder urothelial and smooth muscle cells on small intestinal submucosa: potential applications for tissue engineering technology. J Urol 2000;164:928–34.
[96] Weiser AC, Franco I, Herz DB, Silver RI, Reda EF. Single layered small intestinal submucosa in the repair of severe chordee and complicated hypospadias. J Urol 2003;170:1593–5.
[97] Chen F, Yoo JJ, Atala A. Acellular collagen matrix as a possible "off the shelf" biomaterial for urethral repair. Urology 1999;54:407–10.
[98] Hu YF, Yang SX, Wang LL, Jin HM. Curative effect and histocompatibility evaluation of reconstruction of traumatic defect of rabbit urethra using extracellular matrix. Chin J Traumatol 2008;11:274–8.
[99] Koziak A, Marcheluk A, Dmowski T, Szczesniewski R, Kania P, Dorobek A. Reconstructive surgery of male urethra using human amnion membranes (grafts)--first announcement. Ann Transplant 2004;9:21–4.
[100] Holmes B, Bulusu K, Plesniak M, Zhang LG. A synergistic approach to the design, fabrication and evaluation of 3D printed micro and nano featured scaffolds for vascularized bone tissue repair. Nanotechnology 2016;27.
[101] Mironov V, Kasyanov V, Markwald RR. Organ printing: from bioprinter to organ biofabrication line. Curr Opin Biotechnol 2011;22:667–73.
[102] Farzadi A, Solati-Hashjin M, Asadi-Eydivand M, Abu Osman NA. Effect of layer thickness and printing orientation on mechanical properties and dimensional accuracy of 3D printed porous samples for bone tissue engineering. PLoS One 2014;9.

[103] Rezwan K, Chen QZ, Blaker JJ, Boccaccini AR. Biodegradable and bioactive porous polymer/inorganic composite scaffolds for bone tissue engineering. Biomaterials 2006;27:3413–31.

[104] Eqtesadi S, Motealleh A, Miranda P, Lemos A, Rebelo A, Ferreira JMF. A simple recipe for direct writing complex 45S5 Bioglass (R) 3D scaffolds. Mater Lett 2013;93:68–71.

[105] Liu X, Rahaman MN, Hilmas GE, Bal BS. Mechanical properties of bioactive glass (13-93) scaffolds fabricated by robotic deposition for structural bone repair. Acta Biomater 2013;9:7025–34.

[106] Martinez-Vazquez FJ, Perera FH, Miranda P, Pajares A, Guiberteau F. Improving the compressive strength of bioceramic robocast scaffolds by polymer infiltration. Acta Biomater 2010;6:4361–8.

[107] Johnson AJW, Herschler BA. A review of the mechanical behavior of CaP and CaP/polymer composites for applications in bone replacement and repair. Acta Biomater 2011;7:16–30.

[108] Feng P, Wei PP, Li PJ, Gao CD, Shuai CJ, Peng SP. Calcium silicate ceramic scaffolds toughened with hydroxyapatite whiskers for bone tissue engineering. Mater Charact 2014;97:47–56.

[109] Martinez-Vazquez FJ, Miranda P, Guiberteau F, Pajares A. Reinforcing bioceramic scaffolds with in situ synthesized epsilon-polycaprolactone coatings. J Biomed Mater Res A 2013;101:3551–9.

[110] Miranda P, Pajares A, Saiz E, Tomsia AP, Guiberteau F. Mechanical properties of calcium phosphate scaffolds fabricated by robocasting. J Biomed Mater Res A 2008;85a:218–27.

[111] Eqtesadi S, Motealleh A, Pajares A, Guiberteau F, Miranda P. Improving mechanical properties of 13-93 bioactive glass robocast scaffold by poly (lactic acid) and poly (epsilon-caprolactone) melt infiltration. J Non-Cryst Solids 2016;432:111–9.

[112] Roohani-Esfahani SI, Newman P, Zreiqat H. Design and fabrication of 3D printed scaffolds with a mechanical strength comparable to cortical bone to repair large bone defects. Sci Rep 2016;6:.

[113] Do AV, Khorsand B, Geary SM, Salem AK. 3D printing of scaffolds for tissue regeneration applications. Adv Healthc Mater 2015;4:1742–62.

[114] Raja N, Yun HS. A simultaneous 3D printing process for the fabrication of bioceramic and cell-laden hydrogel core/shell scaffolds with potential application in bone tissue regeneration. J Mater Chem B 2016;4:4707–16.

[115] Wust S, Muller R, Hofmann S. Controlled positioning of cells in biomaterials-approaches towards 3D tissue printing. J Funct Biomater 2011;2:119–54.

[116] Kang HW, Lee SJ, Ko IK, Kengla C, Yoo JJ, Atala A. A 3D bioprinting system to produce human-scale tissue constructs with structural integrity. Nat Biotechnol 2016;34:312–9.

[117] Peltola SM, Melchels FPW, Grijpma DW, Kellomaki M. A review of rapid prototyping techniques for tissue engineering purposes. Ann Med 2008;40:268–80.

[118] Farahani RD, Dube M, Therriault D. Three-dimensional printing of multifunctional nanocomposites: manufacturing techniques and applications. Adv Mater 2016;28:5794–821.

[119] Huang WD, Zheng QX, Sun WQ, Xu HB, Yang XL. Levofloxacin implants with predefined microstructure fabricated by three-dimensional printing technique. Int J Pharm 2007;339:33–8.

[120] Jung JW, Lee H, Hong JM, Park JH, Shim JH, Choi TH, et al. A new method of fabricating a blend scaffold using an indirect three-dimensional printing technique. Biofabrication 2015;7.

[121] Patirupanusara R, Suwanapreuk W, Rubkumintara T, Sulvanprateeb J. Effect of binder content on the material properties of polymethyl methacrylate fabricated by three dimensional printing technique. J Mater Process Tech 2008;207:40–5.
[122] Qiao F, Li DC, Jin ZM, Hao DJ, Liao YH, Gong SH. A novel combination of computer-assisted reduction technique and three dimensional printed patient-specific external fixator for treatment of tibial fractures. Int Orthop 2016;40:835–41.
[123] Sakamoto S, Takaki Y. Three-dimensional print using a one-dimensional screen technique. Jpn J Appl Phys 2008;47:5486–92.
[124] Sun K, Li RX, Jiang WX, Sun YF, Li H. Comparison of three-dimensional printing and vacuum freeze-dried techniques for fabricating composite scaffolds. Biochem Bioph Res Co 2016;477:1085–91.
[125] Tan HT, Yang KQ, Wei PG, Zhang GD, Dimitriou D, Xu L, et al. A novel preoperative planning technique using a combination of ct angiography and three-dimensional printing for complex toe-to-hand reconstruction. J Reconstr Microsurg 2015;31:369–77.
[126] Wu G, Wu WG, Zheng QX, Li JF, Zhou JB, Hu ZL. Experimental study of PLLA/INH slow release implant fabricated by three dimensional printing technique and drug release characteristics in vitro. Biomed Eng Online 2014;13.
[127] Zhang Q, Yan D, Zhang K, Hu GK. Pattern transformation of heat-shrinkable polymer by three-dimensional (3D) printing technique. Sci Rep 2015;5.
[128] Zhang Q, Zhang K, Hu GK. Smart three-dimensional lightweight structure triggered from a thin composite sheet via 3D printing technique. Sci Rep 2016;6.
[129] Billiet T, Vandenhaute M, Schelfhout J, Van Vlierberghe S, Dubruel P. A review of trends and limitations in hydrogel-rapid prototyping for tissue engineering. Biomaterials 2012;33:6020–41.
[130] Kim G, Son J, Park S, Kim W. Hybrid process for fabricating 3D hierarchical scaffolds combining rapid prototyping and electrospinning. Macromol Rapid Commun 2008;29:1577–81.
[131] Lee M, Dunn JCY, Wu BM. Scaffold fabrication by indirect three-dimensional printing. Biomaterials 2005;26:4281–9.
[132] Pataky K, Braschler T, Negro A, Renaud P, Lutolf MP, Brugger J. Microdrop printing of hydrogel bioinks into 3D tissue-like geometries. Adv Mater 2012;24:391–6.
[133] Place ES, Evans ND, Stevens MM. Complexity in biomaterials for tissue engineering. Nat Mater 2009;8:457–70.
[134] Cohen DL, Malone E, Lipson H, Bonassar LJ. Direct freeform fabrication of seeded hydrogels in arbitrary geometries. Tissue Eng 2006;12:1325–35.
[135] Mouser VH, Melchels FP, Visser J, Dhert WJ, Gawlitta D, Malda J. Yield stress determines bioprintability of hydrogels based on gelatin-methacryloyl and gellan gum for cartilage bioprinting. Biofabrication 2016;8:035003.
[136] Levett PA, Melchels FPW, Schrobback K, Hutmacher DW, Malda J, Klein TJ. A biomimetic extracellular matrix for cartilage tissue engineering centered on photocurable gelatin, hyaluronic acid and chondroitin sulfate. Acta Biomater 2014;10:214–23.
[137] Yan YN, Xiong Z, Hu YY, Wang SG, Zhang RJ, Zhang C. Layered manufacturing of tissue engineering scaffolds via multi-nozzle deposition. Mater Lett 2003;57:2623–8.
[138] Pati F, Jang J, Ha DH, Kim SW, Rhie JW, Shim JH, et al. Printing three-dimensional tissue analogues with decellularized extracellular matrix bioink. Nat Commun 2014;5:.
[139] Landers R, Pfister A, Hubner U, John H, Schmelzeisen R, Mulhaupt R. Fabrication of soft tissue engineering scaffolds by means of rapid prototyping techniques. J Mater Sci 2002;37:3107–16.

[140] Gruene M, Pflaum M, Deiwick A, Koch L, Schlie S, Unger C, et al. Adipogenic differentiation of laser-printed 3D tissue grafts consisting of human adipose-derived stem cells. Biofabrication 2011;3.

[141] Armstrong JPK, Burke M, Carter BM, Davis SA, Perriman AW. 3D bioprinting using a templated porous bioink. Adv Healthc Mater 2016;5:1724–30.

[142] Xu T, Rohozinski J, Zhao WX, Moorefield EC, Atala A, Yoo JJ. Inkjet-mediated gene transfection into living cells combined with targeted delivery. Tissue Eng Part A 2009;15:95–101.

[143] Wegman F, Bijenhof A, Schuijff L, Oner FC, Dhert WJA, Alblas J. Osteogenic differentiation as a result of BMP-2 plasmid DNA based gene therapy in vitro and in vivo. Eu Cell Mater 2011;21:230–42.

[144] Hosseinkhani H, Hong PD, Yu DS, Chen YR, Ickowicz D, Farber IY, et al. Development of 3D in vitro platform technology to engineer mesenchymal stem cells. Int J Nanomedicine 2012;7:3035–43.

[145] Hosseinkhani H, Hosseinkhani M, Gabrielson NP, Pack DW, Khademhosseini A, Kobayashi H. DNA nanoparticles encapsulated in 3D tissue-engineered scaffolds enhance osteogenic differentiation of mesenchymal stem cells. J Biomed Mater Res A 2008;85a:47–60.

[146] Loozen LD, Wegman F, Oner FC, Dhert WJA, Alblas J. Porous bioprinted constructs in BMP-2 non-viral gene therapy for bone tissue engineering. J Mater Chem B 2013;1:6619–26.

[147] Cui XF, Boland T. Human microvasculature fabrication using thermal inkjet printing technology. Biomaterials 2009;30:6221–7.

[148] Kolesky DB, Truby RL, Gladman AS, Busbee TA, Homan KA, Lewis JA. 3D bioprinting of vascularized, heterogeneous cell-laden tissue constructs. Adv Mater 2014;26:3124–30.

[149] Miller JS, Stevens KR, Yang MT, Baker BM, Nguyen DHT, Cohen DM, et al. Rapid casting of patterned vascular networks for perfusable engineered three-dimensional tissues. Nat Mater 2012;11:768–74.

[150] Lee KY, Mooney DJ. Hydrogels for tissue engineering. Chem Rev 2001;101:1869–79.

[151] Fidkowski C, Kaazempur-Mofrad MR, Borenstein J, Vacanti JP, Langer R, Wang YD. Endothelialized microvasculature based on a biodegradable elastomer. Tissue Eng 2005;11:302–9.

[152] Khademhosseini A, Langer R, Borenstein J, Vacanti JP. Microscale technologies for tissue engineering and biology. Proc Natl Acad Sci U S A 2006;103:2480–7.

[153] Lee VK, Lanzi AM, Ngo H, Yoo SS, Vincent PA, Dai GH. Generation of multi-scale vascular network system within 3D hydrogel using 3D bio-printing technology. Cell Mol Bioeng 2014;7:460–72.

[154] Bregman BS, Kunkelbagden E, Schnell L, Dai HN, Gao D, Schwab ME. Recovery from spinal-cord injury mediated by antibodies to neurite growth-inhibitors. Nature 1995;378:498–501.

[155] Horner PJ, Gage FH. Regenerating the damaged central nervous system. Nature 2000;407:963–70.

[156] Lee W, Pinckney J, Lee V, Lee JH, Fischer K, Polio S, et al. Three-dimensional bioprinting of rat embryonic neural cells. Neuroreport 2009;20:798–803.

[157] Hsieh FY, Lin HH, Hsu SH. 3D bioprinting of neural stem cell-laden thermoresponsive biodegradable polyurethane hydrogel and potential in central nervous system repair. Biomaterials 2015;71:48–57.

Biopolymer hydrogel bioinks

M. Gelinsky
Dresden University of Technology (TU Dresden), Dresden, Germany

6.1 Bioprinting versus 3D printing and conventional cell seeding

Additive manufacturing (AM) technologies can be used in two different ways for tissue engineering applications. The easier option is to apply methods such as fused deposition modeling (FDM), selective laser sintering (or melting; SLS/SLM), 3D powder printing, or extrusion-based printing (also called 3D plotting, direct writing, robotic dispension, etc.) for scaffold fabrication only, followed by conventional cell seeding and cultivation. The advantage here is that complex geometries, patient-specific size and shape, and highly defined internal porosity can be achieved much easier compared with classical scaffold manufacturing technologies. For this purpose, a huge variety of biomaterials can be utilized, depending on the AM method applied. On the contrary, 3D bioprinting describes the combined printing of cells and suitable jelly-like biomaterials as cells can only be suspended in soft materials with a high water content. For these materials, the term "bioink" has been introduced to the field of AM. Bioprinting offers additional advantages (beside those already mentioned) such as a very high cell seeding efficiency, homogeneous cell distribution, and the possibility to include several cell types which can be positioned in three dimensions with high spatial resolution [1]. This opens up new possibilities like fabrication of complex, multicellular tissues, vascularized tissue constructs, or those consisting of more than one type of tissues. Combined bioprinting of osteoblasts and chondrocytes in a two-layered manner could, for example, lead to artificial tissue constructs, suitable for the therapy of osteochondral defects (more general: for defects located at tissue interfaces).

Even though a variety of natural (this chapter) and synthetic (Chapter 9) bioinks/biomaterials are already available for bioprinting purposes, there are still limitations such as partly weak cell adhesion properties and limited mechanical stability, which also restricts the size of such artificial tissue constructs, producible by bioprinting. In Section 6.4 some strategies will be described to overcome these limitations. Concerning AM technologies, this chapter focuses on extrusion-based methods as those are the only commonly applied ones for bioprinting applications. It has to be mentioned that some inkjet-based methods can also be used for cell printing. But here either single cells, suspended in low viscous liquids such as cell culture media are used, which is not suitable for fabrication of three-dimensional tissue constructs—or spherical cell aggregates are printed (assembled), more or less without additional biomaterials (and therefore bioinks). For the latter, the term "bioassembly" recently has been proposed [2] to clarify the differences between this technology and extrusion-based bioprinting.

3D Bioprinting for Reconstructive Surgery. https://doi.org/10.1016/B978-0-08-101103-4.00008-9

6.2 Advantages and disadvantages of biopolymers for bioprinting

Biopolymers are defined as polymers synthesized by organisms. Biopolymers are present in all types of organisms (and some are secreted by those) such as bacteria, algae, plants, fungi, and animals. In most cases biopolymers fulfill structural functions in the respective animate being—allowing such fascinating properties like up to 100 m high trunks of trees (due to cellulose fibers) or very hard teeth (due to hierarchically organized collagen mineral nanocomposites). The use of of biopolymers (instead of synthetic ones) for biomedical applications has a number of distinct advantages as well as disadvantages. Beneficial is that nearly all biopolymers are biocompatible with mammalian cells and tissues and most can be degraded after implantation, again leading to nontoxic degradation products. It is obvious that biopolymers derived from mammalian tissues such as collagens, hyaluronic acid, or fibrin have the best "biological properties" concerning interaction with human cells. In addition, those materials can be enzymatically degraded such as the natural mammalian extracellular matrix (ECM) components in vivo. In addition, biopolymers are renewable raw materials and therefore classified as ecofriendly. One disadvantage is the difficulty to obtain materials of constant quality and properties: depending on the source biopolymers often suffer from high batch-to-batch variations.

In contrast, synthetic polymers that are often synthesized from petroleum products or other highly defined chemical raw materials are available in high and constant quality and purity. Many properties such as molecular weight (degree of polymerization), degree of branching, chemical composition (in the case of co-polymers), and so on can be controlled during synthesis. Similarly the physical and chemical properties can also be tailored. But as they are artificial materials, they lack binding sites for mammalian cells and biodegradation might lead to problematic degradation products such as lactic acid monomers in the case of polylactide. Many synthetic polymers, used for biomedical applications are thermoplasts, which means that they can be melted without decomposition. They are therefore suitable for AM technologies such as FDM. This is not the case for most of the natural biopolymers which therefore have to be processed as solutions, suspensions, or hydrogels, using extrusion-based AM techniques. Synthetic polymers, suitable for 3D printing and bioprinting applications are described in detail in Chapter 9 including some basic properties of polymers, which are also relevant for biopolymers.

6.3 Biopolymer hydrogels

Biopolymers can be divided into three groups, depending on their monomeric building blocks: polynucleotides (such as DNA and RNA), polypetides/proteins, and polysaccharides. Up to now, polynucleotides do not play an important role as biomaterials whereas a large number of different polypeptides and polysaccharides are used as the scaffold material and for bioprinting applications. In most cases, natural polynucleotides and polypeptides have a highly defined molecular composition. Biomacromolecules having the same sequence and number of monomers also have the same molecular mass: they are monodisperse. In contrast, synthetic polymers are always polydisperse as the chemical synthesis cannot be controlled as good as the biosynthesis. Most polysaccharides are not monodisperse too and their monomeric composition is generally less determined compared with polynucleotides and polypeptides. This will be described using the example of alginate in Section 6.3.1.

As biopolymers are produced by living cells they are adapted to an aqueous environment. But this does not mean that all biopolymers can bind water and form hydrogels—some like highly acetylated chitin (e.g., in exoskeletons of insects and crustaceans) definitely do not. For extrusion-based 3D (bio)printing viscous solutions or suspensions are needed, which can be extruded through nozzles with inner diameters of typically 100–1000 μm. Advantageous for the printing process is a shear thinning behavior, which means that during extrusion the viscosity of the material decreases due to the shear stress, present especially in the nozzle. After deposition of the material strands, viscosity is increasing again, which helps to construct structurally defined, three-dimensional objects. For bioprinting purposes live cells are suspended in the aqueous solution/suspension of the respective (bio)polymer prior to printing. This medium viscous mixture should be called "bioink." After extrusion, the cell-loaded construct has to be further stabilized, which is mostly done by gelation, that is, formation of a hydrogel. The most common example is probably the ionic cross-linking of alginate sols with divalent metal cations such as Ca^{2+}, leading to hydrogels which are stable enough to be kept in an incubator for a couple of weeks. The scheme in Fig. 6.1 clarifies the process of bioprinting and respective terminology.

In principle, most biopolymer hydrogels are suitable for suspending mammalian cells but only in a few cases, cytocompatible stabilization protocols are available. One option to overcome this problem is chemical modification of natural biopolymers, for example, by introducing photo-cross-linkable methacrylate units. Finally, it has to be mentioned that all materials and printer components coming in contact with the bioink have to be sterile, which might lead to difficulties as many sterilization processes such as gamma irradiation significantly alter the properties of biopolymers. For

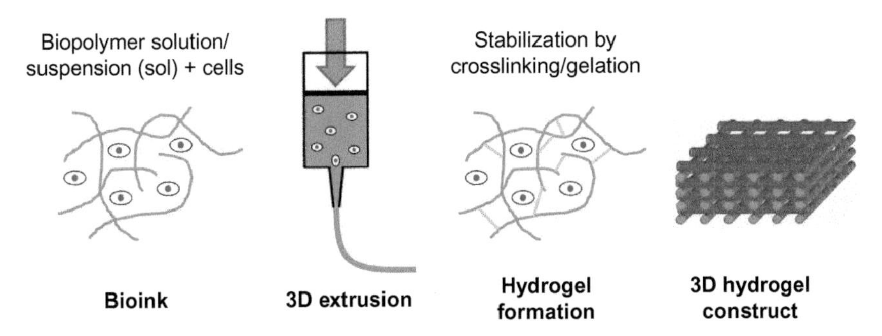

Fig. 6.1 Schematic representation of the bioprinting process, using extrusion-based AM.

every bioink, therefore, a suitable sterilization protocol has to be developed [3]. As 3D bioprinting is an amazingly fast-growing field of research several review papers on bioinks have been published in the last few years, giving a more detailed information compared with this chapter [4–8].

6.3.1 Alginate

Alginate (alginic acid) is a polysaccharide synthesized by brown algae and some bacteria. It is a linear copolymer consisting of the monosaccharide residues β-D-mannuronate (M) and α-L-guluronate (G). The saccharide sequence is not regular and therefore regions of consecutive G-residues (G-blocks), consecutive M-residues (M-blocks), or alternating M- and G-residues (MG-blocks) appear. The molecular weight can be in the range of 10,000–600,000 (typically 50,000–180,000) $g\,mol^{-1}$. Fig. 6.2 shows the general structure of alginate.

Alginate is produced in large scale for several industrial applications, especially in food industry. A number of foods contain alginate as thickener. It is also applied in medicine, especially in wound dressings. Mammals cannot degrade implanted alginate because specific enzymes for cleavage of the polysaccharide chain are not expressed but a very slow degradation should occur due to unspecific enzymatic reactions. Therefore, alginate can be used for immunoprotection of cells which is explored concerning transplantation of allogenic pancreatic islets for diabetes therapy. For bioprinting purposes, alginate is commonly used as aqueous sols at a concentration of c. 3%, possessing acceptable viscosity for extrusion, but lacking shear thinning behavior. After the printing process, ionic cross-linking is induced by immersion in aqueous solutions, containing divalent metal ions such as calcium, strontium, barium, or zinc. By ionic interactions (complexation) of the metal ions with carboxy and hydroxy groups of the alginate chain, especially those of the guluronic acid residues, stable hydrogels are formed. For the resulting chelate complex, the so-called egg box model has been proposed which is shown in Fig. 6.3.

The strength of calcium alginate hydrogels therefore can easily be adjusted by choosing an alginate of suitable molecular weight, ratio of guluronic and manuronic acid residues, and divalent metal ion concentration for cross-linking. As the ionic gelation with calcium ions is a fully cytocompatible process (at least up to Ca^{2+} concentrations of about 100 mM) alginate-based bioinks are currently the most common ones used in bioprinting. Disadvantages of alginates are their limited rheological properties which prevent printing of macroscopic and structurally well-defined 3D

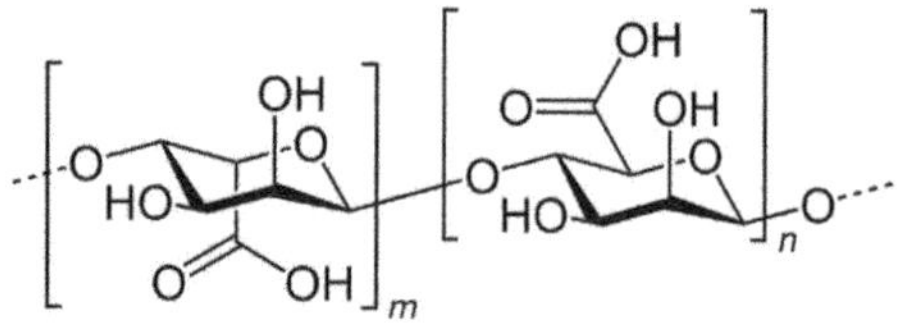

Fig. 6.2 The chemical structure of alginate (here: protonated alginic acid), consisting of mannuronic and guluronic acid monosaccharide residues.
Image from Wikimedia.

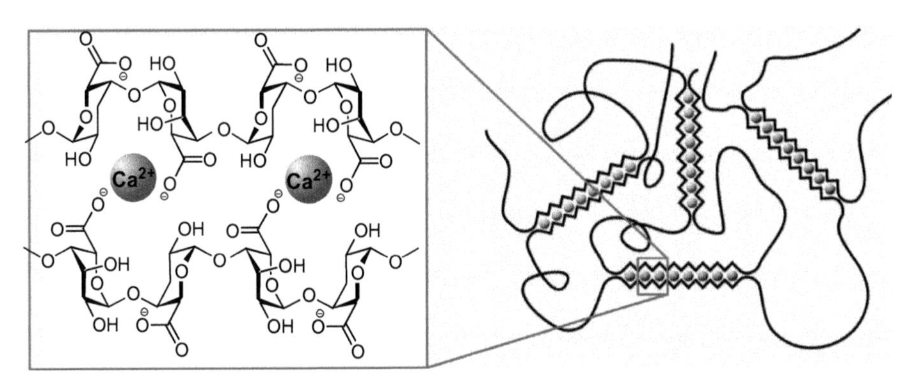

Fig. 6.3 Egg box model for calcium ion complexation by G-block motifs of alginate polysaccharide chains.
From Bruchet M, Melman A. Fabrication of patterned calcium cross-linked alginate hydrogel films and coatings through reductive cation exchange. Carbohydr Polym 2015;131:57–64, with permission.

constructs. Several strategies have been proposed to overcome this limitation. One is to use a slightly precross-linked alginate bioink [9], the other blending with additional (bio)polymers such as for example methylcellulose (MC) [10]. As MC is not affected by calcium cross-linking its addition increases the viscosity of the bioink only temporarily during extrusion but does not alter the stiffness of the finally formed hydrogel. Moreover, the blend possesses distinct shear thinning properties. Fig. 6.4 shows the different outcomes of bioprinting using a pure 3% alginate sol as the bioink compared with a blend consisting of 3% alginate and 9% MC as well as the respective rheological properties.

In this study primary human mesenchymal stem cells (hMSC) were successfully bioprinted with the alginate/MC blend as bioink and similar cell survival rates directly after printing and after 3 weeks of cultivation were found for the blend and a pure 3% alginate bioink. In addition, it could be demonstrated that the MSC still can be differentiated toward the adipogenic lineage after printing which was proved on RNA and protein level and by staining the lipids, produced by adipocytes [10]. Besides soluble components such as MC, insoluble ones such as cellulose nano- or microfibers can be used to increase the viscosity of the alginate, thereby facilitating the printing process as well as stabilizing the final construct. The use of bacterial nanocellulose in combination with alginate was investigated in detail by Paul Gatenholm and coworkers [11], which led to one of the first commercially available bioink products called "cellinc."

An alternative approach for improving the suitability of alginate for bioprinting applications is chemical modification by partial oxidation, leading to the formation of aldehyde groups in the alginate chain. This approach was investigated systematically by Jia et al. [12], who identified a range of bioinks, suitable for printing of different human cell types. Zehnder and coworkers used such alginate aldehydes in combination with methacrylated gelatin (GelMA), leading to a robust ionically and photo-cross-linkable bioink [13]. A variety of other alginate-based bioinks were discussed by Axpe and Oyen [14].

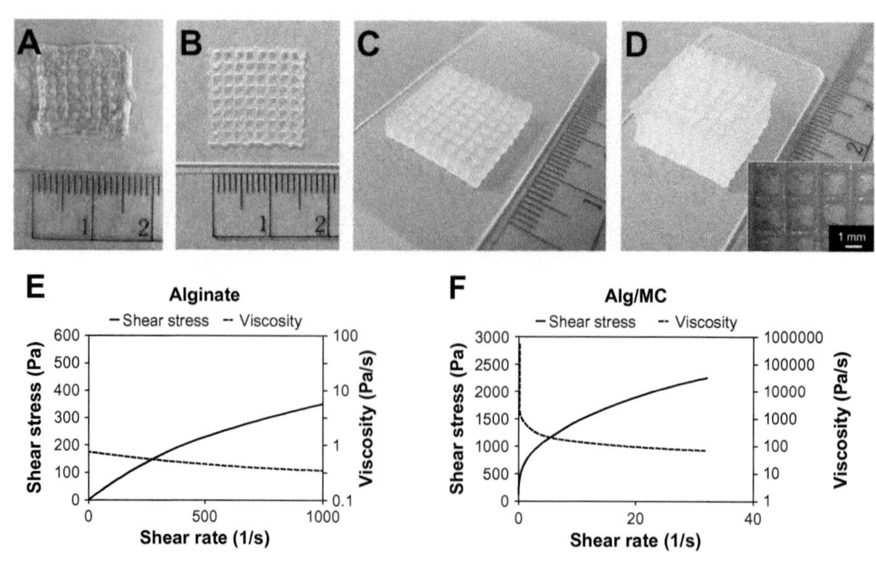

Fig. 6.4 Bioprinting of a 3% alginate bioink (A) in comparison to 3% alginate/9% MC blend (B–D). The scaffolds consist of four layers (A and B), 20 layers (C), and 50 layers (D), respectively. Inset in (D) demonstrates the fully open and well-defined pores between the strands (top view). (E and F) show the rheological properties of both materials.
From Schütz K, Placht A.-M, Paul B, Brüggemeier S, Gelinsky M, Lode A. 3D plotting of a cell-laden alginate/methylcellulose blend: towards biofabrication of tissue engineering constructs with clinically relevant dimensions. J Tissue Eng Regen Med 2015 (in press, doi:10.1002/term.2058), with permission.

6.3.2 Agarose

Like alginate, agarose is a linear polysaccharide, derived from seaweed. It consists of alternating D-galactose and 3,6-anhydro-L-galactopyranose residues, linked by glycosidic bonds. The molecular weight is typically 120,000 Da. It is mostly used for the separation of biomacromolecules such as DNA and RNA by electrophoresis and therefore plays an important role in molecular biology. As bioink agarose is far less frequently used compared with alginate, very few studies on its successful utilization have been published. In a comparative study, conducted by Daly and coworkers four different bioinks were tested concerning bioprinting of articular cartilage constructs [15]. Both alginate and agarose supported the development of hyaline-like cartilage whereas GelMA- and a methacrylated poly(ethylene glycol) (PEGMA)-based bioink led to a more fibrocartilage-like phenotype. The reason for this could be that both polysaccharides do not provide adhesion sites for mammalian cells, which prevent the formation of elongated, fibroblast-like cell morphologies. In another publication, application of a blend consisting of agarose and collagen for bioprinting of human smooth muscle cells was described [16]. It could be shown that a blend consisting of 0.5 wt% agarose and 0.2 wt% collagen type I was well suitable concerning printability and cytocompatibility. In addition, the presence of collagen allowed cells being embedded in the hydrogel matrix to spread.

6.3.3 Gellan gum

Gellan gum is an exopolysaccharide derived from the bacterium *Sphingomonas elodea*. Its molecular composition is quite complicated as it consists of repeating tetrasaccharide units which are made up of D-glucose, L-rhamnose, and D-glucuronic acid residues. The polysaccharide forms typically very long chains with molecular weights of about 500,000 Da. Like alginate, gellan gum form gels in the presence of metal cations; it is used (again like alginate) as a thickener in food industry and also in plant and bacterial cell cultures. As gellan gum forms rather soft gels it needs to be blended with other materials to achieve suitable rheological properties for bioprinting. Kesti et al. have demonstrated very good results with a 3% gellan gum/2% alginate blend, which was cross-linked with strontium ions after extrusion [17]. In this study, primary chondrocytes were successfully printed and afterwards cultivated in vitro for 8 weeks. The constructs showed good structural stability, cell proliferation, and preservation of the chondrogenic phenotype. Together with a pluronic hydrogel used as supportive sacrificial material for the extrusion process anatomically shaped constructs such as a human ear and nose could be fabricated. To improve the rheological properties and thereby the printability several chemical modifications of gellan gum are possible which has been described recently in a review paper [18].

6.3.4 Collagen

Collagen is the most important protein component of mammalian ECMs; it makes up around 25%–35% of the whole protein content of humans. Up to now 28 different collagen variants have been identified of which types I, II, and III are the most important ones. All types of collagen consist of triple-helical molecules with a typical length of 300 nm. For biomedical applications, mostly collagen type I is used as it is the most abundant collagen which can easily be isolated from the dermis or tendon. As the amino acid sequence of collagen type I is highly conserved across the species, collagen isolated from bovine, equine, or porcine sources usually does not lead to immunological rejection when implanted in humans. Collagen-based materials derived from different species are therefore commonly used in a variety of clinical applications. As collagens have both a structural function and also act as adhesion sites for mammalian cells it is obvious that it would be an ideal material for bioprinting. But as the gelation of collagen is based on fibril reassembly and therefore much more difficult to control compared with, for example, ionic gelation of alginate, it is not easy to find suitable process conditions. In addition, stabilization of collagen scaffolds is based on covalent cross-linking which in most cases requires noncytocompatible chemical reactions. Kim and coworkers achieved good results with a 5 wt% collagen bioink and cross-linking with the natural agent genipin [19]. Due to the elevated temperature, needed for collagen fibril reassembly only constructs with a height of up to 4 mm could be printed because of limited heat transfer from the heated building platform through the already deposited layers. Both MG-63 osteosarcoma and hMSC showed good viability after printing and cross-linking with genipin (1 mM for 1 h). Another approach to collagen-based bioinks is utilization of decellularized ECMs (dECM) of various tissues. Pati and coworkers have demonstrated suitability of such materials derived from cartilage, fat, and heart muscle tissue for bioprinting purposes [20] and other groups reported on additional suitable tissue sources such as the liver. Disadvantages of dECM are their weak mechanical properties, which requires co-printing of stiffer supportive materials such as thermoplastic polycaprolactone (PCL), at least if bigger constructs shall be fabricated (see Section 6.4). Of course, collagen was used also for 3D printing of scaffolds for conventional cell seeding. Recently, Lode et al. have published on the manufacturing of mechanically robust pure collagen scaffolds using highly viscous collagen type I dispersions derived from porcine skin [21]. Here, covalent cross-linking with a carbodiimide derivative was applied which excludes integration of cells during printing.

6.3.5 Gelatin

Gelatin can be defined as denatured collagen. Depending on the isolation and purification process used, it consists of collagen polypeptides (i.e., single chains of the triple helical collagen molecules), shorter peptide fragments, but also may still contain some triple-helical fractions. Gelation of gelatin occurs on temperature change: it is readily soluble in hot water but gels upon cooling. As further cultivation of cell-laden, bioprinted constructs includes storage at 37°C which would lead to a softening or even disintegration of pure gelatin constructs, additional stabilization is required. Most groups therefore use GelMA which can be cross-linked photochemically directly after printing. Due to their tunable physical properties, GelMA hydrogels have been used in a variety of biomedical applications including cell encapsulation and bioprinting [22]. As photopolymerization requires addition of photoinitiators and exposure to UV light, the process conditions have to be selected carefully to prevent cell damage. Billiet and coworkers could demonstrate a high cell viability of 97% after bioprinting of HepG2 cells (hepatocarcinoma cell line) in 10–20 wt% GelMA bioinks using VA-086 as photoinitiator [23]. In addition, thorough control of the bioink temperature during and after printing enabled fabrication of structurally well-defined, fully open, porous constructs of relevant dimensions. Fig. 6.5 shows the morphology and high cell viability of printed constructs, using GelMA as a bioink.

Nevertheless, potential DNA damage due to UV irradiation cannot be excluded which limits possible clinical applications of photo-cross-linkable bioinks in general. The use of GelMA/ADA blends for bioprinting has been already mentioned (Section 6.3.1).

6.3.6 Other biopolymer hydrogels

In the last couple of years, a variety of other hydrogel-forming biopolymers have been successfully used as bioinks for 3D printing applications. Of special interest seems to be fibrin—a protein involved in the clotting of blood. Fibrin is formed from fibrinogen by the action of the protease thrombin. Thrombin cleaves the fibrinogen molecule and the resulting fibrin then forms insoluble fibrils. This reaction is also used in “fibrin glue,” a material applied in the clinic as tissue adhesive. As both fibrinogen and thrombin (as prothrombin) are components of blood plasma this bioink

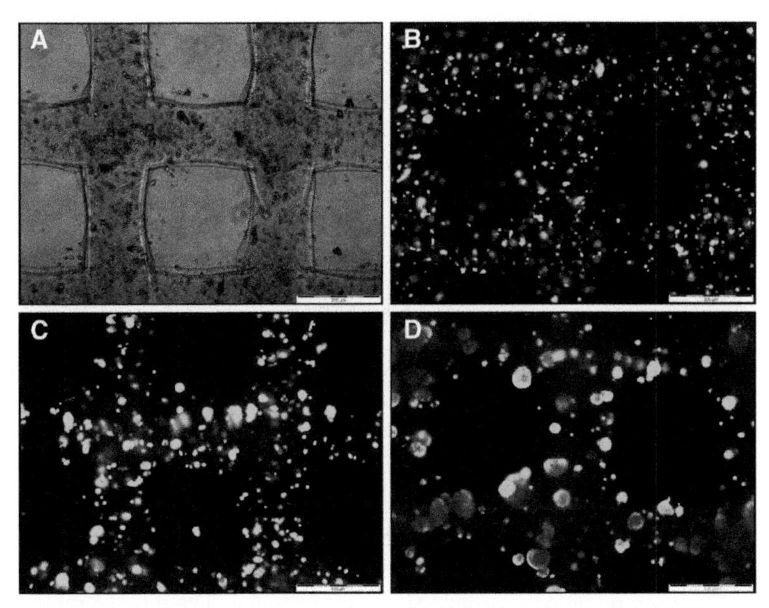

Fig. 6.5 Well-defined 3D bioprinted constructs with HepG2 cells embedded in GelMA bioink. Brightfield image (A) and fluorescence micrographs after live/dead staining taken after 1 day (B), 7 day (C), and 14 day (D). All scale bars represent 500 μm.
From Billiet T, Gevaert E, De Schryver T, Cornelissen M, Dubruel P. The 3D printing of gelatin methacrylamide cell-laden tissue-engineered constructs with high cell viability. Biomaterials 2014;35:49–62, with permission.

could be provided in autologous form. Cubo and coworkers reported on successful bioprinting of human skin using human plasma as bioink; the included fibrinogen was transformed to fibrin upon extrusion [24]. Primary human dermal fibroblasts were embedded during printing and the resulting dermal layer then covered with primary keratinocytes, both isolated from human skin biopsies. The resulting skin constructs were compared with those produced by a similar but manual procedure, being already applied in the clinic—and showed no significant differences. Another biopolymer which attracts attention is silk fibroin, the main protein component of natural silk fibers produced by silk worms and spiders. Aqueous silk fibroin hydrogels possess a strong shear thinning behavior which favors their utilization as bioink. Schacht et al. have used a recombinant, RGD-modified spider fibroin for bioprinting of several cell types and achieved excellent results [25]. Another group proposed a blend consisting of silk fibroin and gelatin as bioink. Here, cytocompatible gelation was induced prior to mixing with the cells either by the addition of a mushroom-derived tyrosinase or sonication [26]. The team could prove high cell viability after printing and structural stability over 3 weeks of cultivation. Only very few studies have been published so far on polynucleotide-based bioinks. Li and coworkers reported about a DNA-based hydrogel which formed upon mixing of two components, a polypeptide-DNA conjugate and the complementary DNA strand which acts as a linker [27]. Despite good cytocompatibility and sufficient strength the limited availability and high price of DNA might prevent broad application of this class of materials for bioprinting applications.

There are many additional studies demonstrating suitability of other biopolymers such as chitosan or hyaluronic acid, either alone, blended with other polymers or chemically modified. Crucial properties are always cytocompatible gelation mechanisms and sufficient mechanical stability. As biopolymer hydrogels in general are soft materials several strategies have been investigated which nevertheless allow bioprinting of macroscopic but still well-defined constructs. A brief overview is given in the following paragraph.

6.4 How to print macroscopic, stable structures with hydrogels

As mammalian cells may only be embedded in low concentrated, soft hydrogels a drastic increase in biopolymer concentration or cross-linking degree would be harmful to them. Therefore, alternative strategies have to be applied if macroscopic and mechanically more reliable constructs shall be fabricated. The most common one is coextrusion of mechanically stable supportive materials such as PCL [28]. PCL seems especially suitable due to its low melting point, but also other materials (thermoplastic polymers or highly concentrated hydrogels) can be used as mechanical support provided that they do not harm the cells. Other options are blending of the hydrogels with either soluble materials (described above using the example of alginate, blended with MC) or fibrous components such as nanocellulose. Also in situ cross-linking during the bioprinting process is an option which improves printability: for example alginate-based bioink layers can be cross-linked by spraying a solution of calcium chloride on top of every layer after it has been deposited to prevent compression of a big hydrogel construct due to gravity effects. Alternatively, the building platform slowly could be lowered during the printing process so that the already extruded layers come into contact with a suitable cross-linking solution. An alternative approach which is intensively investigated at the moment is 3D (bio)printing of strands with a core/shell morphology, so that the soft cell-laden hydrogel as core is mechanically protected by a stiffer material which forms the shell [29]. Strategies for bioprinting of volumetric tissue constructs were recently reviewed by Kilian et al. [30].

6.5 Summary

Biopolymer hydrogels as the natural environment of cells in tissues are highly suitable as bioinks. Many different biopolymers from all classes of organisms (animals, plants, algae, and bacteria) have been successfully used for the development of bioinks. Most biopolymer hydrogels possess good cytocompatibility but limited mechanical strength. Biopolymers isolated from mammalian sources in most cases offer adhesion sites to (mammalian) cells whereas those derived from other sources do not. For every material suitable process conditions and a cytocompatible gelation and/or crosslinking strategy have to be defined. By blending of different biopolymers and also by chemical modification a multitude of novel bioinks can be derived which will open additional new applications in the fast growing field of 3D bioprinting.

References

[1] Malda J, Visser J, Melchels FP, Jüngst T, Hennink WE, Dhert WJ, et al. 25th anniversary article: engineering hydrogels for biofabrication. Adv Mater 2013;25:5011–28.
[2] Groll J, Boland T, Blunk T, Burdick JA, Cho DW, Dalton PD, et al. Biofabrication: reappraising the definition of an evolving field. Biofabrication 2016;8. 013001.
[3] Bernhardt A, Wehrl M, Paul B, Hochmuth Th, Schumacher M, Schütz K, et al. Improved sterilization of sensitive biomaterials with supercritical carbon dioxide at low temperature. PLoS One 2015;10. e0129205.
[4] Chimene D, Lennox KK, Kaunas RR, Gaharwar AK. Advanced bioinks for 3D printing: a materials science perspective. Ann Biomed Eng 2016;44:2090–102.
[5] Jose RR, Rodriguez MJ, Dixon ThA, Omenetto F, Kaplan DL. Evolution of bioinks and additive manufacturing technologies for 3D bioprinting. ACS Biomater Sci Eng 2016;2:1662–78.
[6] Panwar A, Tan LP. Current status of bioinks for micro-extrusion-based 3D bioprinting. Molecules 2016;21:685.
[7] Prendergast ME, Solorzano RD, Cabrera D. Bioinks for biofabrication: current state and future perspectives. J 3D Print Med 2017;1:49–62.
[8] Hospodiuk M, Dey M, Sosnoski D, Ozbolat IT. The bioink: a comprehensive review on bioprintable materials. Biotechnol Adv 2017;35:217–39.
[9] Chung J, Naficy S, Yue Z, Kapsa RM, Quigley AF, Moulton SE, et al. Bio-ink properties and printability for extrusion printing living cells. J Biomater Sci Polym Ed 2013;1:763–73.
[10] Schütz K, Placht A-M, Paul B, Brüggemeier S, Gelinsky M, Lode A. 3D plotting of a cell-laden alginate/methylcellulose blend: towards biofabrication of tissue engineering constructs with clinically relevant dimensions. J Tissue Eng Regen Med 2015. https://doi.org/10.1002/term.2058. in press.
[11] Markstedt K, Mantas A, Tournier I, Martínez Avila H, Hägg D, Gatenholm P. 3D bioprinting human chondrocytes with nanocellulose—alginate bioink for cartilage tissue engineering applications. Biomacromolecules 2015;16:1489–96.
[12] Jia J, Richards DJ, Pollard S, Tan Y, Rodriguez J, Visconti RP, et al. Engineering alginate as bioink for bioprinting. Acta Biomater 2014;10:4323–31.
[13] Zehnder T, Sarker B, Boccaccini AR, Detsch R. Evaluation of an alginate-gelatine crosslinked hydrogel for bioplotting. Biofabrication 2015;7. 025001.
[14] Axpe E, Oyen ML. Applications of alginate-based bioinks in 3D bioprinting. Int J Mol Sci 2016;17:1976.
[15] Daly AC, Critchley SE, Rencsok EM, Kelly DJ. A comparison of different bioinks for 3D bioprinting of fibrocartilage and hyaline cartilage. Biofabrication 2016;8. 045002.
[16] Köpf M, Duarte Campos DF, Blaeser A, Sen KS, Fischer H. A tailored three-dimensionally printable agarose-collagen blend allows encapsulation, spreading, and attachment of human umbilical artery smooth muscle cells. Biofabrication 2016;8. 025011.
[17] Kesti M, Eberhardt Ch, Pagliccia G, Kenkel D, Grande D, Boss A, et al. Bioprinting complex cartilaginous structures with clinically compliant biomaterials. Adv Funct Mater 2015;25:7406–17.

[18] Bacelar AH, Silva-Correia J, Oliveira JM, Reis RL. Recent progress in gellan gum hydrogels provided by functionalization strategies. J Mater Chem B 2016;4:6164–74.
[19] Kim YB, Lee H, Kim GH. Strategy to achieve highly porous/biocompatible macroscale cell blocks, using a collagen/genipin-bioink and an optimal 3D printing process. ACS Appl Mater Interfaces 2016;8:32230–40.
[20] Pati F, Jang J, Ha DH, Won Kim S, Rhie JW, Shim JH, et al. Printing three-dimensional tissue analogues with decellularized extracellular matrix bioink. Nat Commun 2014;5:3935.
[21] Lode A, Meyer M, Brüggemeier S, Paul B, Baltzer H, Schröpfer M, et al. Additive manufacturing of collagen scaffolds by three-dimensional plotting of highly viscous dispersions. Biofabrication 2016;8. 015015.
[22] Yue K, Trujillo-de Santiago G, Alvarez MM, Tamayol A, Annabi N, Khademhosseini A. Synthesis, properties, and biomedical applications of gelatin methacryloyl (GelMA) hydrogels. Biomaterials 2015;73:254–71.
[23] Billiet T, Gevaert E, De Schryver T, Cornelissen M, Dubruel P. The 3D printing of gelatin methacrylamide cell-laden tissue-engineered constructs with high cell viability. Biomaterials 2014;35:49–62.
[24] Cubo N, Garcia M, del Cañizo JF, Velasco D, Jorcano JL. 3D bioprinting of functional human skin: production and in vivo analysis. Biofabrication 2017;9. 015006.
[25] Schacht K, Jüngst T, Schweinlin M, Ewald A, Groll J, Scheibel Th. Biofabrication of cell-loaded 3D spider silk constructs. Angew Chem Int Ed 2015;54:2816–20.
[26] Das S, Pati F, Choi Y-J, Rijal G, Shim J-H, Kim SW, et al. Bioprintable, cell-laden silk fibroin–gelatin hydrogel supporting multilineage differentiation of stem cells for fabrication of three-dimensional tissue constructs. Acta Biomater 2015;11:233–46.
[27] Li C, Faulkner-Jones A, Dun AR, Jin J, Chen P, Xing Y, et al. Rapid formation of a supramolecular polypeptide-DNA hydrogel for in situ three-dimensional multilayer bioprinting. Angew Chem Int Ed 2015;54:3957–61.
[28] Schuurman W, Khristov V, Pot MW, van Weeren PR, Dhert WJ, Malda J. Bioprinting of hybrid tissue constructs with tailorable mechanical properties. Biofabrication 2011;3:21001.
[29] Akkineni AR, Ahlfeld T, Lode A, Gelinsky M. A versatile method for combining different biopolymers in a core/shell fashion by 3D plotting to achieve mechanically robust constructs. Biofabrication 2016;8. 045001.
[30] Kilian D, Ahlfeld T, Akkineni AR, Lode A, Gelinsky M. 3D bioprinting of volumetric tissues and organs. MRS Bull 2017; in press.

Further Reading

[1] Bruchet M, Melman A. Fabrication of patterned calcium cross-linked alginate hydrogel films and coatings through reductive cation exchange. Carbohydr Polym 2015;131:57–64.

Synthetic material bioinks

7

E. Abelardo
Swansea University Medical School, Swansea, United Kingdom

7.1 Introduction

Bioprinting is an exciting technology that propelled the developments in tissue engineering research in the recent decades. By combining the wealth of knowledge on cell physiology, materials science, and engineering principles, this emerging technology can potentially address the unmet needs of laboratory-engineered transplantable organs, 3D cell culture platforms, in vitro models for studying disease processes, drug delivery and screening, and so on.

The early forms of 3D architecture involved cell seeding after scaffold formation, which failed to create homogenous distribution of cells in the tissue constructs. The advent of 3D bioprinting allowed precise layer-by-layer deposition of cells, biologic cues, and scaffolds at desired location and timing. This spatiotemporal control of the assemblies attracted lots of researchers in regenerative medicine and pharmaceutics in the recent decade. So far, there are around 40 types of bioprinting strategies and fused deposition modeling, light-mediated stereolithography, inkjet printing, selective laser sintering, colorjet printing, and direct extrusion appear to be the most popular [1,2]. However, the bottleneck lies in the bioink, which is the polymer matrix, either naturally derived or synthetic, loaded with cells and other bioactive cues that mimic the extracellular matrix (ECM) and provides the appropriate microenvironment for cellular differentiation [3]. Hydrogels as cell-seeded microcarriers, for example, could have very low viscosity, which poses printing design challenge as fluid inks or could be very solid that breaks up on extrusion. The problem lies in adapting the traditional tissue engineering biomaterials in bioprinting processes as opposed to designing a specific bioink with desired characteristics fit for 3D printing [4]. The best strategy would be to develop a bioink that would address the needs of a specific cell and tissue lineage and conform to the design requirements of a particular bioprinting processor. The material scientists need to collaborate with bioengineers and cell physiologists for appropriate bioink design criteria. Fig. 7.1 illustrates the interlocking relations of bioink properties, bioprinting design, and cellular requirements to achieve the ideal 3D-printed tissue/organ constructs. This chapter elaborates on the desired characteristics of bioink and highlights the developments in the field of synthetic polymers as building blocks for 3D printing of connective tissue constructs.

7.2 Bioink properties

7.2.1 *Biocompatibility and biodegradability*

As the material will be transplanted in a patient's body, it is important that the scaffolds do not pose a risk to human health. To obtain regulatory approval, the

3D Bioprinting for Reconstructive Surgery. https://doi.org/10.1016/B978-0-08-101103-4.00009-0

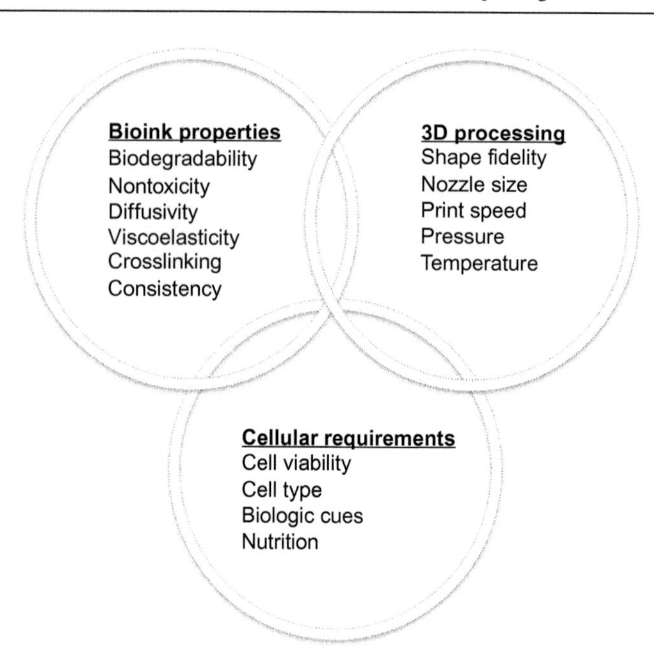

Fig. 7.1 Interlocking relations of bioink properties, bioprinting design, and cellular requirements to achieve the ideal 3D-printed tissue/organ constructs.

chemically defined biomaterials should show no, if not very minimal, immune response to host organ and have known mechanism and controllable rate of biodegradability. The engineered scaffolds are expected to break down and replaced by naturally produced ECM as cells undergo initial remodeling and be resorbed by the body. The native ECM is not the same in all tissues: the biochemical and mechanical cues vary in each organ.

7.2.2 Cross-linking and tensile modulus

Printing by using fluid bioink solution with very low viscosity is technically challenging. To overcome this obstacle, various cross-linking techniques were proposed such as photo-, ion-, electrostatic-, pH-, and temperature-based strategies [5]. Synthetic polymers have the advantage over natural products of being designed to contain cross-linkable components to form a covalently bonded network and a handle over cross-linking mechanisms. One example is the incorporation of methacrylate groups to poly(ethylene glycol) (PEG), which can be photocross-linked to form poly(ethylene glycol) dimethacrylate (PEGDMA) [6]. However, cross-linking is not limited to synthetic biomaterials. Alginate hydrogels are natural-based polymers that are ionically cross-linked with the help of a divalent cation Ca^{2+} and the gel strengths can be controlled by adjusting the concentration of $CaCl_2$ solution [7]. By varying the cross-linking interactions, the tensile modulus of the scaffolds is modified, which influences cellular behavior and stem cell fate via actin-mediated pathways. Other cross-linking methods employ high temperature and toxic reagents that subject the cells to harsh environment and are inappropriate to be used in bioprinting processes.

7.2.3 Viscosity and shear thinning

Ideally, the bioink must possess two phases: fluid-like substance while in the nozzle and solid phase after extrusion [5]. In rheological terms, some polymers possess shear thinning properties where the viscosity of the biomaterial decreases under shear stress and recovers quickly post extrusion thus supporting the solid phase of the scaffold formation [8]. The flow of the bioink through the dispensing tip is crucial. During extrusion, clogging and fracture can occur if the viscosity is too high or cell damage can occur if it is too low. The bioink should protect the encapsulated cells and at the same time support the bioprinting process. Gel-like materials provide more protection to dispersed and encapsulated cells than fluid-like bioinks.

7.2.4 Hydrophilicity

Constructs made up of hydrophobic polymers tend to contract while those with hydrophilic components swell. Water absorption and porosity influence the mechanical strength and nutrient and metabolic waste transport within the polymer matrix.

7.2.5 Cell-matrix interaction

Naturally derived polymers containing biological interaction sites such as adhesion molecules show improved cell growth and differentiation compared to synthetic scaffolds. This can be overcome by adding cell-binding moieties such as Arg-Gly-Asp (RGD) sequences in the construct, either in the backbone design or in blending the synthetic materials with natural polymers [9].

7.3 Synthetic polymers

Synthetic bioinks have the advantage of chemically defined building blocks, improved and tunable mechanical strength, controlled biodegradability, access to further chemical manipulation such as addition of cross-linkage sites or biomimetic molecules, and consistency as opposed to batch-by-batch variability occasionally in naturally derived polymers. However, only around 10% of polymers used in bioprinting are synthetic, which include PEG, poly(lactide-*co*-glycolide), poly(caprolactone) (PCL), and poly(L-lactic acid) [5]. Among these, only a few can be strictly considered as real bioinks because of their inability to encapsulate and incorporate cells in the construct and they mainly act as additive biomaterials. Furthermore, many printing processes such as sterolithography, selective laser sintering, and fused deposition modeling employ high temperature, powder beds, solvent baths, and high radiations, which make the conditions poor for cellular viability [2]. Inkjet and extrusion printing are the popular strategies that can print cell-laden constructs in physiologic environment.

7.3.1 PEG-based bioinks

PEG is a polymer of ethylene oxide (Fig. 7.2). It is FDA-approved that has medical, biological, chemical, industrial, and commercial uses. The polymer exhibits high hydrophilicity, biocompatibility, and very minimal immunogenicity, which make it an attractive material for 3D scaffold designs.

PEG on its own does not form hydrogel and needs chemical modification to form cross-linked strands and 3D architecture. It has a linear and branched structure with two hydroxyl groups that can be modified to functional groups such as acrylate, methyloxyl, azide, thiol, vinyl sulfone, carboxyl, acetylene, and amine for various applications [10]. Acrylation or methacrylation, resulting in PEG diacrylate (PEGDA) or PEG dimethylacrylate (PEGDMA), respectively, for example, paves way for cross-linking in the presence of an appropriate photoinitiator to form the desired three-dimensional network. The acrylate groups cross-link via free radical polymerization requiring between 1 and 10min of photoinduction, and the light intensity can be modulated to avoid harmful effects on cells. The essential conditions are physiological temperature and pH with high solubility to water thus making cell encapsulation with homogenous distribution feasible. PEG-based hydrogels are resistant to protein adsorption, which results in biologically inactive surfaces and where desired biofunctionality can be built. Other strategies for cross-linking include condensation, Michael-type addition, click chemistry, native chemical ligation, and enzymatic reaction [11]. Gel strengths can be tunable and several extrudable bioinks with a wide range of rheological properties have been reported by controlling the polymer contents, geometries (*branched*, *star*, *comb*), and functional groups [12]. Fig. 7.3 illustrates synthesis of four-armed PEG with different chain lengths.

PEG polymer is not degraded naturally but material scientists were able to incorporate degradable segments in the backbone to accelerate degradation. One strategy was adding polyhydroxyacids such as poly(lactic acid), poly(glycolic acid), and PCL, which are known to degrade by hydrolysis [13]. Also, thiol-acrylate reaction has been employed to enhance degradation through the ester bonds attached to PEG [14]. Another approach is to incorporate enzymatically cleavable sequences in the backbone strand, which makes the scaffolds matrix metallo-proteinase (MMP)- or elastase-sensitive [15].

One limitation of any synthetic polymer such as PEG-based hydrogels is the inherent absence of biologic attachments for cellular interaction. The native ECM provides not only structural integrity but also biologic feedback that influences stem cell fate and differentiation, including further production of ECM substrates. Owing to the robust and versatile properties of PEG, bioactive molecules can be tethered to its strand to mimic cell-matrix adhesions and carry signaling cues. Short peptide sequences that

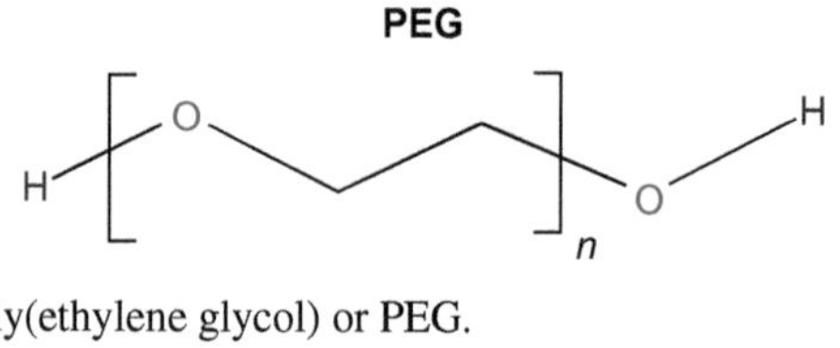

PEG

H O O H n

Fig. 7.2 Subunit of poly(ethylene glycol) or PEG.

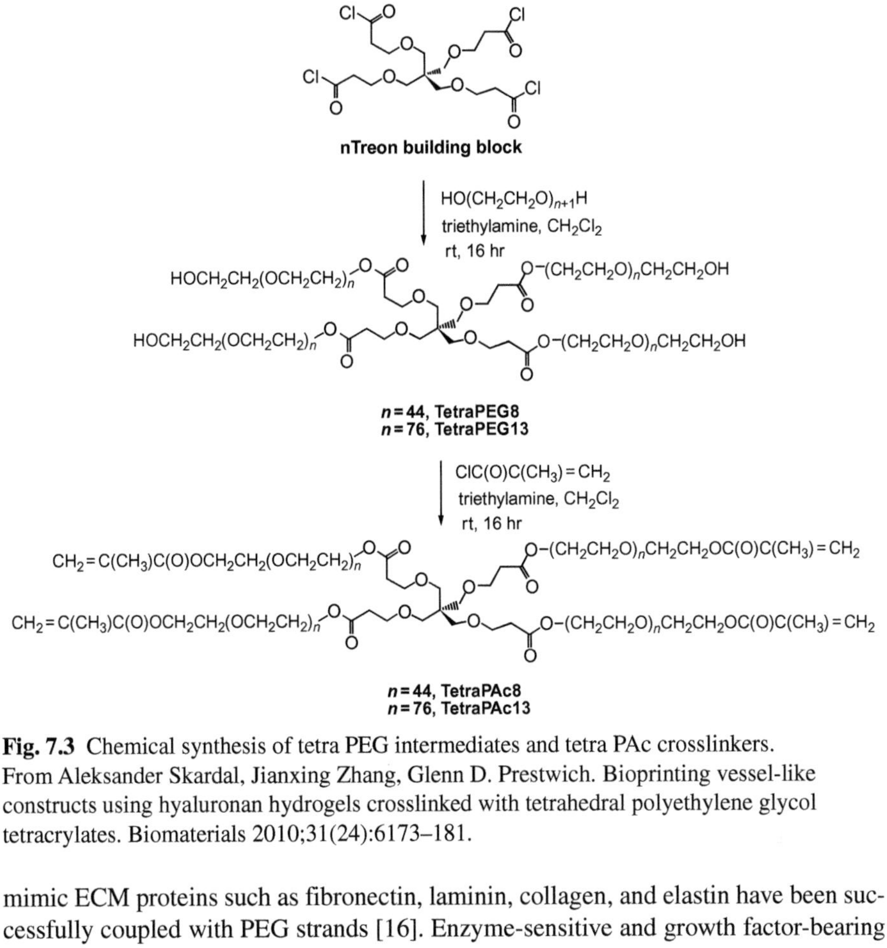

nTreon building block

$HO(CH_2CH_2O)_{n+1}H$
triethylamine, CH_2Cl_2
rt, 16 hr

$n = 44$, TetraPEG8
$n = 76$, TetraPEG13

$ClC(O)C(CH_3) = CH_2$
triethylamine, CH_2Cl_2
rt, 16 hr

$n = 44$, TetraPAc8
$n = 76$, TetraPAc13

Fig. 7.3 Chemical synthesis of tetra PEG intermediates and tetra PAc crosslinkers. From Aleksander Skardal, Jianxing Zhang, Glenn D. Prestwich. Bioprinting vessel-like constructs using hyaluronan hydrogels crosslinked with tetrahedral polyethylene glycol tetracrylates. Biomaterials 2010;31(24):6173–181.

mimic ECM proteins such as fibronectin, laminin, collagen, and elastin have been successfully coupled with PEG strands [16]. Enzyme-sensitive and growth factor-bearing PEG hydrogels have also been reported [11].

7.3.2 Polyhydroxy acid/polyester-based bioinks

Polyhydroxy acids such as poly(lactic-glycolic acid) (PLGA) and poly(L-lactic acid) (PLLA) and polyesters such as PCL are well-established biodegradable and biocompatible synthetic polymers that have been long used in medicine as sutures, implants, drug delivery system, and biofabrication (Fig. 7.3). Their thermoplastic properties and improved mechanical strength are advantageous for bioprinting. Their rate of degradation by hydrolysis is also well understood and controllable. However, their melting temperature at around 60°C limits the use of these polymers as cell-laden bioink as this high temperature compromises cell viability. They have been used for 3D bioprinting but the cells are seeded postextrusion similar to the first generation 3D scaffolds. Furthermore, the high viscosity of PCL necessitates extremely high extrusion pressure, which goes beyond practical limits [5]. These polymers are then not technically bioinks but more as additive bioprinting materials (Fig. 7.4).

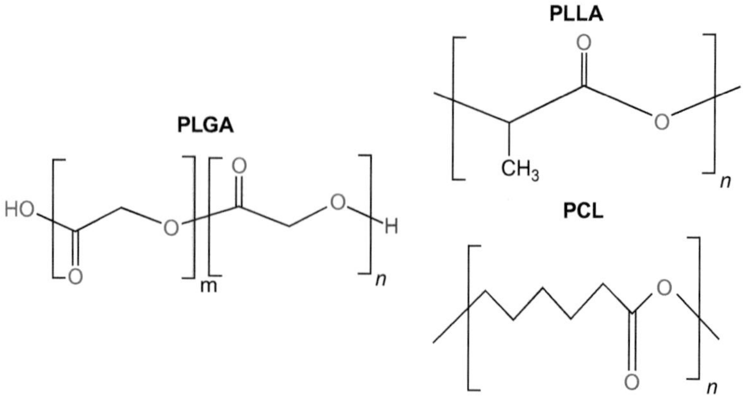

Fig. 7.4 Polyesters used as bioinks. (PLGA, poly(lactic-glycolic acid); PLLA, poly(L-lactic acid); PCL, poly(caprolactone)).

7.3.3 Pluronic acid-based bioinks

Pluronic acid is a triblock copolymer of poly(ethylene glycol)-poly(propylene oxide)-poly(propylene glycol) and has similar properties like PEG as discussed previously. This copolymer has been used as a 3D bioink in omnidirectional printing to mimic microvascular network with the main advantage of having excellent resolution of the printed construct [17]. The weakness of pluronic acid-based hydrogels is their poor mechanical strength despite cross-linking strategies using methylacrylate similar to PEGMDA and poor cell viability due to the lack of cell binding affinity [18,19].

7.3.4 Polyurethane-based bioinks

Water-based polyurethane (PU) has been recently reported as bioinks for cartilage tissue engineering [20]. Similar to PLLA, PLGA, and PCL, PU has good biocompatibility and biodegradability can be engineered by tethering with hydrolysis-prone segments. Unlike polyhydroxy acids and polyesters, PU elastomer is processed in aqueous media and has higher mechanical strength. The main components of PU are PCL diol and a second oligodiol containing amphiphilic blocks [21]. The nanoparticles are dispersed in water by incorporating ionic hydrophilic groups to the hydrophobic backbone. For this specific study by Hung et al., the so-called soft segment composition was designed to match the mechanical properties and degradation rate appropriate for cartilage studies. However, the PU-based bioink did not contain cells and seeding was performed after scaffold processing, which falls short to the strict definition of cell-laden bioink. But another study reported success of neural stem cell-laden thermoresponsive biodegradable PU-based bioink dispersion before gelation [22]. Furthermore, the PU-based construct seems to rescue nerve damage and promote neural reinnervation in in vivo spinal cord injury in a zebrafish model.

7.4 Summary

3D bioprinting is an emerging and promising technology in regenerative medicine, especially on organ fabrication and drug screening. It involves layer-by-layer construction, with high precision and timing, of building blocks, cells, and biologic signals. Synthetic polymers such as bioink provide good structural support and mechanical cues of the tissue construct. They also enjoy the advantage of being chemically defined, high biocompatibility, known and controllable biodegradability, absence of batch-by-batch variability, tunable rheologic properties, and easy handling. The main disadvantage, however, lies in the inherent lack of cell adhesion sites, which compromises cellular viability although strategies such as tethering of biomimetics have been employed. Furthermore, the high temperature involved in synthetic polymer processing creates a harsh environment to cells; thus, these biomaterials act only as additive biomaterials and not necessarily a cell-laden bioink. The real challenge lies in improving bioink printability without compromising cellular viability. This can only be achieved through a concerted collaboration among material scientists, engineers, and cell physiologists.

References

[1] An J, Teoh JEM, Suntornnond R, Chua CK. Design and 3D printing of scaffolds and tissues. Engineering 2015;1(2):261–8.
[2] Panwar A, Tan LP. Current status of bioinks for micro-extrusion-based 3D bioprinting. Molecules 2016;21(6). pii: E685.
[3] Skardal A, Devarasetty M, Kang HW, et al. A hydrogel bioink toolkit for mimicking native tissue biochemical and mechanical properties in bioprinted tissue constructs. Acta Biomater 2015;25:24–34.
[4] Skardal A, Atala A. Biomaterials for integration with 3-D bioprinting. Ann Biomed Eng 2015;43(3):730–46.
[5] Carrow J, Kerativitayanan P, Jaishwal M, et al. Polymers for bioprinting. In: Atala A, Yoo J, editors. Essentials of 3D biofabrication and translation. 1st ed. London: Academic Press; 2015. p. 229–48.
[6] Killion JA, Geever LM, Devine DM, et al. Mechanical properties and thermal behaviour of PEGDMA hydrogels for potential bone regeneration application. J Mech Behav Biomed Mater 2011;4(7):1219–27.
[7] Augst AD, Kong HJ, Mooney DJ. Alginate hydrogels as biomaterials. Macromol Biosci 2006;6(8):623–33.
[8] Skardal A, Zhang J, McCoard L, Xu X, Oottamasathien S, Prestwich GD. Photocrosslinkable hyaluronan-gelatin hydrogels for two-step bioprinting. Tissue Eng Part A 2010;16(8):2675–85.
[9] Daly AC, Critchley SE, Rencsok EM, Kelly DJ. A comparison of different bioinks for 3D bioprinting of fibrocartilage and hyaline cartilage. Biofabrication 2016;8(4). 045002.
[10] Peppas NA, Keys KB, Torres-Lugo M, Lowman AM. Poly(ethylene glycol)-containing hydrogels in drug delivery. J Control Release 1999;62:81–7.
[11] Zhu J. Bioactive modification of poly(ethylene glycol) hydrogels for tissue engineering. Biomaterials 2010;31(17):4639–56.

[12] Skardal A, Devarasetty M, Kang HW, et al. Bioprinting cellularized constructs using a tissue-specific hydrogel bioink. J Vis Exp 2016;(110). e53606.
[13] Sawhney AS, Pathak CP, Hubbell JA. Bioerodible hydrogels based on photopolymerized poly(ethylene glycol)-co-poly(α-hydroxy acid) diacrylate macromers. Macromolecules 1993;26:581–7.
[14] Jo YS, Gantz J, Hubbell JA, Lutolf MP. Tailoring hydrogel degradation and drug release via neighboring amino acid controlled ester hydrolysis. Soft Matter 2009;5:440–6.
[15] Mann BK, Gobin AS, Tsai AT, Schmedlen RH, West JL. Smooth muscle cell growth in photopolymerized hydrogels with cell adhesion and proteolytically degradable domains: synthetic ECM analogs for tissue engineering. Biomaterials 2001;22:3045–51.
[16] Hersel U, Dahmen C, Kessler H. RGD modified polymers: biomaterials for stimulated cell adhesion and beyond. Biomaterials 2003;24:4385–415.
[17] Wu W, DeConinck A, Lewis JA. Omnidirectional printing of 3D microvascular networks. Adv Mater 2011;23(24):H178–83.
[18] Müller M, Becher J, Schnabelrauch M, Zenobi-Wong M. Nanostructured pluronic hydrogels as bioinks for 3D bioprinting. Biofabrication 2015;7(3). 035006.
[19] Smith CM, Stone AL, Parkhill RL, et al. Three-dimensional bioassembly tool for generating viable tissue-engineered constructs. Tissue Eng 2004;10(9-10):1566–76.
[20] Hung KC, Tseng CS, Hsu SH. Synthesis and 3D printing of biodegradable polyurethane elastomer by a water-based process for cartilage tissue engineering applications. Adv Healthc Mater 2014;3(10):1578–87.
[21] Tsai YC, Li S, Hu SG, et al. Synthesis of thermoresponsive amphiphilic polyurethane gel as a new cell printing material near body temperature. ACS Appl Mater Interfaces 2015;7(50):27613–23.
[22] Hsieh FY, Lin HH, Hsu SH. 3D bioprinting of neural stem cell-laden thermoresponsive biodegradable polyurethane hydrogel and potential in central nervous system repair. Biomaterials 2015;71:48–57.

Candidate bioinks for 3D bioprinting soft tissue

S.P. Tarassoli, Z.M. Jessop*,‡, S. Kyle†, I.S. Whitaker†,‡*
*Swansea University Medical School, Swansea, United Kingdom, †Reconstructive Surgery and Regenerative Medicine Research Group, Swansea University Medical School, Swansea, United Kingdom, ‡The Welsh Centre for Burns and Plastic Surgery, Morrison Hospital, Swansea, United Kingdom

8.1 Introduction

Ideality is a sought-after concept in reconstructive surgery. Tissue grafts must meet the immunological and spatial demands of the surgical process or they are likely to fail, leading to detrimental consequences for the patient. Thus, it is imperative that bioinks fulfill the criteria laid out for them during their development and use.

Bioprinting is defined as the precise and accurate deposition of biomaterials within a predesigned space using a computer-aided printer. The process was first described in 1988 [1] as "cytoscribing," which involved depositing colored ink (which are similar in size to biologic materials) inspired by the two-dimensional paper printers that preceded them. Since then, biological materials have been used and increasingly complex printing methods developed to further the potential of the technology [2]. These printable biomaterials are known as bioinks. They can mimic the extracellular matrix (ECM) environment to support cell adhesion, differentiation, and proliferation. These bioinks are dissimilar to other inks used in additive manufacturing (or 3D printing), as they are printed at much lower temperatures, are typically derived naturally, and have mild cross-linking conditions to preserve the cells while preventing unwanted degradation of the biomolecules [3].

The development and discovery of bioinks has had a staggered past; there has been no natural progression of the technique (see Fig. 8.1 [4–12]). The initial advance of bioinks was either inspired or completely derived from the composition of existing natural-based polymers (such as alginate or gelatin). These bioinks are still commonplace in modern practice and some consider them to be the gold standard. There are currently several approaches using chemical synthesis that are seemingly "trial and error" in the form of smart polymers: the advantages of synthetic materials combined with the biological activity of proteins [13]. Although these methods can yield good biomaterials, they are not necessarily compatible with the bioprinters or have poor "printability" characteristics.

8.2 Bioink construct types

Bioinks can be divided into two general types: scaffold-based or scaffold-free. These types are not related to the material that they are made of, but the way they are printed and the cells loaded. As the name suggests, the cells can either be loaded with the aid

3D Bioprinting for Reconstructive Surgery. https://doi.org/10.1016/B978-0-08-101103-4.00026-0

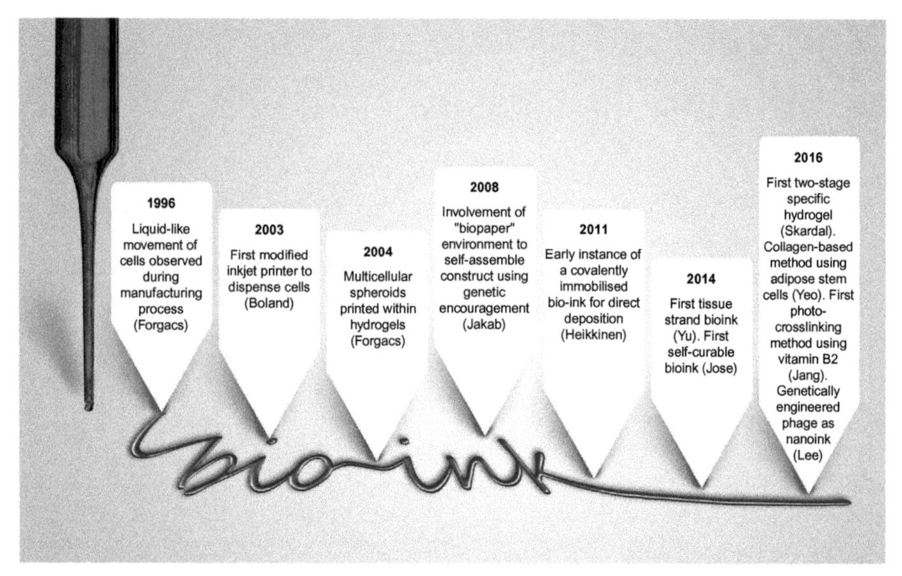

Fig. 8.1 Timeline of bioink development methodology.

of a scaffold (to facilitate tissue formation by allowing cell growth and proliferation) and then printed in the required 3D shape or in a specific pattern allowing them to mature into larger-scale tissues [14] or without the aid of a scaffold at all. These two categories are then subdivided by the derivation of the material that is used, i.e. natural or synthetic (discussed in Section 8.3).

8.2.1 Scaffold-based designs

The aim of the scaffold is seemingly simple on the surface yet remains difficult to master: mimicry of the ECM. This has been attempted using several different approaches. This is either by recreating the cross-linking environment while allowing large volumes of cell encapsulation (hydrogels), designing structures from the ground up with specific porosities (microcarriers) or removing the cells from extracted ECM (decellularized matrices). These methods, although individually have their merits, also have limitations and thus are not suitable for all tissue engineering uses.

8.2.1.1 Hydrogels

Hydrogels are a group of polymers that have a high affinity for water retention and absorption. Due to their versatility they are used in modern medical applications ranging from vision correction [15] to controlled pharmaceutical release [16]. Their high affinity has demonstrated them to take up more than 500 times their original weight [17], which not only has good connotations for cell encapsulation, but also implies a strong mechanical 3D environment and profuse permeability to nutrients for cell growth encouragement. Hydrogels have also been discovered to have unique 3D "niches" available to loaded cells that mimic the environment by promoting mineralization (by the aid of added chemical groups such as phosphate) resulting in enhanced cell-matrix interactions [18].

Despite all these favorable features for tissue engineering, hydrogels are not necessarily "bioprintable." This depends on not only the method of printing and bioprinter used, but the rheological and mechanical attributes of the hydrogel. Hydrogels that are shear thinning, that is to say that their viscosity decreases under shear strain, are deemed to be more suitable for extrusion-based printing methods. The high pneumatic (or mechanical, depending on the bioprinter) pressure that the hydrogel undergoes allows it to reduce its viscosity during bioprinting and recover stability once fully deposited. This principle also applies to droplet-based bioprinting. The hydrogel needs to be free of fibrous consistency to prevent it congesting the printing system. In addition, the bioink should have an optimal surface tension for the bioprinter pressure to allow it to be manipulated at a droplet level. The hydrogels used for laser-based bioprinting require different characteristics and thus should be designed in that fashion. The bioink should have a high affinity for thermal energy transference while exhibiting a sufficient adhesion for uniform layer thickness. If the method includes a photopolymerization step, then the hydrogel should be enhanced with photosensitizers and absorbers [19].

As well as the large biological capacity potential of a hydrogel, its cross-linking ability is equally appealing in the tissue engineering field. The customizability of the cross-linking could allow for increased elasticity of the hydrogel, changes in viscosity and solubility all while increasing its overall toughness [20]. This adds a further element of modification to the development and design of the bioink. Hydrogels can be cross-linked physically, chemically, or enzymatically. Once again, this would be dependent on the candidacy of the bioink and the bioprinting method. Physical cross-linking (involving ionic, hydrogen, and electrostatic bonding interactions) can be adjusted to create temperature-sensitive rheology [21]. Chemical cross-linking, involving covalent polymer bonding, can provide better mechanical stability, but, unlike physical cross-linking, uses exogenous agents that may lead to cytotoxicity of the hydrogel or undesirable effects such as reduced growth factor diffusion [22]. Photopolymerization (a subtype of chemical cross-linking), is a way of circumventing the cytotoxic aspect (when kept under a certain concentration [23]) of chemical cross-linking by using focused radiation. However, this method is currently limited to low-molecular-weight monomers and there are only a few photoinitiators capable of this in circulation [24]. Enzymatic, the most popular being fibrin [25], cross-linked hydrogels are becoming more common in 3D bioprinting practice due to their biocompatibility and nontoxicity. However, the lack of viable options (such as photopolymerization) has stunted its ability to become a leading methodology of cross-linking.

There are some limitations associated with using hydrogels in a printing capacity. Hydrogels, unlike cell aggregates and decellularized designs, do not contain ECM elements and cannot replicate the native environment accurately. In addition, as they encapsulate the cells, the growth of the construct is limited by the size of the hydrogels and a high density of cells cannot be achieved; if the hydrogel concentration is increased to encapsulate more cells, its biological ability decreases as it limits the mobility of the cells. Hydrogels are also known to degrade slower than other options and must be designed specifically to degrade at a certain rate. Their degradation products can also be toxic to cells, but research [26] has demonstrated that this can be removed in the fabrication process.

8.2.1.2 Microcarriers

Scaffolds have inherently been used for their structural support, and though they may bring other features to the fold such as growth facilitation, they are also used for mechanical stability. Microcarriers are specialized "reinforcement blocks" for bioprinting [27]. They have a spheroid architecture with interlaced porosity allowing for cells to proliferate and culture within expansive layers [28]. They can also increase the efficiency of nutrient diffusion due to these specifically engineered pores [29].

This focused structural support they provide has been demonstrated by the improvement of compressive moduli while preserving "phenotypic stability at high concentrations." Microcarriers are typically loaded onto hydrogels (to increase the cell loading capacity further) and then seeded. Studies have shown that the microcarrier-hydrogel hybrids exhibit higher differentiation and cell-matrix interaction levels in comparison with solely bioprinted hydrogels [30].

8.2.1.3 Decellularized matrices

Although mimicry of the ECM is reaching more accurate milestones every day, using actual ECM components remains the most "truthful" approach. Bioinks based on removing all cellular components of extracted ECM and seeding with desired cells have been used in bioprinting practices for several years now [31], but successful (determined to be ~98% removal) decellularization still eludes much of the research conducted in this field [32]. However, of the construct types used in bioprinting, these matrices are shown to have the highest cell viability exhibition and lowest cytotoxicity range. Decellularized ECM can also be manipulated to increase its structural viability if the base is known to lose integrity after being printed [33]. Once seeded, cells have been found to degrade the ECM due to the production of matrix metalloproteinases [34]. The major drawback of this method is the extraction of the ECM itself. It is a complex process and the risk of insufficient procurement is a high one, making it the most expensive technique used in tissue engineering.

8.2.2 Cell aggregate-based constructs

The fabrication of tissues relies on cells assembling themselves within the required 3D space. The development of scaffolds have produced good results, but a perfect mimicry of their "native counterpart" is yet to be achieved [35]. Time will tell if this is a feasible goal, but advances have been made to form tissues from cells organized in specific patterns with the help of genetic and growth factor driving forces [36]. These bioinks are categorized by the shape in which they are printed which also, in turn, have a bearing on their suitability to the bioprinter used.

Spherical tissues are the most common scaffold-free bioink used. They involve the cells to be organized preprinting in a spheroid shape (up to 500 μm) [37]. This is typically done by microwell culturing on some form of cell adhesion mold. Recent endeavors in the field have replaced the manual aspect of this by designing robotic models to fabricate the spheroids [38]. Hydrogels have been a popular choice for this mold due to their bioinertness [39]. Other than microwell deposition, self-assembly driven by gravity, also known as the "hanging-drop method" [40], is one of the other approaches used to instigate cell-cell binding [41]. Once the spheroids have been fabricated, they are loaded as a bioink into a printer and extruded (the only tried and tested bioprinting method for spheroids [42]) onto another biocompatible mold. As the tissue develops and grows, the mold is then removed. Recent work has shown that a needle array can be used in place of the mold support structure [43]. One drawback of tissue spheroids is that they are prone to necrosis at the core, but this can be counteracted if they were fabricated to have a lumen to allow sufficient vascularization [44]. A modified version of the spheroid, the cell pellet, is also used in the 3D bioprinting field. A cell suspension (after significant centrifugation [45]) is collected after aggregation and extruded within a mold. This approach has been popular for the development of nerve constructs [46] and in the maxillofacial field [47]. The major limitation of using the cell pellet is the need for a temporary mold, meaning that the size of tissues is restricted [48] to a certain shape.

Tissue strands are emerging as a good option of a scaffold-free bioink. They are cylindrical, that is strand-like, in shape and have been explicitly designed and engineered for large scalability. It is the most novel of the approaches and involves cells packed at a high density into biocompatible tubes [49]. The tubes are characteristically semipermeable to allow for oxygen and nutrient diffusion. Like spheroid and pellets, the strands can only be extruded mechanically, and if they have not matured optimally they will not be suitable for printing. One drawback of this method is the hypoxic conditions of the core strands that is, those within the center of the tube, but studies have suggested that this would tailor this method toward the cartilage tissue engineering (and the known hypoxic conditions associated with chondrocytes [50]). The major limitation of utilizing tissue strands is that the loading of each strand must be done manually and thus the bioprinting parameters need to be optimized for each one, showing there is clear lack of automation in the process.

8.3 Bioink constituents

Bioinks can be grossly characterized as being nature-derived or synthetic, determined by their derivation or development. Nature-derived bioinks have been classically preferred due to their superior microenvironment mimicry, but as synthetic bioink development has progressed, tunability (a key feature of synthetic bioinks) has become more appealing. Clearly, there is overlap between the two and the use of blends has blurred the initial well-defined classifications in the field. However, the candidacy of a bioink is dependent on the tissue type that they are required to replace.

Bioinks are additionally divided based on their use. The archetypal bioink (used for cell encapsulation) is known as a "matrix bioink." There are also "curing bioinks," also known as photoinitiators, which interact with matrix bioinks to undergo photopolymerization to allow for temporal and spatial control and tunability of the overall bioink gelation. Sacrificial bioinks offer temporary support and are removed once printed. The permanent version of this is known as a support bioink. They should not be confused with matrix bioinks, as the support bioinks are not able to encapsulate cells due to the high temperature they are printed at.

8.3.1 Matrix bioinks

8.3.1.1 Alginate

Alginate is a natural anionic polysaccharide first found in the cell walls of brown algae [51]. Although it has a hydrophilic surface (and therefore unfavorable for cell attachment [52]), it can been modified to provide additional sites for the cells to attach. It is widely used, if not the most used, in bioprinting as it is easily to manipulate, has a high biocompatibility index, versatile cross-linking, and cheap to produce on a large scale [53]. It is mostly used as a matrix bioink due to this, and is accepted as a good option for cell encapsulation without the need for sacrificial material [54]. In addition, its gelation occurs in seconds and requires chemical encouragement. Not only does alginate provide adequate conditions for cell storage, but a good environment for proliferation and differentiation as well [55].

In the bioprinting world, extrusion-based approaches are the most common for alginate. Some would even argue it is the most used combination in the field [56]. This is due to the several significant benefits that are present when using extrusion-based bioprinting with alginate: ease of cross-linking, manipulation of the bioink concentration, stability in a 3D shape, and the ability to print alginate in more than one form [57]. However, studies have shown that viable printing depends on a good balance of the alginate concentration, wettability, and predetermined biolayer thickness [58]. The concentration of the alginate is one of the most important factors influencing the printability and the final product. This is because it affects a combination of the porosity, cross-linking time, and the viscosity. This has connotations for the mechanical aspect of the printing such as clogging the extruder and ultimately distorting the biological printed outcome. Alginate's versatility also allows it to be blended with other biomaterials to accentuate certain factors. Soft tissue constructs were favored when a fluorocarbon solution of alginate and agarose was fabricated [59]. Inkjet printing has also been used to produce similar products [60]. Recent advances have additionally shown alginate to be printed successfully using a modified laser printer to print directly rather than on a glass transfer plate [61].

8.3.1.2 Chitosan

Chitosan is a linear polysaccharide derived from the chitin shells of crustaceans. It is predominantly used for soft tissue engineering, specifically cartilage [62]. It has been used with extrusion-based printers to create vascular channels in conjunction with an ionic cross-linking mechanism. Laser deposition has also been attempted with limited success [63]. Furthermore, chitosan has been printed via an extrusion method, but not largely for soft tissue engineering [64]. Although chitosan is a promising avenue for tissue engineering and cell encapsulation, a limitation of the material is its structural insufficiency for large-scale scaffolds [65]. This could, perhaps, be resolved in further work in a blend design.

8.3.1.3 Collagen (type 1)

Collagen is a helical protein attained from nature and is a major component of connective tissue. With alginate being possibly the most commonly used bioink in bioprinting methodology, collagen is regarded to be the next most extensively used substance, typically for bone tissue engineering [66], but a bilayer skin graft has also been designed with collagen [67]. Due to its biocompatibility, collagen matrices are ideal for both cell adhesion and growth [68]. Collagen has been used as a sole bioink in extrusion bioprinting, but studies have found blending it with a synthetic can produce favorable results, that is, increase proliferation and growth [69,70]. Cross-linking has also been found to be a benefit when using collagen, though this has typically been demonstrated with droplet-based endeavors [71]. Valve-based methods have also been carried out due to collagen's fibrous microarchitecture, but only as a blend with another natural-based protein to culture skin [72]. Laser-based bioprinting has additionally been done successfully to print skin and collagen's adhesive properties were stated to make the plate transference easier [73]. However, there are some disadvantages of using collagen as a bioink. It has a comparatively longer gelation time (up to an hour at 37°C [74]). This makes printing accurate and precise 3D constructs difficult, as the loaded cells' positions are skewed by gravity before the deposition collagen gelates.

8.3.1.4 Fibrin

Fibrin is a matrix hydrogel derived from the enzymatic reaction of fibrinogen and thrombin [75]. It comprises of a network of soft filaments that can withstand a high level of distortion [76]. This allows them to be loaded with cocultures of cell and still exhibit angiogenic behavior [77]. Fibrin has been the target of a range of bioprinting studies. Unfortunately, once cross-linked, fibrin has relatively weak mechanical properties and therefore is not ideal for extrusion-based bioprinting. However, due to the two components of fibrin (fibrinogen and thrombin), droplet-based printing would be more suitable to maximize the probability of a successful scaffold; microvalve methodology has been especially worked on [58]. Considering its biological and chemical composition, one would assume that fibrin is only suitable for designing vascular and connective tissue structures, but viable neural tissue has also been constructed [78]. However, fibrin has its disadvantages. The degradation time of fibrin is quick and thus not ideal for long-term culture of cells. In addition, it is problematic to create an accurate shape due to difficulty in predicting the required thrombin concentration. The lengthy cross-linking duration means that it has a weaker mechanical strength and consequently unsuitable for laser-based bioprinting [79].

8.3.1.5 Hyaluronic acid

Hyaluronic acid is a linear glycosaminoglycan found in abundance in the ECM of cartilage [80], making it one of the most biocompatible bioinks. It is commonly used in tissue engineering methods due to its ability to form versatile hydrogels as a matrix bioink. This is generally done by photopolymerization, but chemical modifications have also been attempted successfully [81]. Hyaluronic acid can be easily manipulated so that any unfavorable characteristics can be resolved with chemical modification. In extrusion-based bioprinting, blending has been used to enhance factors such as solidification time and printability; one group used poly(ethylene glycol) to produce vascular structures [82]. Self-assembly has also been demonstrated using a thiol-modified hydrogel using gold nanoparticles [83]. Due to hyaluronic acid's viscous nature, droplet-based printing is not possible, and there are no ongoing studies. Laser-based bioprinting, however, has been used with hyaluronic acid with a blend to increase mechanical strength and cross-linking time [79].

8.3.1.6 Matrigel

Matrigel is a cell- and tissue-derived substance from natural ECM used as a matrix bioink [84]. It has been shown to encourage cell growth from various cell types and tissue fragments and demonstrated adequate vascularization [85]. Its versatility allows it to observe the growth of complex cells not possible on other 3D surfaces such as the sprouting of capillaries or reaction to radioactive factors [86]. Studies [87] have shown that its mechanical ability allows Matrigel to create constructs with super cell viability numbers compared with other conventional bioinks. The gelation time of Matrigel is rapid and found to be thermally reversible, allowing conservative cross-linking times with the use of a heated plate. Matrigel has been used in extrusion-based bioprinting with endothelial cells (both mammary [86] and progenitor [88]). Matrigel has been utilized as a matrix bioink in droplet-based bioprinting, but as a pseudo-sacrificial bioink [89]; the substrate was deposited to help placement and patterning of the cells rather than promoting growth. Its thermal properties and mechanical strength have made Matrigel a popular option for laser-based bioprinting and has been the target of research involving a wide range of cells such as myoblasts, ductal carcinoma cells, neural stem cells, and fibroblasts [90]. Both matrix-assisted [91] and forward transfer [92] laser printing have been attempted with Matrigel.

8.3.1.7 Poly(ethylene glycol)

Poly(ethylene glycol) (PEG) is a synthetic hydrophilic compound (in hydrogel form) that can conjugate with biological molecules [93,94] used for matrix bioink purposes. It has been classically used as a drug delivery stabilizing molecule [95] and for biosignaling [96]. Due to its hydrophilicity, it can withstand protein adsorption that allows cells that do not require biological encouragement such as osteoblasts, to survive well when encapsulated within PEG [97]. Studies have shown that if additional promotion is required, then precoating PEG can assist this [98]. This resistance to protein adsorption makes PEG a good option for tissue construct bioprinting, and its water solubility allows it to be manipulated easily to add more suitable characteristics [99]; photo-cross-linking with the addition of diacrylate or methacrylate is a popular alteration to produce mechanical and functional adjustments [100]. PEG has been used in all bioprinting modalities. Extrusion-based methods have produced successful aortic valve scaffolds [101] as a matrix bioink as well as is being used as a cross-linker with hyaluronic acid and gelatin to create various tissue constructs [102]. PEG's high mechanical stiffness has caused it to be used widely in droplet-based bioprinting; thermal inkjet printers are classically used in a layer-by-layer fashion [103] in a wide range of studies ranging from producing neocartilage [104,105] to suspending bone marrow-derived human mesenchymal stem cells with bioactive glass [99]. Laser-based bioprinting with PEG was first demonstrated in 2004 [106] and the methodology has been honed over time to use fibroblasts [23], HeLA cells [107], and many others.

8.3.1.8 Miscellaneous matrix bioinks

There are a few other naturally occurring matrix bioinks (Fig. 8.2 [108–114]) that have been used in current techniques, but are not common due to either a lack of viable characteristics or difficulty extracting from the source.

8.3.2 Sacrificial bioinks

Sacrificial bioinks are designed to provide temporary support to the cells. They are printed to create complex shapes and, once cultured, are removed by chemical washing or thermal manipulation. They are best suited for the development of vasculature within a tissue [115]. Sacrificial bioinks typically offer high cytocompatibility and printability with an easy removal method.

8.3.2.1 Agarose

Agarose is a natural polysaccharide, typically used in a hydrogel form that is predominantly a sacrificial bioink [2,116]. Comparatively low temperatures (40°C average) are required of its gelation [117] and maintains its shape once it has undergone the process (despite its brittleness in a solid form [118]). However, some of agarose's bioink counterparts (such as alginate) are more biocompatible with cells [119]. Agarose is not a popular choice in the bioprinting world as it cannot be used in droplet-based bioprinting due to its viscosity plugging the nozzle [120], and it needs temperature control in microvalve printing. It has, however, been used successfully in extrusion-based

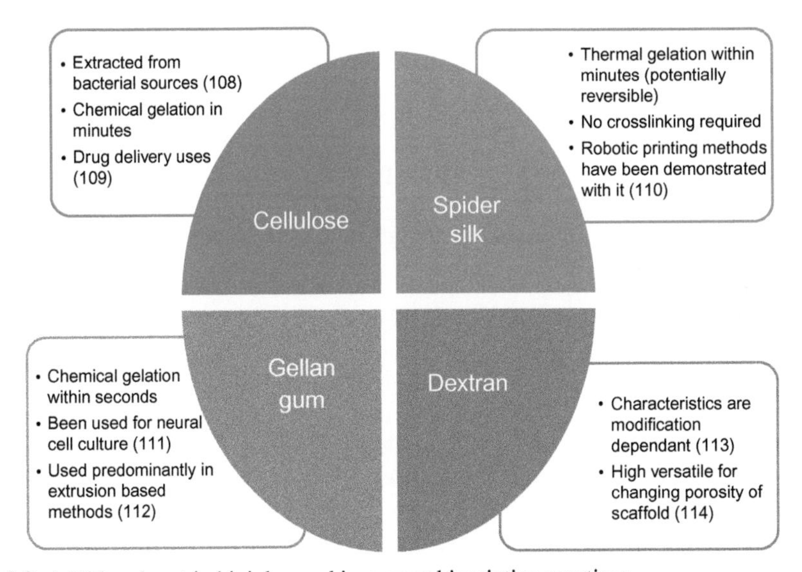

Fig. 8.2 Additional matrix bioinks used in current bioprinting practices.

printing methods and laser bioprinting [121]. Some of the flaws of agarose can be addressed by blending it with another hydrogel that exhibit some of the features it lacks, such as collagen for increased cell viability [122].

8.3.2.2 Gelatin

Gelatin is a denatured protein derived from collagen and obtained from bone and connective tissue [123]. It forms a gel-like state when at low temperatures, but reverts to its "coil confirmation" when the temperature increases [124] and causes complete dissolution. This can be counteracted with chemical cross-linking, but some studies have shown that this results in cytotoxicity for the constructs [125]. Gelatin has been found to be highly biocompatible and keeps cells viability for extensive periods of time, but, curiously, does not encourage cell elongation. In the bioprinting sense, gelatin is classically used as a sacrificial bioink because of its poor mechanical properties (and even then, it is typically cross-linked or designed as a blend). Gelatin has been extruded successfully and loaded with hepatocytes in the desired shape [126]. Very little evidence is available for using gelatin in droplet-based printing [127], but due to its viscoelastic properties it has been used in laser-based bioprinting with favorable results [128,129].

There is a modified version of gelatin also used in bioprinting techniques. It consists of methacrylate group conjugation and allows for adjustable mechanical and functional characteristics [130]. Due to the production steps required to create it, it is technically known as a synthetic hydrogel, but its derivation is from natural sources. Methacrylated gelatin (GelMA) is predominantly a matrix bioink despite its low cell proliferation rates, but research has ameliorated this by blending it with more cell viable bioinks or additional cross-linking [131]. The majority of work conducted has used endothelial cells [132], but GelMA is known to have a high biocompatibility, thus is not considered to be limited by the cell type it can interact with. GelMA has been used extensively in extrusion-based bioprinting [133,134], especially for cartilage tissue engineering [135]. Due to its low room temperature viscosity and modified cross-linking and gelation time, the extrusion of GelMA is easy. Droplet-based printing has also been used to recreate the fibrocartilage microenvironment using various growth factors into a precise and accurate spatial pattern [136]. GelMA is commonly cross-linked with photoinitiators that allows to retain its biocompatibility with minimal cytotoxicity [24]. GelMA's mechanical strength additionally allows it to be printed using laser deposition methods [137], although some studies have shown the cells to be damaged by the laser [19].

8.3.2.3 Pluronic F-127

Pluronic is a synthetically derived polymer comprising hydrophilic and hydrophobic blocks, typically used as a sacrificial bioink. As temperature increases, it reverses its gelation while also cross-linking [138]. Pluronic, as a sole construct, has been shown to erode quickly within several hours, but when blended it has been used as a controlled drug release scaffold [139] due to its surfactant properties (made possible by the combination of hydrophilic and hydrophobic blocks [140]).

Pluronic can be cured using ultraviolet light for bioink purposes, but initial studies have shown that this has a detrimental effect on cell viability. Extrusion-based bioprinting is possible with Pluronic after 20°C as it becomes more viscous and is more easily extruded, creating a need for a heating platform for the printer. This is potentially troublesome for cell viability, thus it is advised that Pluronic is kept in a semi-liquid suspension state. Pluronic's ability to liquefy at low temperatures (~4°C) has been exploited to fabricate complex structures with diffusible channels and high cell-loading numbers [141], although this highly viscoelastic characteristic of Pluronic has meant that it is not possible to be printed using either droplet-based or laser-based techniques.

8.3.3 Curing bioinks

Curing bioinks are photoinitiators designed specifically for bioprinting. These bioinks are used in conjunction with matrix bioinks to photopolymerize the construct for full control of its gelation. Curing bioinks are chosen depending on their cross-linking mechanisms, that is, how quick and cytocompatible they are. There are four main purely curing bioinks used in the bioprinting industry (described in Table 8.1).

8.3.4 Support bioinks

Support bioinks are used in a similar fashion to sacrificial bioinks, but offer a more permanent option. They provide improved mechanical stability when blended with a matrix bioink, making them a good selection for cartilage tissue engineering. Most support bioinks are synthetic polymers, which are thermosensitive providing biodegradability control. Fig. 8.3 [146–150] shows the most common support bioinks used.

8.3.5 Nanostructured bioinks

Nanostructured bioinks are biomaterials that have been specifically designed for 3D bioprinting and comprise of at least one nanodimensional element [151]. The design is to improve the cytochemical potential and diffusion ability of the bioink. One study [152] could increase the print resolution using a blend of nanofibers. An increase of the storage modulus was also reported. All the work carried out with these types of bioinks have only been carried out with extrusion-based methods for cartilage tissue engineering [153], and thus should be considered when determining the candidacy of a bioink.

The introduction of nanostructured cavities into bioinks have also been the focus of some work in field. One group [154] used nanostructured Pluronic within hyaluronic acid for chondrocyte cell printing. The cell viability was found to be more than 50% higher than the nonnanostructured control. They theorize that the nanoporosity of the structure improved permeability. This branch of research is novel, and therefore work is still ongoing, but steps have been made to design and develop nanostructured hybrids [12]. The work still states that further investigation is needed to qualify the nanostructure effect on cell function and viability.

Table 8.1 Curable bioinks currently used in 3D bioprinting techniques

Bioink commercial names	Chemical names	Cross-linking wavelength	Printing method	References
Biokey	Lithium phenyl-2,4,6-trimethylbenzoylphosphinate	UV 365 nm Blue light 405 nm	Extrusion	[142,143]
Eosin Y	Eosin Y	Green light 519 nm	Stereolithography	[24]
Igracure	[(2-hydroxy-1-[4-(2-hydroxyethoxy) phenyl]-2-methyl-1-propanone	UV 365 nm	Laser-based deposition	[144]
VA-086	(2,2′-azo[2-methyl-N-(2-hydroxyethyl) propionamide]	UV 365 nm	Extrusion (pneumatic)	[145]

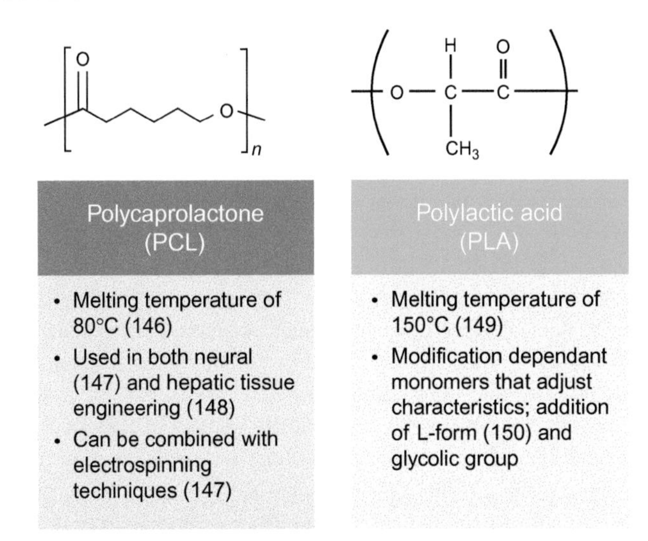

Fig. 8.3 Synthetically fabricated polymers typically used as support bioinks.

8.3.6 Self-assembled bioinks

The assembly of these bioinks is driven by the noncovalent interaction made possible by either the inherent ability of the bioink itself or by chemically manipulating the bioink to gain these abilities (done with moieties such as DNA or cyclodextrins). Studies have also shown that noncovalent interaction plays a crucial role in maintaining tissue integrity within the ECM [155], implying a better environment for nutrient absorption. Most self-assembled hydrogels have been designed for drug delivery [156], but there are a few that have been used for 3D printing applications.

The self-assembled bioinks have the advantage of self-healing properties as well as good cell guarding while being temperature sensitive. The tunability of the design is also an appealing aspect of the development of these types of bioinks. However, the print resolution possible is very low [157] and results is poorly constructed scaffolds. This leads to the other disadvantage of using self-assembled bioinks: shape fidelity. This can be counteracted by using a photo-cross-linking method [158] to eliminate the use of a chemical one (which would use covalent elements). It is a promising avenue for 3D bioprinting, but more work is required to optimize the self-assembly.

8.4 Determinant bioink moduli

Suitability is the most crucial element in 3D bioprinting. There is a wealth of biomaterials that can be used in the field, but not all are appropriate for the variability in requirement. Each bioink has been evaluated on characteristics such as biocompatibility, bioprintability, mechanical integrity, and many more. With this evaluation, a choice is then made to use a bioink or blend (to accentuate or remove characteristics).

The compatibility of the bioink with the printing modality is perhaps the most important factor. Extrusion-based bioprinting is known to be the most versatile of the methods due to its compatibility with both scaffold-free and scaffold-based construct designs. Droplet-based bioprinting and laser-based bioprinting have only been found to successfully print hydrogels. This is partly due to the lack of research in the field and the technological limitation; highly viscous materials are needed for droplet-based methods and adequate mechanical strength and lightweight material is required for laser-based bioprinting. When designing a technique and construct for clinical use, scalability also becomes a consideration. Most hydrogels can be obtained in abundance compared with the scaffold-free options; tissues spheroids are especially difficult to scale up [49]. Decellularized ECM is yielded in a notably small amount with current methods. The scalability and compatibility of the bioink influences the bioprinting time. This also poses questions of the affordability aspect of certain bioinks (though it is difficult to quantify fully when the technology is novel) when many of them are not available commercially. Hydrogels are generally the quickest to print, though, if additional blending or cross-linking is required this will increase the process duration. However, hydrogels are recognized [159] as taking a much longer time for the cells to fully grow and proliferate in comparison with the scaffold-free alternatives.

The bioprintability of the bioink is also a key feature for determining the choice and has more discrepancy in comparison with the printer compatibility. Printability is, technically, a subjective term, depending on the required application and the printing method, but also the compressive and storage moduli possible of the bioink with or without cell involvement [160]. Although a bioink may in theory "print," the product may not be appropriate for use. For example, tissue spheroids are known to be problematic for bioprinting. The use of a custom-made extruder [161] attempts to counteract that, but studies have found that it can still clog the mechanism resulting in unwanted gaps within the structure. The variability in the bioprintability of bioink types is also present in the bioink materials as well. Decellularized ECM, despite its high biocompatibility, requires a structural bioink until gelation [33]. The size of the construct being printed needs to be considered as well; microcarriers [162] have been found to disrupt the nozzle when higher densities are loaded. In relation to bioprintability, the resolution of the process has a significant effect on the printed product, but the resolution is dependent on the modality as well. Extrusion-based bioprinting can print at the highest resolution in comparison with droplet-based and laser-based techniques. It should be noted that though it is possible to print cell aggregates [especially cell pellets [163]] at high resolutions, it is not advised as this reduction in nozzle diameter causes the cells to undergo higher shear stress than normal, causing a distortion or disruption in the construct shape.

The formation of the ECM after cell encapsulation is vital to the success of the printed construct. Hydrogels are a preferred bioink as they can secrete their own ECM, which is shown to provide a good structural basis. However, this is not always the case. To produce a biomimetic environment, lower hydrogel concentrations are procured to encourage better cell growth and proliferation, but this is at the detriment of the construct's structural and mechanical integrity. Cell aggregation models have been found to develop integrity over time by the "cadherin mediated cell-cell adhesion followed by ECM deposition" [164]. Further work has also shown that the mechanical properties of the construct develops differently over time per the cell type used; stromal cells reported to be much stronger mechanically in comparison with parenchymal cells. In addition, there are inherent limitations with the cell interactions within a construct, more specifically hydrogels, as the cells are unable to migrate and manipulate themselves into the required shape [14]. Cell aggregation blends have made some progress to negate this and mimic the native environment due to their lack of matrix immobilization and higher possible cell density. Decellularized ECMs are known to be the "most" biomimetic of the bioink types available as they contain beneficial matrix proteins [165]. The degradability of the bioink is also a factor to consider in the process. Many hydrogels are thermosensitive and therefore lose their shape easily. This is an appealing characteristic for sacrificial bioinks, but not for their matrix counterparts. Studies have shown that hydrogels can be adapted to tune the degradability using different components [166]. The degradation rate is one factor that is entirely reliant on the application of the designed construct. The immune reaction that could potentially be induced should be understood during the creation process. Bioinks from different species have been demonstrated [167] to illicit a low-level immune response in the form of lymphocyte migration, but the group does theorize that this can perhaps be offset by decellularized manipulation. Autologous cell aggregates, however, should not result in an immunogenic response and thus, should be considered when designing constructs that involve highly vascular or complex architecture.

8.5 Clinical application and selection

Despite the breakthroughs in tissue engineering technologies in the past decade, the clinical application is still ambiguous. This is partly as the clinical basis of the technologies relies on the intended outcome. This, in turn, has connotations for the development and opting of the bioink. Biocompatibility, bioinertness, and biodegradability are all characteristics that are required of a bioink, but may have different weightings depending on the use of the biomaterial. When designing a bioink, biodegradability, for example, may be sacrificed for a higher mechanical strength as the bioprinting modality may require that for optimal printing. This applies to the tissue type attempting to be recreated as well as the required clinical use. Some bioinks are known as the "gold standard" for building a certain tissue-type construct, and therefore research and protocols have been honed to create an adequate standard.

In the field of reconstructive surgery, this generally applies to repairing wounds and gross defects. Thus, not only is the cytocompatibility of the construct crucial, but the aesthetic result is just as key. Most of the designs will be used superficially and therefore will not require the most extensive immunogenics. However, the spatial requirements will be high so the resolution and the printability of the bioink will be a more important factor.

Thus, it is important to systematically categorize bioinks depending on their potential utilization. Fig. 8.4 [32,68,72,121,168–171] demonstrates a method of choosing the bioink from the list of candidates discussed. It aids the selection process by

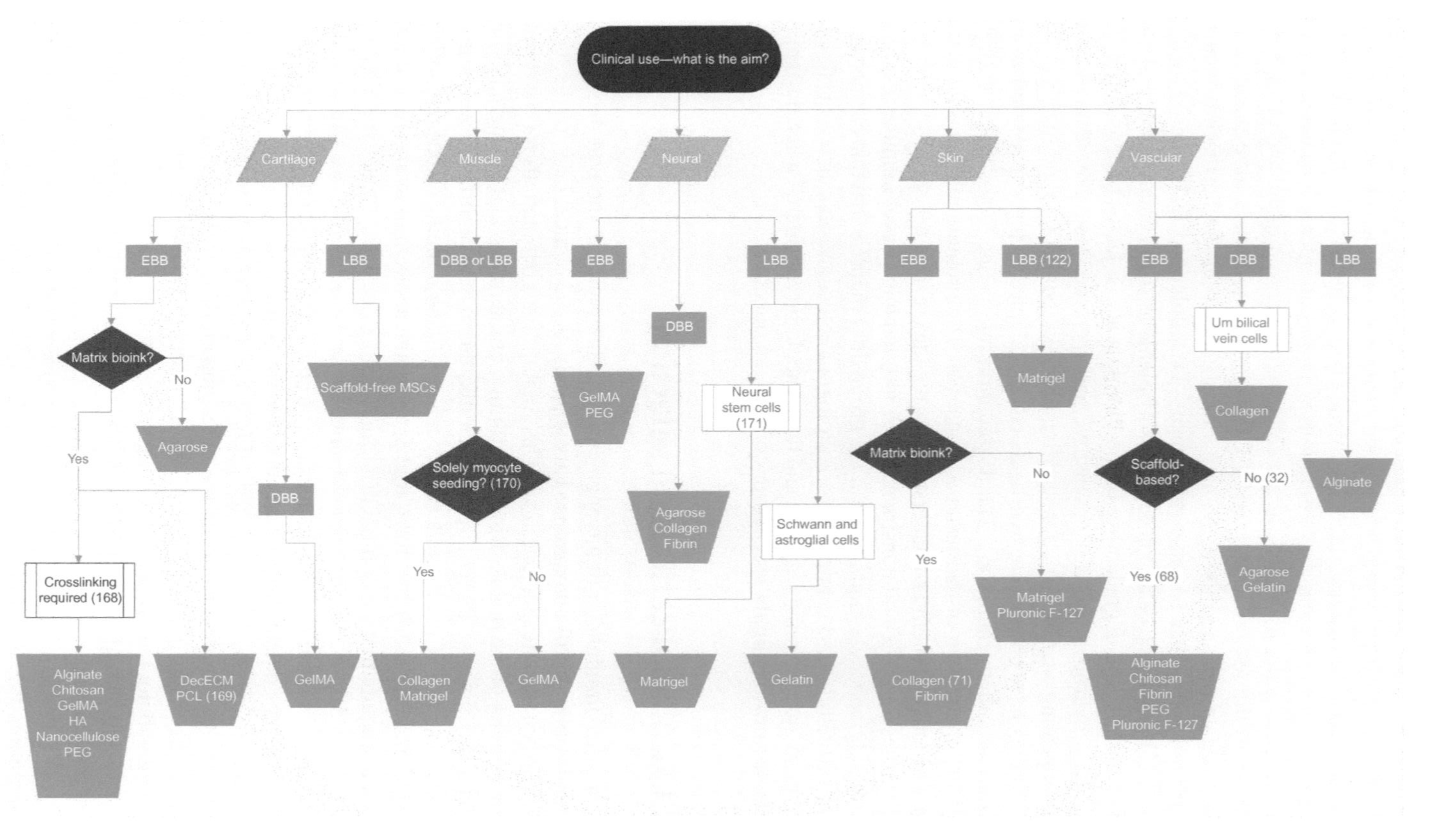

Fig. 8.4 Diagrammatical flowchart of candidate bioinks for 3D bioprinting soft tissues (DBB, droplet-based bioprinting; DecECM, decellularized extracellular matrix; EBB, extrusion-based bioprinting; GelMA, methacrylated gelatin; HA, hyaluronic acid; LBB, laser-based bioprinting; MSC, mesenchymal stem cells; PCL, polycaprolactone; PEG, poly(ethylene glycol)).

cataloguing the bioinks that could be printed successfully (from evidence) depending on the tissue type needed and bioprinting modality available. It should be noted that when blends are considered to alter characteristics, a systematic formula may not apply.

8.6 Future potential

Although the field of bioprinting and bioink design is flourishing, there are still many ongoing endeavors to better the research in the field. There are currently many biomaterials used in modern medicine, but very few of them are specifically designed for the 3D bioprinting process. These materials have been refined or combined to hone them for their printing use, but the need for more suitable bioinks is apparent. However, this will depend on many aspects such as the printing methodology, clinical outcome required, and cell seeding modality.

The development of new bioinks will become a crucial field in the future. Refining biomaterials will only get the technology so far; as printing modalities become more advanced, the need for explicitly designed bioinks will increase. Many researchers have begun work on developing what are known as "smart polymers" [172]. These materials marry the advantageous characteristics of synthetic polymers with the biological abilities of natural proteins. These polymers are able to undergo "large conformational changes in response to small environmental changes" [13] which can then exploited for required gelation or degradability in a construct design. Research has shown that these conformational changes are, in most cases, reversible. The possibilities of biological applications have depth and width. Drug delivery, gene therapy, and enzymatic manipulation [173] are some of the aspects currently being worked on. With this versatility in development, the customizability with a bioprinting process would be probable and is likely to produce favorable results. Incorporating these smart biomaterials with growth factors and ECM components would revolutionize future tissue engineering approaches. This has been attempted in part [174] by conjugate proteins in a "site-specific" way that has demonstrated to change the state of the polymer (if near a ligand-binding domain [175]). This could permit accurate and precise spatial delivery of cells to a construct, a major limitation of current regenerative medicine techniques. Another emergence in this field has been that of "programmable biomaterials." It involves merging genetic engineering with regenerative medicine to design scaffold with a DNA-base which could, in theory, exhibit certain characteristics depending on their environment. For example, adapting its mechanical property per what tissue type is needed. Now, these biomaterials have not been used in 3D bioprinting [176], but could be a potential future approach.

Designing biomaterials will, no doubt, be a key aspect in future work, but the development of new bioinks altogether will also see an increase. Efforts will be made to seek out biomaterials in present natural sources and, the chemical synthesis of new materials. With the development of advanced combinatorial libraries combined with large-scale research methodologies [177], an uncountable number of candidate bioinks can be developed. However, their genuine suitability for scaffold design will still need to be evaluated [178]. Perhaps, a unity of nanotechnology and tissue engineering could be an output for the revolutions in the bioink field such as developing new self-assembling nanoarchitecture in the form of nanorods and nanofilaments [179]. However, incorporating the biocompatibility aspect would still be a challenge, although one group could coat a scaffold with bioactive sequences via nanofiber gels [180] so there is research ongoing to investigate the feasibility of this. Natural sources, especially keratin, have been investigated for their potential use in tissue engineering approaches. It contains large amounts of the arginine-glycine-aspartate (RGD) motif that are indicated in cell adhesion and ECM development. Clearly, more refinement of keratin is needed before it could be considered a candidate for scaffold design, but there is research currently addressing this [181].

8.7 Conclusion

This chapter discussed the candidate bioinks used in 3D bioprinting soft tissues. There are two different approaches when deciding the bioink that should be used: scaffold-based and scaffold-free, although the bioinks are categorized in certain ways (such as matrix, sacrificial, etc.) using blends and cross-linking merges the characteristics together and makes the decision more complex. Systematically selecting the bioink is a good method for producing the best results, but it should be noted that the clinical need is just as important as the material that is to be used. Future efforts involve both the refinement of current bioinks and the development of new ones. Incorporating genetic engineering and nanotechnology will be the focus of many groups in the future due to their increased tunable capability and versatility. However, full mimicry of the ECM is still not possible and should be a priority for current research. Despite this, the field of 3D bioprinting is exciting, and future technology will be key in regenerative medicine and reconstructive surgery.

References

[1] Klebe RJ. Cytoscribing: a method for micropositioning cells and the construction of two- and three-dimensional synthetic tissues. Exp Cell Res 1988;179(2):362–73.
[2] Lee W, Lee V, Polio S, Keegan P, Lee JH, Fischer K, et al. On-demand three-dimensional freeform fabrication of multi-layered hydrogel scaffold with fluidic channels. Biotechnol Bioeng 2010;105(6):1178–86.
[3] Malda J, Visser J, Melchels FP, Jüngst T, Hennink WE, Dhert WJA, et al. 25th anniversary article: engineering hydrogels for biofabrication. Adv Mater 2013;25(36):5011–28.
[4] Foty RA, Pfleger CM, Forgacs G, Steinberg MS. Surface tensions of embryonic tissues predict their mutual envelopment behavior. Development 1996;122(5):1611–20.
[5] Wilson WC, Boland T. Cell and organ printing 1: protein and cell printers. Anat Rec A: Discov Mol Cell Evol Biol 2003;272A(2):491–6.
[6] Jakab K, Neagu A, Mironov V, Markwald RR, Forgacs G. Engineering biological structures of prescribed shape using self-assembling multicellular systems. Proc Natl Acad Sci U S A 2004;101(9):2864–9.

[7] Jakab K, Norotte C, Damon B, Marga F, Neagu A, Besch-Williford CL, et al. Tissue engineering by self-assembly of cells printed into topologically defined structures. Tissue Eng A 2008;14(3):413–21.
[8] Heikkinen JJ, Kivimaki L, Maatta JA, Makela I, Hakalahti L, Takkinen K, et al. Versatile bio-ink for covalent immobilization of chimeric avidin on sol-gel substrates. Colloids Surf B: Biointerfaces 2011;87(2):409–14.
[9] Yu Y, Ozbolat IT. Tissue strands as "bioink" for scale-up organ printing. Conf Proc IEEE Eng Med Biol Soc 2014;2014:1428–31.
[10] Yeo M, Lee JS, Chun W, Kim GH. An innovative collagen-based cell-printing method for obtaining human adipose stem cell-laden structures consisting of core-sheath structures for tissue engineering. Biomacromolecules 2016;17(4):1365–75.
[11] Jang J, Kim TG, Kim BS, Kim SW, Kwon SM, Cho DW. Tailoring mechanical properties of decellularized extracellular matrix bioink by vitamin B2-induced photo-crosslinking. Acta Biomater 2016;33:88–95.
[12] Lee DY, Lee H, Kim Y, Yoo SY, Chung WJ, Kim G. Phage as versatile nanoink for printing 3-D cell-laden scaffolds. Acta Biomater 2016;29:112–24.
[13] Galaev IY, Mattiasson B. "Smart" polymers and what they could do in biotechnology and medicine. Trends Biotechnol 1999;17(8):335–40.
[14] Ozbolat IT. Scaffold-based or scaffold-free bioprinting: competing or complementing approaches? J Nanotechnol Eng Med 2015;6(2):024701–6.
[15] White CJ, McBride MK, Pate KM, Tieppo A, Byrne ME. Extended release of high molecular weight hydroxypropyl methylcellulose from molecularly imprinted, extended wear silicone hydrogel contact lenses. Biomaterials 2011;32(24):5698–705.
[16] Nie S, Hsiao WLW, Pan W, Yang Z. Thermoreversible Pluronic(®) F127-based hydrogel containing liposomes for the controlled delivery of paclitaxel: in vitro drug release, cell cytotoxicity, and uptake studies. Int J Nanomedicine 2011;6:151–66.
[17] Ahmed EM. Hydrogel: preparation, characterization, and applications: a review. J Adv Res 2015;6(2):105–21.
[18] Nuttelman CR, Tripodi MC, Anseth KS. Synthetic hydrogel niches that promote hMSC viability. Matrix Biol 2005;24(3):208–18.
[19] Ovsianikov A, Muhleder S, Torgersen J, Li Z, Qin XH, Van Vlierberghe S, et al. Laser photofabrication of cell-containing hydrogel constructs. Langmuir 2014;30(13):3787–94.
[20] Wong RS, Ashton M, Dodou K. Effect of crosslinking agent concentration on the properties of unmedicated hydrogels. Pharmaceutics 2015;7(3):305–19.
[21] Jeong B, Kim SW, Bae YH. Thermosensitive sol-gel reversible hydrogels. Adv Drug Deliv Rev 2002;54(1):37–51.
[22] Hennink WE, van Nostrum CF. Novel crosslinking methods to design hydrogels. Adv Drug Deliv Rev 2002;54(1):13–36.
[23] Arcaute K, Mann BK, Wicker RB. Stereolithography of three-dimensional bioactive poly(ethylene glycol) constructs with encapsulated cells. Ann Biomed Eng 2006;34(9):1429–41.
[24] Wang Z, Abdulla R, Parker B, Samanipour R, Ghosh S, Kim K. A simple and high-resolution stereolithography-based 3D bioprinting system using visible light crosslinkable bioinks. Biofabrication 2015;7(4). 045009.
[25] Ye Q, Zund G, Benedikt P, Jockenhoevel S, Hoerstrup SP, Sakyama S, et al. Fibrin gel as a three dimensional matrix in cardiovascular tissue engineering. Eur J Cardiothorac Surg 2000;17(5):587–91.
[26] Islam A, Yasin T, Gull N, Khan SM, Sabir A, Munawwar MA, et al. Fabrication and performance characteristics of tough hydrogel scaffolds based on biocompatible polymers. Int J Biol Macromol 2016;92:1–10.

[27] Turner AE, Flynn LE. Design and characterization of tissue-specific extracellular matrix-derived microcarriers. Tissue Eng Part C Methods 2012;18(3):186–97.
[28] Goh TK, Zhang ZY, Chen AK, Reuveny S, Choolani M, Chan JK, et al. Microcarrier culture for efficient expansion and osteogenic differentiation of human fetal mesenchymal stem cells. Biores Open Access 2013;2(2):84–97.
[29] Malda J, Frondoza CG. Microcarriers in the engineering of cartilage and bone. Trends Biotechnol 2006;24(7):299–304.
[30] Levato R, Visser J, Planell JA, Engel E, Malda J, Mateos-Timoneda MA. Biofabrication of tissue constructs by 3D bioprinting of cell-laden microcarriers. Biofabrication 2014;6(3). 035020.
[31] Ott HC, Matthiesen TS, Goh SK, Black LD, Kren SM, Netoff TI, et al. Perfusion-decellularized matrix: using nature's platform to engineer a bioartificial heart. Nat Med 2008;14(2):213–21.
[32] Zhao L, Chen HL, Xie LQ, Wang M, Li XF, Feng ZW, et al. Modification of decellularized vascular scaffold with conditioned medium to enhance cell reseeding. J Biomater Sci Polym Ed 2016;27(11):1115–25.
[33] Pati F, Jang J, Ha D-H, Won Kim S, Rhie J-W, Shim J-H, et al. Printing three-dimensional tissue analogues with decellularized extracellular matrix bioink. Nat Commun 2014;5:3935.
[34] Banyard DA, Borad V, Amezcua E, Wirth GA, Evans GR, Widgerow AD. Preparation, characterization, and clinical implications of human decellularized adipose tissue extracellular matrix (hDAM): a comprehensive review. Aesthet Surg J 2016;36(3):349–57.
[35] Rivron NC, Rouwkema J, Truckenmuller R, Karperien M, De Boer J, Van Blitterswijk CA. Tissue assembly and organization: developmental mechanisms in microfabricated tissues. Biomaterials 2009;30(28):4851–8.
[36] Jorgenson AJ, Choi KM, Sicard D, Smith KM, Hiemer SE, Varelas X, et al. TAZ activation drives fibroblast spheroid growth, expression of profibrotic paracrine signals, and context-dependent ECM gene expression. Am J Physiol Cell Physiol 2017;312(3):C277–85.
[37] Czajka CA, Mehesz AN, Trusk TC, Yost MJ, Drake CJ. Scaffold-free tissue engineering: organization of the tissue cytoskeleton and its effects on tissue shape. Ann Biomed Eng 2014;42(5):1049–61.
[38] Moldovan NI, Hibino N, Nakayama K. Principles of the Kenzan method for robotic cell spheroid-based three-dimensional bioprinting. Tissue Eng B Rev 2017;23(3):237–44.
[39] Wheeldon I, Ahari AF, Khademhosseini A. Microengineering hydrogels for stem cell bioengineering and tissue regeneration. JALA Charlottesv Va 2010;15(6):440–8.
[40] Lin B, Miao Y, Wang J, Fan Z, Du L, Su Y, et al. Surface tension guided hanging-drop: producing controllable 3D spheroid of high-passaged human dermal papilla cells and forming inductive microtissues for hair-follicle regeneration. ACS Appl Mater Interfaces 2016;8(9):5906–16.
[41] Beauchamp P, Moritz W, Kelm JM, Ullrich ND, Agarkova I, Anson BD, et al. Development and characterization of a scaffold-free 3D spheroid model of induced pluripotent stem cell-derived human cardiomyocytes. Tissue Eng Part C Methods 2015;21(8):852–61.
[42] Mironov V, Visconti RP, Kasyanov V, Forgacs G, Drake CJ, Markwald RR. Organ printing: tissue spheroids as building blocks. Biomaterials 2009;30(12):2164–74.
[43] Itoh M, Nakayama K, Noguchi R, Kamohara K, Furukawa K, Uchihashi K, et al. Scaffold-free tubular tissues created by a bio-3D printer undergo remodeling and endothelialization when implanted in rat aortae. PLoS One 2015;10(9). e0136681.

[44] Fleming PA, Argraves WS, Gentile C, Neagu A, Forgacs G, Drake CJ. Fusion of uniluminal vascular spheroids: a model for assembly of blood vessels. Dev Dyn 2010;239(2):398–406.
[45] Niyama K, Ide N, Onoue K, Okabe T, Wakitani S, Takagi M. Construction of osteochondral-like tissue graft combining beta-tricalcium phosphate block and scaffold-free centrifuged chondrocyte cell sheet. J Orthop Sci 2011;16(5):613–21.
[46] Owens CM, Marga F, Forgacs G, Heesch CM. Biofabrication and testing of a fully cellular nerve graft. Biofabrication 2013;5(4). 045007.
[47] Na S, Zhang H, Huang F, Wang W, Ding Y, Li D, et al. Regeneration of dental pulp/dentine complex with a three-dimensional and scaffold-free stem-cell sheet-derived pellet. J Tissue Eng Regen Med 2016;10(3):261–70.
[48] Norotte C, Marga FS, Niklason LE, Forgacs G. Scaffold-free vascular tissue engineering using bioprinting. Biomaterials 2009;30(30):5910–7.
[49] Yu Y, Moncal KK, Li J, Peng W, Rivero I, Martin JA, et al. Three-dimensional bioprinting using self-assembling scalable scaffold-free "tissue strands" as a new bioink. Sci Rep 2016;6:28714.
[50] Schrobback K, Klein TJ, Crawford R, Upton Z, Malda J, Leavesley DI. Effects of oxygen and culture system on in vitro propagation and redifferentiation of osteoarthritic human articular chondrocytes. Cell Tissue Res 2012;347(3):649–63.
[51] Yagi H, Fujise A, Itabashi N, Ohshiro T. Purification and characterization of a novel alginate lyase from the marine bacterium Cobetia sp. NAP1 isolated from brown algae. Biosci Biotechnol Biochem 2016;80(12):2338–46.
[52] Rowley JA, Madlambayan G, Mooney DJ. Alginate hydrogels as synthetic extracellular matrix materials. Biomaterials 1999;20(1):45–53.
[53] Purcell EK, Singh A, Kipke DR. Alginate composition effects on a neural stem cell-seeded scaffold. Tissue Eng Part C Methods 2009;15(4):541–50.
[54] Hill E, Boontheekul T, Mooney DJ. Designing scaffolds to enhance transplanted myoblast survival and migration. Tissue Eng 2006;12(5):1295–304.
[55] Galateanu B, Dimonie D, Vasile E, Nae S, Cimpean A, Costache M. Layer-shaped alginate hydrogels enhance the biological performance of human adipose-derived stem cells. BMC Biotechnol 2012;12:35.
[56] Lee H, Ahn S, Bonassar LJ, Chun W, Kim G. Cell-laden poly(ε-caprolactone)/alginate hybrid scaffolds fabricated by an aerosol cross-linking process for obtaining homogeneous cell distribution: fabrication, seeding efficiency, and cell proliferation and distribution. Tissue Eng Part C Methods 2013;19(10):784–93.
[57] Zhang Y, Yu Y, Chen H, Ozbolat IT. Characterization of printable cellular micro-fluidic channels for tissue engineering. Biofabrication 2013;5(2). 025004.
[58] Gudapati H, Dey M, Ozbolat I. A comprehensive review on droplet-based bioprinting: past, present and future. Biomaterials 2016;102:20–42.
[59] Blaeser A, Duarte Campos DF, Weber M, Neuss S, Theek B, Fischer H, et al. Biofabrication under fluorocarbon: a novel freeform fabrication technique to generate high aspect ratio tissue-engineered constructs. Biores Open Access 2013;2(5):374–84.
[60] Xu T, Zhao W, Zhu JM, Albanna MZ, Yoo JJ, Atala A. Complex heterogeneous tissue constructs containing multiple cell types prepared by inkjet printing technology. Biomaterials 2013;34(1):130–9.
[61] Guillotin B, Souquet A, Catros S, Duocastella M, Pippenger B, Bellance S, et al. Laser assisted bioprinting of engineered tissue with high cell density and microscale organization. Biomaterials 2010;31(28):7250–6.

[62] Croisier F, Jérôme C. Chitosan-based biomaterials for tissue engineering. Eur Polym J 2013;49(4):780–92.
[63] Kingsley DM, Dias AD, Corr DT. Microcapsules and 3D customizable shelled microenvironments from laser direct-written microbeads. Biotechnol Bioeng 2016;113(10):2264–74.
[64] Kozhikhova K, Ivantsova M, Tokareva M, Shulepov I, Tretiyakov A, Shaidarov L, et al. Preparation of chitosan-coated liposomes as a novel carrier system for the antiviral drug Triazavirin. Pharm Dev Technol 2016;1–28.
[65] Geng L, Feng W, Hutmacher DW, San Wong Y, Tong Loh H, Fuh JYH. Direct writing of chitosan scaffolds using a robotic system. Rapid Prototyp J 2005;11(2):90–7.
[66] Ferreira AM, Gentile P, Chiono V, Ciardelli G. Collagen for bone tissue regeneration. Acta Biomater 2012;8(9):3191–200.
[67] Yanez M, Rincon J, Dones A, De Maria C, Gonzales R, Boland T. In vivo assessment of printed microvasculature in a bilayer skin graft to treat full-thickness wounds. Tissue Eng A 2015;21(1-2):224–33.
[68] Lee VK, Lanzi AM, Haygan N, Yoo S-S, Vincent PA, Dai G. Generation of multi-scale vascular network system within 3D hydrogel using 3D bio-printing technology. Cell Mol Bioeng 2014;7(3):460–72.
[69] Smith CM, Stone AL, Parkhill RL, Stewart RL, Simpkins MW, Kachurin AM, et al. Three-dimensional bioassembly tool for generating viable tissue-engineered constructs. Tissue Eng 2004;10(9-10):1566–76.
[70] Homenick CM, de Silveira G, Sheardown H, Adronov A. Pluronics as crosslinking agents for collagen: novel amphiphilic hydrogels. Polym Int 2011;60(3):458–65.
[71] Deitch S, Kunkle C, Cui X, Boland T, Dean D. Collagen matrix alignment using inkjet printer technology. MRS Proc 2011;1094:.
[72] Skardal A, Mack D, Kapetanovic E, Atala A, Jackson JD, Yoo J, et al. Bioprinted amniotic fluid-derived stem cells accelerate healing of large skin wounds. Stem Cells Transl Med 2012;1(11):792–802.
[73] Michael S, Sorg H, Peck CT, Koch L, Deiwick A, Chichkov B, et al. Tissue engineered skin substitutes created by laser-assisted bioprinting form skin-like structures in the dorsal skin fold chamber in mice. PLoS One 2013;8(3). e57741.
[74] Lee V, Singh G, Trasatti JP, Bjornsson C, Xu X, Tran TN, et al. Design and fabrication of human skin by three-dimensional bioprinting. Tissue Eng Part C Methods 2014;20(6):473–84.
[75] Hakam MS, Imani R, Abolfathi N, Fakhrzadeh H, Sharifi AM. Evaluation of fibrin-gelatin hydrogel as biopaper for application in skin bioprinting: an in-vitro study. Biomed Mater Eng 2016;27(6):669–82.
[76] Janmey PA, Winer JP, Weisel JW. Fibrin gels and their clinical and bioengineering applications. J R Soc Interface 2009;6(30):1–10.
[77] Nakatsu MN, Sainson RC, Aoto JN, Taylor KL, Aitkenhead M, Perez-del-Pulgar S, et al. Angiogenic sprouting and capillary lumen formation modeled by human umbilical vein endothelial cells (HUVEC) in fibrin gels: the role of fibroblasts and Angiopoietin-1. Microvasc Res 2003;66(2):102–12.
[78] Xu T, Gregory CA, Molnar P, Cui X, Jalota S, Bhaduri SB, et al. Viability and electrophysiology of neural cell structures generated by the inkjet printing method. Biomaterials 2006;27(19):3580–8.
[79] Gruene M, Pflaum M, Hess C, Diamantouros S, Schlie S, Deiwick A, et al. Laser printing of three-dimensional multicellular arrays for studies of cell-cell and cell-environment interactions. Tissue Eng Part C Methods 2011;17(10):973–82.

[80] Zhang L, Hu J, Athanasiou KA. The role of tissue engineering in articular cartilage repair and regeneration. Crit Rev Biomed Eng 2009;37(1-2):1–57.
[81] Burdick JA, Prestwich GD. Hyaluronic acid hydrogels for biomedical applications. Adv Mater 2011;23(12):H41–56.
[82] Skardal A, Zhang J, Prestwich GD. Bioprinting vessel-like constructs using hyaluronan hydrogels crosslinked with tetrahedral polyethylene glycol tetracrylates. Biomaterials 2010;31(24):6173–81.
[83] Skardal A, Zhang J, McCoard L, Oottamasathien S, Prestwich GD. Dynamically crosslinked gold nanoparticle—hyaluronan hydrogels. Adv Mater 2010;22(42):4736–40.
[84] Fan R, Piou M, Darling E, Cormier D, Sun J, Wan J. Bio-printing cell-laden Matrigel-agarose constructs. J Biomater Appl 2016;31(5):684–92.
[85] Kleinman HK, Martin GR. Matrigel: basement membrane matrix with biological activity. Semin Cancer Biol 2005;15(5):378–86.
[86] Snyder JE, Hamid Q, Wang C, Chang R, Emami K, Wu H, et al. Bioprinting cell-laden matrigel for radioprotection study of liver by pro-drug conversion in a dual-tissue microfluidic chip. Biofabrication 2011;3(3). 034112.
[87] Melchels FPW, Domingos MAN, Klein TJ, Malda J, Bartolo PJ, Hutmacher DW. Additive manufacturing of tissues and organs. Prog Polym Sci 2012;37(8):1079–104.
[88] Fedorovich NE, Wijnberg HM, Dhert WJ, Alblas J. Distinct tissue formation by heterogeneous printing of osteo- and endothelial progenitor cells. Tissue Eng A 2011;17(15–16):2113–21.
[89] Horváth L, Umehara Y, Jud C, Blank F, Petri-Fink A, Rothen-Rutishauser B. Engineering an in vitro air-blood barrier by 3D bioprinting. Sci Rep 2015;5:7974.
[90] Hong S, Song S-J, Lee JY, Jang H, Choi J, Sun K, et al. Cellular behavior in micropatterned hydrogels by bioprinting system depended on the cell types and cellular interaction. J Biosci Bioeng 2013;116(2):224–30.
[91] Barron JA, Spargo BJ, Ringeisen BR. Biological laser printing of three dimensional cellular structures. Appl Phys A 2004;79(4):1027–30.
[92] Schiele NR, Koppes RA, Corr DT, Ellison KS, Thompson DM, Ligon LA, et al. Laser direct writing of combinatorial libraries of idealized cellular constructs: biomedical applications. Appl Surf Sci 2009;255(10):5444–7.
[93] Alexander A, Ajazuddin, Khan J, Saraf S, Saraf S. Poly(ethylene glycol)-poly(lactic-co-glycolic acid) based thermosensitive injectable hydrogels for biomedical applications. J Control Release 2013;172(3):715–29.
[94] Pasut G, Sergi M, Veronese FM. Anti-cancer PEG-enzymes: 30 years old, but still a current approach. Adv Drug Deliv Rev 2008;60(1):69–78.
[95] Tarassoli SP, de Pinillos Bayona AM, Pye H, Mosse CA, Callan JF, MacRobert A, et al. Cathepsin B-degradable, NIR-responsive nanoparticulate platform for target-specific cancer therapy. Nanotechnology 2017;28(5). 055101.
[96] Sharma S, Johnson RW, Desai TA. XPS and AFM analysis of antifouling PEG interfaces for microfabricated silicon biosensors. Biosens Bioelectron 2004;20(2):227–39.
[97] Benoit DS, Durney AR, Anseth KS. Manipulations in hydrogel degradation behavior enhance osteoblast function and mineralized tissue formation. Tissue Eng 2006;12(6):1663–73.
[98] Vladkova T, Krasteva N, Kostadinova A, Altankov G. Preparation of PEG-coated surfaces and a study for their interaction with living cells. J Biomater Sci Polym Ed 1999;10(6):609–20.
[99] Gao G, Schilling AF, Yonezawa T, Wang J, Dai G, Cui X. Bioactive nanoparticles stimulate bone tissue formation in bioprinted three-dimensional scaffold and human mesenchymal stem cells. Biotechnol J 2014;9(10):1304–11.

[100] Pereira RF, Bártolo PJ. 3D bioprinting of photocrosslinkable hydrogel constructs. J Appl Polym Sci 2015;132(48).
[101] Hockaday LA, Kang KH, Colangelo NW, Cheung PY, Duan B, Malone E, et al. Rapid 3D printing of anatomically accurate and mechanically heterogeneous aortic valve hydrogel scaffolds. Biofabrication 2012;4(3). 035005.
[102] Skardal A, Devarasetty M, Kang HW, Seol YJ, Forsythe SD, Bishop C, et al. Bioprinting cellularized constructs using a tissue-specific hydrogel bioink. J Vis Exp 2016;110. e53606.
[103] Soman P, Kelber JA, Lee JW, Wright TN, Vecchio KS, Klemke RL, et al. Cancer cell migration within 3D layer-by-layer microfabricated photocrosslinked PEG scaffolds with tunable stiffness. Biomaterials 2012;33(29):7064–70.
[104] Cui X, Breitenkamp K, Finn MG, Lotz M, D'Lima DD. Direct human cartilage repair using three-dimensional bioprinting technology. Tissue Eng A 2012;18(11-12):1304–12.
[105] Cui X, Breitenkamp K, Lotz M, D'Lima D. Synergistic action of fibroblast growth factor-2 and transforming growth factor-beta1 enhances bioprinted human neocartilage formation. Biotechnol Bioeng 2012;109(9):2357–68.
[106] Dhariwala B, Hunt E, Boland T. Rapid prototyping of tissue-engineering constructs, using photopolymerizable hydrogels and stereolithography. Tissue Eng 2004;10(9-10):1316–22.
[107] Huang TQ, Qu X, Liu J, Chen S. 3D printing of biomimetic microstructures for cancer cell migration. Biomed Microdevices 2014;16(1):127–32.
[108] Mohite BV, Patil SV. A novel biomaterial: bacterial cellulose and its new era applications. Biotechnol Appl Biochem 2014;61(2):101–10.
[109] Zhang Q, Lin D, Yao S. Review on biomedical and bioengineering applications of cellulose sulfate. Carbohydr Polym 2015;132:311–22.
[110] Schacht K, Jüngst T, Schweinlin M, Ewald A, Groll J, Scheibel T. Biofabrication of cell-loaded 3D spider silk constructs. Angew Chem Int Ed 2015;54(9):2816–20.
[111] Lozano R, Stevens L, Thompson BC, Gilmore KJ, Gorkin 3rd R, Stewart EM, et al. 3D printing of layered brain-like structures using peptide modified gellan gum substrates. Biomaterials 2015;67:264–73.
[112] Kirchmajer DM, Gorkin III R, in het Panhuis M. An overview of the suitability of hydrogel-forming polymers for extrusion-based 3D-printing. J Mater Chem B 2015;3(20):4105–17.
[113] Pescosolido L, Schuurman W, Malda J, Matricardi P, Alhaique F, Coviello T, et al. Hyaluronic acid and dextran-based semi-IPN hydrogels as biomaterials for bioprinting. Biomacromolecules 2011;12(5):1831–8.
[114] Lévesque SG, Lim RM, Shoichet MS. Macroporous interconnected dextran scaffolds of controlled porosity for tissue-engineering applications. Biomaterials 2005;26(35):7436–46.
[115] Kolesky DB, Truby RL, Gladman AS, Busbee TA, Homan KA, Lewis JA. 3D bioprinting of vascularized, heterogeneous cell-laden tissue constructs. Adv Mater 2014;26(19):3124–30.
[116] Mandrycky C, Wang Z, Kim K, Kim D-H. 3D bioprinting for engineering complex tissues. Biotechnol Adv 2016;34(4):422–34.
[117] Griess GA, Guiseley KB, Serwer P. The relationship of agarose gel structure to the sieving of spheres during agarose gel electrophoresis. Biophys J 1993;65(1):138–48.
[118] Tako M, Nakamura S. Gelation mechanism of agarose. Carbohydr Res 1988;180(2):277–84.

[119] Xu T, Jin J, Gregory C, Hickman JJ, Boland T. Inkjet printing of viable mammalian cells. Biomaterials 2005;26(1):93–9.
[120] Yamada M, Imaishi H, Morigaki K. Microarrays of phospholipid bilayers generated by inkjet printing. Langmuir 2013;29(21):6404–8.
[121] Koch L, Kuhn S, Sorg H, Gruene M, Schlie S, Gaebel R, et al. Laser printing of skin cells and human stem cells. Tissue Eng Part C Methods 2010;16(5):847–54.
[122] Duarte Campos DF, Blaeser A, Korsten A, Neuss S, Jakel J, Vogt M, et al. The stiffness and structure of three-dimensional printed hydrogels direct the differentiation of mesenchymal stromal cells toward adipogenic and osteogenic lineages. Tissue Eng A 2015;21(3-4):740–56.
[123] Aldana AA, Abraham GA. Current advances in electrospun gelatin-based scaffolds for tissue engineering applications. Int J Pharm 2016;523(2):441–53.
[124] Chiou B-S, Avena-Bustillos RJ, Bechtel PJ, Jafri H, Narayan R, Imam SH, et al. Cold water fish gelatin films: effects of cross-linking on thermal, mechanical, barrier, and biodegradation properties. Eur Polym J 2008;44(11):3748–53.
[125] Liang H-C, Chang W-H, Liang H-F, Lee M-H, Sung H-W. Crosslinking structures of gelatin hydrogels crosslinked with genipin or a water-soluble carbodiimide. J Appl Polym Sci 2004;91(6):4017–26.
[126] Wang X, Yan Y, Pan Y, Xiong Z, Liu H, Cheng J, et al. Generation of three-dimensional hepatocyte/gelatin structures with rapid prototyping system. Tissue Eng 2006;12(1):83–90.
[127] Pataky K, Braschler T, Negro A, Renaud P, Lutolf MP, Brugger J. Microdrop printing of hydrogel bioinks into 3D tissue-like geometries. Adv Mater 2012;24(3):391–6.
[128] Schiele NR, Chrisey DB, Corr DT. Gelatin-based laser direct-write technique for the precise spatial patterning of cells. Tissue Eng Part C Methods 2011;17(3):289–98.
[129] Raof NA, Schiele NR, Xie Y, Chrisey DB, Corr DT. The maintenance of pluripotency following laser direct-write of mouse embryonic stem cells. Biomaterials 2011;32(7):1802–8.
[130] Nichol JW, Koshy S, Bae H, Hwang CM, Yamanlar S, Khademhosseini A. Cell-laden microengineered gelatin methacrylate hydrogels. Biomaterials 2010;31(21):5536–44.
[131] Benton JA, DeForest CA, Vivekanandan V, Anseth KS. Photocrosslinking of gelatin macromers to synthesize porous hydrogels that promote valvular interstitial cell function. Tissue Eng A 2009;15(11):3221–30.
[132] Nikkhah M, Eshak N, Zorlutuna P, Annabi N, Castello M, Kim K, et al. Directed endothelial cell morphogenesis in micropatterned gelatin methacrylate hydrogels. Biomaterials 2012;33(35):9009–18.
[133] Bertassoni LE, Cardoso JC, Manoharan V, Cristino AL, Bhise NS, Araujo WA, et al. Direct-write bioprinting of cell-laden methacrylated gelatin hydrogels. Biofabrication 2014;6(2). 024105.
[134] Mingchun D, Bing C, Qingyuan M, Sumei L, Xiongfei Z, Cheng Z, et al. 3D bioprinting of BMSC-laden methacrylamide gelatin scaffolds with CBD-BMP2-collagen microfibers. Biofabrication 2015;7(4). 044104.
[135] Schuurman W, Levett PA, Pot MW, van Weeren PR, Dhert WJ, Hutmacher DW, et al. Gelatin-methacrylamide hydrogels as potential biomaterials for fabrication of tissue-engineered cartilage constructs. Macromol Biosci 2013;13(5):551–61.
[136] Gurkan UA, El Assal R, Yildiz SE, Sung Y, Trachtenberg AJ, Kuo WP, et al. Engineering anisotropic biomimetic fibrocartilage microenvironment by bioprinting mesenchymal stem cells in nanoliter gel droplets. Mol Pharm 2014;11(7):2151–9.

[137] Gauvin R, Chen YC, Lee JW, Soman P, Zorlutuna P, Nichol JW, et al. Microfabrication of complex porous tissue engineering scaffolds using 3D projection stereolithography. Biomaterials 2012;33(15):3824–34.
[138] Wang S, Lee JM, Yeong WY. Smart hydrogels for 3D bioprinting. Int J Bioprinting 2015;1(1):3–14.
[139] Gong CY, Shi S, Dong PW, Zheng XL, Fu SZ, Guo G, et al. In vitro drug release behavior from a novel thermosensitive composite hydrogel based on Pluronic f127 and poly(ethylene glycol)-poly(ε-caprolactone)-poly(ethylene glycol) copolymer. BMC Biotechnol 2009;9:8.
[140] Mesa M, Sierra L, Patarin J, Guth J-L. Morphology and porosity characteristics control of SBA-16 mesoporous silica. Effect of the triblock surfactant Pluronic F127 degradation during the synthesis. Solid State Sci 2005;7(8):990–7.
[141] Wu W, DeConinck A, Lewis JA. Omnidirectional printing of 3D microvascular networks. Adv Mater 2011;23(24):H178–83.
[142] Fairbanks BD, Schwartz MP, Bowman CN, Anseth KS. Photoinitiated polymerization of PEG-diacrylate with lithium phenyl-2,4,6-trimethylbenzoylphosphinate: polymerization rate and cytocompatibility. Biomaterials 2009;30(35):6702–7.
[143] Huber B, Borchers K, Tovar GE, Kluger PJ. Methacrylated gelatin and mature adipocytes are promising components for adipose tissue engineering. J Biomater Appl 2016;30(6):699–710.
[144] Ovsianikov A, Deiwick A, Van Vlierberghe S, Dubruel P, Moller L, Drager G, et al. Laser fabrication of three-dimensional CAD scaffolds from photosensitive gelatin for applications in tissue engineering. Biomacromolecules 2011;12(4):851–8.
[145] Billiet T, Gevaert E, De Schryver T, Cornelissen M, Dubruel P. The 3D printing of gelatin methacrylamide cell-laden tissue-engineered constructs with high cell viability. Biomaterials 2014;35(1):49–62.
[146] Kundu J, Shim JH, Jang J, Kim SW, Cho DW. An additive manufacturing-based PCL-alginate-chondrocyte bioprinted scaffold for cartilage tissue engineering. J Tissue Eng Regen Med 2015;9(11):1286–97.
[147] Lee SJ, Nowicki M, Harris B, Zhang LG. Fabrication of a highly aligned neural scaffold via a table top stereolithography 3D printing and electrospinning. Tissue Eng A 2017;23(11–12):491–502.
[148] Lee JW, Choi YJ, Yong WJ, Pati F, Shim JH, Kang KS, et al. Development of a 3D cell printed construct considering angiogenesis for liver tissue engineering. Biofabrication 2016;8(1). 015007.
[149] Bandyopadhyay A, Bose S, Das S. 3D printing of biomaterials. MRS Bull 2015;40(2):108–15.
[150] Carrow JK, Kerativitayanan P, Jaiswal MK, Lokhande G, Gaharwar AK. Polymers for bioprinting. In: Atala A, Yoo JJ, editors. Essentials of 3D Biofabrication and Translation. Cambridge, MA: Academic Press; 2015.
[151] Panwar A, Tan LP. Current status of bioinks for micro-extrusion-based 3D bioprinting. Molecules 2016;21(6).
[152] Markstedt K, Mantas A, Tournier I, Martinez Avila H, Hagg D, Gatenholm P. 3D bioprinting human chondrocytes with nanocellulose-alginate bioink for cartilage tissue engineering applications. Biomacromolecules 2015;16(5):1489–96.
[153] Muller M, Ozturk E, Arlov O, Gatenholm P, Zenobi-Wong M. Alginate sulfate-nanocellulose bioinks for cartilage bioprinting applications. Ann Biomed Eng 2017;45(1):210–23.
[154] Muller M, Becher J, Schnabelrauch M, Zenobi-Wong M. Nanostructured pluronic hydrogels as bioinks for 3D bioprinting. Biofabrication 2015;7(3). 035006.

[155] John SA, Revel JP. Connexon integrity is maintained by non-covalent bonds: intramolecular disulfide bonds link the extracellular domains in rat connexin-43. Biochem Biophys Res Commun 1991;178(3):1312–8.
[156] Khodaverdi E, Gharechahi M, Alibolandi M, Tekie FS, Khashyarmanesh BZ, Hadizadeh F. Self-assembled supramolecular hydrogel based on PCL-PEG-PCL triblock copolymer and gamma-cyclodextrin inclusion complex for sustained delivery of dexamethasone. Int J Pharm Investig 2016;6(2):78–85.
[157] Li C, Faulkner-Jones A, Dun AR, Jin J, Chen P, Xing Y, et al. Rapid formation of a supramolecular polypeptide-DNA hydrogel for in situ three-dimensional multilayer bioprinting. Angew Chem Int Ed Engl 2015;54(13):3957–61.
[158] Highley CB, Rodell CB, Burdick JA. Direct 3D printing of shear-thinning hydrogels into self-healing hydrogels. Adv Mater 2015;27(34):5075–9.
[159] Mellati A, Fan CM, Tamayol A, Annabi N, Dai S, Bi J, et al. Microengineered 3D cell-laden thermoresponsive hydrogels for mimicking cell morphology and orientation in cartilage tissue engineering. Biotechnol Bioeng 2017;114(1):217–31.
[160] Guvendiren M, Molde J, Soares RM, Kohn J. Designing biomaterials for 3D printing. ACS Biomater Sci Eng 2016;2(10):1679–93.
[161] Tan Y, Richards DJ, Trusk TC, Visconti RP, Yost MJ, Kindy MS, et al. 3D printing facilitated scaffold-free tissue unit fabrication. Biofabrication 2014;6(2). 024111.
[162] Jakob PH, Kehrer J, Flood P, Wiegel C, Haselmann U, Meissner M, et al. A 3-D cell culture system to study epithelia functions using microcarriers. Cytotechnology 2016;68(5):1813–25.
[163] Mehta G, Hsiao AY, Ingram M, Luker GD, Takayama S. Opportunities and challenges for use of tumor spheroids as models to test drug delivery and efficacy. J Control Release 2012;164(2):192–204.
[164] Takeichi M. The cadherins: cell-cell adhesion molecules controlling animal morphogenesis. Development 1988;102(4):639–55.
[165] Jang J, Park HJ, Kim SW, Kim H, Park JY, Na SJ, et al. 3D printed complex tissue construct using stem cell-laden decellularized extracellular matrix bioinks for cardiac repair. Biomaterials 2017;112:264–74.
[166] Nicodemus GD, Bryant SJ. Cell encapsulation in biodegradable hydrogels for tissue engineering applications. Tissue Eng B Rev 2008;14(2):149–65.
[167] Murphy SV, Skardal A, Atala A. Evaluation of hydrogels for bio-printing applications. J Biomed Mater Res A 2013;101(1):272–84.
[168] Daly AC, Critchley SE, Rencsok EM, Kelly DJ. A comparison of different bioinks for 3D bioprinting of fibrocartilage and hyaline cartilage. Biofabrication 2016;8(4). 045002.
[169] Zhang ZZ, Wang SJ, Zhang JY, Jiang WB, Huang AB, Qi YS, et al. 3D-printed poly (epsilon-caprolactone) scaffold augmented with mesenchymal stem cells for total meniscal substitution. Am J Sports Med 2017;45:1497–511. 363546517691513.
[170] Gao L, Kupfer M, Jung J, Yang L, Zhang P, Sie Y, et al. Myocardial tissue engineering with cells derived from human induced-pluripotent stem cells and a native-like, high-resolution, 3-dimensionally printed scaffold. Circ Res 2017;120(8):1318–25.
[171] Zhu W, George J, Sorger V, Zhang L. 3D printing scaffold coupled with low level light therapy for neural tissue regeneration. Biofabrication 2017;9(2). Article No. 025002.
[172] Furth ME, Atala A, Van Dyke ME. Smart biomaterials design for tissue engineering and regenerative medicine. Biomaterials 2007;28(34):5068–73.
[173] Roy I, Gupta MN. Smart polymeric materials: emerging biochemical applications. Chem Biol 2003;10(12):1161–71.

[174] Hoffman AS. Bioconjugates of intelligent polymers and recognition proteins for use in diagnostics and affinity separations. Clin Chem 2000;46(9):1478–86.
[175] Stayton PS, Shimoboji T, Long C, Chilkoti A, Chen G, Harris JM, et al. Control of protein-ligand recognition using a stimuli-responsive polymer. Nature 1995;378(6556):472–4.
[176] Jung H, Pena-Francesch A, Saadat A, Sebastian A, Kim DH, Hamilton RF, et al. Molecular tandem repeat strategy for elucidating mechanical properties of high-strength proteins. Proc Natl Acad Sci 2016;113(23):6478–83.
[177] Simon Jr. CG, Yang Y, Dorsey SM, Ramalingam M, Chatterjee K. 3D polymer scaffold arrays. Methods Mol Biol 2011;671:161–74.
[178] Anderson DG, Levenberg S, Langer R. Nanoliter-scale synthesis of arrayed biomaterials and application to human embryonic stem cells. Nat Biotechnol 2004;22(7):863–6.
[179] Wagner DE, Phillips CL, Ali WM, Nybakken GE, Crawford ED, Schwab AD, et al. Toward the development of peptide nanofilaments and nanoropes as smart materials. Proc Natl Acad Sci U S A 2005;102(36):12656–61.
[180] Guler MO, Hsu L, Soukasene S, Harrington DA, Hulvat JF, Stupp SI. Presentation of RGDS epitopes on self-assembled nanofibers of branched peptide amphiphiles. Biomacromolecules 2006;7(6):1855–63.
[181] Placone JK, Navarro J, Laslo GW, Lerman MJ, Gabard AR, Herendeen GJ, et al. Development and characterization of a 3D printed, keratin-based hydrogel. Ann Biomed Eng 2017;45(1):237–48.

Assessing printability of bioinks

9

S. Kyle*, Z.M. Jessop*,†, S.P. Tarassoli‡, A. Al-Sabah*, I.S. Whitaker*,†
*Reconstructive Surgery and Regenerative Medicine Research Group, Swansea University Medical School, Swansea, United Kingdom, †The Welsh Centre for Burns and Plastic Surgery, Morrison Hospital, Swansea, United Kingdom, ‡Swansea University Medical School, Swansea, United Kingdom

9.1 Introduction

Congenital or acquired tissue loss and dysfunction is an enormous socioeconomic burden for healthcare systems globally [1]. Tissue engineering holds promise to repair or replace human tissues and organs in order to restore normal function. This requires interplay between various technologies and multidisciplinary fields such as cell, molecular and developmental biology, materials and computer sciences, engineering and clinical medicine.

Bioprinting is a computer-assisted technology which has emerged as an attractive method for assembling living and nonliving biological materials into spatially controlled, precise, and well-defined three-dimensional (3D) structures. Bioprinting aims to fabricate engineered tissues or organs by designing and guiding specific cell types and materials into organized geometries that mimic native tissue architecture [2]. Changes and control of advanced geometry can be introduced by 3D printing in order to mimic the mechanical property in native tissues [3]. Groll et al. have defined bioprinted constructs to be "when living single cells, bioactive molecules, biomaterials, or cell-aggregates small enough to be printed are used for fabrication" [4]. Moreover, bioprinting devices have the capability to print cells and biomaterials in specific locations in a repeatable and high throughput manner [5]. A combination of the "biopaper," which provides structural and mechanical support, and specialized cells are referred to as the "bioink" (discussed in Chapter 9). The central dogma to the creation of 3D constructs lies in the careful selection of the bioink and bioprinter so that fully functional tissue can be fabricated (Fig. 9.1). It is therefore essential that bioinks have the desired functional and mechanical properties that closely match the tissue it is to replace. Ultimately, this means that the bioink and cell integrity in the printed construct must be maintained. Hence, there is a need to correlate process and printing parameters and factors that affect printability with the shape fidelity and print resolution of the final fabricated construct. This chapter focuses on understanding the printability of bioinks in 3D bioprinting.

9.2 *The* biofabrication window *and printability*

Printability can be defined as the relationship between bioinks and substrates that results in printing an accurate, high-quality pattern [6]. The *biofabrication window*

3D Bioprinting for Reconstructive Surgery. https://doi.org/10.1016/B978-0-08-101103-4.00027-2

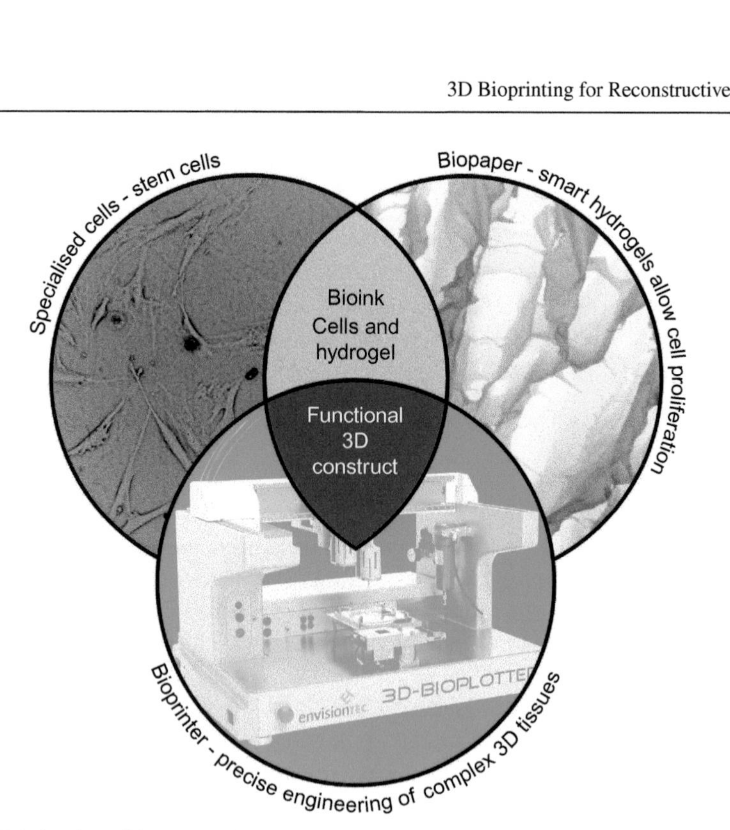

Fig. 9.1 Fabrication of functional 3D constructs can be achieved through careful selection and control of (1) specialized cells, (2) biopaper, and (3) a suitable bioprinter.
From Kyle S, Jessop ZM, Al-Sabah A, Whitaker IS. 'Printability' of candidate biomaterials for extrusion based 3D printing: state-of-the-art. Adv Healthc Mater 2017;6(16). https://doi.org/10.1002/adhm.2017002642017.

paradigm (Fig. 9.2) necessitates compromise between suitability for fabrication (printability) and the ability to encapsulate cells and maintain cell viability [1,7,8]. Traditionally, there was a trade-off between the ideal bioink and cell culture environments. Often, bioinks had inadequate mechanical support in the printed construct with limited bioactivity and functionality to maintain appropriate cell biology. A new generation of advanced bioinks are now being designed to significantly improve shape and print fidelity, and biocompatibility. To design and utilize such advanced bioinks, careful control of printing parameters, in addition to physicochemical and biological properties are required.

9.3 Nozzle/printing parameters and filament dimensions

To successfully translate bioprinting technology toward the clinical setting, careful control of printing accuracy, and cell viability is required. Many studies have reported better relationships between manipulation of printing parameters and fabricated tissue outcomes. It has been established that nozzle diameter, printing speed, material flow

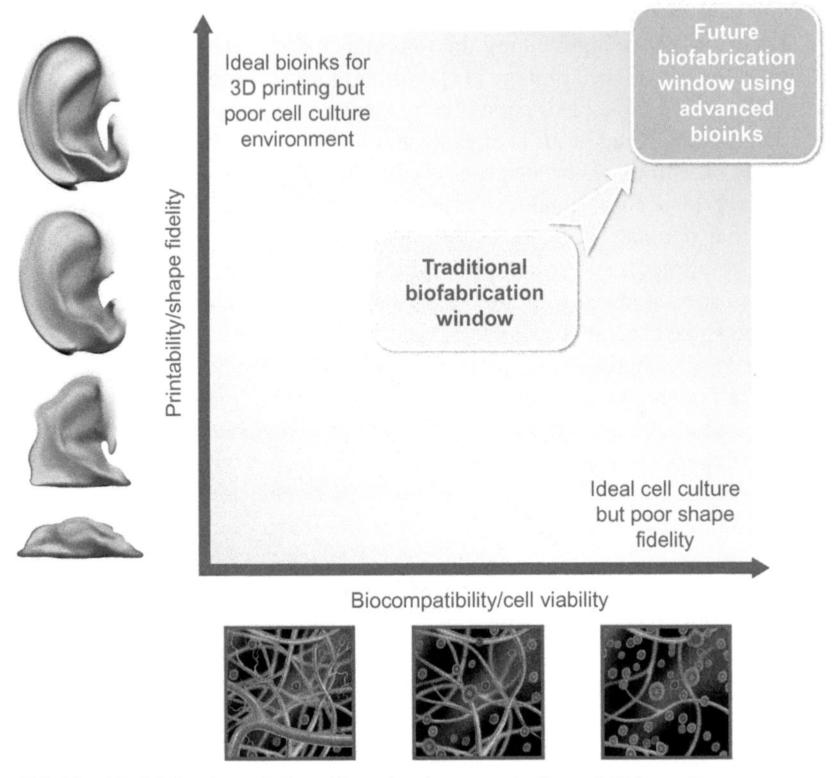

Fig. 9.2 The *biofabrication window*. There has been a paradigm shift from the more traditional bioinks which have seen a standoff between shape fidelity and biocompatibility (top left and bottom right).
From Kyle S, Jessop ZM, Al-Sabah A, Whitaker IS. 'Printability' of candidate biomaterials for extrusion based 3D printing: state-of-the-art. Adv Healthc Mater 2017;6(16). https://doi.org/10.1002/adhm.2017002642017; Chimene D, Lennox KK, Kaunas RR, Gaharwar AK. Advanced bioinks for 3D printing: a materials science perspective. Ann Biomed Eng 2016; 44(6): 2090–2102; Malda J, Visser J, Melchels FP, Jungst T, Hennink WE, Dhert WJ, et al. 25th anniversary article: engineering hydrogels for biofabrication. Adv Mater 2013; 25(36): 5011–5028.

rate, and printed filament height are fundamental when determining widths and heights of printed hydrogel filaments [9]. Chang et al. [9] found that printing viscous hydrogels through narrow nozzles resulted in thinner print widths which required a higher dispensing pressure to overcome resistance by the nozzle diameter. In another earlier study, Hockaday et al. [10] implemented an onboard printing/photocross-linking technique for rapidly engineering complex, heterogeneous aortic valve scaffolds. The 3D printing geometric accuracy/shape fidelity was quantified and compared using micro-computed tomography (micro-CT). They found that the 3D printing strategy generated scaffolds with high geometric precision, but accuracy decreased to some extent with reduced size. They proposed that print resolution may be enhanced with smaller nozzle diameters and lower extrusion rates. More recently, smaller diameter nozzles and valves have significantly compromised chondrocyte cell proliferation following bioink extrusion, hence highlighting the importance of nozzle geometry not only on print fidelity but also on cell biology [11]. Furthermore, cell spreading properties were maintained with the lowest extrusion pressure and shear stress.

In a later study, Kang et al. [12] evaluated the effects of printing parameters on the accuracy of various 3D printed hydrogels. They used quantitative image analysis to determine print accuracy and resolution along the length, width, and height. They concluded that optimal nozzle pressure, path height (which determines how many layers will be printed), and path width (which determines the distance between adjacent paths for each layer) were increased linearly with nozzle diameter. Moreover, the predicted value generated accurate 3D geometries while poorly chosen parameters yielded inaccurate, unpredictable geometries.

Clearly, a better understanding of assessing printability of 3D bioprinted constructs is required. Some groups have tried to assess the printability of bioinks based on bioink composition and rheology [13], whereas others have reported on the physical and rheological properties of bioinks under different printing conditions using different printing instruments [8]. Correlations have also been made between printability and bioink mechanical properties [14]. However, it is not until recently that research has moved toward assessing relationships between printing parameters and printing fidelity. He et al. [15] investigated printability of hydrogels from printing lines in 1D to printing lattices/films in 2D and then printing 3D structures. Hydrogels were created using alginate and gelatin, with calcium chloride as the cross-linking agent. They investigated the influence of air pressure, feed rate, printing distance, and printing sharp angles on the structural integrity of printed lines, lattices, and 3D structures. Printing quality was also found to be worse with acute angles compared with obtuse and right angles, but could be improved when the extrusion rate was controlled by the motor speed. Furthermore, they highlighted that diffusion rates had a direct impact on the shape fidelity, whereby line distance and intersection area of the lines affected lattice quality. Diffusion does not appear to play an important role with larger structures with larger line distances, when compared with smaller, more intricate structures.

It is evident from earlier reports on 3D printing that a trial and error approach was mostly adopted and the assessment of printability appeared less important. Research has definitely moved in the direction of not only being able to fabricate a 3D construct with encapsulated cells but also more importantly that this construct must have a degree of structural integrity over time. In fact, excellent progress has been made in trying to define printability and research now is paving the way into trying to quantify the concept so that 3D constructs can be fabricated with meticulous precision in a reproducible manner on a large scale. Ouyang et al. [16] have made an excellent attempt in semiquantifying the process of printability by understanding the influence of bioink properties and printing parameters. Using various mathematical equations, they were able to predict and confirm that 3D hydrogel constructs demonstrated excellent filament morphology and mechanical stability. They also found that both printability and cell viability were influenced by three main factors: (1) printing temperature, (2) bioink concentration, and (3) holding time.

Another factor that has been shown to be important when assessing printability is surface tension, which is measured by the contact angles between two media [17]. Surface tension and wettability have been shown to be important in droplet formation in inkjet printing since the affinity of the nanofluid for the substrate can be formulated from the solid surface tensions [18]. Surface tension of the bioink can influence printing quality, print resolution, and resulting dimensions of printed filaments [19]. It has been proposed that filaments should be printed thinner than the target size so that the expanded final state matches the native model [10]. An understanding of the mechanics of surface tension would therefore be critical in controlling print fidelity and resolution. To build constructs in 3D, bioinks should not be printed too flat on the substrate. This means that there must be sufficient surface tension in the vertical direction and have a large contact angle with the substrate [17]. Hence, the actual substrate by which bioinks are to be printed plays an important role in helping to create 3D constructs with structural integrity. Owing to glass being a commonly employed substrate with a poor contact angle, research has moved toward better surface coatings in order to improve printability [20]. Contact angles above and below 90 degrees correspond to hydrophobic and hydrophilic surfaces, respectively. Studies have exploited this so that surfaces can also be improved for better cell attachment/interaction in 3D constructs [21].

9.4 Viscosity and rheology

An understanding of the rheological properties of bioinks is critical for successful 3D bioprinting. The ink should be sufficiently fluid to be extruded through smaller diameter nozzles without requiring high pressures (>4 bars). In the absence of shear, filamentary deformation can be maintained when the bioink has a high enough elastic modulus (> few kPa) and yield stress (~few 10^2 Pa) [22]. It has been suggested that the rheological requirements can be achieved by careful design of bioinks that display a non-Newtonian viscoelastic response, generally evident by its high storage modulus over the loss modulus at low shear stresses [23,24].

It is well documented that lower viscosity bioinks can often result in collapse post-printing. Initially, studies manipulated bioink concentration and molecular weights in order to change viscosity, but with little improvement in shape fidelity of printed 3D constructs. For this reason, many studies have simultaneously printed bioinks with other materials. Alginates isolated from brown algae have been used as bioinks for 3D bioprinting. The main benefit for using alginate is the ease of cross-linking which can be achieved using divalent cations, such as calcium or barium chloride. Even with careful manipulation of bioink concentration, the required viscosity is often not achieved. Printability of alginate has often been improved with sacrificial polymers [25], gelatin [13], chitosan [26], and agarose [27].

Nanocellulose is another material that many groups have employed owing to its unique physicochemical and biological properties, which include high stiffness and strength, flexibility, chemical modification capacity, high surface area and aspect ratio, and excellent biocompatibility. Nanocellulose has been combined with alginate

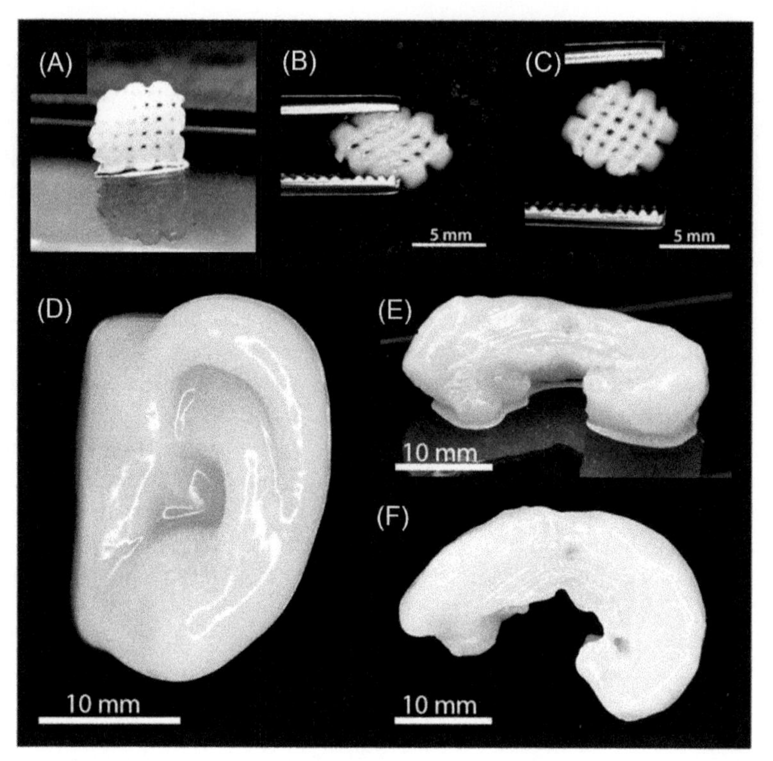

Fig. 9.3 3D bioprinting of nanocellulose-alginate bioinks. (A) Printed small grids with nanocellulose:alginate 80:20 (w/w) after cross-linking with calcium chloride. Shape of grid (B) deformed while squeezing and (C) restored after squeezing. 3D printed (D) human ear, and (E) top and (F) side views of sheep meniscus.
From Markstedt K, Mantas A, Tournier I, Martinez Avila H, Hagg D, et al. 3D bioprinting human chondrocytes with nanocellulose-alginate bioink for cartilage tissue engineering applications. Biomacromolecules 2015; 16(5): 1489–1496.

to prepare sponges used for adipose tissue [28], cell encapsulation [29], and potential applications in wound dressings and biosensing [30]. Nanocellulose and alginate composites have increasingly been used for cartilage tissue engineering applications. This was first highlighted by Markstedt et al. [31] who found that nanocellulose alone was shear thinning at higher viscosities which led to excellent shape fidelity, but this was destroyed when mechanical force was applied. By combining alginate and nanocellulose, a bioink with the rheological properties of nanocellulose and cross-linking ability of alginate was formulated. A variety of 3D bioprinted structures is shown in Fig. 9.3. This group more recently went on to show that 3D bioprinted constructs composed of nanofibrillated cellulose:alginate (60:40 w/w%) initially maintained pluripotency of human-derived pluripotent stem cells, and after 5 weeks, hyaline-like cartilaginous tissue with collagen type II expression and lacking tumorigenic Oct4 expression was observed [11].

Nanocellulose can also be utilized as rheological modifiers for the bioinks. This can be done not only by ensuring the viscoelastic response required for filament printing, but also by their specific mechanical properties, their ability to support cells, and the possibility to accurately control pore structure and shape in scaffolds and implants. Higher viscosity (due to a higher concentration, molecular weight, or cross-linking) ensures a better postprinting structural stability but at the same time lowers the cell viability due to increased shear force. Thus, optimizing viscosity and cross-linking is essential to prepare a bioink which is cell compatible and provides high fidelity scaffold structures [32].

Other rheological modifiers or thickening agents have also been employed in order to improve print and shape fidelity. Gellan gum, an anionic polysaccharide has been used in tissue engineering because of its versatile thermo-reversible gelling properties. Good print fidelity is also achieved with high viscosity, shear thinning behavior, and high yield stress [33]. The stabilization procedure after printing is also important in maintaining print fidelity [8]. Gellan gum at lower concentrations has been used in 3D bioprinting for scaffold-based tissue engineering, but a lower viscosity has often generated constructs with poor print fidelity. Studies have therefore used either higher polymer concentrations to attain sufficient viscosity for printing [34], or used multifunctional, cross-linked blends of polymers [35]. Recently, alginate and gellan gum (cross-linked with calcium chloride) have been used for 3D printing hydrogel scaffolds, as alginate has excellent processability and gellan gum gives enhanced stability [33]. Characterization of the resulting composite scaffolds revealed an increased stiffness and lower volume swelling in cell culture medium compared to pure alginate scaffolds. In addition, human mesenchymal stem cells showed better cell attachment and proliferation than pure alginate. Calcium phosphate scaffolds combined with vascular endothelial growth factor (VEGF)-laden alginate-gellan gum hydrogel strands have also been shown to be suitable as VEGF delivery systems in bone tissue regeneration [36]. Scaffold shape fidelity was assessed using stereomicroscopic analysis and micro-CT, and was not compromised in the fabrication process.

What is particularly evident when assessing printability of bioinks in 3D printing is most research groups rarely manipulate only one printing parameter in order to improve print/shape fidelity of the final fabricated construct. It is the strategic control of various parameters, as discussed in this chapter, that helps create 3D constructs with excellent shape fidelity and biocompatibility. For example, Pluronic (or poloxamer) is a block copolymer comprising poly(propylene oxide) and poly(ethylene oxide) blocks. It has been used in a variety of applications, including drug delivery, wound dressings, 3D printing, and sacrificial molding. It has received much attention due to its excellent chemical, physical, and biological properties. Pluronic gels have been found to be shear thinning and show good shear recovery which is essential for 3D extrusion printing. Muller et al. [37] have extensively investigated Pluronic composition and concentrations, rheology, nozzle pressures and feed rates, and cell viability in order to show good printability of 3D constructs. They found one composition to have both shear thinning in its gelled state and Newtonian behavior below its gelation temperature, which may simplify handling with the bioinks. It was found that this composition exhibited quick shear recovery which meant that the printed structures did not flow when printed and provided a high initial printing fidelity. Long-term shape fidelity was further enhanced by the fact that the Pluronic composition had a yield point rather than a zero-shear viscosity.

Rheology of bioinks is directly related to printing parameters. Shear stress, in particular, is very much influenced by nozzle diameters, printing pressures, and viscosity of the dispensed bioink [38,39]. The ability to control shear stress is paramount in 3D printing so that the microenvironment is able to optimize cell biology and maintain or enhance printing resolution and shape/print fidelity of the final construct. A study by Blaeser et al. [40] showed that hydrogel viscosity and nozzle size directly affect shear stress. Using a microvalve-based bioprinting system and applying fluid-dynamic models, they were able to precisely control shear stress at the nozzle site, with adjustments in hydrogel viscosity, printing pressure, and nozzle diameter. They were able to generate multilayered hydrogel structures with high resolution at a low shear stress. This was a pioneering study that highlighted how better control of shear stress in 3D bioprinting has a direct impact on the printing resolution and stem cell integrity.

9.5 Cross-linking and sacrificial/support materials

This chapter has focused on an array of biofabrication strategies. These include direct bioprinting whereby bioinks are printed as a result of predetermined shapes and configurations. The bioink formulations are optimized to achieve desired physicochemical properties and mechanical integrity for fast shear recovery during and after printing [8,41–43]. In addition, a quasiscaffold structure that supports cell differentiation, migration, and/or proliferation postprinting must be fabricated. These bioinks can be optimized via cross-linking using chemicals [44], heat [45], ultraviolet (UV) light [46], or a mixture of cross-linking agents [47]. Other methods of cross-linking include depositing the bioink into a cross-linking bath [48], sequentially depositing the cross-linking agent and bioink through careful control of printing parameters [49], or using multiple bioinks with various cross-linking mechanisms to enhance printability [50].

In fact, novel cross-linking systems have been employed for improved print fidelity. One study highlighted the importance of oxygen inhibition on print fidelity of 3D biofabricated and photopolymerizable gelatin-methacryloyl hydrogel constructs [51]. The system was composed of water-soluble photoinitiators ruthenium (Ru) and sodium persulfate (SPS) compounds which absorb photons in the visible light (Vis) range. They found that compared to the UV light + Irgacure 2959 (I2595) system, the Vis + Ru/SPS system yielded cell viability >85% at high Ru/SPS concentrations and Vis irradiation intensities for up to 21 days. In addition, 3D biofabricated constructs showed high shape fidelity as shown in Fig. 9.4.

Physiological stability of printed constructs has been achieved by a number of different cross-linking methods. In cartilage tissue engineering, hyaluronan and chondroitin sulfate have often been used in combination with thermoresponsive polymers to enhance printability and cell biology [52]. Kesti et al. [53] developed a dual cross-linked bioink composed of methacrylated hyaluronan (HA-MA) and poly(*N*-isopropylacrylamide)

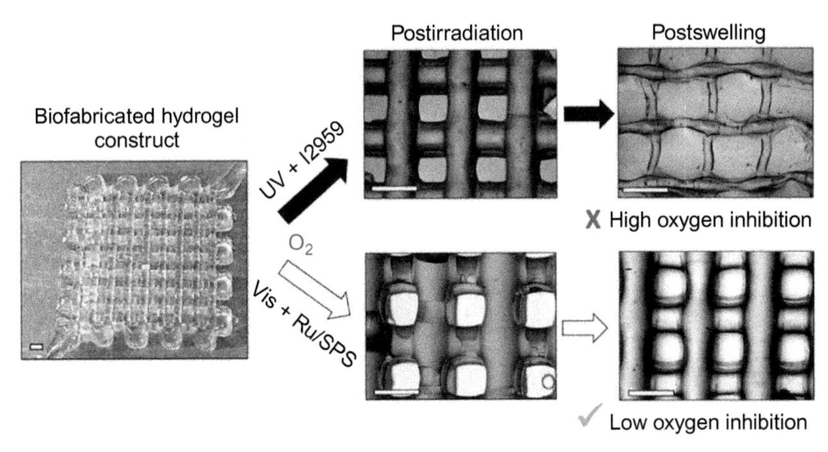

Fig. 9.4 3D plotting of gelatin + methacryloyl + collagen-based bioink, photopolymerized with UV + I2959 and Vis + Ru/SPS systems. Shape fidelity can be seen postirradiation and postswelling in high and low oxygen inhibition environments.
From Lim KS, Schon BS, Mekhileri NV, Brown GCJ, Chia CM, Prabakar S, et al. New visible-light photoinitiating system for improved print fidelity in gelatin-based bioinks. ACS Biomater Sci Eng 2016; 2(10): 1752–1762.

grafted hyaluronan (HA-pNIPAAM). These hydrogels were liquid at room temperature and formed a gel at body temperature. After stabilization of the biopolymer with free radical polymerization of the HAMA, the HA-pNIPAAM was eluted leaving a glycosaminoglycan-based scaffold. The thermoresponsive nature of the HA-pNIPAAM component provided rapid gelation and enhanced shape and print fidelity in the post-printed construct.

To improve versatility and tunability, and optimize structural integrity and biological performance while maintaining printability, Rutz et al. [54] designed a bioink from gelatin methacrylate (GelMA) and multifunctional polyethylene glycol (PEGX, where X is succinimidyl valerate). PEGX was used to loosely cross-link the gelatin backbone and provide the necessary viscosity for bioprinting. PEGX cross-linking allowed for chemical and physical properties to be carefully tuned, and may further expand the number of 3D printable bioinks available. The resulting constructs showed high shape and structural fidelity, without compromising cytocompatibility. This novel cross-linking system shows great potential toward developing tailorable platforms for studying cell-cell signaling and tissue morphogenesis in 3D, in addition to creating more customized, biomimetic 3D printed tissue constructs [54].

Bioprinted constructs have also been fabricated in tandem with support or sacrificial materials which provide extra mechanical support and strength [47,55,56]. In some cases, support baths [57,58], often in the form of cross-linking agents, have been employed which help to improve the final shape fidelity of the printed construct. These baths can then be removed postprocessing. Bioink printability has been assessed when extruded into a gelatin support bath [57]. This support bath maintains the intended structure during the print process and significantly improves print fidelity. They used multimaterial bioinks to build complex 3D structures by extruding the material into a

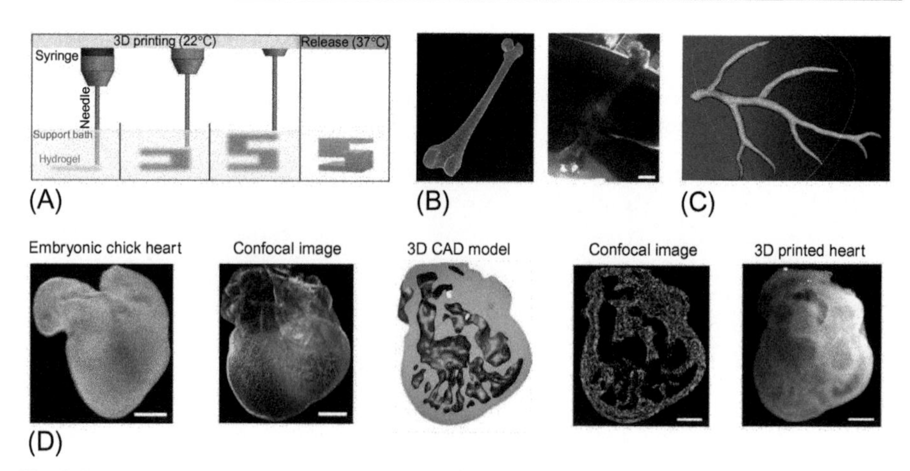

Fig. 9.5 Multimaterial bioinks for 3D bioprinting using sacrificial support baths. (A) Hydrogel bioinks printed within a thermoreversible gelatin support bath. Using computer models, complex tissue structures can be printed, including (B) a human femur (scale bar = 4 mm), (C) a human right coronary arterial tree, and (D) an explanted embryonic chick heart (scale bar = 1 mm).
Modified from Hinton TJ, Jallerat Q, Palchesko RN, Park JH, Grodzicki MS, Shue HJ, et al. Three-dimensional printing of complex biological structures by freeform reversible embedding of suspended hydrogels. Sci Adv 2015; 1(9): e1500758.

gelatin slurry. Once the 3D structure is printed, the temperature of the bath is raised to 37°C, causing the gelatin support bath to melt in a nondestructive manner. This study highlighted how complex structures such as a human femur, right coronary arterial tree, and an explanted embryonic chick heart can be printed with exquisite structural fidelity (Fig. 9.5).

Owing to the complex shapes and compositions of various tissues in the body, 3D printing technology has been employed in combination with sacrificial layers in order to improve print resolution, shape fidelity, and ultimately tissue regeneration. This is particularly evident in ear reconstruction for the regeneration of auricular cartilage and adipose tissue. Lee et al. [25] printed poly-caprolactone and cell-laden alginate hydrogel, with a poly-ethylene-glycol (PEG) sacrificial layer to support the main structure. Upon fabrication, the PEG could be easily removed in aqueous solutions, and this removal was shown to have no detrimental impact on cell viability.

Alginate has been used as a sacrificial template to generate polymeric network templates that physically entrap various prepolymer solutions. Tamayol et al. [59] mixed alginate with a bioactive polymer to generate pure polymeric fibers using sacrificial polymeric networks. Following gelation using calcium chloride, the polymer replicates the structure of the alginate template. Upon selective dissolution of the alginate by a calcium chelating agent, a homogeneous and polymeric structure remains. The resulting fiber networks has shown promise in vascular tissue engineering [14].

In the absence of perfusable vascular networks, 3D engineered tissues densely populated with cells quickly develop a necrotic core [55]. Yet creating such networks remains a major challenge for 3D tissue culture [60]. The 3D printed fugitive materials removed postprocessing have been later shown to be highly effective for predefining void regions within a bulk gel for microvascular networks [61]. Studies have used the thermosensitive Pluronic F-127 ink or water-soluble carbohydrate glass, casted on an ECM gel based on a blend of gelatin and fibrinogen around the printed structure. The sacrificial bioink is removed to produce vascular-like tubular structures [62,63]. This approach is translatable to other 3D void environments that are otherwise difficult to fabricate, such as convoluted proximal tubules in the kidney [64] and vascular lattice networks [55].

9.6 Computer simulation and modeling

Computer and complex mathematic models are advancing the fabrication of tunable, responsive, and multifunctional materials in 3D bioprinting. In moving forward from a trial and error approach to fabrication, it is clear that printing parameters must be carefully tuned in order to create revolutionary materials. Computational approaches have also helped to design scaffolds with gradient porosity [65], and have been used to study the interactions between scaffold pore morphology, mechanical stimuli, and cell culture conditions [66].

To fabricate constructs with better shape fidelity, many groups have employed deposition models to understand the printing quality of filaments. Various mathematical models based on the Hagen-Poiseuille, Bernoulli, and Ostwald-de Waele laws have been utilized to understand the effects of various nozzle parameters on the structural integrity of the printed construct. These models help shed light on relationships between filament dimensions and printing parameters.

It is clearly evident that computer simulation can help design better 3D constructs. Yet what is critical is that these constructs are functional and remain functional in their targeted applications. Therefore, computer simulation should help predict changes in the structure and function of bioprinted materials. Monte Carlo kinetic studies have been employed to study cell self-assembly and cellular aggregate fusion of multicellular systems in postprinting constructs [67]. The model was developed to describe and predict time evolution of postprinting morphological structure formation during morphogenesis of tissues or organs. It was evident that much research is needed to study interactions between cells and materials in order to achieve better simulation data.

Experimental, theoretical, and computational formalisms have been developed to rationalize complex, self-assembly processes to produce large-scale functional tissue. Shape evolution during tissue fusion [68] has been shown to be analogous to merging of liquid drops [69,70]. This analogy was adopted by Shafiee et al. [71] who used cellular particles dynamics (CPD) to predict shape evolution of multicellular systems that undergo shape-changing biomechanical relaxation. This has been translated to make 3D bioprinting more predictive, reproducible, and time efficient. They found that complex, mathematical modeling via CPD simulations can be used to predict postprinting structure formation even in the case of volume-changing bioink units.

Rapid prototyping technologies help build 3D constructs in a layer-by-layer manner. In combination with computational models and simulation, they have been indispensable tools in diagnosis and treatment planning in reconstructive surgery. A reverse engineering approach was adopted by Shahbazian et al. [72] who employed a cone beam CT-based stereolithographic tooth replica and surgical guide in order to reduce injury to the periodontal ligament of the donor tooth in the autotransplantation procedure. These techniques provide an accurate method to produce models of donor teeth that serve as prototypes for recipient site preparation in tooth autotransplantation procedures.

9.7 Challenges and future outlook

The current bottleneck in designing complex tissue structures and organs using 3D bioprinting technologies is the limited availability of versatile bioinks [7]. Multimaterial bioinks have been extensively explored; however, further evaluation of polymer combinations with tunable and versatile physicochemical and biological properties is required. Smart, adaptive, and multifunctional bioinks consisting of interpenetrating networks, nanoparticle composites, and/or supramolecular hydrogels may pave the way in providing facile and effective methods for attaining desirable properties. These rationally designed materials must have excellent printability and structural/print fidelity, and superior mechanical, physical, and biological properties. In addition, the potential to biofunctionalize bioinks and incorporate bioactivity may enhance physicochemical and biological properties of next generation biomaterials. Advanced bioinks also have the potential to be stimuli responsive where cells and materials interact in various physicochemical and biological environments. Over time, these materials will have the ability to change shape and adapt to these evolving environments. This will be achieved through advanced engineering and technology by carefully designing state-of-the-art multihead bioprinters which have the ability to produce complex, heterogeneous, multicellular tissue constructs at clinically relevant scales and in a time efficient manner.

The ability to distribute materials and cells in a temporo-spatially controlled, homogenous manner within viscous hydrogels must also be a focus for future studies. Simultaneously, the 3D constructs must be printed with high precision and resolution. Hence, they pave the way for emerging 4D printing methods. Long-term effects of 3D bioprinted constructs on cell differentiation, proliferation, and maturation must also be addressed. This will require consideration of nutrient transport and vascularization networks within hydrogels. Careful selection of bioinks and better manipulation of printing parameters will be critical for achieving enhanced properties. Merging multimaterial bioinks with multiple process/printing parameters will prove essential for improving biological functional and complexity.

9.8 Conclusions

3D bioprinting is an integrated field that brings collaboration between engineering, materials science, cell biologists, and clinicians with the goal of addressing current problems with tissue/organ dysfunction and failure. Early studies focused on the assessment of printability of bioinks while investigating different biofabrication platforms. Printability was assessed based on mechanical properties and biocompatibility of 3D printed constructs. Advanced technology and better bioprinter designs now allow for more precise control of process and print parameters during the printing process so that fabricated constructs have accurate shape fidelity and exquisite print resolution. These systems will provide advanced motion control and metrological validation, and produce complex 3D structures with nanoliter precision. Studies have shown that key factors that should be controlled in order to achieve better printability, high shape fidelity, and reproducibility include bioink concentration and composition, nozzle/print and process parameters, rheology, and use of computational modeling. In fact, computational and mathematical models may prove essential for future studies in helping to predict and design complex 3D geometries that evolve over time due to cell-material remodeling.

Worldwide there is an huge shortage of organ donors. It is therefore in the interests of patients, scientists, clinicians, and industry to see translation of 3D bioprinting to clinical applications. This technology can only be successful if 3D tissue constructs are produced in a scalable, cost-effective and reproducible manner, and overcome regulatory/legal obstacles. Advances in personalized medicine also mean that products will be made to order rather than available "off the shelf." 3D bioprinting technologies offers great hope in producing a paradigm shift in clinical practice.

References

[1] Kyle S, Jessop ZM, Al-Sabah A, Whitaker IS. 'Printability' of candidate biomaterials for extrusion based 3D printing: state-of-the-art. Adv Healthc Mater 2017;6(16):https://doi.org/10.1002/adhm.201700264.

[2] Derby B. Printing and prototyping of tissues and scaffolds. Science 2012;338(6109):921–6.

[3] Wu DJ, Bouten CVC, Dankers PYW. From molecular design to 3D printed life-like materials with unprecedented properties. Curr Opin Biomed Eng 2017;2:43–8.

[4] Groll J, Boland T, Blunk T, Burdick JA, Cho DW, Dalton PD, et al. Biofabrication: reappraising the definition of an evolving field. Biofabrication 2016;8(1):013001.

[5] Skardal A, Atala A. Biomaterials for integration with 3-D bioprinting. Ann Biomed Eng 2015;43(3):730–46.

[6] Murphy SV, Atala A. 3D bioprinting of tissues and organs. Nat Biotechnol 2014;32(8):773–85.

[7] Chimene D, Lennox KK, Kaunas RR, Gaharwar AK. Advanced bioinks for 3D printing: a materials science perspective. Ann Biomed Eng 2016;44(6):2090–102.

[8] Malda J, Visser J, Melchels FP, Jungst T, Hennink WE, Dhert WJ, et al. 25th anniversary article: engineering hydrogels for biofabrication. Adv Mater 2013;25(36):5011–28.

[9] Chang CC, Boland ED, Williams SK, Hoying JB. Direct-write bioprinting three-dimensional biohybrid systems for future regenerative therapies. J Biomed Mater Res B Appl Biomater 2011;98(1):160–70.

[10] Hockaday LA, Kang KH, Colangelo NW, Cheung PY, Duan B, Malone E, et al. Rapid 3D printing of anatomically accurate and mechanically heterogeneous aortic valve hydrogel scaffolds. Biofabrication 2012;4(3):035005.

[11] Nguyen D, Hagg DA, Forsman A, Ekholm J, Nimkingratana P, Brantsing C, et al. Cartilage tissue engineering by the 3D bioprinting of iPS cells in a nanocellulose/alginate bioink. Sci Rep 2017;7(1):658.
[12] Kang KH, Hockaday LA, Butcher JT. Quantitative optimization of solid freeform deposition of aqueous hydrogels. Biofabrication 2013;5(3):035001.
[13] Chung JHY, Naficy S, Yue Z, Kapsa R, Quigley A, Moulton SE, et al. Bio-ink properties and printability for extrusion printing living cells. Biomater Sci 2013;1(7):763–73.
[14] Bertassoni LE, Cardoso JC, Manoharan V, Cristino AL, Bhise NS, Araujo WA, et al. Direct-write bioprinting of cell-laden methacrylated gelatin hydrogels. Biofabrication 2014;6(2):024105.
[15] He Y, Yang F, Zhao H, Gao Q, Xia B, Fu J. Research on the printability of hydrogels in 3D bioprinting. Sci Rep 2016;6:29977.
[16] Ouyang L, Yao R, Zhao Y, Sun W. Effect of bioink properties on printability and cell viability for 3D bioplotting of embryonic stem cells. Biofabrication 2016;8(3):035020.
[17] Mandrycky C, Wang Z, Kim K, Kim DH. 3D bioprinting for engineering complex tissues. Biotechnol Adv 2016;34(4):422–34.
[18] Vafaei S, Wen D, Borca-Tasciuc T. Nanofluid surface wettability through asymptotic contact angle. Langmuir 2011;27(6):2211–8.
[19] Vafaei S, Tuck C, Ashcroft I, Wildman R. Surface microstructuring to modify wettability for 3D printing of nano-filled inks. Chem Eng Res Des 2016;109:414–20.
[20] Chen F, Zhang D, Yang Q, Yong J, Du G, Si J, et al. Bioinspired wetting surface via laser microfabrication. ACS Appl Mater Interfaces 2013;5(15):6777–92.
[21] Wang W, Caetano G, Ambler SW, Blaker JJ, Frade AM, Mandal P, et al. Enhancing the hydrophilicity and cell attachment of 3D printed PCL/graphene scaffolds for bone tissue engineering. Materials 2016;9(12):992.
[22] Sultan S, Siqueira G, Zimmermann T, Mathew AP. 3D printing of nano-cellulosic biomaterials for medical applications. Curr Opin Biomed Eng 2017;2:29–34.
[23] Studart AR. Additive manufacturing of biologically-inspired materials. Chem Soc Rev 2016;45(2):359–76.
[24] Compton BG, Lewis JA. 3D-printing of lightweight cellular composites. Adv Mater 2014;26(34):5930–5.
[25] Lee JS, Hong JM, Jung JW, Shim JH, Oh JH, Cho DW. 3D printing of composite tissue with complex shape applied to ear regeneration. Biofabrication 2014;6(2):024103.
[26] Demirtas TT, Irmak G, Gumusderelioglu M. A bioprintable form of chitosan hydrogel for bone tissue engineering. Biofabrication 2017;9(3):035003.
[27] Daly AC, Critchley SE, Rencsok EM, Kelly DJ. A comparison of different bioinks for 3D bioprinting of fibrocartilage and hyaline cartilage. Biofabrication 2016;8(4):045002.
[28] Krontiras P, Gatenholm P, Hagg DA. Adipogenic differentiation of stem cells in three-dimensional porous bacterial nanocellulose scaffolds. J Biomed Mater Res B Appl Biomater 2015;103(1):195–203.
[29] Park M, Shin S, Cheng J, Hyun J. Nanocellulose based asymmetric composite membrane for the multiple functions in cell encapsulation. Carbohydr Polym 2017;158:133–40.
[30] Leppiniemi J, Lahtinen P, Paajanen A, Mahlberg R, Metsa-Kortelainen S, Pinomaa T, et al. 3D-printable bioactivated nanocellulose-alginate hydrogels. ACS Appl Mater Interfaces 2017;9(26):21959–70.
[31] Markstedt K, Mantas A, Tournier I, Martinez Avila H, Hagg D, Gatenholm P. 3D bioprinting human chondrocytes with nanocellulose-alginate bioink for cartilage tissue engineering applications. Biomacromolecules 2015;16(5):1489–96.

[32] Datta P, Ayan B, Ozbolat IT. Bioprinting for vascular and vascularized tissue biofabrication. Acta Biomater 2017;51:1–20.
[33] Akkineni RA, Ahlfeld T, Funk A, Waske A, Lode A, Gelinsky M. Highly concentrated alginate-gellan gum composites for 3D plotting of complex tissue engineering scaffolds. Polymers 2016;8(5):170.
[34] Luo Y, Lode A, Akkineni AR, Gelinsky M. Concentrated gelatin/alginate composites for fabrication of predesigned scaffolds with a favorable cell response by 3D plotting. RSC Adv 2015;5(54):43480–8.
[35] Hart LR, Li S, Sturgess C, Wildman R, Jones JR, Hayes W. 3D printing of biocompatible supramolecular polymers and their composites. ACS Appl Mater Interfaces 2016;8(5):3115–22.
[36] Ahlfeld T, Akkineni AR, Förster Y, Köhler T, Knaack S, Gelinsky M, et al. Design and fabrication of complex scaffolds for bone defect healing: combined 3D plotting of a calcium phosphate cement and a growth factor-loaded hydrogel. Ann Biomed Eng 2017;45(1):224–36.
[37] Muller M, Becher J, Schnabelrauch M, Zenobi-Wong M. Nanostructured pluronic hydrogels as bioinks for 3D bioprinting. Biofabrication 2015;7(3):035006.
[38] Chang R, Nam J, Sun W. Effects of dispensing pressure and nozzle diameter on cell survival from solid freeform fabrication-based direct cell writing. Tissue Eng Part A 2008;14(1):41–8.
[39] Nair K, Gandhi M, Khalil S, Yan KC, Marcolongo M, Barbee K, et al. Characterization of cell viability during bioprinting processes. Biotechnol J 2009;4(8):1168–77.
[40] Blaeser A, Duarte Campos DF, Puster U, Richtering W, Stevens MM, Fischer H. Controlling shear stress in 3D bioprinting is a key factor to balance printing resolution and stem cell integrity. Adv Healthc Mater 2016;5(3):326–33.
[41] Lee JM, Yeong WY. Design and printing strategies in 3D bioprinting of cell-hydrogels: a review. Adv Healthc Mater 2016;5(22):2856–65.
[42] Kesti M, Eberhardt C, Pagliccia G, Kenkel D, Grande D, Boss A, et al. Bioprinting complex cartilaginous structures with clinically compliant biomaterials. Adv Funct Mater 2015;25(48):7406–17.
[43] Panwar A, Tan LP. Current status of bioinks for micro-extrusion-based 3D bioprinting. Molecules 2016;21(6):685.
[44] Lode A, Meyer M, Bruggemeier S, Paul B, Baltzer H, Schropfer M, et al. Additive manufacturing of collagen scaffolds by three-dimensional plotting of highly viscous dispersions. Biofabrication 2016;8(1):015015.
[45] Dang Q, Liu K, Zhang Z, Liu C, Liu X, Xin Y, et al. Fabrication and evaluation of thermosensitive chitosan/collagen/alpha, beta-glycerophosphate hydrogels for tissue regeneration. Carbohydr Polym 2017;167:145–57.
[46] He Y, Tuck CJ, Prina E, Kilsby S, Christie SDR, Edmondson S, et al. A new photocrosslinkable polycaprolactone-based ink for three-dimensional inkjet printing. J Biomed Mater Res B Appl Biomater 2017;105(6):1645–57.
[47] Wust S, Godla ME, Muller R, Hofmann S. Tunable hydrogel composite with two-step processing in combination with innovative hardware upgrade for cell-based three-dimensional bioprinting. Acta Biomater 2014;10(2):630–40.
[48] Gao Q, He Y, Fu JZ, Liu A, Ma L. Coaxial nozzle-assisted 3D bioprinting with built-in microchannels for nutrients delivery. Biomaterials 2015;61:203–15.
[49] Li C, Faulkner-Jones A, Dun AR, Jin J, Chen P, Xing Y, et al. Rapid formation of a supramolecular polypeptide-DNA hydrogel for in situ three-dimensional multilayer bioprinting. Angew Chem Int Ed Engl 2015;54(13):3957–61.

[50] Schuurman W, Levett PA, Pot MW, van Weeren PR, Dhert WJ, Hutmacher DW, et al. Gelatin-methacrylamide hydrogels as potential biomaterials for fabrication of tissue-engineered cartilage constructs. Macromol Biosci 2013;13(5):551–61.

[51] Lim KS, Schon BS, Mekhileri NV, Brown GCJ, Chia CM, Prabakar S, et al. New visible-light photoinitiating system for improved print fidelity in gelatin-based bioinks. ACS Biomater Sci Eng 2016;2(10):1752–62.

[52] D'Este M, Sprecher CM, Milz S, Nehrbass D, Dresing I, Zeiter S, et al. Evaluation of an injectable thermoresponsive hyaluronan hydrogel in a rabbit osteochondral defect model. J Biomed Mater Res A 2016;104(6):1469–78.

[53] Kesti M, Müller M, Becher J, Schnabelrauch M, D'Este M, Eglin D, et al. A versatile bioink for three-dimensional printing of cellular scaffolds based on thermally and photo-triggered tandem gelation. Acta Biomater 2015;11:162–72.

[54] Rutz AL, Hyland KE, Jakus AE, Burghardt WR, Shah RN. Multimaterial bioink method for 3D printing tunable, cell-compatible hydrogels. Adv Mater 2015;27(9):1607–14.

[55] Miller JS, Stevens KR, Yang MT, Baker BM, Nguyen DH, Cohen DM, et al. Rapid casting of patterned vascular networks for perfusable engineered three-dimensional tissues. Nat Mater 2012;11(9):768–74.

[56] Kang HW, Lee SJ, Ko IK, Kengla C, Yoo JJ, Atala A. A 3D bioprinting system to produce human-scale tissue constructs with structural integrity. Nat Biotechnol 2016;34(3):312–9.

[57] Hinton TJ, Jallerat Q, Palchesko RN, Park JH, Grodzicki MS, Shue HJ, et al. Three-dimensional printing of complex biological structures by freeform reversible embedding of suspended hydrogels. Sci Adv 2015;1(9):e1500758.

[58] Wu W, DeConinck A, Lewis JA. Omnidirectional printing of 3D microvascular networks. Adv Mater 2011;23(24):H178–83.

[59] Tamayol A, Najafabadi AH, Aliakbarian B, Arab-Tehrany E, Akbari M, Annabi N, et al. Hydrogel templates for rapid manufacturing of bioactive fibers and 3D constructs. Adv Healthc Mater 2015;4(14):2146–53.

[60] Griffith LG, Swartz MA. Capturing complex 3D tissue physiology in vitro. Nat Rev Mol Cell Biol 2006;7(3):211–24.

[61] Hinton TJ, Lee A, Feinberg AW. 3D bioprinting from the micrometer to millimeter length scales: size does matter. Curr Opin Biomed Eng 2017;1:31–7.

[62] Kolesky DB, Homan KA, Skylar-Scott MA, Lewis JA. Three-dimensional bioprinting of thick vascularized tissues. Proc Natl Acad Sci U S A 2016;113(12):3179–84.

[63] Kolesky DB, Truby RL, Gladman AS, Busbee TA, Homan KA, Lewis JA. 3D bioprinting of vascularized, heterogeneous cell-laden tissue constructs. Adv Mater 2014;26(19):3124–30.

[64] Homan KA, Kolesky DB, Skylar-Scott MA, Herrmann J, Obuobi H, Moisan A, et al. Bioprinting of 3D convoluted renal proximal tubules on perfusable chips. Sci Rep 2016;6:34845.

[65] Leong KF, Chua CK, Sudarmadji N, Yeong WY. Engineering functionally graded tissue engineering scaffolds. J Mech Behav Biomed Mater 2008;1(2):140–52.

[66] Olivares AL, Marsal E, Planell JA, Lacroix D. Finite element study of scaffold architecture design and culture conditions for tissue engineering. Biomaterials 2009;30(30):6142–9.

[67] Sun Y, Wang Q. Modeling and simulations of multicellular aggregate self-assembly in biofabrication using kinetic Monte Carlo methods. Soft Matter 2013;9(7):2172–86.

[68] Perez-Pomares JM, Foty RA. Tissue fusion and cell sorting in embryonic development and disease: biomedical implications. Bioessays 2006;28(8):809–21.

[69] Jakab K, Norotte C, Damon B, Marga F, Neagu A, Besch-Williford CL, et al. Tissue engineering by self-assembly of cells printed into topologically defined structures. Tissue Eng Part A 2008;14(3):413–21.

[70] Jakab K, Neagu A, Mironov V, Markwald RR, Forgacs G. Engineering biological structures of prescribed shape using self-assembling multicellular systems. Proc Natl Acad Sci U S A 2004;101(9):2864–9.
[71] Shafiee A, McCune M, Forgacs G, Kosztin I. Post-deposition bioink self-assembly: a quantitative study. Biofabrication 2015;7(4):045005.
[72] Shahbazian M, Wyatt J, Willems G, Jacobs R. Clinical application of a stereolithographic tooth replica and surgical guide in tooth autotransplantation. Virtual Phys Prototyp 2012;7(3):211–8.

Bioreactor processes for maturation of 3D bioprinted tissue

*J. Rosser**, *D.J. Thomas*†
*Welsh Centre for Printing and Coating, Swansea University, Swansea, United Kingdom,
†3Dynamic Systems Group, Swansea, United Kingdom

10.1 Introduction

A bioreactor is defined as a device that uses mechanical means to influence biological processes. In tissue engineering bioreactors can be used to aid in the in vitro development of a new tissue by providing biochemical and physical regulatory signals to cells and encouraging them to undergo differentiation and/or to produce extracellular matrix before in vivo implantation. This chapter discusses the necessity for bioreactors in tissue engineering, the numerous types of bioreactors, and the means by which they stimulate cells and how their functionality is governed by the requirements of the specific tissue being engineered and the cell type undergoing stimulation.

10.2 Process parameters

During normal activities within a person's lifetime, the articular cartilage experiences various stresses, strains, and pressures. Cartilaginous tissues are hypersensitive to this mechanical environment and during their development, in vivo mechanical factors are vital for the establishment of the tissue's functions [1,2]. During joint loading, chondrocytes vary their metabolite activity and produce proteoglycans and collagen to maintain the homeostasis of the extra cellular matrix (ECM). In this regard, research has focused on the effects of mechanical stimulus application on chondrogenesis in vitro with variations in scaffold and culture environment. In this section, current literature has been reviewed and its key findings are highlighted.

10.2.1 Hydrostatic pressure

During daily activities, articular cartilage is predominantly subjected to compression. Therefore, most studies investigating the effects of mechanical stimulus on chondrogenesis have considered this factor. During loading, the hydrostatic pressure within the cartilaginous tissues increases as they retain synovial fluid within the extracellular matrix due to the presence of hydrophilic proteoglycans [3]. Hydrostatic pressure induces a uniform stress throughout the cells (Fig. 10.1), without resulting in tissue deformation due to the incompressibility of water and the extracellular matrix [4].

3D Bioprinting for Reconstructive Surgery. https://doi.org/10.1016/B978-0-08-101103-4.00010-7

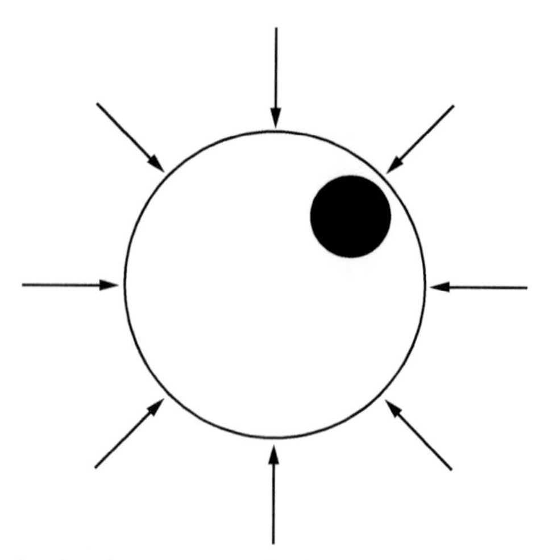

Fig. 10.1 Illustration of a chondrocyte exposed to hydrostatic pressure as the cell experiences a uniform normal stress, without any measurable tissue strain.

Within the synovial joints, articular cartilage is typically exposed to stresses between 0.1 and 4 MPa during walking and 1 MPa during static standing, although peak stresses as high as of 20 MPa in the hip joint when jumping have been reported [5–8]. Additionally, a loading frequency of up to 1 Hz has been reported during human walking cadence [5]. With knowledge of the physiologic levels of fluid pressurization and frequency, multiple studies have been conducted to study the effects on cartilaginous tissue using these values.

Elder et al. [9] studied the response of scaffold-less bovine cartilaginous tissues with the application of temporal hydrostatic pressure in vitro. In the first study, a pressure of 10 MPa was applied for 1 h per day to three separate samples at various times (6–10, 10–14, and 14–18 days) during the culture period. The tensile and compressive mechanical properties, collagen, and glycosaminoglycan (GAG) production were assessed. The study revealed that temporal effects were most beneficial when applied at 10–14 days of the culture period. Significant increases in the production of GAG and collagen content were observed, resulting in increased aggregate and Young's modulus, therefore enhancing the constructs mechanical properties. In a second study, Elder et al. [10] studied the effects of various hydrostatic pressure values when applied between 10 and 14 days during the culture process. Pressures of 1, 5, and 10 MPa were applied for 1 h per day where application of 5 MPa resulted in an increased aggregate and Young's modulus by 95% and 101% respectively, with corresponding increases in GAG and collagen content. Application of 10 MPa resulted in an increased aggregate and Young's modulus by 96% and 92%, respectively, with corresponding increases in GAG and collagen content.

Suh et al. [11] cultured chondrocytes in monolayer acquired from young bovine while exposing them to 10 cycles of 0.8 MPa, 5 min on and 30 min off. This treatment had an insignificant effect on the collagen content, but increased proteoglycan synthesis by 40% and enhanced aggrecan messenger ribonucleic acid (mRNA). Jortikka et al. [12] applied 5 MPa of hydrostatic pressure at 1 Hz for 20 h to cultured chondrocytes in monolayer and the results were compared to a continuous application of 5 MPa for 20h. Cyclic loading significantly increased sulfated glycosaminoglycan (sGAG) incorporation whereas an application of continuous load resulted in no significant effects on sGAG incorporation.

Smith et al. [13] studied the time-dependent effects of intermittent hydrostatic pressure. A monolayer culture of chondrocyte cells was pressurized by 10MPa at 1Hz for periods of 2, 4, 8, 12, and 24h. An optimum time of 4h was identified as type II collagen and aggrecan mRNA signal levels increased by five- and threefold, respectively. Using this optimum time, the load profile was then changed to 4h per day for 4 days. This further increased the type II collagen and aggrecan mRNA signal levels by 9- and 20-fold. Smith et al. [14] also studied the effects on mRNA signal levels with the application of continuous pressure of 10MPa for 4h, but this proved disadvantageous as it resulted in decreased collagen mRNA signal levels.

Hu et al. [15] applied 10MPa of hydrostatic pressure at 1Hz to chondrocytes on self-assembled constructs for 4h per day, 5 days a week for up to 8 weeks. This treatment resulted in a significant increase in collagen synthesis. Toyoda et al. [16] investigated the effects on mRNA of specific ECM proteins in chondrocyte cultures in 2% agarose constructs during applications of hydrostatic pressure. A pressure of 5MPa was applied for 4h, resulting in a 400% and 50% increase in aggrecan and type II collagen mRNA, respectively, along with a 11% increase in GAG production.

The effects of hydrostatic pressure application ranging within physiological values have proven to be beneficial to chondrocyte protein synthesis, gene expression, and biomechanical properties; these effects dependent on the hydrostatic pressure frequency, magnitude, regimens, and application time. Studies indicate that hydrostatic pressure can be both beneficial and nonbeneficial, depending on the type of culture. It seems that when considering a monolayer culture, application of dynamic hydrostatic pressure seems beneficial, whereas static hydrostatic pressure can cause suppressive effects. On the contrary, static hydropressure application to chondrocytes in explants or 3D culture has been reported to be advantageous.

10.2.2 Fluid-induced shear stress

During joint loading, interstitial fluid flows around the compressed articular cartilage and through the pores of its ECM. As a result, fluid-induced shear stresses are generated from the increased mass transfer [17]. Due to the avascular nature of cartilage, this mass transfer is of high importance as it supplies the chondrocytes with the required nutrients [18]. As well as nutrient transportation, sufficient gaseous exchange and waste removal is achieved through interstitial fluid flow, promoting cellular proliferation and growth [19]. Gharravi et al. studied the effects of flow-induced shear stress on chondrocyte constructs by using a perfusion bioreactor [20] where shear stress values were predicted by using the computational fluid dynamics (CFD) approach. The fluid properties used were assumed to be equivalent to water, due to the medium being water-based. Chondrocytes were first expanded within a static culture before being transferred to the perfusion bioreactor. After culture, cells were analyzed histologically and a homogenous distribution of cells was found throughout the scaffold. CFD model estimated a maximum shear stress of 0.00482Pa, which was determined advantageous to the developing cartilage.

Tarng et al. [21] investigated the effects of fluid flow and chondrogenesis by using a spinner flask. Periosteal explants were harvested and attached to a poly-ε-caprolactone scaffold before being suspended inside a spinner-flask bioreactor. While suspended in the culture medium, the constructs were exposed to varying flow. Flows of 0, 20, 60, and 150 rpm were applied for 4 h each day for 6 weeks. During static culture, the cartilage yield in periosteal explants was 17%. When fluid flow was introduced, the yield increased to 65%–75%. Explants exposed to flows of 60 rpm showed a greater Young's modulus to those that were exposed to flows of 20 and 150 rpm. Applied flows enhanced cell organization, chondrogenic differentiation, biomechanical properties, and overall cell proliferation. It was also concluded that due to the nature of the spinner flask, turbulent flows produced nonuniform tissues as seen in Fig. 10.2.

John S et al. [22] employed a spinner flask to study the integration effects of flow on bovine cartilage. Static culture and a dynamic culture (90 rpm) were studied after 2 or 4 weeks of culture and evaluated, biomechanically, biochemically, and histologically. Strength and extent of integration between native and tissue-engineered cartilage improved significantly over time due to the increase in collagen at the integration zones. Highest integration was demonstrated in constructs cultured for 4 weeks. It was concluded that this was caused by both the mechanical stimulation and improved nutrient diffusion caused by the flow.

Smith et al. [23,24] examined the effects on monolayer cultures of chondrocytes during application of fluid-induced shear using a cone viscometer. Shear stresses of 0.16 and 0.6 Pa were applied to a monolayer culture for 3 days. This caused elongation

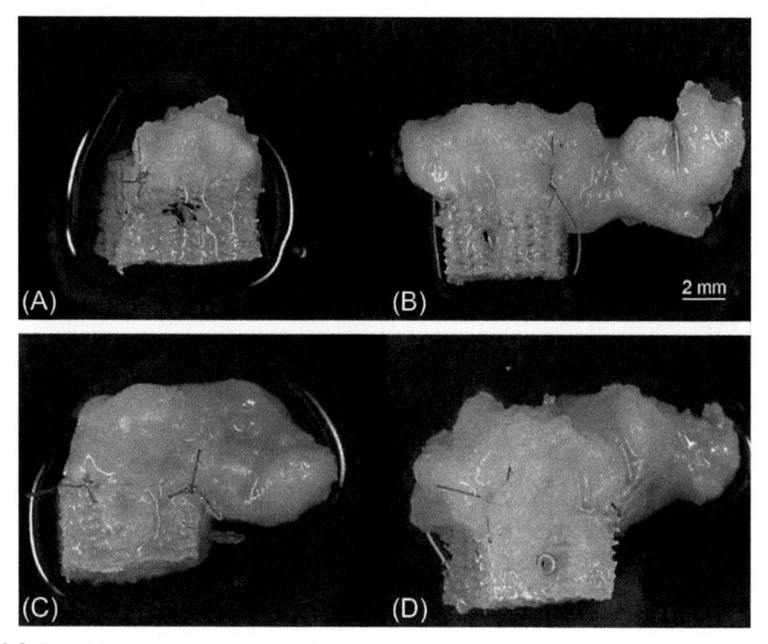

Fig. 10.2 Resulting engineered cartilaginous tissue after spinner flask culturing at 0 (A), 20 (B), 60 (C), and 150 (D) RPM [21].

of the chondrocytes and resulted in an enhanced proteoglycan synthesis, although higher values of shear stress had adverse effects on aggrecan and collagen II mRNA expression. Contrary to this, Germmiti et al. used increased shear stresses of 2 Pa on a cartilaginous construct and saw no significant change in proteoglycan synthesis. However, up-regulation of type II collagen incorporation was observed [25].

It is clear that fluid flow can be a beneficial mechanical stimulant on chondrogenesis due to resulting shear stresses and improved nutrient diffusion. The amount of shear stress on the tissue during culture is proportional to the fluid's flow rate and large amounts of shear stress on the construct can have an adverse effect on the cellular metabolism, damaging the cells and even detaching them from the construct [17]. Separately, both shear stress and hydrostatic are vital to the healthy development of cartilaginous tissues, but mechanical stimulus applied to native cartilage in vivo is more complicated and loading is applied synergistically [13]. To apply these loads to cartilage tissues, several types of bioreactor have been constructed.

10.3 Bioreactor types

To apply the previously described mechanical stimulus to cartilaginous tissues, bioreactors have been employed. In the context of tissue engineering, bioreactors are used to reproduce the well-defined physiological environment of specific tissues found in vivo [26]. This relies on the control of the biological, biochemical (nutrient and oxygen concentration), biophysical (mechanical regulations), and physiologically relevent environmental conditions such as temperature, pH, and CO_2 [3,26–28]. Within most cell culture systems, incubators are employed to maintain constant temperature and CO_2 values of 37°C and 5.1%, respectively, as seen in Fig. 10.3.

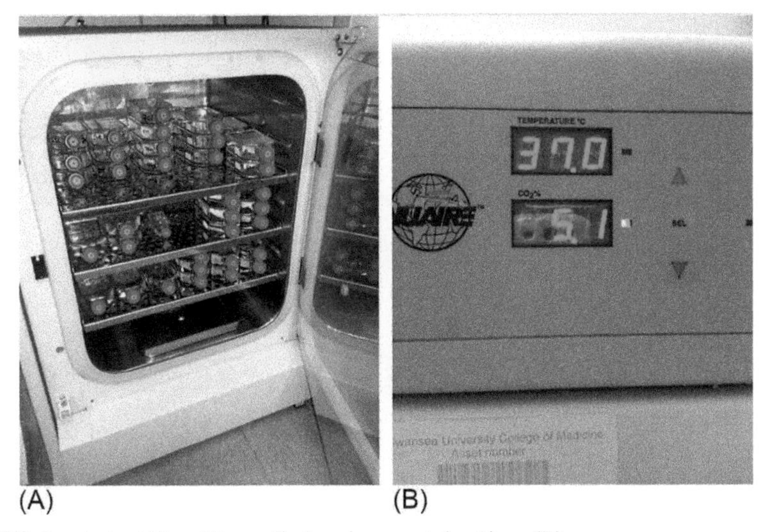

Fig. 10.3 Incubator (A) and its applied environmental settings (B).

The requirements of a bioreactor are dependent on the type of the tissue culture and are therefore designed and manufactured for its tissue-specific purpose. Ideally, a bioreactor system would be able to reliably and reproducibly form, store, and deliver engineered tissues that exactly replicate every property of the native tissue for successful implantation and sustained function in vivo [29]. Over the years, a variety of bioreactor systems has been developed and employed for different types of physical stimulation [30]. In the context of cartilaginous tissue engineering, several bioreactors have been implemented including static, spinner flask, and perfusion systems. In this section, current literature relating to the design of bioreactors to employ the cellular requirement of cartilaginous tissues has been addressed.

10.3.1 Static culture systems

Static culturing systems provide tissues with the required nutrients in a static fluid environment. As a result, the media containing the essential nutrients has to be changed more often than other techniques employed due to waste build-up and nutrient expenditure. Waldman et al. tailored the Mech-1 mechanical loading device to apply cyclic compressive loading to a static culture (Fig. 10.4). This allowed the application of various compressive loads and frequencies to the cultures [31].

Correia V, et al. [32] designed and validated another static culture bioreactor. It allowed the application of both static and cyclic mechanical stimulations to multiple three-dimensional scaffolds, a schematic representation of which is depicted in Fig. 10.5. Compressive forces were applied by using vertically dynamic Teflon pistons controlled via a screw and computer-controlled step motor. The hermetically enclosed bioreactor was made for use within an incubator and compassed a user-friendly design with application of components for disassembly and sterilization. The materials of the piston and hermetic enclosure were Teflon and acrylic, respectively; all other materials involved were of stainless steel.

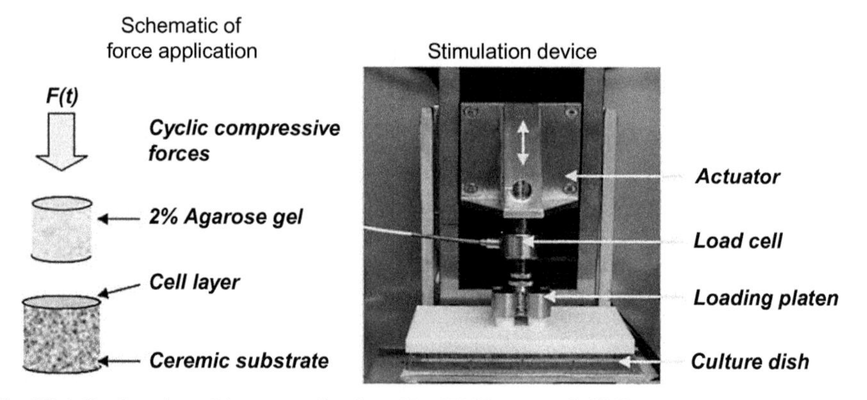

Fig. 10.4 Static culture bioreactor developed by Waldman et al. [31].

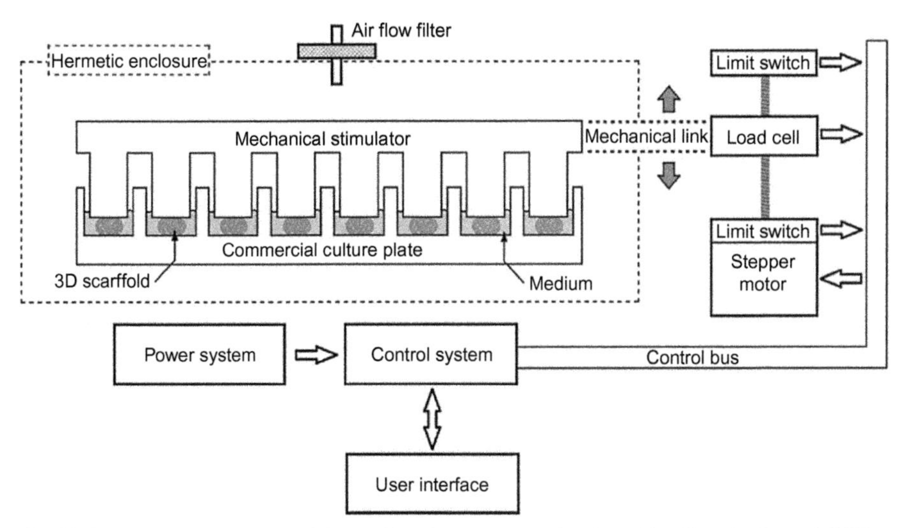

Fig. 10.5 Schematic diagram of the static culture system developed by Correia V, et al. [32].

10.3.2 Spinner-flask systems

Spinner-flask systems apply fluid-induced shear stresses to submerged constructs by circulation of nutrient-rich media. As with static systems, spinner-flask systems also require frequent media exchange as the nutrient levels within the fluid are not replenished autonomously.

Sucosky et al. [33] developed a simple spinner-flask bioreactor to mimic the dynamic environment of articular cartilage. It was a simple construction that induced fluid flow to submerged constructs by using a magnetic stirrer. The hermetically sealed container was made of glass and allowed gaseous exchange through two small microfilters for use in an incubator (Fig. 10.6).

Shahin et al. [34] modified a spinner flask to implement mechanical compression and shear stress on cartilaginous constructs to simulate the rolling action of articular joints. Cyclic and shear compressions were applied by rotation and height alteration (respectively) of the three Teflon wheels in conjunction with the base plate. Each movement used an independent motor as depicted in Fig. 10.7. The Teflon wheels and stainless steel base plate containing the cartilage constructs were submerged in a bath of media inside a 1.5-L hermetically sealed vessel. Fluid flow was also applied with the implementation of a magnetic stirrer.

10.3.3 Perfusion systems

Perfusion bioreactors employ laminar flow through the culture area, providing the scaffold with the required nutrients, oxygen, and waste removal. By introducing flow through the culture area, perfusion bioreactors are capable of applying fluid-induced shear stress to the surfaces of the scaffold. As previously mentioned this can be advantageous when controlled.

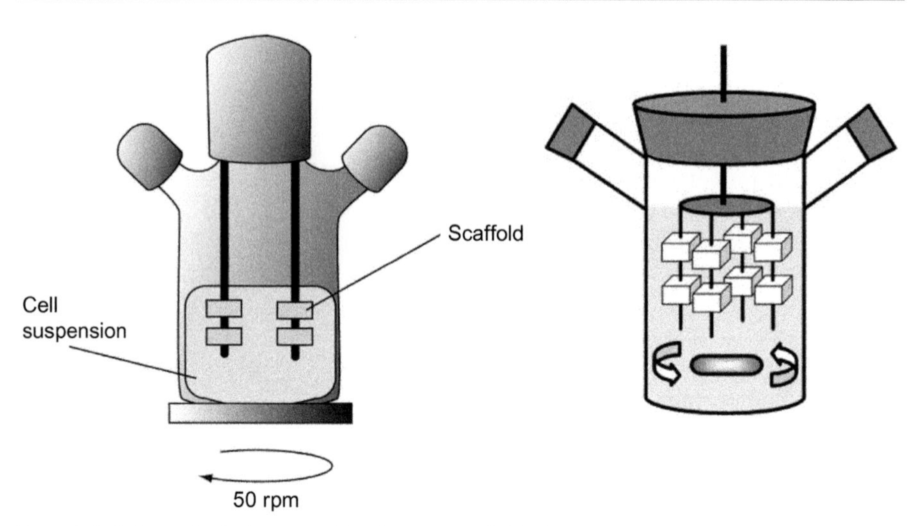

Fig. 10.6 Spinner flask developed by Sucosky et al. [33].

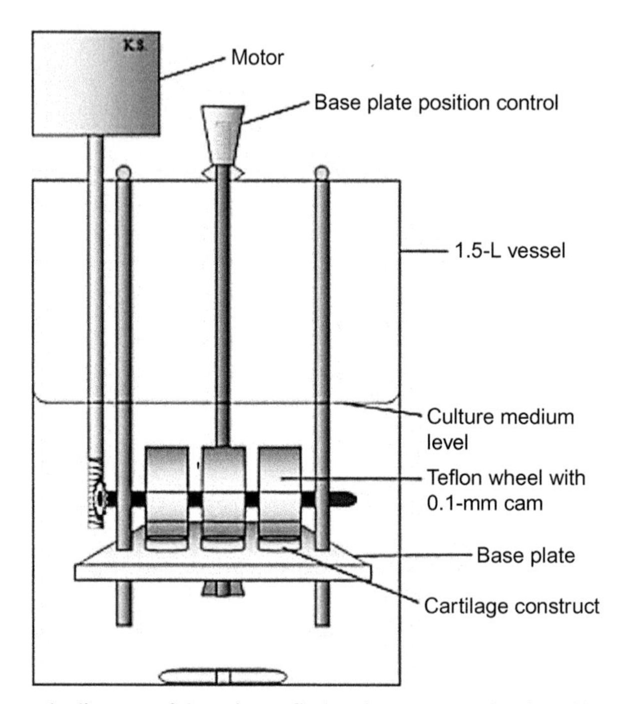

Fig. 10.7 Schematic diagram of the spinner-flask culture system developed by Shahin et al. [34].

Gilbert E. et al. [35] developed a "through-thickness" perfusion bioreactor for monolayer culture of scaffold-free cartilaginous tissues. The bioreactor system (Fig. 10.8) was for use inside a CO_2 incubator whereby the temperature, CO_2, and pH were monitored and controlled over time. The cylindrical polysulfone container

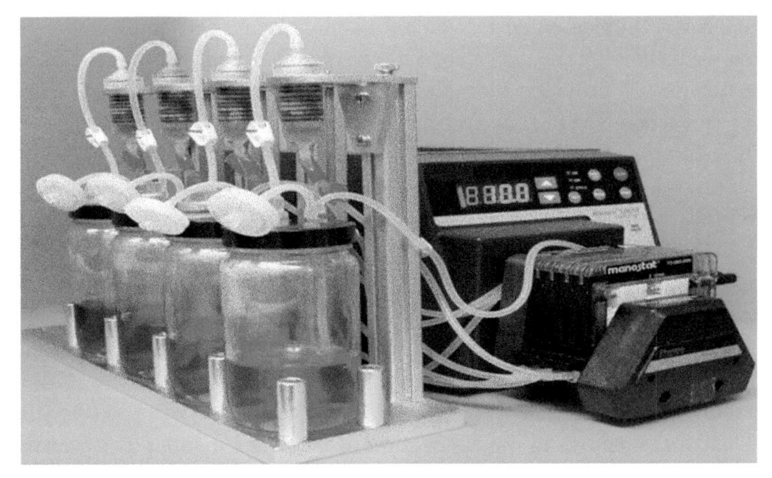

Fig. 10.8 Perfusion bioreactor system developed by Gilbert E. et al. [35].

that was used to house the culture process was tailored to allow for a 2.4-cm-diameter transwell insert capable of multidirectional flow. Flow was induced at various flow rates using a peristaltic pump and transported by silicon tubing. Unlike most other perfusion bioreactors, the flow was restricted and forced through the construct itself. The system was completely sealed apart from 0.22-μm membrane filter to facilitate gaseous exchange. A media reservoir was used for containment and exchange of media. Depending on the direction of flow, this bioreactor could produce both shear and compressive forces evenly across the construct according to the CFD conducted.

Gharravi A. et al. [36] developed a portable perfusion bioreactor (Fig. 10.9) for use with alginate gel moulds in a three-dimensional culture system. A piston controlled by an electromotor allowed biomimicry of the temporomandibular joint. Mechanical

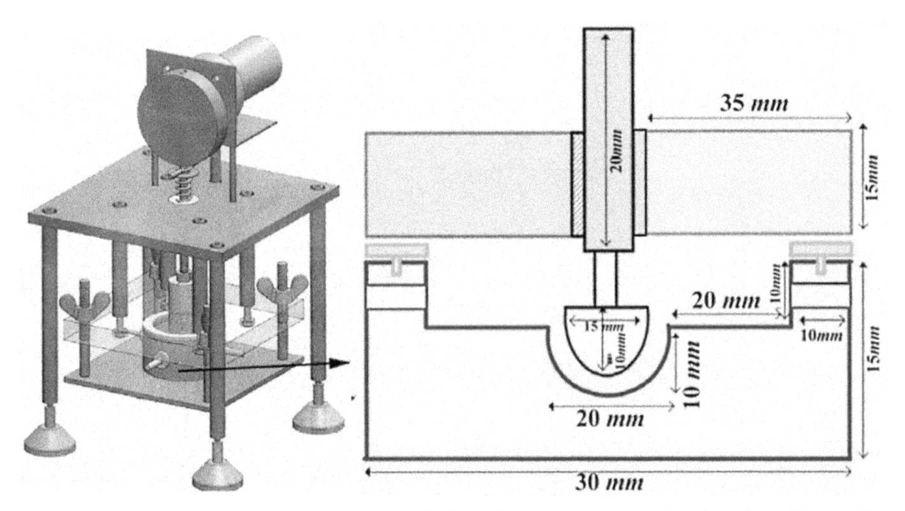

Fig. 10.9 Assembled model of portable perfusion bioreactor system developed by Gharravi A. et al. [36].

compression was implemented by vertical displacement (between 2 and 10 mm) of the piston at various frequencies. The culture chamber and piston were of stainless steel. Perfusion was provided by multichannel peristaltic pump that was capable of producing flows between 0.01 and 0.5 L/min, resulting in fluid-induced shear stresses.

Yih-Wen Tarng et al. [37] developed a recirculating flow-perfusion bioreactor that offers oscillating fluid-induced shear stress and hydrodynamic pressure to simulate the dynamics of the arthrodial joint. The system consisted of a culture chamber, four electronic control valves (D1, D2, E1, and E2 in Fig. 10.10), two reservoirs containing the nutrient-rich media, an oil-free air compressor, and a programmable logic controller (PLC). Fluid flow was induced via the air compressor and two separate control valves allowing bidirectional flow from reservoir A or B. Hydrodynamic pressure (0–15 Psi) could be applied with air compressor D1 active, valve E1 open, and valve E2 closed (Fig. 10.10A). Shear stresses (up to 250 dyne/cm^2) could be applied with D1 active, and both E1 and E2 open (Fig. 10.10B).

It is clear that methods in the application of mechanical stimulus to cartilaginous cultures are varied. The most disadvantageous of these methods seems to be static culture systems. Owing to the nondynamic nature of static cultures, they require a frequent media change due to the exhaustion of nutrient supply and build-up of waste products through metabolic activities. The same has been said for spinner-flask

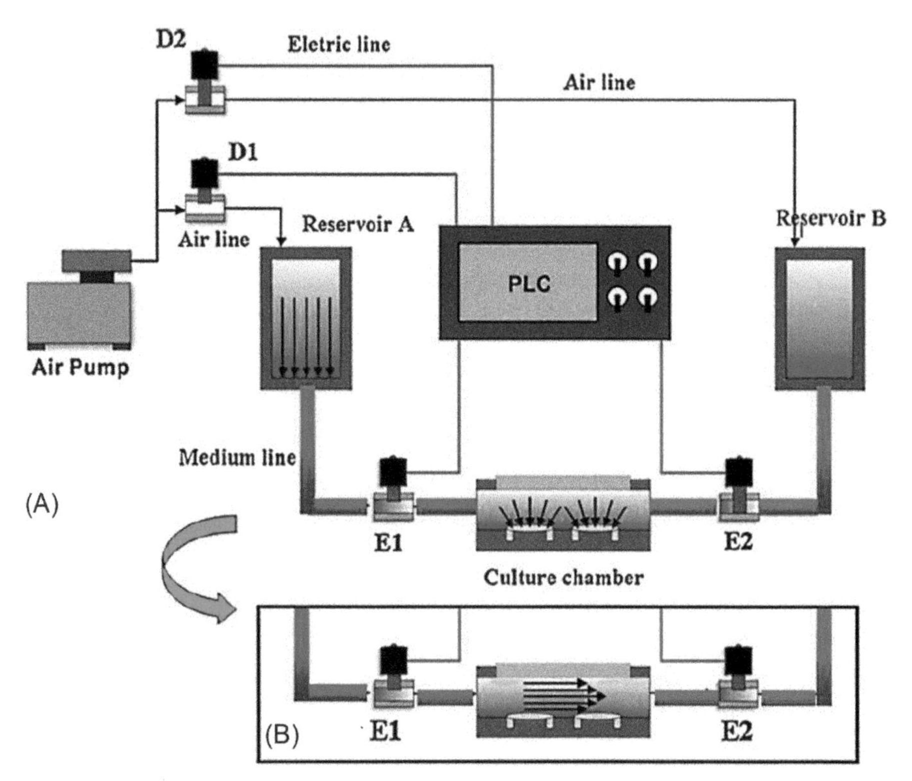

Fig. 10.10 Schematic representation of the recirculating flow-perfusion system developed by Yih-Wen Tarng et al. depicting (A) application of hydrodynamic pressure and (B) fluid-induced shear stress [37].

bioreactors, although the addition of fluid flow allows fluid-induced shear stresses and distribution of waste products within the media, thereby extending the culture time capable without media exchange.

Perfusion bioreactors have proven to be the most appropriate choice when considering the culture of cartilaginous tissues. The dynamic flow within each perfusion system permitted the application of shear stresses on each explant as well as even distribution of media and prolonged time between media. Additionally, both Gharravi and Yih-Wen Tarng et al. developed systems capable of applying pressure by physical contact and media compression.

10.4 Building a perfusion bioreactor system for 3D tissue culture

Perfusion bioreactors employ circulatory flow throughout the system with a singular input and output to provide sufficient nutrient transfer, removal of waste products, and application of mechanical stimuli. To establish such a system, three major components have been identified by review of literature: the culture encasing bioreactor, a vessel containing the oxygenated nutrient-rich medium, and a pump capable of generating flow throughout the system. A system schematic representation is seen in Fig. 10.11.

10.4.1 Computational modeling

Using SolidWorks, initial designs were developed and modeled with consideration to the design requirements. Along with each design, fluid flow analysis was conducted using CFD. By using CFD techniques a geometric configuration could be determined to maximize flow uniformity around the tissue scaffold.

Flow phenomena occurring within the bioreactor were approximated by solution of the *Navier-Stokes equations* for the applicable boundary conditions in conjunction with the *Continuity equation*. The *Navier-Stokes* and *Continuity equations* represent the conservation of momentum and mass, respectively.

Water-based media flowing throughout the system was regarded as a three-dimensional inviscid and incompressible Newtonian fluid modeled by Eqs. (10.1), (10.2).

Resulting Navier-Stokes equation for inviscid and incompressible fluids

$$\rho\left(\frac{\delta v}{\delta t}+v\cdot\nabla v\right)=-\nabla p+f \tag{10.1}$$

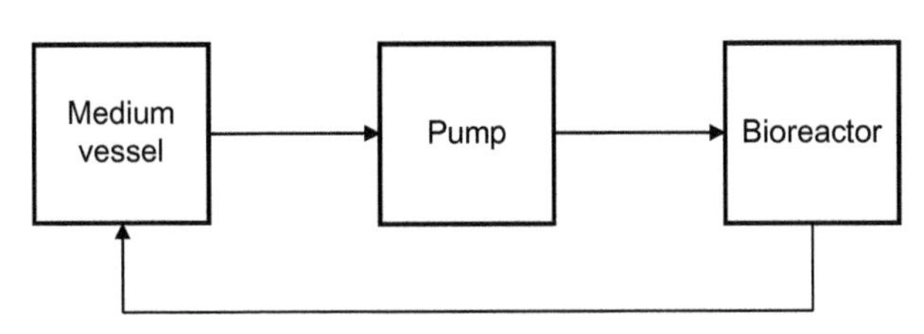

Fig. 10.11 Major components of the perfusion bioreactor system.

where ρ, v, t, p and f are fluid density, fluid velocity, time, pressure, and external forces within the system, respectively.

Continuity equation

$$\frac{\delta \rho}{\delta t} + \nabla \cdot (\rho v) = 0 \tag{10.2}$$

where ρ, v, t, p and f are fluid density, fluid velocity, and time, respectively.

Initially, a model was created representing the internal geometry of the bioreactor (disregarding any cavities without flow conditions). These were analyzed by using the SolidWorks' Flow Simulation package. Initially, each concept was meshed using automated meshing tools. A level 5 quadrilateral mesh with advanced narrow channel refinement was applied to each model. This tool allowed automated mesh refinement in regions of narrow geometry by reducing cell size when approaching gaps. With each mesh generated, the following boundary conditions were applied:

- Gravitational interactions were applied in the *y*-direction with a value of $-9.81\ \text{m/s}^2$ and flow types included both laminar and turbulent flows
- Environmental pressure and temperatures were valued at 101,325 Pa (ambient pressure) and 310.15 K (required culturing temperature), respectively
- Water (properties listed in Table 10.1) was identified as the fluid and was assumed to have equal properties to that of the water-based medium [38]
- Both the inlet and outlet had a diameter of 4 mm
- A flow rate of 0.1 L/min was applied at the inlet
- An opening at environmental pressure was applied at the outlet

By conducting flow analysis by application of CFD, the fluid flow around the scaffold was analyzed; for the bioreactor to function effectively there must be sufficient fluid flow around the scaffold. The objective of this investigation identified any regions of slow moving or stagnant fluid. Ideally, flow inside the culture zone would be uniform to optimize nutrient delivery, waste removal, and application of mechanical stimuli to the scaffold. Areas of low velocity flow can lead to areas of nutrient deficiency, waste build-up, and variation in mechanical stimulus.

A cylinder (20 mm diameter, 10 mm height) was modeled to represent the scaffold used throughout this chapter. The orientation of the scaffold was first optimized by using CFD analysis. It was concluded that a vertical orientation maximizes both uniform flow around the scaffold and fluid–scaffold contact (Fig. 10.12). To accommodate this scaffold positioning, four support pins were also implemented.

Table 10.1 Fluid properties of water at 37°C (310.15 K)

Property	Value	Unit
Density	993.27564	kg/m^3
Dynamic viscosity	0.0006928	Pa s
Specific heat	4179.5398	J/kg K
Thermal conductivity	0.6262054	W/m K

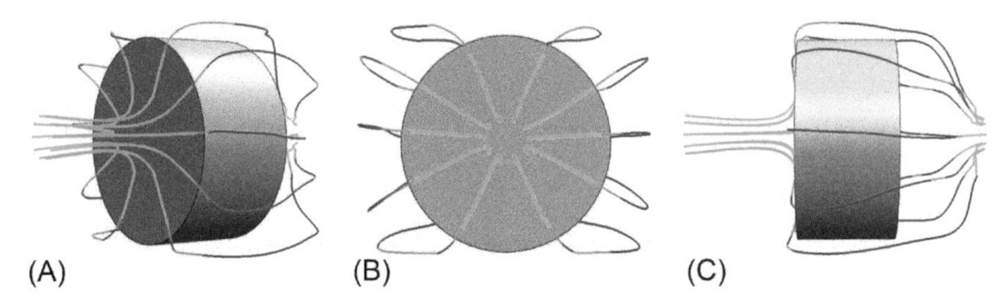

Fig. 10.12 Flow trajectories around the vertical scaffold; (A) isometric, (B) front, and (C) side views.

Several models were created to simulate fluid flow within the bioreactor to achieve an optimized design. Each model simulated the following variables independently: internal shape, inlet/outlet positioning, and scaffold fixture positioning.

To conduct the analysis, the computational setup and an inlet flow rate of 0.1 L/min were applied. Flows were analyzed by using the resulting 2D velocity contours that were studied at three heights: 1 mm above the internal base, 1 mm below the internal surface of the lid, and at the central height of the bioreactor. The 2D velocity contours contain a spectrum of colors from blue to red representing regions of minimum to maximum flow, respectively. The optimal configuration would result in contours with minimal color differentiation.

Through the combination of each independent study a final design was defined with the following configuration: a cylindrical internal geometry, vertically central inlet/outlet, and a white scaffold fixture positioning. The resulting contours of the final configuration are found in Fig. 10.13. Furthermore, the symmetry of the design also proves advantageous as it allows flow reversal for further flow uniformity during the culture process.

10.4.2 Bioreactor

With the internal geometry optimized the bioreactor was designed to suit the configuration: the drawing of which is found in Fig. 10.14. The base of the bioreactor included a 1.5 mm recess to house a silicon rubber seal, six bolt holes for lid–base assembly, and four 2-mm recesses in the internal base.

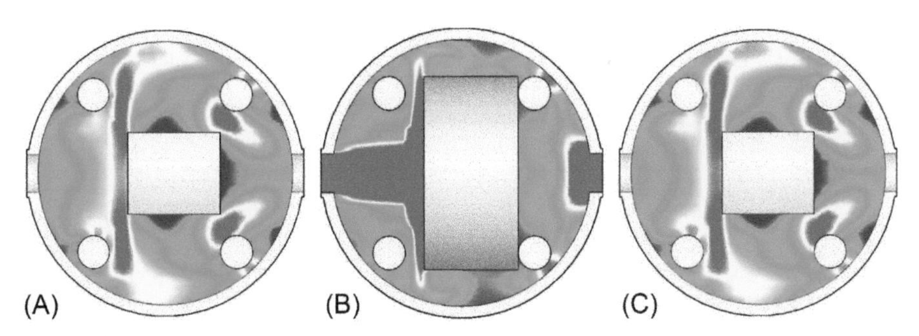

Fig. 10.13 (A) Lower, (B) middle, and (C) upper velocity contours with application-wide scaffold supports.

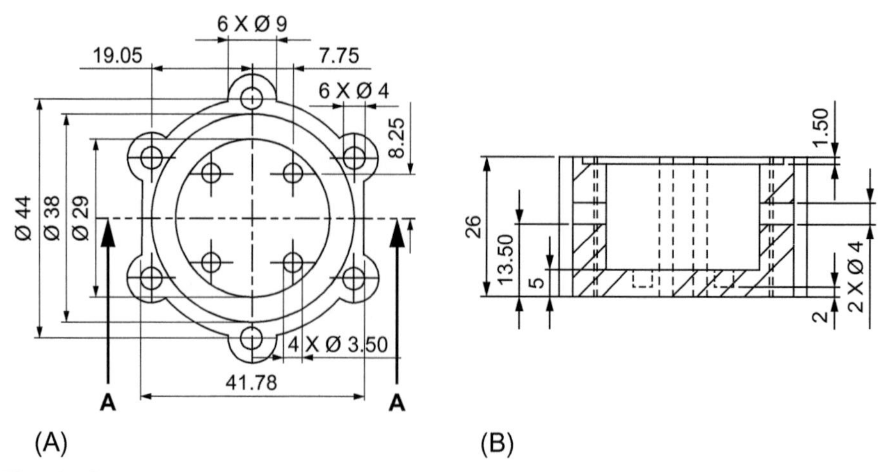

Fig. 10.14 Bioreactor base dimensions (mm): (A) top and (B) A-A sectional views.

Initially, the model was first converted to a stereolithographic (STL) file format and split into 72 layers (0.25-mm-layer thickness). The base of the bioreactor was then manufactured by using fused deposition modeling (FDM, a 3D printing technique) with Makerbot's "The Replicator". A feedstock containing acrylonitrile butadiene styrene (ABS) was fed into the printer with a diameter of 1.75 mm, which was then extruded to a diameter of 0.35 mm. Printing parameters used are found in Table 10.2.

The total printing time of the bioreactor base was 3 h. During the first prototype test the bioreactor began to leak due to the highly porous structure caused by the nature of this manufacturing technique; to combat this issue the bioreactor was cast. First, an advanced room temperature vulcanizing silicon rubber mould was produced by using the 3D-printed bioreactor before filling it with a two part polyurethane casting resin. The filled mould was then placed into a 60°C oven for 1 h to set the mixture. Using the same mould the bioreactor could be reproduced multiple times. Each hole within the bioreactor was re-bored by using a pillar drill to allow smooth bolt input.

Each system component was to be interconnected using flexible silicon tubing (4 mm internal diameter), for ease of use two inexpensive ABS tubing connectors were fixed to the input and output of the bioreactor. The scaffold supports were then manufactured and placed in the four recesses within the bioreactor's base. Each support consisted of a steel tube that was 3.5 mm in diameter and 24 mm in length. With the

Table 10.2 **Prototype printing parameters**

Printing parameter	Value	Unit
Layer thickness	0.25	mm
Speed	50	mm/s
Number of shells	5	–
Temperature (extruder)	220	°C
Temperature (base plate)	120	°C
Hatch pattern	Honeycomb	–

Fig. 10.15 Finished bioreactor base component.

lid attached to the base of the bioreactor, the scaffold supports were fixed (due to their length). This allowed support removal and replacement when necessary. This concluded the manufacture of the base of the bioreactor (Fig. 10.15).

Following the base, the lid was then manufactured. To monitor the culture process, a transparent lid was essential. The lid was manufactured by using transparent acrylic (10 mm thick) in conjunction with laser beam machining. First, the laser cutting parameters of the Helix Epilog Laser were optimized. Next, the bioreactor's lid was drawn by using Coral Draw X7 and cut (Fig. 10.16).

This concluded the component manufacturer of the base (Fig. 10.17). Finally, the scaffold was inserted and the lid was attached with a silicon rubber seal, six stainless steel bolts and nuts, and six zinc-plated washers.

10.4.3 Medium vessel

A polypropylene vessel was used to house the medium with an approximate internal volume of 330 mL. The vessel's outlet tubing was submerged within the nutrient-rich medium,

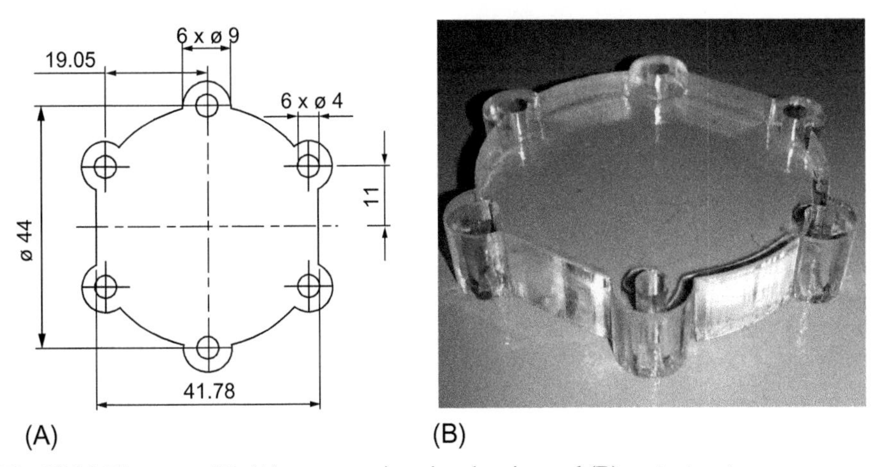

Fig. 10.16 Bioreactor lid: (A) precut engineering drawing and (B) postcut part.

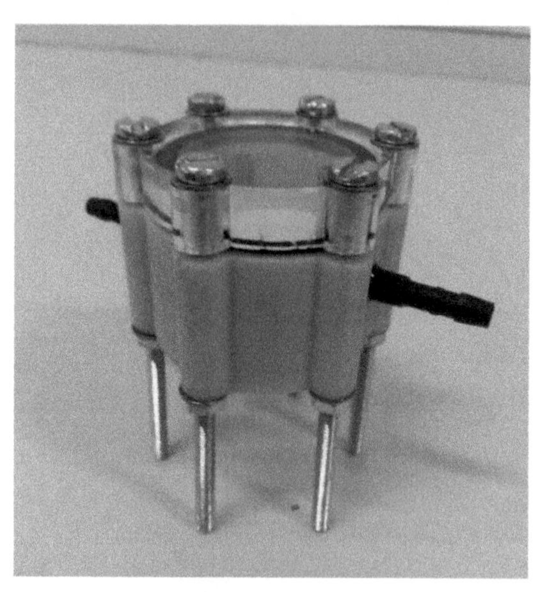

Fig. 10.17 Fully constructed bioreactor unit.

whereas the inlet tubing was left out of the medium to allow the fluid to fall from a height, inducing oxygenation to the fluid. The vessel outlet and inlet were connected to the pump inlet and bioreactor outlet, respectively. An acrylic sterile syringe filter was attached to the vessel's lid; the filter contained hydrophilic polyethersulfone (PES) membrane with a 0.22 μm pore size. This allowed filtration during gaseous exchange (of CO_2 and O_2) while in the incubator, inhibiting any airborne entities that may otherwise enter the system.

Using a 3D printer, two connectors were designed and printed to manufacture an inlet and outlet attachment for the medium vessel to accommodate the silicon tubing used in this system (Fig. 10.18). Each inlet and outlet component was then fixed and sealed to the medium vessel insuring that all unwanted airborne entities could not enter the system.

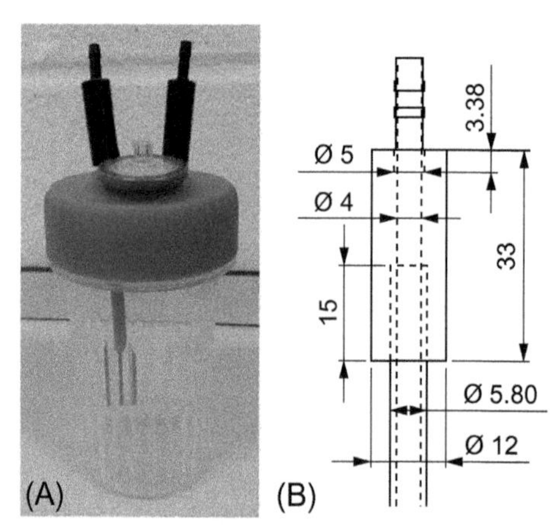

Fig. 10.18 (A) Medium vessel assembly and (B) dimensional drawing of the inlet/outlet component.

10.4.4 Pump

A 12V 4.8W 0.8A brushless DC pump was used to induce flow within the system. It was a small and inexpensive turbine pump controlled by a variable voltage DC power source (Fig. 10.19). The pump was capable of disassembly using four screws, allowing the removal of all electrical components. This allowed the sterilization using an autoclave without damaging any electronic components. Two tailored attachments were also fixed to the pump for tubing attachments (Fig. 10.19B and C).

Flow rates of the 12V DC pump were analyzed to approximate the flow rates achievable and determine its voltage–flow rate correlation. The flow rates were determined by timing how long it took to pump a specified volume of water. Results of this experiment are graphed in Fig. 10.20.

10.4.5 System setup and operation

The final system consisted of four major components: the medium vessel, pump, bioreactor, and manual valve. A system schematic diagram is seen in Fig. 10.21.

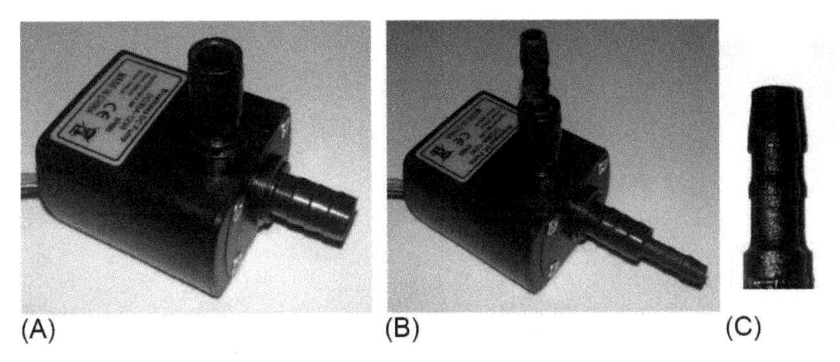

Fig. 10.19 (A) Pump, (B) adapted pump, and (C) connector used.

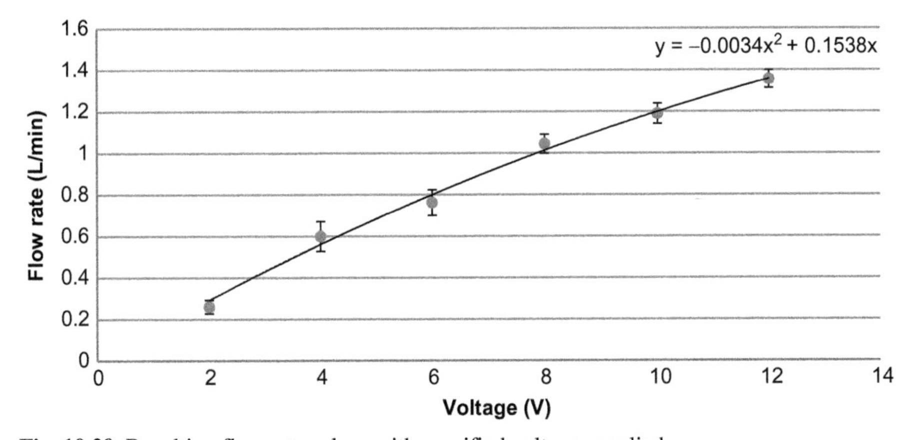

Fig. 10.20 Resulting flow rate values with specified voltages applied.

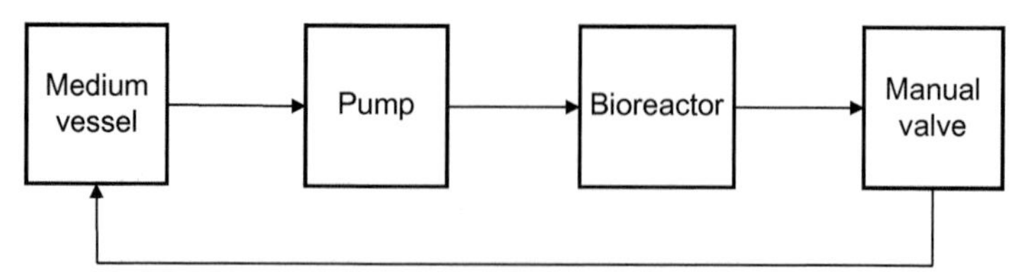

Fig. 10.21 System schematic diagram of the finalized system.

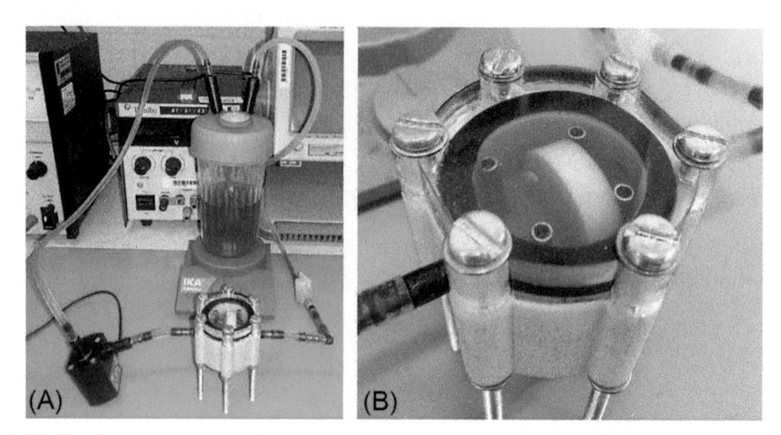

Fig. 10.22 Perfusion system during operation: (A) overview and (B) bioreactor.

After system setup, power was applied using a variable voltage meter. This drew the nutrient-rich media from the vessel, through the pump and into the bioreactor (Fig. 10.22). As the media flows around the scaffold, shear stresses occur on the surfaces, nutrients are supplied, and waste is extracted from the culture area. The media output area was reduced by then turning the manual valve, therefore increasing the hydrostatic pressure inside the system (discussed in Fig. 10.5). When passed through the manual valve, the waste contaminated media would then return to the medium vessel thereby completing the perfusion loop. In addition, the medium vessel was placed on top of an IKA Topolino S2 magnetic stirrer. This ensured thorough mixing of the media, even distribution of waste, and time extension before medium change. Second, it promotes gaseous exchange between the fluid and air, allowing oxygenation of the media.

10.5 Hydrostatic pressure analysis

Before conducting further CFD analysis on the final model, a mesh sensitivity study was performed to maximize result accuracy. The final mesh was concluded as suitable as soon as the resulting values produced a percentage difference of $<1\%$; this resulted in a structured quadrahedral mesh containing 348,649 elements.

Application and control of hydrostatic pressure were accomplished with the use of a manual valve. By turning the manual valve, the fluid outlet area of the bioreactor was decreased. During a constant flow rate, this decrease in area would cause an increase in fluid velocity to satisfy the continuity equation (Eq. 10.3). Bernoulli's equation (Eq. 10.4) states that both pressure and velocity of the fluid are dependent on the energy (head) supplied to the system. The only energy supplied to the system was provided by the pump (h_{Pump}); the gravitational energy (h) effects are assumed to be negligible due to no differentials in height and any energy losses (h_l) within the system were not considered. This results in an increase in pressure to support the increase in fluid velocity to satisfy Bernoulli's equation.

Continuity equation

$$Q = V \cdot A \tag{10.3}$$

where Q, V, and A is the volume flow rate, velocity and area respectively.

Bernoulli's equation

$$\frac{P}{\rho g} + \frac{V^2}{2g} + h + h_{\text{Pump}} + h_l = 0 \tag{10.4}$$

where $P, \rho, g, V, h, h_{\text{Pump}}$ and h_l are Pressure, density, gravitational acceleration, fluid velocity, gravitational head, pump head and head loss respectively.

To analyze the achievable hydrostatic pressures within the system, several models of the final design were simulated using the CFD settings mentioned in the previous section and the newly optimized mesh. In each model the outlet diameter was reduced in increments from 4 to 0.1 mm and hydrostatic pressure in the system was determined.

In each model the flow rate was kept constant at 0.1 L/min and the resulting hydrostatic pressures are graphed and tabulated in Fig. 10.23 and Table 10.3, respectively. The resulting values of pressure verified that during a constant flow rate, a decrease in outlet diameter increased the pressure within the system. When the diameter of the outlet equalled that of the inlet, a predicted pressure of 101.39 kPa was

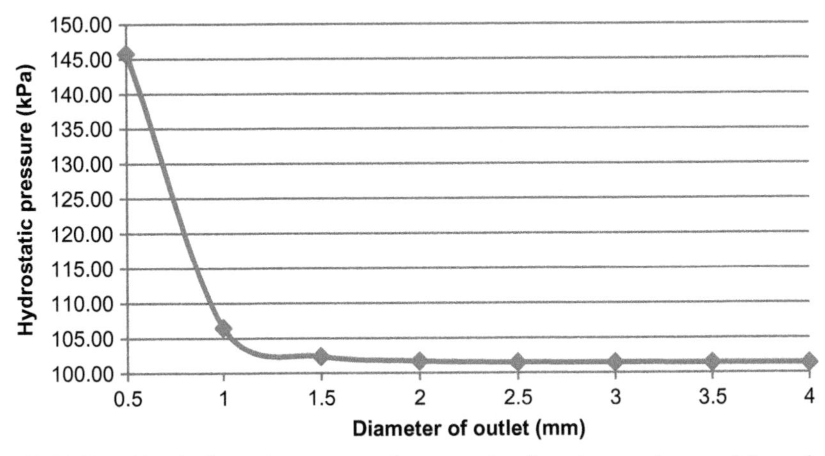

Fig. 10.23 Resulting hydrostatic pressure values as outlet diameter was decreased through CFD analysis.

Table 10.3 Resulting hydrostatic pressure values (P, kPa) as outlet diameter (D_o, mm) was decreased through CFD analysis

D_o	4	3.5	3	2.5	2	1.5	1	0.5
P	101.39	101.41	101.46	101.50	101.70	102.42	106.50	145.82

achieved, 0.067 kPa above environmental pressure (101.32 kPa). As the diameter was decreased, the pressure increased exponentially. No significant pressure change was observed until an outlet diameter of 0.5 mm was applied causing an increased pressure of 145.82 kPa within the system.

The resulting pressure values acquired through the CFD simulation did not reach the physiological values mentioned in related literature. Unfortunately, due to a lack of computational power, further investigation of smaller outlet diameters could not be conducted through use of CFD. This was due to the additional mesh density required when implementing the narrower channel formed by the small outlet diameter. To further investigate the effects on pressure during changes in outlet diameter values, the Continuity equation and Bernoulli's equation were used. As the diameter was further decreased, the pressure increased exponentially and some values achieved were within the physiological range of values (Fig. 10.24 and Table 10.4). By applying an outlet diameter of 0.2 mm, the pressure increased to 1.51 MPa. A maximum pressure of 22.62 MPa was achieved by implementing an outlet diameter of 0.1 mm; this value is beyond physiological levels.

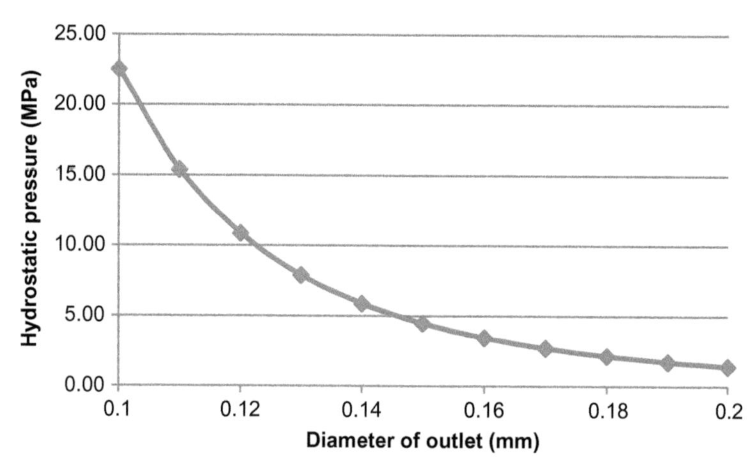

Fig. 10.24 Hand-calculated results of hydrostatic pressure with further reductions in outlet diameter.

Table 10.4 Hand-calculated results of hydrostatic pressure (P, MPa) with further reductions in outlet diameter (D_o, mm)

D_o	0.2	0.19	0.18	0.17	0.16	0.15	0.14	0.13	0.12	0.11	0.1
P	1.51	1.83	2.25	2.80	3.54	4.55	5.96	7.99	10.9	15.5	22.6

10.6 Fluid-induced shear stress analysis

Shear stress is caused by the flow of fluid across the surface and its value is directly proportional to the velocity of the surrounding fluid [38]. With the lack of sensors, shear stress could only be approximated with use of CFD techniques. Several simulations were conducted with variations in flow rate to estimate shear stress values on the surfaces of the scaffold. Maximum shear stress values were determined at flow rates of 0.01, 0.12, 0.23, 0.34, 0.45, 0.56, 0.67, 0.78, 0.89, and 1 L/min.

First, the distributions of shear stresses on all surfaces of the scaffold were evaluated with an applied flow rate of 0.12 L/min. From this study, it was found that the front surface was most affected by the imposed flow (Fig. 10.25A). Therefore, it was concluded that the shear stresses were to be evaluated on the front face of the scaffold.

Proceeding with this knowledge, various flow rates were then applied at the inlet to study the correlation flow rate and shear stress. The maximum shear stress values were evaluated linearly from the center to the edge of the front face of the scaffold. From the results, the correlation between flow rate and shear stress were verified (Fig. 10.26; Table 10.5).

The distribution of shear stress on the front surface varied greatly between the center and edge of the scaffold. Between the center and 2 mm from the center, the shear stress on the face of the scaffold increased significantly. From 2 to 4 mm from the center, the maximum shear stress on the scaffold was observed, progressing from 4 to 6 mm from the center, the shear stress then decreased. Finally, between 6 and 10 mm from the center, the shear stress increased slightly, before decreasing to a minimal value.

10.7 Discussion

The requirements of the perfusion system were determined through an in depth literature review regarding bioreactors. A bioreactor design was developed by conceptual design and study of the geometry–flow relations determined by CFD. It was

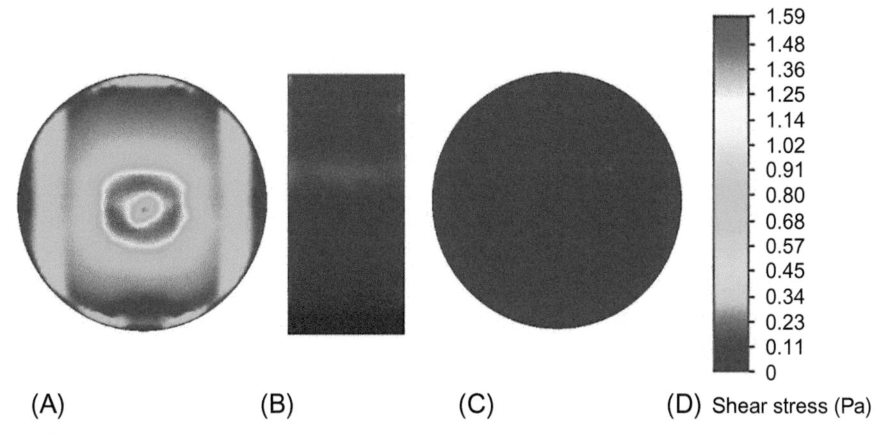

Fig. 10.25 Shear stress surface contours on scaffold: (A) front, (B) side, (C) rear views, and (D) shear stress scale bar.

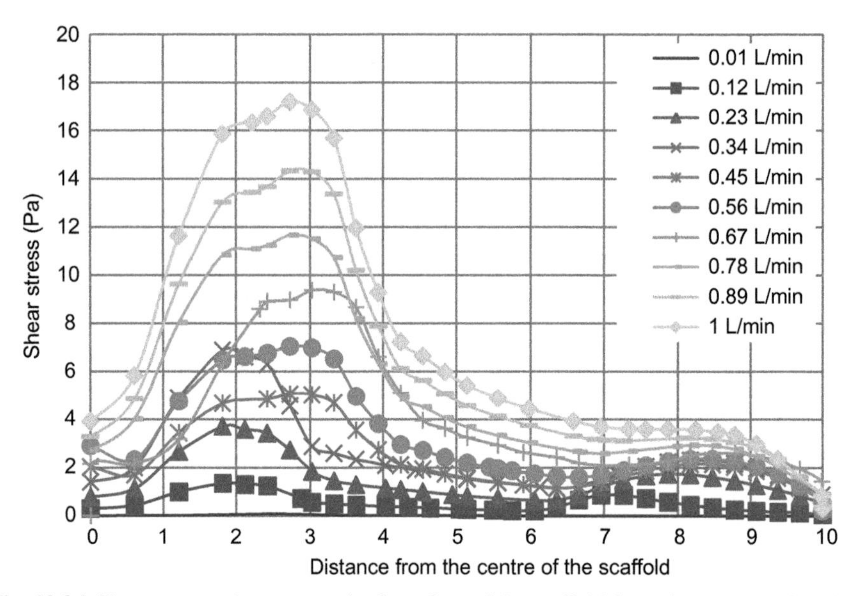

Fig. 10.26 Shear stress values across the front face of the scaffold from the center to the edge.

Table 10.5 **Shear stress (SS, Pa) values on front, side, and rear faces of scaffold during variations in volume flow rates (V, L/min)**

V	0.01	0.12	0.23	0.34	0.45	0.56	0.67	0.78	0.89	1.00
Front	0.08	1.59	4.42	5.19	6.03	8.42	11.21	13.92	17.31	20.57
Side	0.02	0.53	1.29	1.95	2.17	4.17	3.01	4.27	4.55	13.21
Back	0.16	8.91	3.67	4.19	6.58	5.55	7.11	8.34	10.47	12.10

concluded that a circular-shaped bioreactor with vertically central inlet/outlet holes produced maximum flow uniformity around a vertically supported scaffold. The initial prototype was constructed to validate the sufficiency of the materials used and manufacturing techniques employed. It was discovered that due to the porous nature of the 3D printing, that technique was to be avoided and the base of the bioreactor was cast. Next, laser beam machining produced a transparent acrylic bioreactor lid. By evaluating the final prototype, it was realized that the pump in use was not sufficient to induce the pressure required to reach the physiological range. To approximate the achievable hydrostatic pressures within the system using a pump capable of producing such stimulus, CFD was once again employed (following mesh sensitivity study). CFD analysis resulted in a maximum hydrostatic pressure of 145.82 kPa, which could be achieved with an outlet diameter of 0.5 mm. In pursuit of achieving values of hydrostatic pressure within the physiological range, further investigation was conducted using hand calculations. It was estimated that an outlet diameter between 0.2 and 0.1 mm during a flow rate of 0.1 L/min would provide hydrostatic pressures between 1.51 and 22.62 MPa, respectively (values within the physiological range). Analysis of shear stress within the system revealed that the largest effect was on the front face of the scaffold. Minimum and maximum values of shear stress were 0.08 and 20.57 Pa during flow rates of 0.01 and 1 L/min, showing a strong correlation between the value of shear stress and flow rate.

10.8 Conclusions

Bioprinted tissues are highly sensitive to mechanical environment during maturation processes in vivo mechanical factors are vital for the establishment of the tissue's functions. These factors include hydrostatic pressure and shear stress. To supply mechanical stimuli and other requirements to the cartilaginous tissues, a novel perfusion bioreactor was designed and manufactured. This bioreactor was designed to hold a scaffold and using fluid flow, a mechanical stimulus was applied to the culture while continuously supplying the cells with fresh, oxygenated media. Meanwhile the waste products that are produced by continuous cellular metabolic activity are removed. Using computation fluid dynamics the flow with the system was observed for design optimization. In addition to this, the hydrostatic pressure, shear stress, and turbulence were monitored during variations in flow rate to correlate the relations and approximate the final values achievable.

References

[1] Khan IM, Redman SN, Williams R, Dowthwaite GP, Oldfield SF, Archer CW. The development of synovial joints. Current topics in developmental biology. London: Academic Press; 2007. p. 1–36.

[2] O'Conor CJ, Case N, Guilak F. Mechanical regulation of chondrogenesis. Stem Cell Res Ther 2013;4(4):61.

[3] Grad S, Eglin D, Alini M, Stoddart MJ. Physical stimulation of chondrogenic cells in vitro: a review. Clin Orthop Relat Res 2011;469(10):2764–72.

[4] Bachrach NM, Mow VC, Guilak F. Incompressibility of the solid matrix of articular cartilage under high hydrostatic pressures. J Biomech 1998;31(5):445–51 [Research Support, U.S. Gov't, P.H.S.].

[5] Afoke NY, Byers PD, Hutton WC. Contact pressures in the human hip joint. J Bone Joint Surg (Br) 1987;69(4):536–41.

[6] Gooch K, Tennant C. Chondrocytes. In: Gooch K, Tennant C, editors. Mechanical forces: their effects on cells and tissues. Berlin Heidelberg: Springer; 1997. p. 79–100.

[7] Hall AC, Urban JP, Gehl KA. The effects of hydrostatic pressure on matrix synthesis in articular cartilage. J Orthop Res 1991;9(1):1–10 [Research Support, Non-U.S. Gov't].

[8] Soltz M, Ateshian G. Interstitial fluid pressurization during confined compression cyclical loading of articular cartilage. Ann Biomed Eng 2000;28(2):150–9.

[9] Elder BD, Athanasiou KA. Effects of temporal hydrostatic pressure on tissue-engineered bovine articular cartilage constructs. Tissue Eng Part A 2009;15(5):1151–8.

[10] Elder BD, Athanasiou KA. Synergistic and additive effects of hydrostatic pressure and growth factors on tissue formation. PLoS One 2008;3(6). e2341, [Research Support, N.I.H., Extramural].

[11] Suh J-K, Baek GH, Arøen A, Malin CM, Niyibizi C, Evans CH, et al. Intermittent sub-ambient interstitial hydrostatic pressure as a potential mechanical stimulator for chondrocyte metabolism. Osteoarthr Cartil 1999;7(1):71–80.

[12] Jortikka MO, Parkkinen JJ, Inkinen RI, Karner J, Jarvelainen HT, Nelimarkka LO, et al. The role of microtubules in the regulation of proteoglycan synthesis in chondrocytes under hydrostatic pressure. Arch Biochem Biophys 2000;374(2):172–80 [Research Support, Non-U.S. Gov't].

[13] Smith RL, Lin J, Trindade MC, Shida J, Kajiyama G, Vu T, et al. Time-dependent effects of intermittent hydrostatic pressure on articular chondrocyte type II collagen and aggrecan mRNA expression. J Rehabil Res Dev 2000;37(2):153–61 [Research Support, U.S. Gov't, Non-P.H.S.].

[14] Smith RL, Rusk SF, Ellison BE, Wessells P, Tsuchiya K, Carter DR, et al. In vitro stimulation of articular chondrocyte mRNA and extracellular matrix synthesis by hydrostatic pressure. J Orthop Res 1996;14(1):53–60 [Research Support, U.S. Gov't, Non-P.H.S.].

[15] Hu JC, Athanasiou KA. The effects of intermittent hydrostatic pressure on self-assembled articular cartilage constructs. Tissue Eng 2006;12(5):1337–44 [Comparative Study].

[16] Toyoda T, Seedhom BB, Yao JQ, Kirkham J, Brookes S, Bonass WA. Hydrostatic pressure modulates proteoglycan metabolism in chondrocytes seeded in agarose. Arthritis Rheum 2003;48(10):2865–72.

[17] Wendt DJ. Effects of fluid-induced shear stress on articular cartilage regeneration under simulated aspects of microgravity. Retrospective Theses and Dissertations, 461, 2001. p. 1–125.

[18] Lee C, Grad S, Wimmer M, Alini M. The influence of mechanical stimuli on articular cartilage tissue engineering. Topics Tissue Eng 2006;2:1–32.

[19] Gaspar DA, Gomide V, Monteiro FJ. The role of perfusion bioreactors in bone tissue engineering. Biomatter 2012;2(4):167–75.

[20] Gharravi AM, Orazizadeh M, Hashemitabar M. Fluid-induced low shear stress improves cartilage like tissue fabrication by encapsulating chondrocytes. Cell Tissue Bank 2016;17(1):117–22. https://doi.org/10.1007/s10561-015-9529-2.

[21] Tarng YW, Casper ME, Fitzsimmons JS, Stone JJ, Bekkers J, An KN, et al. Directional fluid flow enhances in vitro periosteal tissue growth and chondrogenesis on poly-epsilon-caprolactone scaffolds. J Biomed Mater Res A 2010;95(1):156–63 [Research Support, N.I.H., Extramural Research Support, Non-U.S. Gov't].

[22] Theodoropoulos JS, DeCroos AJ, Petrera M, Park S, Kandel RA. Mechanical stimulation enhances integration in an in vitro model of cartilage repair. Knee Surg Sports Traumatol Arthrosc 2016;24(6):2055–64. https://doi.org/10.1007/s00167-014-3250-8.

[23] Smith RL, Donlon BS, Gupta MK, Mohtai M, Das P, Carter DR, et al. Effects of fluid-induced shear on articular chondrocyte morphology and metabolism in vitro. J Orthop Res 1995;13(6):824–31 [Research Support, Non-U.S. Gov't Research Support, U.S. Gov't, Non-P.H.S.].

[24] Smith RL, Carter DR, Schurman DJ. Pressure and shear differentially alter human articular chondrocyte metabolism: a review. Clin Orthop Relat Res 2004;427:S89–95.

[25] Gemmiti CV, Guldberg RE. Fluid-induced shear stress up-regulates collagen type ii incorporation at the surface of tissue-engineered cartilage, 50th Annual meeting of the Orthopaedic Research Society, Poster No. 0841.

[26] Chen HC, Hu YC. Bioreactors for tissue engineering. Biotechnol Lett 2006;28(18):1415–23 [Research Support, Non-U.S. Gov't Review].

[27] Schulz RM, Wüstneck N, van Donkelaar CC, Shelton JC, Bader A. Development and validation of a novel bioreactor system for load-and perfusion-controlled tissue engineering of chondrocyte-constructs. Biotechnol Bioeng 2008;101(4):714–28.

[28] Martin I, Wendt D, Heberer M. The role of bioreactors in tissue engineering. Trends Biotechnol 2004;22(2):80–6.
[29] Korossis SA, Bolland F, Kearney JN, Fisher J, Ingham E. Bioreactors in tissue engineering. Topics Tissue Eng 2005;2(8):1–23.
[30] Hansmann J, Groeber F, Kahlig A, Kleinhans C, Walles H. Bioreactors in tissue engineering—principles, applications and commercial constraints. Biotechnol J 2013;8(3):298–307.
[31] Waldman SD, Couto DC, Grynpas MD, Pilliar RM, Kandel RA. A single application of cyclic loading can accelerate matrix deposition and enhance the properties of tissue-engineered cartilage. Osteoarthritis Cartilage 2006;14(4):323–30 [Research Support, Non-U.S. Gov't.
[32] Correia V, Panadero JA, Ribeiro C, Sencadas V, Rocha JG, Gomez Ribelles JL, et al. Design and validation of a biomechanical bioreactor for cartilage tissue culture. Biomech Model Mechanobiol 2015;8:1–8.
[33] Sucosky P, Osorio DF, Brown JB, Neitzel GP. Fluid mechanics of a spinner-flask bioreactor. Biotechnol Bioeng 2004;85(1):34–46.
[34] Shahin K, Doran PM. Tissue engineering of cartilage using a mechanobioreactor exerting simultaneous mechanical shear and compression to simulate the rolling action of articular joints. Biotechnol Bioeng 2012;109(4):1060–73 [Research Support, Non-U.S. Gov't].
[35] Gilbert E, Mosher M, Gottipati A, Elder S. A Novel Through-Thickness Perfusion Bioreactor for the Generation of Scaffold-Free Tissue Engineered Cartilage. Processes 2014;2(3):658–74.
[36] Gharravi AM, Orazizadeh M, Ansari-Asl K, Banoni S, Izadi S, Hashemitabar M. Design and fabrication of anatomical bioreactor systems containing alginate scaffolds for cartilage tissue engineering. Avicenna J Med Biotechnol 2012;4(2):65–74.
[37] Tarng YW, Huang BF, Su FC. A novel recirculating flow-perfusion bioreactor for periosteal chondrogenesis. Int Orthop 2012;36(4):863–8 [Research Support, Non-U.S. Gov't].
[38] Hutmacher DW, Singh H. Computational fluid dynamics for improved bioreactor design and 3D culture. Trends Biotechnol 2008;26(4):166–72 [Review].

3D bioprinting: Regulation, innovation, and patents

P. Li
University of Sussex, Brighton, United Kingdom

With the technological breakthrough in additive manufacturing (AM, aka, three-dimensional (3D) printing, 3D printing), 3D bioprinting technologies have emerged as an efficient tool for tissue engineering and regenerative medicine. The disruptive nature of the bioprinting technologies not only weaves a dream of "printed-organ-on-demand," but also challenges the traditional regulatory framework. The product development of 3D bioprinted technologies involved "re-distributed manufacturing (RDM)," which means that the manufacturing process has taken place at several technical stages at multiple sites. This includes procurement and initial processing, processing and cell manipulation, scaffold fabrication and product assembly, packaging, testing, shipping, and end-user delivery and implantation [1]. To ensure quality, it is essential to have clear characterization of each process step:

- Preprinting: Designing process; software workflow phase (design manipulation software is a computer program that enables a medical device to modify the design for customization)
- Printing (build)
- Postprocessing
- Testing and characterization

Quality assurance and testing is required at every stage before moving onto the next site for further processing. Hence, this chapter analyzes the issues associated with each process step of manufacturing. There is currently not a sui generis regulatory regime governing the whole bioprinting process but piecemeal legislations in relation to tissue engineering and regenerative medicine. Regenerative medicine using cell or tissue engineering is named advanced therapies in the European Commission (EC) Regulation on Advanced Therapy Medicinal Products (ATMP) Regulation. The principles developed in the ATMP Regulation are applicable to 3D bioprinting. Relevant legal instruments include the ATMP regulation, the EC Cells and Tissues Directive, and the new Medical Device Regulation (MDR), all are applicable at different stages of production.

This chapter considers the effects of existing laws on bioprinting, particularly through the lens of risk regulation and IP. The first section introduces relevant EU and UK regulatory instruments at their respective stages, the second section will examine the bioprinting patent regime which acts as a corresponding regulatory tool for innovation.

3D Bioprinting for Reconstructive Surgery. https://doi.org/10.1016/B978-0-08-101103-4.00020-X

11.1 Introduction

The 3D bioprinting technology is particularly beneficial to regenerative medicine. It is primarily driven by three factors: the quality of bioprinters and source materials, computer-aided design (CAD) software, and advances in regenerative medicine. Regenerative medicine means using self-healing processes of the human body via enhancement of medicine by therapeutic options of tissue engineering, which deals with the process of generating repair tissues and organs (bioartificial tissues) using cells from an individual patient (using cells to create tissues outside the body and then implanting them). The advantage of using a patient's autologous cells is that it decreases tissue rejection.

The following are the steps involved in using a bioprinter with autologous cells:

- Collect and grow stem cells or cells from patient biopsies.
- Make bioink from enough cells and load into the cartridge.
- Use the bioink and hydrogel to print tissues with software and a bioprinter.
- The printed tissues are left to mature for several weeks.

After maturation, the printed tissues can be used in medical research, new drug testing, or as a transplant material. It is expected that in the near future such a bioprinter could be an essential tool for in-vivo printing in the hospital operating theater. 3D bioprinted organs are expected to address the shortage of transplantable organs thus preventing the complications of sources and commercialization of human organs, which the existing Human Organ Transplants Act 1989 seeks to regulate.

Researchers have long been reproducing cells in laboratories, including skin tissues, blood vessels, and other cells from various organs. There is a wide range of materials which could be used in bioprinting. In addition to cells and collagen, stem cells are an attractive option to be used as bioink for printing different organs and tissues as they adapt easily to host tissues. Though 3D bioprinting opens a door to printing a whole organ, currently the printing of the complicated networks of veins is a major obstacle to reproducing a human organ.

11.2 Regulation

Bioprinting is a subcategory of 3D printing but the same regulation rationale cannot be applied due to the inherently different policy consideration from the perspective of human health and safety. These range from the fundamental philosophical and bioethical issues to practical risk, biosafety, and security concerns. The UK Nuffield Council on Bioethics defines biosafety as "the safe handling and containment of infectious microorganisms and hazardous biological materials" [2, p. 137]. 3D bioprinting presents the recurrent risks and challenges arising from cell therapy, stem cell research, and organ transplantation. There may be questionable sources of biomaterials, unhealthy donors, and posttransplant infections. Bioethical concerns include treatment safety; animal testing; human enhancement; moralization; "playing God" objection; justice and patients' access. The concerns regarding clinical trials include product safety, standardization, customization, quality control, surveillance and traceability, quality system (QS), good manufacturing practices (GMP), ISO, standard review pathways, abbreviated review pathways; product standardization, and classifications.

The current legal regime on cell therapy and stem cell research lacks clarity, which also impacts the bioprinting regulation. The legal complexities of bioprinting are further compounded by the multiple actors involved in the production chain. Specifically, the adoption of computer-aided manufacturing (CAM) through which CAD software is used to customize the product and to trigger the bioprinting process. Actors or stakeholders of 3D bioprinting include printer manufacturers, 3D model designers, surgeons, other healthcare professionals, hospital, ethics committee, and insurance companies. Expertise from legislature and health professionals; legal experts; engineers; biologists, and computer programmers are necessary for multistakeholder collaboration in order to shape a suitable pathway for bioprinting progression.

Among the key aspects of bioprinting governance are risk regulation, product liability, and IP issues. One of the pending fundamental issues is the definition of a "manufacturer"—who will be liable for product failure? Are 3D model designer, 3D printer producer, and health professional expected to share liability? And what is defined as a "product," and under what category within the current legal classification?

11.2.1 Preprinting: Procurement and initial processing, processing and cell manipulation

The 3D bioprinted product using bioink developed from the autologous cells would avoid patient's rejection rate, yet in certain circumstances it is also possible to use allogeneic cells for fabrication. Cells and tissues are first acquired from a patient or a donor before the initial processing of cell releasing from tissue matrix, cell isolation/enrichment, cell culture, harvest, and encapsulation of cells in hydrogel [1]. To ensure the quality and safety of cell and tissue material, the early stage of donation, procurement, and testing of tissue and cells is governed by the Human Tissue Act 2004 (following the EU Tissues and Cells Directive 2004/23); while the later stage of manufacture, storage, and distribution of tissues and cells is under the auspice of the EU Regulation on Advanced Therapeutic Medicinal Products [3].

Regulation of biotechnology is based on principles of bioethics, such as do no harm; confidentiality; informed consent; and benefit sharing. According to the Human Tissue Act 2004, only a licensed person is allowed to remove a living person's transplantable material. Such removal need to be noncommercial. A full informed consent process will minimize the risk of harm and possible violation of ethical considerations. Express consent from the donor is required to remove, store, and use his tissues. Consent could be given by the donor during his lifetime or after death if such consent is expressed by will, a nominated person, or the next of kin. A child is to give consent himself, but in situations where a child is incapable of giving consent or involves the mentally incapacitated, a parent or guardian is to express consent. Consent will not be required if the donor or the next of kin could not be traced. The ethics committee may approve the research without consent. Donors should be informed of the current and future use of his cells and tissues. Complete details about the composition of the bioprinted product, the implantation process, all conflicts of interests, and all potential outcomes and adverse effects, are suggested to be noted in the consent form [4, p. 288]. Furthermore, it is noteworthy that a "presumed consent" (opt-out) model of the deceased is applied in Wales.

In preparation for the materials for fabrication, the use of human embryonic stem cells (HESCs) as bioink, once again touches upon the social and religious taboos of cloning and the interpretation of humanity and human dignity. The ATMP Regulation states that the regulation of ATMP at the European Community level should not interfere with decisions made by individual Member States on whether to allow the use of any specific type of human cells, such as embryonic stem cells or animal cells. In view of potential objections, researchers may pursue "induced pluripotent stem cells" (iPSC) technologies or to collect multipotent stem cells (adult/somatic stem cells) to produce pluripotent stem cells for 3D tissue engineering in order to bypass the destruction of human embryos.

The ATMPs Regulation also expresses the respect of fundamental rights and makes reference to the Council of Europe Convention for the Protection of Human Rights and Dignity of the Human Being with regard to the Application of Biology and Medicines: Convention on Human Rights and Biomedicine. Particularly, it is stated that "[R]emoval of organs or tissue from a living person for transplantation purposes may be carried out solely for the therapeutic benefit of the recipient and where there is no suitable organ or tissue available from a deceased person and no other alternative therapeutic method of comparable effectiveness" [5].

It is recognized that the same regulatory principles for other types of biotechnology medicinal product should be applicable to ATMP, and specific technical requirements with regard to quality and clinical data to ensure quality, safety, and efficacy, may also be required. Following the principle of altruism in the Cell and Tissue Directive, the ATMP Regulation also stresses that the procurement of human cells or tissues used in ATMP should be voluntary and unpaid donation, the anonymity of the donor and the recipient and solidarity between the donor and the recipient should be respected. Specifically, the ATMP Regulation urges Member States to encourage a strong public and nonprofit sector involvement in the procurement of human cells or tissues.

11.2.2 Printing: Scaffold fabrication and product assembly

The printing process is triggered by the instructions generated by CAD or CAM systems. The customized CAD file is generated by 3D scanning and image segmentation. The medical use of 3D printers involves a digital file generated from the 3D scanning of a human body part, and reprinting the wound for repair. In medical use, researchers could use the digital data acquired from a CT scan to make a 3D design model of an organ. The next step is to convert the design or data into a Stereolithography (STL) file. The STL file is then imported into the printer software package to set up the machine for printing. It is then reviewed by a surgeon before the CAM process or biofabrication.

Mass digitization and dissemination of STL file poses risks to the regulatory framework of consumer safety, product liability (quality control), data protection and confidentiality, and safety. A major issue is that the design involves multiple translation and compilation steps between multiple software applications, each having different error sensitivities. Hence errors in customization settings are likely to occur. The current computer-aided process planning still relies on human expertise and manual interaction that the manufacturer will need to verify and account for the accuracy of the image [1]. These errors would consequently have direct impacts on the mechanical strength of the finished 3D bioprinted product. A key question is that how could the translation and compilation steps ensure reproducibility and comparability of printed products?

It is also noteworthy that if the software and hardware set up for bioprinting are regarded as "medical devices" that fall under the auspice of EU MDRs, then it will need certification before being put on the market. The new EU MDR states that "It is necessary to clarify that software in its own right, when specifically intended by the manufacturer to be used for one or more of the medical purposes set out in the definition of a medical device, is qualified as a medical device, while software for general purposes, even when used in a healthcare setting, or software intended for well-being application is not a medical device. The qualification of software, either as device or accessory, is independent of its location or type of interconnection between the software and a device" (MDR Article 18a).

Stand-alone software, which drives a device or influence the use of a device, falls automatically in the same class as the device. If stand-alone software is independent of any other device, it is classified in its own right (MDR Annex VII Sec II.3).

Once a software is qualified as a medical device, it will fall within the safety and performance requirement when used for diagnosis, prevention, monitoring, treatment or alleviation of disease, injury or disability and the "investigation, replacement or modification of the anatomy or of a physiological or pathological process or state."

11.2.3 Postprinting: End-user delivery and implantation

The ATMP Regulation classifies tissue-based or cell-based products as medicinal products including gene therapies, cell therapies, and tissue engineering under the regulation of regenerative medicines. Cells or tissues that are subject to "substantial manipulation" are considered "engineered." The ATMP regulation involves one centralized marketing authorization pathway. The European Medicines Agency (EMA) is responsible for the regulatory framework, while the Committee for Advanced Therapies (CAT) act as the scientific committee dealing with individual submission. The CAT will decide the classification and evaluate the quality, safety, and efficacy of an ATMP product in accordance with the scientific information submitted to the committee.

The key issue of postprinting is the classification of a bioprinted product: as organs and tissues are not "original" body parts… are they to be regulated as "products," medical devices, drugs, ATMP, or a new category? When the printing process takes place at clinics or hospitals—are they put "on the market" or "put into service"?

(The criteria for ATMPs are set out in Article 17 of Regulation (EC) No 1394/2007.)

The dealings of ATMPs should follow the principles of GMP and good clinical practice during the conduct of clinical trials. Clinical trials on ATMPs are required

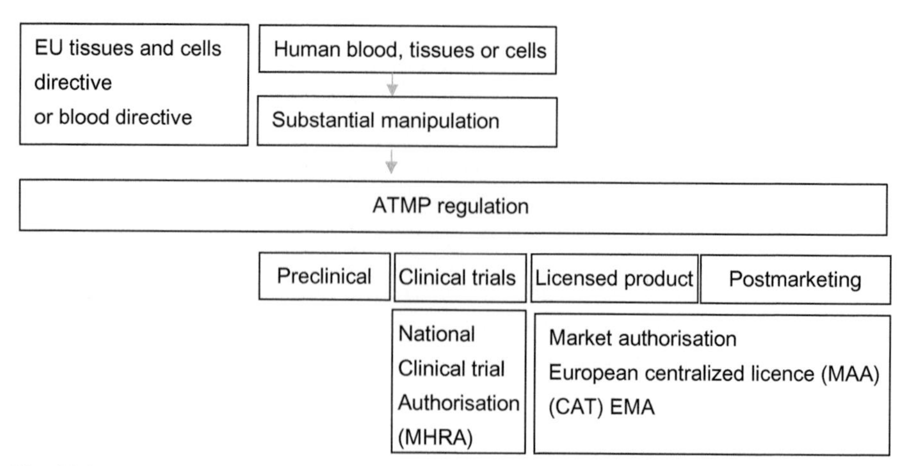

Fig. 11.1 ATMP regulation.

for market authorization. It is also noteworthy that a "combined" ATMP comprised of both medical device and viable cells or tissues will require premarket approval from the EMA, which involves assessment of both the medical device part and the medicinal product part.

The ATMP regulation waives the requirement of premarket approval under the "hospital exemption" where the ATMPs are prepared on a nonroutine basis according to specific quality standards. Due to the experimental nature, the bioprinted products are used via the "hospital exemption" (it is called "Specials" in the United Kingdom). It takes place at a hospital premise under the exclusive professional responsibility of a medical practitioner in order to comply with an individual medical prescription for a custom-made product for an individual patient.

After the bioprinting process, traceability of the patient and the ATMP is required to monitor the safety of ATMP. The holder of marketing authorization is required to keep the relevant data for a minimum of 30 years after the expiry date of the product for follow-up of efficacy and risk management post authorization. Extra caution needs to be taken to protect the personal data. Information for documentation should include: manufacturing parameters, risks identified for each step of the manufacturing process, mitigations of these risks, software steps, validation and testing steps, and acceptance activities (Fig. 11.1).

11.3 Patents as an actor for regulation

The protection of IP plays an important role in the innovation of 3D bioprinting. 3D bioprinting technologies may attract a wide array of IP rights including patents, copyright, design rights, trademark and passing off, and confidentiality. This section will focus on the role of patents in bioprinting innovation and regulation.

The patent system is a social contract of give and take. Patentees are required to register and disclose sufficient information in relation to the invention to society, in exchange for market exclusivity for 20 years. Thus the patent system is an incentive for inventors to share their knowhow, as opposed to the "trade secret" or "confidentiality" regime in which technologies are kept within a small group of people. It is anticipated the progress of science and technology could be promoted after the monopoly.

Patent rights provide market monopolies over an invention for up to 20 years on products, manufacturing processes, or "product by process." A patent may be granted for an invention if it is new, involves an inventive step, is capable of industrial application, and is not excluded in the UK Patent Act 1977. Some scientific breakthrough will not be deemed patentable due to special public policy. Exclusions to patentability mainly fall under three strands: contrary to morality or public policy; animal or plant variety and essential biological processes for the production of animals or plants (not being a microbiological or other technical process or the product of such a process); and methods of medical or veterinary treatment. Specifically, the UK Patent Act 1977 explicitly identifies that the following categories are not inventions and thus nonpatentable (s. 1(2) UK PA 1977):

- a discovery, scientific theory, or mathematical method;
- a literary, dramatic, musical, or artistic work or any other esthetic creation;
- a scheme, rule, or method for performing a mental act, playing a game or doing business, or a program for a computer;
- the presentation of information

Biological materials, animals and plants and parts of the human body are potentially patentable, except for the following:

- the human body, at the various stages of its formation and development, and the simple discovery of one of its elements, including the sequence or partial sequence of a gene;
- processes for cloning human beings;
- processes for modifying the germ line identity of human beings;
- uses of human embryos for industrial or commercial purposes;
- processes for modifying the genetic identity of animals which are likely to cause them suffering without any substantial medical benefit to man or animal, and also animals resulting from such processes

(Para. 3 Schedule A2 to the UK PA 1977 implementing the EU Biotechnology Directive)

Patent monopolies would create exclusive market advantage, yet at the same time it would also lead to artificial barriers to dissemination of knowledge. In recent years, there has been initiatives to reflect upon the downsides of patents, such as the call for open innovation and open licensing, the access to medicines, and the access to knowledge campaigns. The scope of patentability has been constantly refiguring by the development of case laws in order to strike a fine balance between public and private interests [6].

Bioprinting patents are composed of materials, manufacturing process, software, and printed products (see Fig. 11.2, source: UK IPO). Patent protection may cover bioprinter, computer software, 3D scanner, 3D scanner driver, CAD file software, materials, and printed products (see Fig. 11.2, source: UK IPO).

The following paragraphs will introduce the key issues in patents including patentability, morality and medical treatment exception, infringement, enforcement,

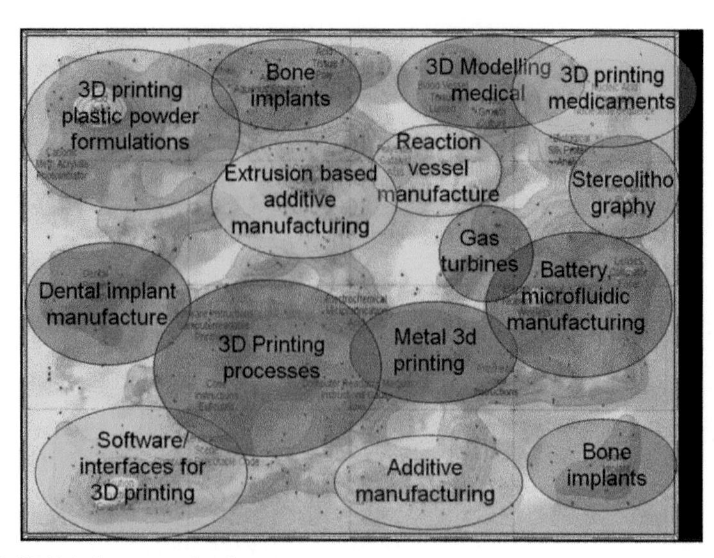

Fig. 11.2 3D Printing patent landscape.
Source: UK Intellectual Property Office 2013.

licensing, and accessibility of patents. As discussed above, patents provide an economic incentive to innovation. It is built upon the hypothesis that creating short-term (20 years) market monopoly on the condition of disclosure of technology to the public will overall promote innovation in society. Yet at the same time, overprotection of technologies creating "patent thickets" which will lead to the adverse effects of stifling innovation. It is therefore vital to strike a balance between patent monopoly and the public interests of accessing the invention.

In the bioprinting scenario, patent protection will insure the inventor to recoup his investment in research and development of the technology. However, the multilayered protection of various patents on this technology inevitably increases the cost of using this technology, thus may possibly restrict the majority from accessing the scientific benefit brought by bioprinting. It is essential to consider both the need to protect the inventor's IP right as well as the need to protect patients' right in benefiting from this invention. The following section will take a number of patents, for example, examining the current scope of protection in the EU, including patent protection on a bioprinter, a software program, and a bioprinted vascular vessel.

11.3.1 Patents on 3D bioprinter hardware

The following is an example of a patent for the inkjet printing with viable cell technology (see Fig. 11.3). The patent "Multilayered Vascular Tubes," provides protection for the invention on engineered multilayer vascular tubes, methods for producing such tubes, and therapeutic, diagnostic, and research uses for such tubes. Such a patent claim directed to 3D bioprinting technologies would exclude others from making, using, offering to sell, selling, or importing any products that meet the claim. A license

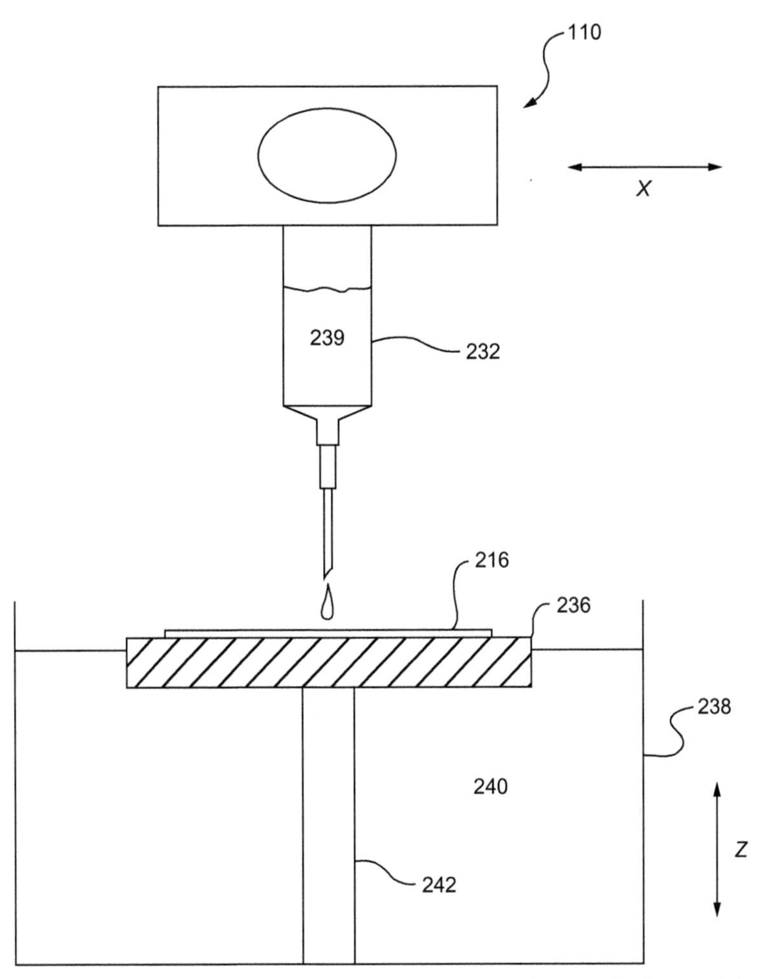

Fig. 11.3 A schematic view of a printer head that may be used in on embodiment of the invention.
Source: United States Patent Application Publication, US 2004/0237822 A1.

needs to be acquired from the patentee before utilizing this invention. Thus the consideration of patentability will have significant impacts on access to these technologies.

11.3.2 Patents on 3D printing software

Computer software in the form of source code and object code is a set of statements or instructions to be used in a computer to perform a certain operation, it directly deals with both the literal aspects of copyright and the functional aspects of patents. There is a dichotomy between copyright and patents: while copyright protects creative esthetic works, patents protect practical functional inventions. The literal aspects of computer programs are protected as literary works under the copyright regime, and are explicitly excluded from patentability in Europe. Yet the appropriateness of using copyright to protect software has been questioned due to the inherent purpose of giving computer instructions to perform certain functions. Thus there has been preference to protect software via patents.

Despite the explicit exclusion from patentability, in practice the EU/UK has shifted toward software patents as long as a "hardware" or a "technical effect" exists in the invention. The UK Court expressed in *Symbian Ltd v Comptroller General of Patents* that exclusion to patentability would not automatically apply merely on the ground that the use of a computer program was involved.

The *Aerotel/Macrossan* approach is the current position taken by the UK courts on software patents, which is deemed equivalent to the prior case law test of "technical contribution." The *Aerotel/Macrossan* four-step approach refers to: properly construe the claim; identify the actual contribution; ask whether it falls solely within the excluded subject matter; and check whether the actual or alleged contribution is actually technical in nature.

Currently, the 3D printing industry uses open source software. There is an example involving a software patent developed to prevent the printing of unauthorized 3D designs. This "Manufacturing Control System" may act to control illegitimate dissemination of 3D CAD files, on the basis of risk regulation, through a recorded protection system which would check whether a licensing agreement has been acquired before printing (US Patent 8286236, "Manufacturing Control System"). Online platforms have been described as facilitating illegal file sharing. If the same software is to be considered for patentability in the United Kingdom, it would need to pass the *Aerotel/Macrossan* test as discussed above, examining whether the claim falls within the patent subject matter and that the contribution of which is technical in nature.

However, IP infringement will probably be not a key issue at professional healthcare institutions, instead, issues such as informed consent, privacy, and data protection will be the main concerns in the bioprinting scenario. Images generated from CT or MRI for further use in CAD and STL files are considered to be personal data that should not be disclosed, distributed, or uploaded without a patient's informed consent.

Another example is the patent on "Method and system for registration/alignment of 3D digital models" (WO/2014/049049). It involves a collection of systems and methods, devices, and software for use in medical imaging applications that require registration of digital images. This system can automatically find correspondence between multiple digital images of multiple 3D objects with identical or similar geometry. The advantage of the system is that it could ignore noise, scatter, occlusion, or clutter in order to find correspondence of the images.

Following the European precedents, a "hardware" or an "apparatus" included in the application would be required to acquire patentability; while the UK Aerotel/Macrossan approach expects the system to involve a technical effect or technical contribution. The software system is composed of the following steps: volumetric scan; load image data into the computer; segmentation; low- and high-resolution models of definition; registration of low- and high-resolution images; select optional tools;

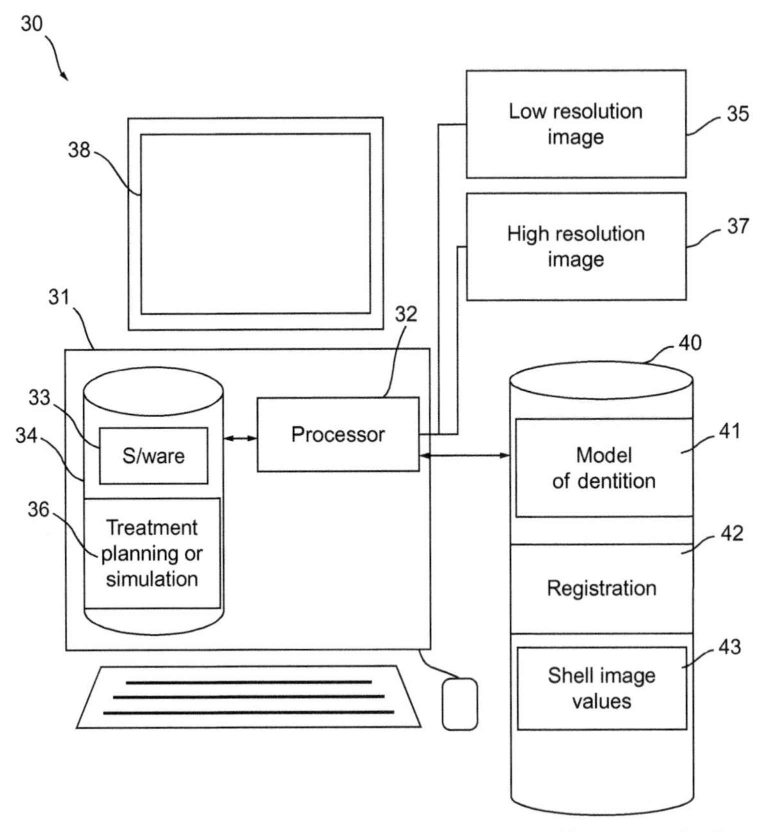

Fig. 11.4 A schematic representation of a computer system according to an embodiment of the present invention.
Source: WO/2014/049049.

image and value calculation; convert to array of values; optional normalization; comparison of correspondences, and registration (Fig. 11.4).

On application, the International Searching Authority then examines the novelty, inventive step against the state of the art, and industrial applicability. The clarity of the major and dependent claims was also scrutinized. Any ambiguity or imprecise statement of the patent claims is requested to be rectified.

11.3.3 Patents on printed tissues or organs

3D printed tissues or organs particularly relate to the field of regenerative medicine and tissue engineering. In the case of the patent on "Multilayered Vascular Tubes" (Fig. 11.5), patent protection covering the invention of engineered multilayer vascular tubes includes methods for producing such tubes, and therapeutic, diagnostic, and research uses for such tubes. As discussed above, exclusion to patentability include the *morality* clause and medical treatment exception.

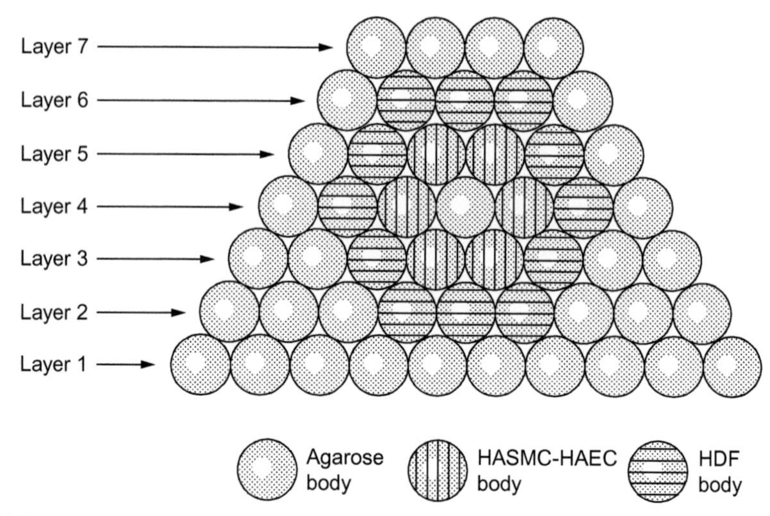

Fig. 11.5 Geometrical arrangement for multilayered vascular tubes.
Source: UK Patent 2478801.

11.3.3.1 The morality clause

The debate over the legitimacy of granting monopolies on human body parts could be traced back to the *Moore* case in the United States and the *Yearwood* case in the United Kingdom. The commercial exploitation of inventions should not be endorsed if such activities are deemed to offend against the social perceptions of morality. It demonstrates that the patent system allows a margin for embedding citizens' culture and preferences.

The European Directive on the Legal Protection of Biotechnological Inventions (Biotech Directive) specifically sets out commercialization of certain technologies as immoral and unpatentable: processes for cloning human beings; processes for modifying the germ line genetic identity of human beings; uses of human embryos for industrial or commercial purposes; and processes for modifying the genetic identity of animals where their suffering outweighs substantial medical benefits to man or animal (Article 6.2). Yet it also provides that "An element isolated from the human body or otherwise produced by means of a technical process, including the sequence or partial sequence of a gene, may constitute a patentable invention, even if the structure of that element is identical to that of a natural element" (Article 5(2)). This implies that printed tissues or organs which have been isolated from the human body and reproduced by means of a technical process may be considered patentable, even if the structure is identical to that of a natural element. 3D-printed tissues using hydrogel and cultured in a scaffold will develop their own structure layer by layer and thus are distinct from the original cell, but patentability is still valid even if a 3D-printed organ produced by digital fabrication is identical to the original organ. Thus, the 3D-printed tissues and organs may satisfy the requirement in Article 5(2) of the Biotech Directive.

However, the invention will still need to pass the examination of the "morality" test in Article 53(a) of the European Patent Convention (EPC). The morality clause is embedded in patent law for achieving a balance and coherence between science and culture. The interpretation of the morality clause has been developed from case law. It was held in the case of the *Harvard/Onco-mouse* that the threshold of patentability could be satisfied as long as the potential benefit to mankind outweighed the suffering of the animals used in scientific experiments. The EPO further elaborated that the morality provision would only apply where universal outrage is involved and that the "abhorrent" threshold is developed to measure immorality: "A fair test to apply [the morality provision] is to consider whether it is probable that the public in general would regard the invention as so abhorrent that the grant of patent rights would be inconceivable."

The recent case rulings further elaborate the morality clause through the definition of human life and embryo. The *Brüstle v Greenpeace* case suggests that an invention is excluded from patentability if the process involves the destruction of embryos, yet it may be patentable if the production could be achieved without destruction of embryos. The *International Stem Cell Corporation (ISCO) v Comptroller General of Patents* further defined a human embryo as having the "inherent capacity to develop into a human being." As pluripotent stem cells are not considered to be embryos due to their incapacity to develop into human beings, patents may still be granted if the claims relate to pluripotent stem cells instead of totipotent ones, on the condition that their isolation does not involve any destruction of human embryos. Therefore, researchers may then opt to pursue iPSC technologies or to collect multipotent stem cells (adult/somatic stem cells) to bypass the destruction of human embryos for producing pluripotent stem cells for 3D tissue engineering. Researchers can still seek patents for iPSC-base inventions.

11.3.3.2 Medical treatment exception

In addition to the morality clause, the EPC also draws limits to patentability presented by the medical treatment exemption. It notes that European patents shall not be granted in respect of inventions of methods for treating the human or animal body by surgery or therapy and diagnostic methods practised on the human or animal body; this provision does not apply to products such as substances or compositions for use in these methods (Article 53(c)). The UK Patents Act (UKPA) also articulates a special exclusion to patentability for medical treatments. It states that a patent shall not be granted for the invention of methods for treatment of the human or animal body by surgery or therapy, or a diagnosis method practised on the human or animal body (Section 4A). The purpose of this provision is "to free from restraint non-commercial and non-industrial medical and veterinary activities."

The purpose of the provision is to insure that "actual use, by practitioners, of methods for medical treatment when treating patients should not be subject to restriction or restraint by patent monopolies." The patentability of an invention in this field is reliant on the scope of the interpretation of what constitutes a method of surgery, therapy, or diagnosis practiced on the human or animal body. It is noteworthy that the exclusion of methods of diagnosis is only applicable to in vivo diagnosis; therefore, in vitro diagnosis is still patentable. Careful attention needs to be paid to the drafting of patent claims. It is necessary to differentiate between method claims about the operation of a device used in healthcare and methods involving treatment, surgery, or diagnosis. Patent claims relating to "purely technical methods" will still be granted.

11.4 Conclusion and future trends

The emergence of 3D bioprinting presents a useful tool for tissue engineering and regenerative medicine, yet the novel production methods and products blur the existing boundaries and classifications in regulation and IP. It remains to be seen whether the EU ATMP Directive and the new MDR is sufficient for regulation. Bioprinting patents facilitate the development of the industry, but it also creates undesirable downsides of high prices and create barriers to the technologies. While granted market monopolistic rights, at the same time, inventors are also expected to discharge their obligations of sharing the benefits of scientific progress to society. A broad or open licensing scheme will promote the accessibility of 3D bioprinting technologies. It remains to be seen how the product liability regime and the insurance policy will adapt to meet the challenges of bioprinting. It is expected that public debates would take place in the near future on the development pathway of the technology. Policymakers will need to make an informed decision on whether to have bioprinting products and services covered by national or private health insurances. Hopefully, this fantastic technology will not broaden but minimize the social divide.

Sources of further information

Li, P. (2014) '3D bioprinting technologies: patents, innovation and access' 6(2) Law, Innovation and Technology 282-304.

Alex Faulkner, 'Tissue engineered technologies: regulatory pharmaceuticalization in the European Union' The European Journal of Social Sciences Research (2012) 25:4; pp 389-408.

Paul Hourd, Nicholas Medcalf, Joel Segal, and David Williams, 'A 3D-bioprinting exemplar of the consequences of the regulatory requirement on customized process' Regenerative Medicine (2015) 10(7) 863-883.

Acknowledgments

This work is based on my previous publication: "3D bioprinting technologies: patents, innovation and access," and findings from the UK EPSRC (*Engineering and Physical Sciences Research Council, EPSRC*, Institute for Manufacturing at Cambridge University) funded project: "A feasibility study of mass customization governance: regulation, liability, and IP of RDM in 3D printing." I would like to express my gratitude to the funder and research team member: Alex Faulkner, Nick Medcalf, Dan Thomas, James Griffin, Richard Everson, and Katy Joyce.

References

[1] Hourd P, Medcalf N, Segal J, Williams D. A 3D-bioprinting exemplar of the consequences of the regulatory requirement on customized process. Regen Med 2015;10(7):863–83.
[2] Nuffield Council on Bioethics (2012) Emerging biotechnologies: technology, choice and the public good [online]. Available at https://nuffieldbioethics.org/wp-content/uploads/2014/07/Emerging_biotechnologies_full_report_web_0.pdf.
[3] Council Regulation (EC) 1394/2007 on Advanced Therapy Medicinal Products, amending Council Directive (EC) 2001/83/EC (ATMP Regulation). Para 1 Preamble ATMP Regulation.
[4] Varkey M, Atala A. Organ bioprinting: a closer look at ethics and policies. Wake Forest J Law Policy 2015;5:275–98.
[5] Council of Europe (1997) 'Convention for the protection of human rights and dignity of the human being with regard to the application of biology and medicine: convention on human rights and biomedicine', European Treaty Series—No. 164.
[6] Li P. 3D bioprinting technologies: patents, innovation and access. Law Innov Technol 2014;6(2):282–304.

Further Reading

[1] Symbian Ltd v Comptroller General of Patents [2008] EWCA Civ 1066.
[2] UK Intellectual Property Office, Manual of Patent Practice: 'Patentable Inventions', (2014, para. 1.08); Aerotel v Telco [2006] EWCA Civ 1371; Macrossan's Patent Application [2006] EWHC 705 (Ch).
[3] Art 2 of the Council Regulation (EC) 1394/2007 on Advanced Therapy Medicinal Products, amending Council Directive (EC) 2001/83/EC (ATMP Regulation). See also: Council Directive (EC) 2009/120 amending Directive 2001/83/EC of the European Parliament and of the Council on the Community Code relating to medicinal products for human use as regards advanced therapy medicinal products [2009] OJ L242/3.
[4] Douglas JF, Cronin AJ. The Human Transplantation (Wales) Act 2013: an Act of encouragement, not enforcement. Mod Law Rev 2015;78(2):324–48.
[5] Faulkner A. Tissue engineered technologies: regulatory pharmaceuticalization in the European Union. Eur J Soc Sci Res 2012;25(4):389–408.
[6] U.S. Department of Health and Human Services Food and Drug Administration, Technical considerations for additive manufactured devices: Draft guidance for industry and Food and Drug Administration Staff, Draft Guidance, 10 May 2016.

PART B

3D Bioprinting Applications

The clinical need for 3D printed tissue in reconstructive surgery

T.H. Jovic[*,†], Z.M. Jessop[*,†], A. Al-Sabah[*], I.S. Whitaker[*,†]*
[*]Reconstructive Surgery and Regenerative Medicine Research Group, Swansea University Medical School, Swansea, United Kingdom, [†]The Welsh Centre for Burns and Plastic Surgery, Morrison Hospital, Swansea, United Kingdom

The form and function of the human body is maintained through a network of connective tissues, derived largely from embryological mesoderm such as bone, cartilage, muscle, tendons and ligaments (Fig. 12.1), and ectodermal derivatives such as skin and nerves (Fig. 12.2).

Injury or disease to any of these mesenchymal or ectodermal derivatives can occur due to an array of congenital, neoplastic, or traumatic aetiologies and evoke a subsequent presentation to clinical practice (Fig. 12.3). It is the goal, therefore, of the reconstructive surgeon to address both congenital and acquired defects in tissue architecture and to endeavor to restore their structural and functional integrity [1].

The field of plastic and reconstructive surgery has been revolutionized by advancements in microsurgery and transplantation [2]. However, despite these developments reconstruction remains hindered by the availability of donor sites, the morbidity associated with autologous tissue harvest [3–6] and the potential complications that arise posttransplantation [7] (Fig. 12.3).

Considering the associated complications, and a finite supply of appropriate donor tissue, advancements in the regenerative approach to connective tissue repair have been a strong focus of research interest [8]. 3D biomanufacturing of tissue has the potential to evade the morbidity associated with autologous tissue and the need for immunosuppression with allografts [9,10]. As a clinical specialty, the innovative and reconstructive characteristics of plastic surgery lend themselves to lead developments in 3D biomanufacturing [9].

12.1 Bioprinting principles

A combined effort of engineering, rheology, computer science, and cell biology has made the possibility of 3D bioprinting a reality. 3D printing software can be used to extract digital data from patient images such as computed tomography, magnetic resonance imaging, or laser scanning to yield custom-made and personalized implants [10,11]. The printer hardware deposits a combination of biomaterials and cells in the form of a bioink to create the structure required.

3D Bioprinting for Reconstructive Surgery. https://doi.org/10.1016/B978-0-08-101103-4.00002-8

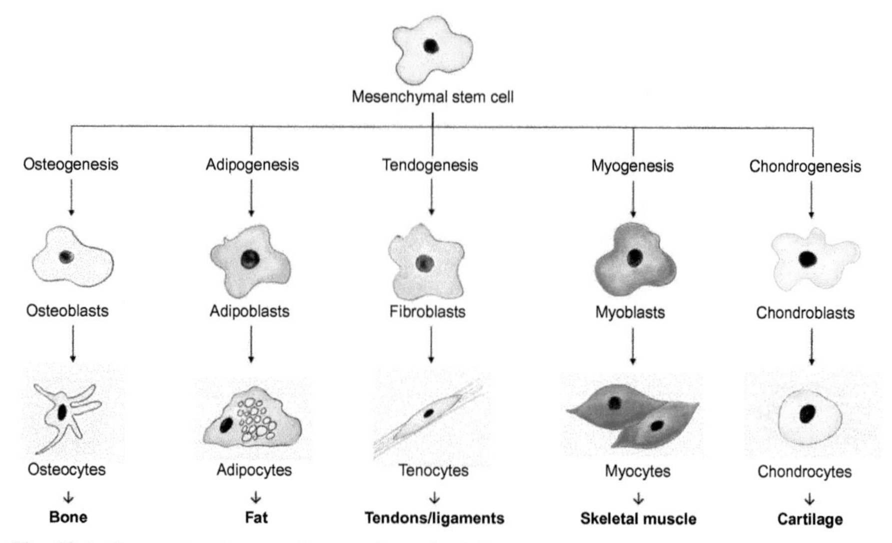

Fig. 12.1 Connective tissues of mesenchymal origin.

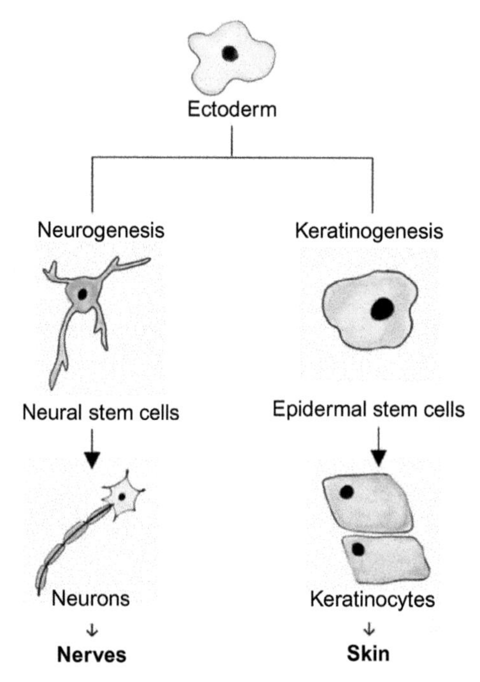

Fig. 12.2 Tissues of ectodermal origin.

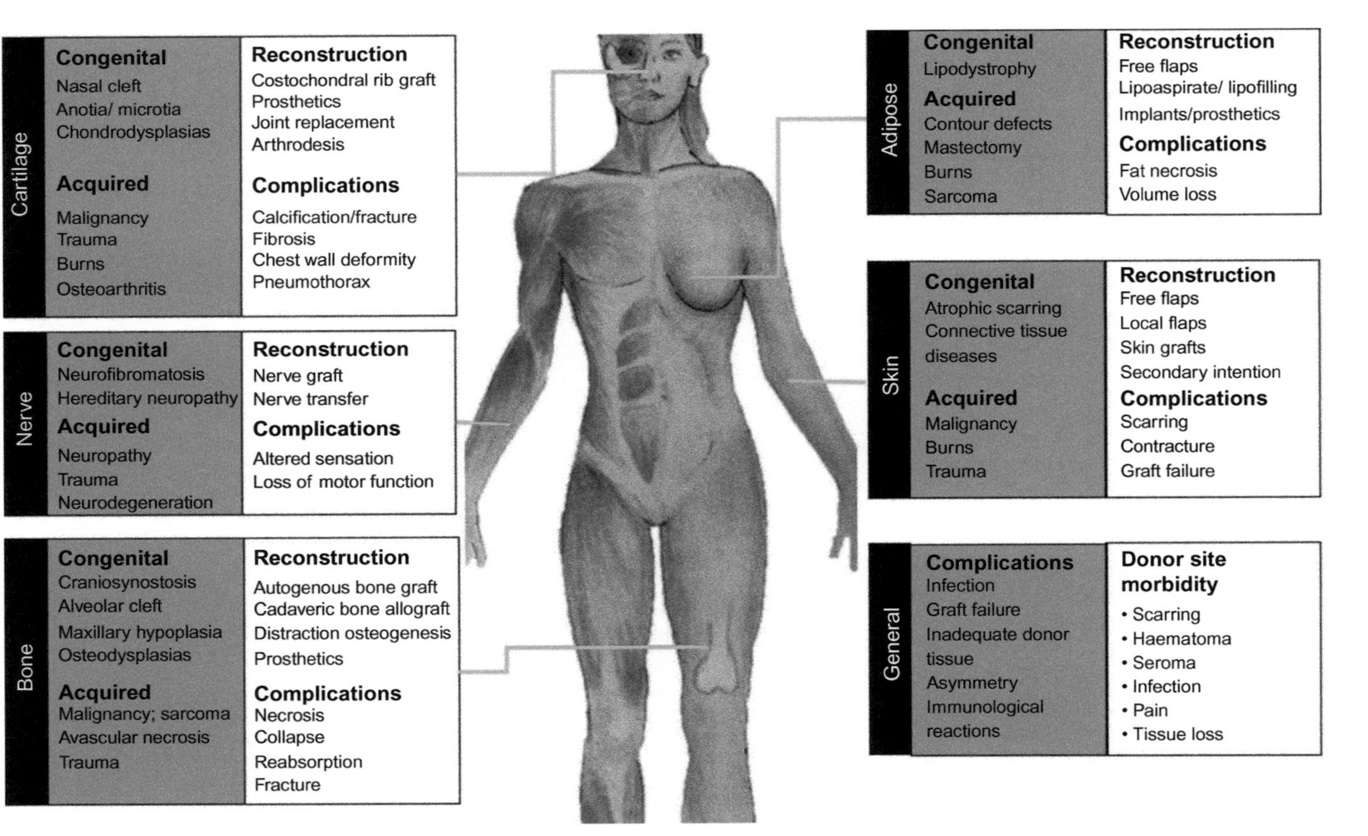

Fig. 12.3 Reconstruction of tissue defects and the associated complications.ART: Please replace present image of Fig. 12.3 to the attached file "Chapter 2 Figure 3"

12.2 Bioprinting for tissue assembly

3D bioprinting offers the potential to replicate the native tissue anisotropy of biological structures through the biofabrication of a tailored macrostructure, microstructure and nanostructure to optimize patient specificity, extracellular matrix (ECM) composition, cellular interactions, and topography [12,13]. The traditional approach to tissue assembly was through the use of solid, porous scaffolds seeded with a cell suspension [14,15]. The porosity is achieved by particulate leeching or electrospinning, and allows for control over pore dimensions, shape, and interconnectivity: key determinants of cell migration and proliferation [16]. Alternative methods of tissue assembly have included layering, rolling [17,18], cell encapsulation within hydrogels [19], and scaffold-free cell patterning approaches [20,21]. Each of these approaches is hindered by limited control over cel-cell contact and microarchitecture [22]: limitations that ultimately restrict the final size of the constructs due to a demand for adequate nutrient supply, cell function, and waste product removal [23,24]. Bioprinting has advantages over alternative means of tissue assembly in its replicability, customization, potential for vascularization, automation, and high-resolution manufacture [25,26]. These properties render this approach a frontrunner for successful clinical translation.

12.3 Connective tissue reconstruction and the role of bioprinting

There are currently three bioprinting approaches used in reconstructive surgery. Biomimicry seeks to match the tissue micro- and macroarchitecture to the native tissue [27,28]; autonomous self-assembly encourages cell-driven tissue formation based on embryological processes of tissue development [29], whereas biopatterning can be employed to yield small structural and functional blocks of tissue that can be assembled into larger constructs through biomimicry or self-assembly [26,30]. Current approaches have utilized these methods either alone or in combination to produce connective tissues such as bone [31,32], cartilage [33–35], muscle [36,37], fat [38,39], and skin [40]. The production of cells, scaffolds, and biomolecules in a spatially controlled manner is not sufficient to produce native-like tissue for reconstruction: successful clinical translation remains hindered by the transition of weak 3D printed neo-tissue into surgically appropriate tissue for transplantation [26,41].

12.3.1 Bone reconstruction

Bone displays a remarkable capacity for regeneration, through the sequential processes of osteoconduction, osteoinduction, and osteogenesis [42]. This innate regenerative capacity may be hindered by large defects in the bony architecture, disabling the process of osteoconduction [43]. As such, the demands on the reconstructive surgeon may traverse a number of congenital bone disorders including the management of craniosynostosis, alveolar cleft, or maxillary hypoplasia (Treacher Collins Syndrome) (Fig. 12.3). Furthermore, in the aftermath of trauma, malignancy, complex upper limb fractures, osteonecrosis, and postoperative bone defects may require a reconstructive approach to restore the structural integrity and locomotive capacity of the skeleton.

In the instance of successful osteogenesis, the bone construct must be able to integrate into host tissues, and be capable of load bearing and remodeling to serve the structural functions of bone [44]. The current gold standard of bone reconstruction is the bone graft [45]. Commonly harvested from an autologous or cadaveric iliac crest, bone grafts present a significant risk of postoperative complication including bowel herniation, nerve injury, pain, and haematoma in addition to recipient site complications such as graft failure, immunoreaction, and infection [45–48] (Fig. 12.3).

The main inorganic component of bone is hydroxyapatite, and most scaffolds currently used for osteogenesis are polymers of calcium phosphate or sulfate, or chitosan and collagen [49]. Scaffolds should also possess the capacity to resorb as new bone grows to enable further osteogenesis and osseointegration [49]. Natural scaffold materials such as collagen have demonstrated angiogenesis and osteogenesis, but at the expense of subsequent bone strength [50]. Further efforts will be required to ensure the optimal combination of resorption and strength produce the most durable scaffold for bone tissue engineering and bioprinting.

12.3.2 Cartilage reconstruction

Cartilage tissue is unique in its flexibility, elasticity, and resilience [51]. These properties render it an ideal material for forming a protective lining of joint surfaces and providing a stiff yet pliable structure to the ears, nose, rib cage, and airways. There are three types of cartilage: elastic, hyaline and fibrocartilage, the properties of which reflect invarying degrees of proteoglycan and collagen fibers in the ECM [52].

Cartilage tissue engineering has been a focus of translational regenerative medicine research, in particular due to the increasing burden of osteoarthritis in the aging population [53]. Cartilaginous defects may also present to the Plastic Surgeon due to congenital causes, such as microtia or anotia, or secondary to trauma, malignancy, and burns affecting the cartilaginous structures of the head and neck. There is a finite amount of suitable cartilage that can be harvested for reconstruction. Cartilage from the costochondral component of the rib cage is commonly used, but at the expense of significant morbidity, ranging from pneumothorax [54] and chest wall deformity [55] to pain [56] and problematic scarring [57]. The quality of the cartilage yielded from the rib cage is also inconsistent, and becomes prone to ossification with age [58], rendering it a brittle material for reconstruction and poorly equipped to restore the necessarily elastic function of nasal and auricular cartilage.

Cartilage is an ideal tissue to bioengineer as it is avascular, aneural, and devoid of the extensive cell-cell connections that currently challenge the attainment of construct anisotropy and longevity [12,59].

12.3.3 Adipose reconstruction

The use of adipose tissue autografts for soft tissue defects, such as the breast, is becoming an increasing focus of interest [60,61]. The success of conventional implant-based reconstruction has been hindered by complications such as capsular contracture, infection, rupture, and incompatibility with radiotherapy [62–64]. As such, adipose tissue, in the form of a free flap, is becoming an increasingly favored reconstructive modality.

Fat graft survival is one of the major issues with reconstruction, with up to 70% volume loss postoperatively [65]. Inadequate revascularization underlies this process of fat resorption, particularly where larger volumes of fat are required such as breast reconstruction. Consequently, multiple procedures may be required, increasing the risk of operative and postoperative complications in addition to the associated financial and emotional implications. Currently, free-flap reconstruction is the gold standard. Tissue engineering and stem cell research could revolutionize soft tissue reconstruction, obviating the requirements for microsurgical expertise, lengthy intraoperative periods and the potential for donor site morbidity [66–68].

12.3.4 Skin reconstruction

Acute and chronic wounds constitute an enormous proportion of the plastic surgery workload. The skin serves an array of important structural roles in the human body, providing a barrier against injury and microorganisms in addition to physiological roles in thermoregulation and sensation. The skin has a remarkable capacity for healing in the face of minor insults or breaks to this protective barrier, however, large wounds predispose to prolonged and potentially impaired healing in addition to increased infection risk. It is in these larger defects, therefore, that the role of the reconstructive surgeon is paramount.

The reconstructive toolkit contains an array of means by which to close wounds. Large skin defects can occur following skin cancer excision, particularly due to the need for adequate margins [69]. Local cutaneous flaps, split thickness, and full thickness skin grafts are commonly used, but require a donor site, and can be complicated by failure to integrate, necrosis and infection, and may heal with significant scarring, contracture, and fibrosis [70,71] with cosmetic and functional implications for the patient. In instance of burn injuries, another major limiting factor is the availability of donor tissue. This has increased the interest in tissue-engineered solutions, and in certain settings, tissue-engineered structures such as acellular dermal matrices are used currently in clinical practice [72]. The use of human tissue allografts carries the risk of disease transmission and immunogenicity [72,73]. As such, the ability to bioprint autogenous skin for reconstruction has the potential to bypass the restrictions of donor site availability and the complications associated with harvest and implantation.

12.4 Conclusion

Disorders of skin, nerves, and connective tissue underpin the role of the reconstructive surgeon. The current gold standard approaches to reconstructing tissue defects depend largely on the use of autologous tissue transfer to restore form and function. Limitations in the availability of suitable donor tissue restrict the validity of this approach, particularly in cases such as severe burns, where the defect size surpasses the area of donor skin available for grafting. Even in less extensive cases, the harvest of donor tissue carries a significant risk of donor site morbidity, and the potential for failure, infection, and degradation at the host site. 3D bioprinting has the potential to surpass these limitations, offering personalized and durable tissue without the need for extensive tissue harvest.

Future research will need to surmount the biological, technological, and regulatory challenges associated with 3D bioprinting to facilitate translational developments in the field and holds the potential to revolutionize the treatment of congenital and acquired tissue defects.

References

[1] Dunkin CSJ, Pleat JM, Jones SAM, Goodacre TEE. Perception and reality—a study of public and professional perceptions of plastic surgery. Br J Plast Surg 2003;56(5):437–43.
[2] Le Fanu J. Rise and fall of modern medicine. Lancet 1999;354(9177):518.
[3] Blondeel PN, Vanderstraeten GG, Monstrey SJ, Van Landuyt K, Tonnard P, Lysens R, et al. The donor site morbidity of free DIEP flaps and free TRAM flaps for breast reconstruction. Br J Plast Surg 1997;50(5):322–30.
[4] Laurie SWS, Kaban LB, Mulliken JB, Murray JE. Donor-site morbidity after harvesting rib and Iliac bone. Plast Reconstr Surg 1984;73(6):933.
[5] Hallock GG. Relative donor-site morbidity of muscle and fascial flaps. Plast Reconstr Surg 1993;92(1):70–6.
[6] Swan MC, Goodacre TEE. Morbidity at the iliac crest donor site following bone grafting of the cleft alveolus. Br J Oral Maxillofac Surg 2006;44(2):129–33.
[7] Suh JD, Sercarz JA, Abemayor E, Calcaterra TC, Rawnsley JD, Alam D, et al. Analysis of outcome and complications in 400 cases of microvascular head and neck reconstruction. Arch Otolaryngol Head Neck Surg 2004;130(8):962–6. American Medical Association.
[8] D'Amico RA, Rubin JP. Regenerative medicine and the future of plastic surgery. Plast Reconstr Surg 2014;133(6):1511–2.
[9] Jessop ZM, Al-Himdani S, Clement M, Whitaker IS. The Challenge for Reconstructive Surgeons in the Twenty-First Century: Manufacturing Tissue-Engineered Solutions. Front Surg 2015;2(3):3. 3rd ed. Frontiers.
[10] Radenkovic D, Solouk A, Seifalian A. Personalized development of human organs using 3D printing technology. Med Hypotheses 2016;87:30–3.
[11] Horn TJ, Harrysson OLA. Overview of current additive manufacturing technologies and selected applications. Sci Prog 2012;95(3):255–82. Science Reviews 2000 Ltd.
[12] Guillotin B, Guillemot F. Cell patterning technologies for organotypic tissue fabrication. Trends Biotechnol 2011;29(4):183–90.
[13] Mironov V. The second international workshop on bioprinting, biopatterning and bioassembly. Expert Opin Biol Ther 2005;5(8):1111–5. Taylor & Francis.
[14] Leong KF, Cheah CM, Chua CK. Solid freeform fabrication of three-dimensional scaffolds for engineering replacement tissues and organs. Biomaterials 2003;24(13):2363–78.
[15] Yeong W-Y, Chua C-K, Leong K-F, Chandrasekaran M. Rapid prototyping in tissue engineering: challenges and potential. Trends Biotechnol 2004;22(12):643–52.
[16] Lee MH, Arcidiacono JA, Bilek AM, Wille JJ, Hamill CA, Wonnacott KM, et al. Considerations for tissue-engineered and regenerative medicine product development prior to clinical trials in the United States. Tissue Eng Part B Rev 2009;16(1):41–54.

[17] Boland T, Mironov V, Gutowska A, Roth EA, Markwald RR. Cell and organ printing 2: fusion of cell aggregates in three-dimensional gels. Anat Rec 2003;272A(2):497–502. Wiley Subscription Services, Inc., A Wiley Company.

[18] Yang J, Yamato M, Shimizu T, Sekine H, Ohashi K, Kanzaki M, et al. Reconstruction of functional tissues with cell sheet engineering. Biomaterials 2007;28(34):5033–43.

[19] McGuigan AP, Sefton MV. Vascularized organoid engineered by modular assembly enables blood perfusion. Proc Natl Acad Sci 2006;103(31):11461–6. National Acad Sciences.

[20] Nichol JW, Khademhosseini A. Modular tissue engineering: engineering biological tissues from the bottom up. Soft Matter 2009;5(7):1312.

[21] Norotte C, Marga FS, Niklason LE, Forgacs G. Scaffold-free vascular tissue engineering using bioprinting. Biomaterials 2009;30(30):5910–7.

[22] Hurtley S. Location, location, location. Science 2009;326(5957):1205. American Association for the. Advancement of Science.

[23] Kaully T, Kaufman-Francis K, Lesman A, Levenberg S. Vascularization—the conduit to viable engineered tissues. Tissue Eng Part B Rev 2009;15(2):159–69.

[24] Lovett M, Lee K, Edwards A, Kaplan DL. Vascularization strategies for tissue engineering. Tissue Eng Part B Rev 2009;15(3):353–70.

[25] Sun W, Darling A, Starly B, Nam J. Computer-aided tissue engineering: overview, scope and challenges. Biotechnol Appl Biochem 2004;39(1):29–47. Blackwell Publishing Ltd.

[26] Mironov V, Visconti RP, Kasyanov V, Forgacs G, Drake CJ, Markwald RR. Organ printing: tissue spheroids as building blocks. Biomaterials 2009;30(12):2164–74.

[27] Huh D, Torisawa Y-S, Hamilton GA, Kim HJ, Ingber DE. Microengineered physiological biomimicry: organs-on-Chips. Lab Chip 2012;12(12):2156.

[28] Ingber DE. Mechanical control of tissue growth: function follows form. Proc Natl Acad Sci 2005;102(33):11571–2. National Acad Sciences.

[29] Marga F, Neagu A, Kosztin I, Forgacs G. Developmental biology and tissue engineering. Birth Defects Res C Embryo Today 2008;81(4):320–8. Wiley Subscription Services, Inc., A Wiley Company.

[30] Kelm JM, Lorber V, Snedeker JG, Schmidt D, Broggini-Tenzer A, Weisstanner M, et al. A novel concept for scaffold-free vessel tissue engineering: self-assembly of microtissue building blocks. J Biotechnol 2010;148(1):46–55.

[31] Tamimi F, Torres J, Gbureck U, Lopez-Cabarcos E, Bassett DC, Alkhraisat MH, et al. Craniofacial vertical bone augmentation: a comparison between 3D printed monolithic monetite blocks and autologous onlay grafts in the rabbit. Biomaterials 2009;30(31):6318–26.

[32] Shim J-H, Moon T-S, Yun M-J, Jeon Y-C, Jeong C-M, Cho D-W, et al. Stimulation of healing within a rabbit calvarial defect by a PCL/PLGA scaffold blended with TCP using solid freeform fabrication technology. J Mater Sci Mater Med 2012;23(12):2993–3002. Springer US.

[33] Shanjani Y, Hu Y, Toyserkani E, Grynpas M, Kandel RA, Pilliar RM. Solid freeform fabrication of porous calcium polyphosphate structures for bone substitute applications: in vivo studies. J Biomed Mater Res B Appl Biomater 2013;101B(6):972–80. Wiley Subscription Services, Inc., A Wiley Company.

[34] Yen H-J, Tseng C-S, Hsu S-H, Tsai C-L. Evaluation of chondrocyte growth in the highly porous scaffolds made by fused deposition manufacturing (FDM) filled with type II collagen. Biomed Microdevices 2008;11(3):615–24.

[35] Gruene M, Deiwick A, Koch L, Schlie S, Unger C, Hofmann N, et al. Laser printing of stem cells for biofabrication of scaffold-free autologous grafts. Tissue Eng Part C

Methods 2011;17(1):79–87. Mary Ann Liebert, Inc. 140 Huguenot Street, 3rd Floor New Rochelle, NY 10801 USA.

[36] Phillippi JA, Miller E, Weiss L, Huard J, Waggoner A, Campbell P. Microenvironments engineered by inkjet bioprinting spatially direct adult stem cells toward muscle- and bone-like subpopulations. Stem Cells 2008;26(1):127–34. John Wiley & Sons, Ltd.

[37] Merceron TK, Burt M, Seol Y-J, Kang H-W, Lee SJ, Yoo JJ, et al. A 3D bioprinted complex structure for engineering the muscle–tendon unit. Biofabrication 2015;7(3):035003. IOP Publishing.

[38] Williams SK, Touroo JS, Church KH, Hoying JB. Encapsulation of adipose stromal vascular fraction cells in alginate hydrogel spheroids using a direct-write three-dimensional printing system. BioRes Open Access 2013;2(6):448–54. Mary Ann Liebert, Inc. 140 Huguenot Street, 3rd Floor New Rochelle, NY 10801 USA.

[39] Gruene M, Pflaum M, Hess C, Diamantouros S, Schlie S, Deiwick A, et al. Laser printing of three-dimensional multicellular arrays for studies of cell–cell and cell–environment interactions. Tissue Eng Part C Methods 2011;17(10):973–82. Mary Ann Liebert, Inc. 140 Huguenot Street, 3rd Floor New Rochelle, NY 10801 USA.

[40] Lee W, Debasitis JC, Lee VK, Lee J-H, Fischer K, Edminster K, et al. Multi-layered culture of human skin fibroblasts and keratinocytes through three-dimensional freeform fabrication. Biomaterials 2009;30(8):1587–95.

[41] Hajdu Z, Mironov V, Mehesz AN, Norris RA, Markwald RR, Visconti RP. Tissue spheroid fusion-based in vitro screening assays for analysis of tissue maturation. J Tissue Eng Regen Med 2010;4(8):659–64. John Wiley & Sons, Ltd.

[42] Oryan A, Alidadi S, Moshiri A, Maffulli N. Bone regenerative medicine: classic options, novel strategies, and future directions. J Orthop Surg Res 2014;9(1):18. BioMed Central.

[43] Albrektsson T, Johansson C. Osteoinduction, osteoconduction and osseointegration. Eur Spine J 2001;10:S96–101.

[44] Bhumiratana S, Vunjak-Novakovic G. Concise review: personalized human bone grafts for reconstructing head and face. Stem Cells Transl Med 2011;1(1):64–9. AlphaMed Press.

[45] Rogers GF, Greene AK. Autogenous bone graft. J Craniofac Surg 2012;23(1):323–7.

[46] Ebraheim NA, Elgafy H, Xu R. Bone-graft harvesting from iliac and fibular donor sites: techniques and complications. J Am Acad Orthop Surg 2001;9(3):210–8.

[47] Arrington ED, Smith WJ, Chambers HG, Bucknell AL, Davino NA. Complications of iliac crest bone graft harvesting. Clin Orthop Relat Res 1996;329:300–9.

[48] Delloye C, Cornu O, Druez V, Barbier O. Bone allografts: what they can offer and what they cannot. J Bone Joint Surg Br 2007;89-B(5):574–80. Bone and Joint Journal.

[49] Owen SC, Shoichet MS. Design of three-dimensional biomimetic scaffolds. J Biomed Mater Res A 2010;116(Part 12). Wiley Subscription Services, Inc., A Wiley Company.

[50] Kruger EA, Im DD, Bischoff DS, Pereira CT, Huang W, Rudkin GH, et al. In vitro mineralization of human mesenchymal stem cells on three-dimensional type I Collagen versus PLGA scaffolds: a comparative analysis. Plast Reconstr Surg 2011;127(6):2301–11.

[51] Fung YC. Biomechanics: mechanical properties of living tissues 1993. CrossRef Google Scholar.

[52] Zambrano NZ, Monies GS, Shigihara KM, Sanchez EM, Junqueira LCU. Collagen arrangement in cartilages. Cells Tissues Organs 1982;113(1):26–38. Karger Publishers.

[53] Vinatier C, Guicheux J. Cartilage tissue engineering: from biomaterials and stem cells to osteoarthritis treatments. Ann Phys Rehabil Med 2016;59(3):139–44.

[54] Ochi JW, Evans JNG, Bailey CM. Pediatric airway reconstruction at great ormond street: a ten-year review. Ann Otol Rhinol Laryngol 2016;101(6):465–8. SAGE PublicationsSage CA: Los Angeles, CA.

[55] Thomson HG, Kim T-Y, Ein SH. Residual problems in chest donor sites after microtia reconstruction. Plast Reconstr Surg 1995;95(6):961–8.
[56] Cervelli V, Bottini DJ, Gentile P, Fantozzi L, Arpino A, Cannat C, et al. Reconstruction of the nasal dorsum with autologous rib cartilage. Ann Plast Surg 2006;56(3):256–62.
[57] Brent B. Auricular repair with autogenous rib cartilage grafts. Plast Reconstr Surg 1992;90(3):355–74.
[58] McCormick WF. Mineralization of the costal cartilages as an indicator of age: preliminary observations. J Forensic Sci 1980;25(4):736–41. ASTM International.
[59] Perera JR, Jaiswal PK, Khan WS. The potential therapeutic use of stem cells in cartilage repair. Curr Stem Cell Res Ther 2012;7(2):149–56.
[60] Chan CW, McCulley SJ, Macmillan RD. Autologous fat transfer—a review of the literature with a focus on breast cancer surgery. J Plast Reconstr Aesthet Surg 2008;61(12):1438–48.
[61] Ross RJ, Shayan R, Mutimer KL, Ashton MW. Autologous fat grafting. Ann Plast Surg 2015;74(5):633–4.
[62] Holmes JD. Capsular contracture after breast reconstruction with tissue expansion. Br J Plast Surg 1989;42(5):591–4.
[63] Virden CP, Dobke MK, Stein P, Parsons CL, Frank DH. Subclinical infection of the silicone breast implant surface as a possible cause of capsular contracture. Aesthet Plast Surg 1992;16(2):173–9.
[64] Pajkos A, Deva AK, Vickery K, Cope C, Chang L, Cossart YE. Detection of subclinical infection in significant breast implant capsules. Plast Reconstr Surg 2003;111(5):1605–11.
[65] Eto H, Kato H, Suga H, Aoi N, Doi K, Kuno S, et al. The fate of adipocytes after nonvascularized fat grafting. Plast Reconstr Surg 2012;129(5):1081–92.
[66] Fuchs E, Segre JA. Stem cells. Cell 2000;100(1):143–55.
[67] Mimeault M, Hauke R, Batra SK. Stem cells: a revolution in therapeutics-recent advances in stem cell biology and their therapeutic applications in regenerative medicine and cancer therapies. Clin Pharmacol Ther 2007;82(3):252–64.
[68] Kim HY, Jung BK, Lew DH, Lee DW. Autologous fat graft in the reconstructed breast: fat absorption rate and safety based on sonographic identification. Arch Plast Surg 2014;41(6):740–7.
[69] Marsden JR, Newton-Bishop JA, Burrows L, Cook M, Corrie PG, Cox NH, et al. Revised UK guidelines for the management of cutaneous melanoma 2010. J Plast Reconstr Aesthet Surg 2010;63(9):1401–19.
[70] Ragnell A. The secondary contracting tendency of free skin grafts. Br J Plast Surg 1952;5(1):6–24.
[71] Corps BVM. The effect of graft thickness, donor site and graft bed on graft shrinkage in the hooded rat. Br J Plast Surg 1969;22(2):125–33.
[72] Nyame TT, Chiang HA, Leavitt T, Ozambela M, Orgill DP. Tissue-engineered skin substitutes. Plast Reconstr Surg 2015;136(6):1379–88.
[73] Tomford WW. Transmission of disease through transplantation of musculoskeletal allografts. J Bone Joint Surg 1995;77(11):1742–54.

3D bioprinting bone

13

A. Ibrahim
Institute of Child Health, London, United Kingdom

13.1 Introduction

Bone is the second most transplanted tissue after blood. Due to the risks and constraints of reconstruction of defects with foreign-body implants, autologous bone grafts and allogeneic tissue transfer, tissue engineering provides the promise of generating autologous bone tissue free of the limitations and morbidities associated with current treatment options.

13.1.1 Bone

Bone is a complex tissue that has mechanical, hemopoeitic, and metabolic functions. It provides structure and protection to the surrounding soft tissues and is necessary for metabolic regulation of calcium and phosphate as well as hemopoeisis. Additionally, the organization and function of each bone are highly site specific and related to its embryological origin as well as its ability to adapt (e.g., the weight-bearing long bones that are of mesenchymal origin and facilitate locomotion compared with the neural crest-derived flat bones of the face, which provide support and structure).

To date, many different approaches have been taken to tissue engineer bone using osteoprogenitor cells. In general, the reconstructive principles of finding the optimal cell type scaffold capable of providing biomechanical support as well as perfusion and biochemical environment underpin the basic methodology of most research in this area. A multitude of different options are associated with each of these variables so there is no consensus regarding the optimal method for engineering bone at present. Bioprinting bone implants hold the potential to generate a life-long solution without the risks associated with current options such as bone grafts, flaps, or foreign-body implants.

This chapter provides an overview of bone bioprinting. We consider the methods and tools currently available and present the progress made to date in bioprinting bone tissues. With regard to clinical translation and biological considerations, we discuss the limitations and future directions of this important area of research.

13.2 The clinical need for bioprinted bone and limitations of current treatments

Bone defects may arise due to a number of congenital or acquired conditions. Congenital bone anomalies are commonly due to the absence of or maldevelopement of bones (such as zygomatic hypoplasia in Treacher Collins syndrome (TCS) or fibular hemimelia). Acquired bone defects often occur acutely because of trauma, infection,

3D Bioprinting for Reconstructive Surgery. https://doi.org/10.1016/B978-0-08-101103-4.00015-6

neoplasm, or surgical resection. Osteodegenerative diseases such as osteoarthritis are also responsible for bone loss in weight-bearing regions over time. In the context of an aging population, the number of patients suffering from these osteodegenerative diseases and the associated costs of treatment are expected to double by 2020 in the developed and developing countries [1,2].

Foreign-body implants are the mainstay of reconstruction of large-segmental defects by providing a replacement to missing bone and are particularly useful for areas requiring mechanical strength and structural support. Furthermore, the surfaces of these materials can be modified to improve osseointegration, promote periprosthetic bone regeneration, and reduce microbial activity [3,4]. These synthetic materials are still however associated with risks of infection and extrusion and do not grow with the skeleton, which is an important consideration in children who will require multiple revisions as a result [5,6].

Bone grafting provides a life-long regenerative solution by delivering tissue to the defect that can integrate, vascularize, and promote local bone repair. After blood, bone is the second most transplanted tissue worldwide with an estimated 500,000 procedures performed per year in the United States alone requiring bone grafts [7]. At present, the main regenerative treatment options for bone defects are autologous, allogeneic, or the use of bone graft substitutes for the repair of large defects [8].

Autologous bone grafting involves the transfer of bone from one site to reconstruct a defect within the same patient. This has the advantage of providing a cellularized nonimmunogenic tissue that can be revascularized, engraft and permit osteoinduction at the defect [9]. Due to the donor-site morbidity (including pain and leaving a defect at the site of extraction) and associated blood loss, this option is limited to reconstructing smaller defects. Allografting permits the transplantation of bone from one person to another. Due to the immunogeneic response to fresh allografts, these bone grafts are usually processed then used as demineralized bone matrix or lyophilized tissues and thus work mainly through osteoinduction [9]. The main limitation to widespread clinical translation of allografts is potential for transmitting infectious diseases and immunological rejection as well as a deterioration of the material properties of bone if demineralized and/or decellularized [10,11]. In addition to the limitations discussed above, bone defect repair is associated with significant financial costs with autologous bone graft procedures costing up to $9000 per case [2]. Furthermore, bone grafts are limited by high-resorption rates that often necessitate further surgery and thus propagating the risk of clinical complications and financial burden [12].

Synthetic bone graft substitutes are increasingly seen as an attractive cost-effective, biocompatible, and osteoinductive alternative to bone with the potential for overcoming the limitations associated with autologous and allogeneic bone grafts. The most commonly studied and clinically available bone substitute products include hydroxyapatite, calcium phosphates, calcium sulfate, and bioactive glasses [13]. These are either available alone or complexed with titanium, collagen, growth factors, and/or demineralized bone matrix to mimic the structural, mechanical, and functional properties of bone [7]. Despite their safety profile and ease of manufacture, these products have still shown reduced mechanical strength and osteoinduction thus proving useful only when used as an adjunct to implants or scaffolds in the repair of critical defects [11,13] (Fig. 13.1).

The increasing global burden of bone-associated diseases requiring reconstructive surgery cannot be optimally managed with the current treatment options due to the constraints

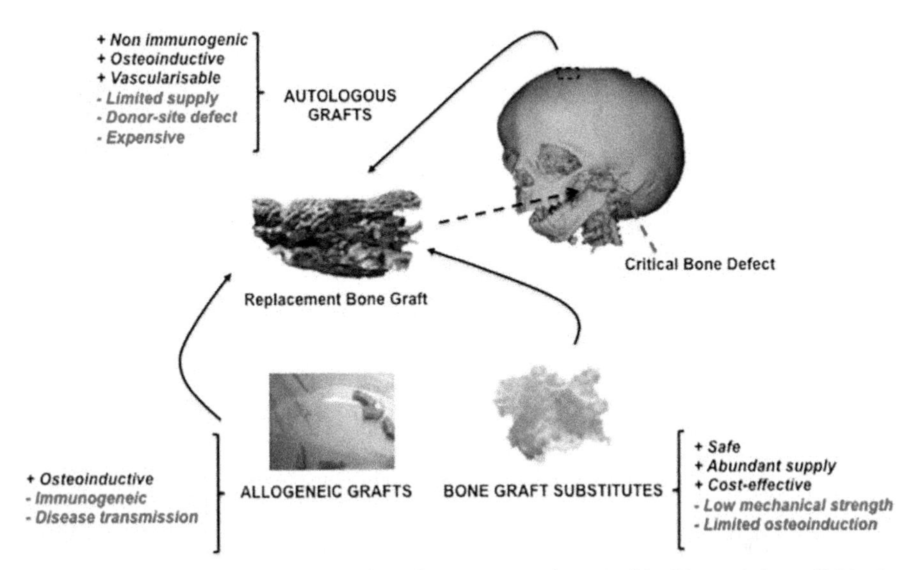

Fig. 13.1 Available bone-grafting options for reconstruction of critical bone defects. Critical bone defects require bone replacement to restore form and function. The gold-standard autologous bone grafts are harvested from sites such as the calvarium. Other less effective sources include allogeneic bone grafts and bone graft substitutes. All three options are associated with a range of limitations that make a tissue engineered alternative necessary.

associated with complications, poor functional, and esthetic outcomes and limited the availability of tissues (in the case of bone grafts) and high-financial costs. Bone tissue engineering offers the potential to create personalized bone tissues that can incorporate and grow with the patient thus providing a life-long solution. These bioengineered bones can also be preshaped and designed to mimic the structure and function of the type of bone required while negating the risks of morbidity and immunogenicity associated with autologous and allogeneic grafts, respectively. Additionally, with the aid of three-dimensional (3D) printing, these bones can be fabricated in a reproducible manner to mass produce large-scale custom-shaped tissues tailored to the patient's needs and meet the growing demand for functionalized bone tissues [14]. For successful clinical translation of 3D bioprinting of bones to occur, first it is important to understand the anatomy and physiology of the desired tissue and design engineering protocols that recreate the osteogenic microenvironment and hence support the creation of functional bone tissues [2].

13.3 Biology, structure and functions of bone

13.3.1 Bone anatomy and physiology

Bone is a hierarchical tissue composed of multiple subunits of organic and inorganic matrix containing cellular elements [15]. Osteoblasts, osteocytes, osteoclasts, and osteoprogenitor cells comprise the cellular constituents of bone [16]. There is extensive crosstalk between these cells, which in concert regulate osteogenesis and remodeling [17]. The acellular components of bone comprise a network of proteins and minerals that provide a scaffold to support cells structurally as well as facilitate their growth, proliferation, differentiation, and signaling. The complexity of bone is further demonstrated by its multiple mechanical and biological functions as well as its dependence on a rich blood supply.

Osteoprogenitors are a population of mesenchymal cells located in the bone "stem-cell niche" that are capable of self-renewal, proliferation, and osteogenic differentiation as well as potentially play a role in regulating angiogenesis [18]. Osteoblasts promote bone formation and mature into osteocytes that act as mechanosensors, control mineral hemostasis, and regulate both osteoclast and osteoblast function [19]. Osteoclasts, alongside osteoblasts, and matrix metalloproteinases play an important role in bone turnover and remodeling [20]. These cells are widely thought to share a common precursor to macrophages and regulate bone turnover through matrix degradation using a hydrogen ion pump mechanism. This allows control of new bone formation and even shaping by creating cavities required for vasculature or bone marrow. Although separate precursor origins, signaling on and from osteoblasts plays a role in stimulating osteoclast recruitment differentiation. This provides a balance between osteoblast-mediated bone formation and osteoclastic resorption in response to the local and systemic environment.

In all, 80%–90% of bone mass is composed of the acellular component of bone, which is largely made up of extracellular matrix proteins the most prominent of which is type I collagen (and to a lesser extent type III collagen and noncollagenous proteins such as osteonectin, bone sialoprotein, and osteocalcin) and inorganic minerals such as calcium and phosphate [21]. Type I collagen comprises 80% of all proteins found in bone matrix, regulates the mechanical properties of bone by imparting the ability of bone to absorb energy (toughness), and balances the stiffness generated by osteoblast-mediated crystal deposition in and around the collagen fibrils [22]. The inorganic component of bone matrix is largely composed of hydroxyapatite crystals, which are deposited as plates within and around the collagen fibrils [23]. This process is regulated by osteoblasts and results in mineralization of bone tissues [24,25].

Bone is a highly functional tissue whose structure is well adapted to its purpose [26]. This is especially evident in its mechanical roles, whereby the difference of extracellular matrix organization in highly organized weight-bearing cortical long bones can be contrasted with the disordered arrangements seen in the flat cancellous bones of the midface, which only have a protective but not load-bearing purpose. The key metabolic function of bone is in the regulation of calcium and phosphate, whereby it acts as a reservoir for these minerals and releases or stores them in response to systemic cues. Bone is also the main site of hemopoesis in adults and children, which occurs in the bone marrow cavities, and thus is responsible for maintenance and differentiation of hemopoeitic stem cells in response to systemic signaling.

A major feature of bone is that it relies on a rich blood supply for its survival and functions [27,28]. Previously maturation of preosteoblast differentiation has been shown to be most vulnerable to oxygen deprivation at the stage of matrix maturation [29]. Recently, vasculature has also been demonstrated to provide a template and guide bone morphogenesis in mouse embryos [30]. In addition, vascularization of bone grafts have shown improved osteogenesis and neoangiogenesis in rabbit studies [31] (Fig. 13.2).

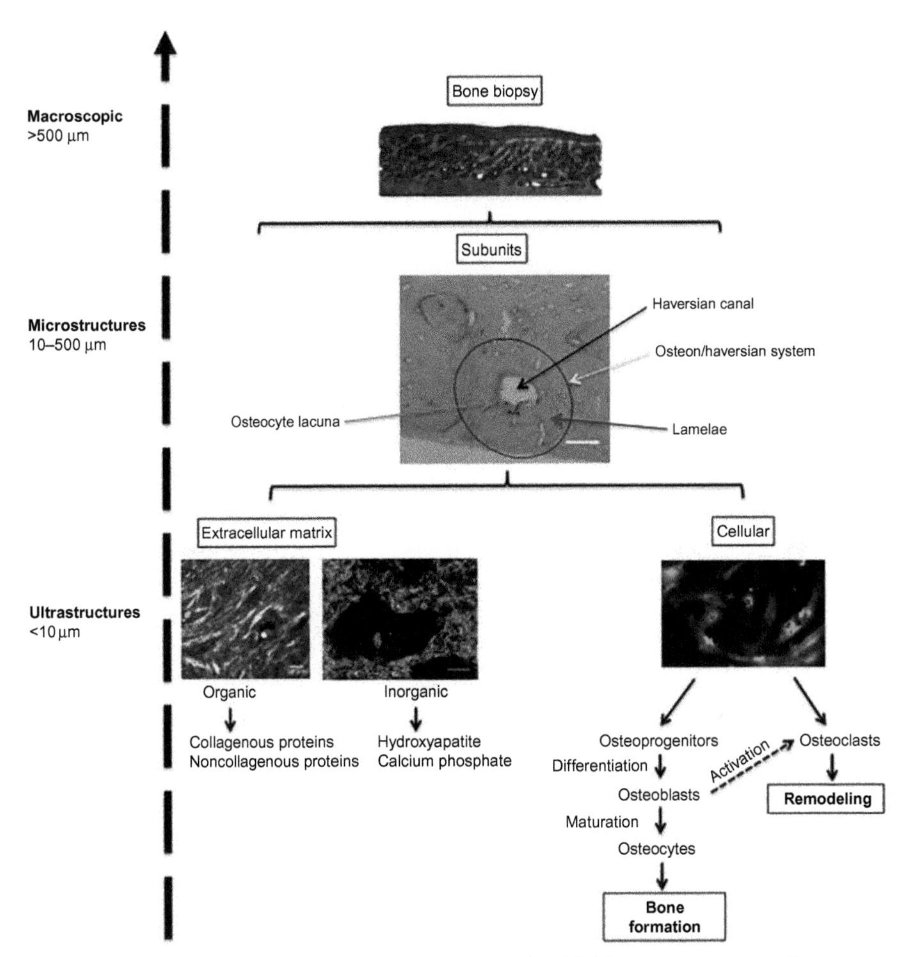

Fig. 13.2 Hierarchical structure of bone tissues. Bone is a highly porous structure that contains an inner meshwork of interconnected trabecular processes and an outer densely packed periosteal layer. Microscopically, bone is composed of repeating subunits called haversian systems, which comprise concentric lamellar rings that contain cells, osteoid, and an inner canal in which reside the neurovascular structures. At the ultrastructural level, bone composes of a cellular component and extracellular matrix. The extracellular matrix is made up of organic materials that are primarily collagenous and are mineralized with hydroxyapatite as well as calcium phosphate crystals. The cellular constituents are namely comprised of an osteoprogenitor reserve, osteoblasts, and osteocytes, which lay down the extracellular matrix and osteoclasts that remodel bone. These cells also regulate the activity of each other.

13.3.2 Development and growth of human bones

Bone can be categorized broadly according to the embryological germ layer from which they develop. The neural tube-derived neural crest cells give rise to the facial and anterior cranial bones while mesenchymal cells from the mesoderm form the trunk, long bones, and the rest of the skull [32,33]. Osteogenesis then occurs through migration of these cells to the relevant anatomic site of the future skeleton in the developing embryo and form cellular condensations to initiate osteogenesis. Skeletogenesis then occurs through one of two distinct pathways depending on the type of bone required to provide osteoblasts that then mature and generate bone tissue. Intramembranous ossification produces bone formation directly through mineralization of the cell condensations and occurs in the development of the facial, anterior cranial, and clavicular bones. The rest of the skeleton is formed indirectly through enchondral ossification, which forms a cartilage template of the desired bone that is gradually replaced with mineralized bone [33].

The stages of osteoblast maturation involve a sequence of up/downregulation of specific genes as the cells progress from an early stage of proliferation to extracellular matrix deposition, matrix maturation, and finally mineralization [34]. Osteoblasts exhibit a distinct pattern of gene expression that corresponds with each stage of maturation. Namely, collagen I and alkaline phosphatase (ALP) gene expression peak in the early stages of osteogenesis and downregulate with increasing maturation, whereas osteopontin and osteocalcin gene expression is maximal at the later stages of bone formation [34]. Runt-related transcription factor 2 (Runx2) is the master regulator of osteogenic differentiation whose absence in mice results in reduced expression of osteogenic genes such as ALP, osteocalcin and osteopontin, abnormal osteoblast morphology and failure of intramembranous, and enchondral ossification of embryonic skeletons [35]. Runx2 has a bimodal distribution and is upregulated both early on in differentiation of osteoprogenitors to preosteoblasts, as well as again expressed at the later stages of maturation.

Postnatally, bone growth and repair are mediated by osteoprogenitor cells, which may be site specific (located in the bone) or recruited from the neighboring bone marrow, adipose, or muscle tissues. The osteoprogenitor cells are recruited through a complex and highly regulated local and systemic signaling pathways [36] involving the bone morphogenetic proteins, Runx2, and osterix pathways. These signals coordinate activation and maturation of the osteoprogenitor cells to differentiate into osteoblasts which in turn generate and lay down the extracellular bone matrix [37]. This collagen-proteoglycan matrix binds calcium crystals to permit calcification. Osteoblasts are separated from this region of calcification by the osteoid matrix that they continually produce although eventually around 15% become buried in their own matrix and form osteocytes. Ongoing ossification eventually results in the formation of bone spicules that are then surrounded by mesenchymal cells forming a membrane (the periosteum). The cells on the inner surface of the periosteum then become osteoblasts and begin depositing osteoid matrix in parallel to the preexisting bony spicules. Furthermore, in the postnatal skeleton, bone is highly adapted to its location. It responds to environmental stresses such as mechanical loading or hormonal changes through a continuous process of modeling and remodeling, whereby new bone is deposited or resorbed [33] (Fig. 13.3).

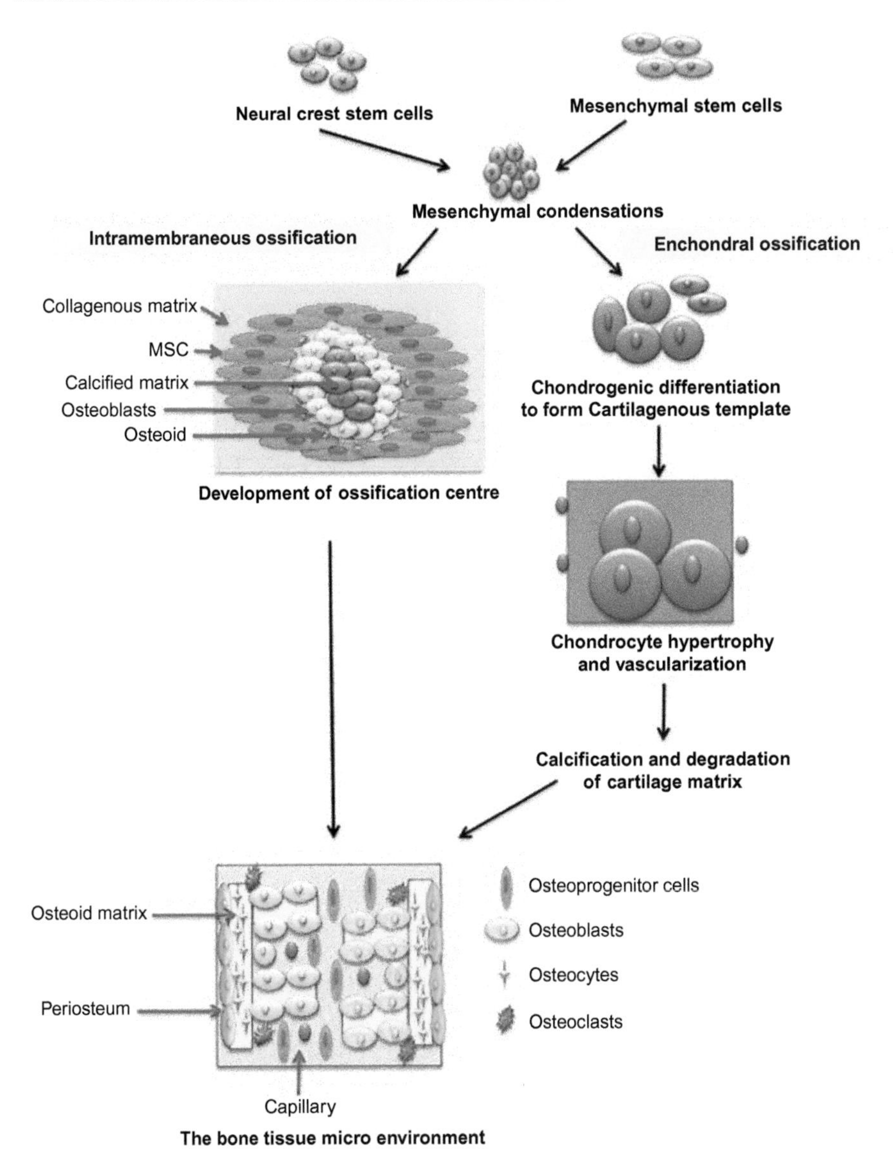

Fig. 13.3 Ossification pathways in bone tissue development. Migrating neural crest cells or mesenchymal stem cells proliferate and aggregate to form mesenchymal condensations. Depending on lineage, ossification can occur either directly through the intramembranous route or indirectly via a cartilage intermediary in enchondral ossification. The outcome of both is the creation of a multicellular and highly vascularized issue. Osteoblasts lay down osteoid matrix and periosteum before becoming buried in their own calcified matrix and mature into osteocytes. Osteoclasts contribute to bone remodeling through break down of the extracellular matrix.

13.4 Approaches to bone bioengineering

A number of different approaches have been investigated for bioengineering bone tissues with varying degrees of success although the fabrication of mature functional autologous bone is yet to be achieved. The general principles of finding the optimal osteoprogenitor cell type scaffold capable of providing biomechanical support and recreation of the bone microtissue environment form the mainstay of investigation.

13.4.1 Cellular

Mesenchymal stem cells (MSC) can be recruited from a number of autologous sites such as the bone, bone marrow, periosteum, fat, and muscle to obtain osteoprogenitor cells. Other potential but less readily accessible sources of osteoprogenitor cells include embryonic [38], amniotic fluid [39], and induced pluripotent stem cells [40]. Due to their ease of harvest and abundance, adipose, bone marrow, and bone-derived osteoprogenitor cells have been the most extensively studied and characterized thus are likely to represent the most suitable cell source for bone tissue engineering.

13.4.1.1 Adipose-derived stem cells

Human adipose-derived stem cells (hADSC) can be easily and safely retrieved from relatively small volumes of fat in adults and children. These cells can be cultured relatively easily and are capable of self-renewal and possess multiline age differentiation potential [41,42]. Another advantage is that they have also shown to confer immune modulatory effects and promote vascularization thus making their clinical translation more attractive [43,44]. In vivo, hADSC have been successfully used to promote critical bone defect repair in rats [45,46]. There is also one case report of their use to reconstruct zygomatic hypoplasia in a TCS patient [47].

13.4.1.2 Bone marrow-derived stem cells

Another well-studied source of stem cell for bone tissue engineering is human bone marrow-derived stem cells (BMSC). These cells possess multipotency, can self-renew, are easily derived from a small amount of donor tissue, and can expand rapidly in vitro [44,48]. BMSC are capable of differentiation along the osteogenic lineage in vitro and when compared with other somatic sources of osteoprogenitor cells, have demonstrated enhanced osteogenic marker expression and mineralization when differentiated down the osteogenic lineage [49]. Similar to hADSC, these cells have also promoted vascularization in vitro [50]. In vivo, MSC have recently shown promise in repair of subacute critical sized long bone defects in a preclinical large animal model and to regenerate successfully a maxillary oral defect in a clinical study of a single patient [51,52]. Unlike hADSC, the method of harvest is invasive in that a bone marrow biopsy or bone specimen is required to isolate these cells, which can leave a donor-site defect.

13.4.1.3 Osteoblast precursor cells

Explant cultures or enzymatic digestion of bone tissue can be used to generate osteoblast-like cells that, when placed in osteoinductive media, upregulate markers of mature bone differentiation and deposit minerals. These human osteoblast precursors have been obtained from a variety of bone sources including intraoral and midface tissues [53,54]. Traditionally these cells have been considered as of being more lineage restricted and thus not considered to have multilineage potential, but recent evidence suggests that they may possess a similar multipotent and proangiogenic profile to hBMSC in children [55,56]. Although these cells are relatively easy to obtain in the case of trauma or neoplasm where bone debridement/excision is necessary, in correcting chronic or congenital defects, harvesting of bone would be associated with an additional invasive surgical procedure and resultant donor site defect. As such, their usefulness in clinical practice maybe limited to cases where bone resection is already required.

13.4.2 Scaffolds for bone tissue engineering

Historically, a wide variety of different scaffolds have been proposed for bone tissue engineering ranging from natural materials such as collagen [57] and fibrin [58] to synthetic biomaterials in the form of polymers [59], all of which have demonstrated distinct advantages and disadvantages [60]. Increasingly, biomaterials that attempt to recreate and stimulate the extracellular matrix and stem cell niche are being generated with the aim of enhancing cell adhesion, proliferation, and differentiation [61,62].

Several studies have attempted to identify the optimal physiochemical and structural properties of biomaterials required to promote osteogenesis [63,64]. The most basic requirement for a scaffold is to promote cell adhesion, survival, and differentiation. In the case of bone, the ability to support vascularization is also critical. In addition, bioabsorbable scaffolds are more desirable in children whose skeleton continues to grow as multiple surgeries are required to remove and/or replace permanent implants.

The current trend in biomaterial design for bone tissue engineering leans toward creating bioactive scaffolds decorated with extracellular matrix components, which have been shown to enhance osteoinduction. The most commonly studied scaffolds for bone bioengineering can be broadly categorized as hydrogels, decellularized matrix, bioceramics polymers, and nanomaterials.

13.4.2.1 Hydrogels

Hydrogels such as fibrin and collagen present an attractive option for use as a scaffold in bone tissue engineering because they are relatively inexpensive, easily manufactured through reproducible protocols, bioabsorbable, and already licensed for clinical use with an established safety profile.

Fibrin is a key constituent of the coagulation cascade and has angiogenic properties, which makes it a desirable scaffold for bone tissue engineering [65]. As a scaffold, it has been repeatedly shown to promote survival, proliferation, and ostegenesis of osteoprogenitor cells in 3D culture as well as support vascularization [66,67]. It can also be decorated with osteoinductive components such as hydroxyapatite crystals or growth factors to stimulate osteogenesis when implanted in rat defects [58]. It has also been used to deliver osteogenically differentiated hBMSC in combination with bone graft to patients with nonhealing upper limb fractures, which were shown to unite on follow-up [68]. The poor mechanical strength and structural support afforded by fibrin is a key limitation. Biomimetic scaffolds that combine fibrin and hydroxyapatite-b-tricalcium have been shown to harness the effect of MSC proliferation and osteogenesis associated with fibrin with the additional benefit of increasing the mechanical strength [69]. This is in keeping with previous studies, which found that addition of fibrin to a mineralized scaffold not only enhanced osteogenesis but also induced secretion of vascular factors by hBMSC [67]. Fibrin-based matrices were also shown to induce secretion of angiogenic factors and lead to vasculogenesis of hADSC in vitro [70].

Collagen I is a basement membrane constituent, biodegradable, and already used in human applications. As discussed earlier, interaction between osteoprogenitors and the secreted collagen in the basement membrane promotes osteogenic differentiation. In bone tissue engineering, collagen has shown promise as a scaffold as it can facilitate osteoprogenitor cell differentiation to repair critical calvarial defects in vivo and promote vascularization [71]. Like fibrin, the strength afforded by collagen hydrogels is limited so, organic-inorganic scaffolds combining minerals with hydrogels are increasingly proving to provide the necessary mechanical strength and biological environment required for bone tissue engineering [72,73].

13.4.2.2 Decellularized matrix

A key advantage of using decellularized matrix as a scaffold is that it provides the extracellular matrix proteins and organization found in mature bone and as such is thought to mimic the bone tissue microenvironment that has been shown to direct MSC behavior [74]. These scaffolds are capable of recreating the stem cell niche to allow expansion of MSC while maintaining their capacity for self-renewal without loss of differentiation potential or induce osteoinduction of MSC without the need for additional osteogenic cues [75,76]. Despite these promising results, the disadvantages associated with traditional bone grafts such as limited supply, complex and potentially expensive decellularization protocols, potential immunogenicity, and infection transmission are a barrier to the suitability for use of these biomaterials in large-scale bone bioengineering for clinical use.

13.4.2.3 Bioceramics

Inorganic biomaterials such as bioceramics provide an attractive scaffold for bone bioengineering due to their mechanical strength, osteoinductive potential, and established safety profile from their preexisting clinical use in maxilla-facial and orthopedic implants. Furthermore, bioceramics can be modified to generate bioabsorbable porous scaffolds such as tricalcium phosphate and biologically active surfaces in the form of bioactive glasses [77]. In addition to promoting proliferation and osteogenic differentiation of human osteoblasts, bioceramics scaffolds have also been shown to activate murine osteoclasts in coculture and thus provide potential for mimicking the bone microenvironment [78]. Coculture of human BMSC and human umbilical vein endothelial cells on bioceramics scaffolds permits interaction between these cell types and results in enhanced osteogenesis and vascularization both in vitro and in vivo [79].

Despite the versatility of bioceramics and promising results in bone bioengineering, wider use of bioceramics scaffolds is hindered by the lack of standardized fabrication and characterization protocols [80].

13.4.2.4 Polymeric scaffolds and nanomaterials

A number of biodegradable polymers have been studied widely in bone tissue engineering due to their ability to mimic the extracellular matrix architecture, biocompatibility, versatility with respect to modification of their material properties, and availability in a number of forms including hydrogels, electrospun fibers, or porous sponges. Natural polymers such as collagen and chitosan have been used separately to induce osteogenic differentiation of MSC [81,82] and in combination to optimize the mechanical strength and overall osteoinductive effect on MSC as well as achieve vascularized bone implants when used to support osteoblast-endothelial cell cocultures [83,84]. An additional advantage of natural polymeric scaffolds is that it can be activated with the addition of inorganic matrices and/or growth factors to improve cell adherence, survival, and osteogenic behavior [85]. The main barrier to natural polymers is limited availability, low mechanical strength, potential to induce immunogenic reaction or transmit pathogens, and inability to standardize the "manufacturing" process [86].

Synthetic polymers bypass the limitations of their natural alternatives as they can be generated using scalable and easily standardized fabrication techniques. The most commonly studied synthetic polymers in bone bioengineering include poly(glycolic) acid, poly(lactic) acid, and polycaprolactone (PCL) all of which can be altered to affect their rate of biodegradation, mechanical strength, and structural properties as well as preferentially direct MSC behavior [87]. These polymers have been shown to promote osteoinduction of MSC in vitro and in vivo [88–91]. Similar to the scaffolds discussed previously, these polymers can be biofunctionalized to enhance cell attachment and migration as well as their osteoinductive and angiogenic properties [86,92].

Nanotechnology has enabled the fabrication of fully customized scaffolds that can be designed to mimic the architectural, biological, and mechanical properties of the desired microtissue environment to direct and regulate MSC behavior and promote vascularization [59,60]. Nanomaterials can be fabricated as nanofibers, self-assembling particles, and nanoceramics all of which have been shown to promote osteoinduction of MSC in vitro [93–95]. Combination of nanomaterials with natural or synthetic polymers can be used to enhance the biomimetic properties of traditional scaffolds to induce osteogenic differentiation and promote mature bone formation in vitro [96,97]. Although the versatility and biocompatibility demonstrated by nanomaterials makes them ideal for designing bone biomimetic scaffolds, further long-term in vivo and clinical studies are required to assess their safety and efficacy.

The complexity of bone structure, function, and physiology poses a number of challenges with regard to finding the ideal scaffold. This is further complicated by the fact that vascularization is essential for bone survival and maturation. Given that there has not been a single biomaterial to date, which has fulfilled all these requirements, it is likely that a combination of available scaffolds may provide the optimal biomaterial for bone bioengineering (Table 13.1).

Table 13.1 Commonly used scaffolds for tissue engineering

Category	Scaffold	Advantages	Disadvantages
Natural	Alginate	Licensed for clinical use Easily incorporates growth factors, peptides, or other scaffolds to enhance bioactivity	
	Collagen	Can be modified through incorporation of growth factors or peptides Licensed for clinical use Easily combined with other scaffolds Main ECM constituent	Rapid degradation Low-mechanical strength
	Fibrin	Licensed for clinical use Osteoinductive Angiogenic	
	Gelatin	Relatively inexpensive Easily printable	
	Decellularized matrix	Provides extracellular architecture and components required to recreate bone niche	Lengthy and complicated fabrication Risk of immunogenicity
Synthetic	Polyglycolic acid (PGA)	Mass producible Can be altered to optimize topography or functionalize with groups/motifs	Degradation products induce inflammation
	Polycaprolactone (PCL)	Versatile structure and easily incorporate bioactive particles Clinically established safety profile Provides mechanical strength	Unstable degradation
	Poly-L-lactic acid (PLA)	Easily modified structure Mechanical strength	Rapid degradation
	Bioceramics	Mechanical strength Bioactive Clinically established safety profile	Not suitable for cell-laden bioinks Complex fabrication protocols

The advantages and limitations of the most commonly used bioabsorbable biomaterials as applied to bone bioengineering. Adapted from Ibrahim A, Bulstrode NW, Whitaker IS, Eastwood DM, Dunaway D, Ferretti P. Nanotechnology for stimulating osteoprogenitor differentiation. Open Orthop J 2016;10:849–61.

13.4.3 The osteogenic microenvironment

The tissue microenvironment has been previously shown to play a crucial role in directing and regulating stem cell survival, proliferation, and differentiation [98,99]. This effect is dependent on cell type and as such has resulted in tissue engineers moving from the use of generic scaffolds toward recreating the niche specific to the stem cell population being studied [100,101]. For bone bioengineering, the 3D microenvironment has been shown to be vital in maintenance of osteogenic differentiation capacity and inducing extracellular matrix deposition by MSC [102,103]. Although a number of pathways can be targeted to drive osteogenic differentiation, those activating the canonical Wnt/β-catenin pathway are generally thought to be of greatest importance in driving osteogenic differentiation and mineralization [104].

Bone formed in vivo by implanted MSCs is often insufficient to repair defects, thus expansion and osteoinduction of osteoprogenitor cells before their implantation into a defect has been postulated to increase bone formation [105–107]. Most differentiation media will contain a combination of dexamethasone, ascorbic acid, and betaglycerophosphate, which are thought to increase Runx2 expression, enhance collagen I secretion, and promote hydroxyapatite formation in vitro, respectively [108].

Many studies have attempted to drive osteogenic differentiation using growth factors such as bone morphogeneic protein 2 [109], fibroblast growth factor 2 (FGF-2) [110,111], vascular endothelial growth factor [112], genetic modification [113], coculture with other cells [114], and using mechanical stimulation [115]. In addition, the role of the musculoskeletal unit (bone-tendon-muscle) provides an important biomechanical stimulus that regulates normal bone development and maintenance [116]. The issue of vascularization and perfusion is also of paramount importance when attempting bone tissue engineering as it not only affects the function of the bone but also the size of tissue that can be generated. Similarly, 3D coculture of MSC with endothelial precursors prompted early vascularization and osteogenic differentiation [117]. Increasingly, bioreactors have proved useful in tailoring the physical environment as well as enhancing perfusion of tissue engineered to optimize mineralization and mechanical strength in animal and human MSC in vitro [118–120].

13.5 3D bioprinting of bone tissue

The ultimate aim of bioengineering is to create functionally and anatomically relevant bone tissues, which can be used as to reconstruct bone defects. 3D bioprinting utilizes the significant progress made over the last 30 years in computer-aided design and computer-aided manufacture that have aided surgical training and planning and allowed generation of custom-shaped surgical tools and implants. Bone bioprinting also builds on previous evidence, which has shown that 3D printing can influence MSC behavior and preferentially direct osteogenic differentiation through spatial patterning of growth factors or pore geometry to recreate the biochemical and physical cues found in native bone microtissue environment [121–123]. Hence, 3D bone bioprinting offers the potential for fabrication of shaped constructs based on the input of computer models whose specifics can be defined entirely by the user. In addition, this approach has the potential for mass scalability and reproducibility increasing the feasibility for clinical translation.

13.5.1 Clinical and technical considerations for bone bioprinting

13.5.1.1 Computer-aided design

3D modeling using routine radiographic imaging modalities such as computed tomography and magnetic resonance imaging has detailed understanding of the macrostructure and anatomy of bone tissues as well as permitted accurate evaluation of bone defects and deformities. These data have already proved clinically useful in planning reconstructive surgery and in the production of patient-specific foreign-body implants and templates, which have been shown to improving operative outcomes [124,125].

Several groups have optimized computer-aided design protocols based on data from routine medical imaging to design bone substitutes and tissues, which have been successfully 3D printed and maintained their complex geometric details and structural integrity [126,127]. Modeling of perfusion patterns in native bone has also enabled the design and 3D printing of microchannel containing scaffolds that mimic these complex vascular networks and allow engineering of vascularized bone grafts [128].

13.5.1.2 Overview of bioprinting approaches

In addition to the challenge of identifying the optimal osteoprogenitor cell type, scaffold, and culture environment required for bone tissue engineering, additional technical challenge are posed when attempting to bioprint bone tissues such as ensuring cell survival during the printing process as well as accurately reproducing the complex structural organization of the extracellular matrix environment and cellular distributions to permit the generation of functional bone [129]. The main bone bioprinting approaches utilized in bone bioprinting are biomimicry, autonomous self-assembly, and mini tissues.

Biomimicry seeks to replicate accurately the cellular and extracellular components and arrangement found in native tissues. It has already been used to successfully bioprint rudimentary cartilage in vitro [130] and has the potential to generate complete bone tissues. This approach is heavily dependent on detailed understanding of the biological, structural, and mechanical properties of bone tissue. In addition, bioprinting methods will need to be able to reproduce accurately the complex arrangement and interactions of the cellular constituents.

Autonomous self-assembly approaches seek to utilize the developmental pathways in embryological bone development and bioprint the cellular components in an environment with that replicates the necessary biochemical and spatial signaling to prompt bone organogenesis. This approach is simpler in that it relies on cells and may not

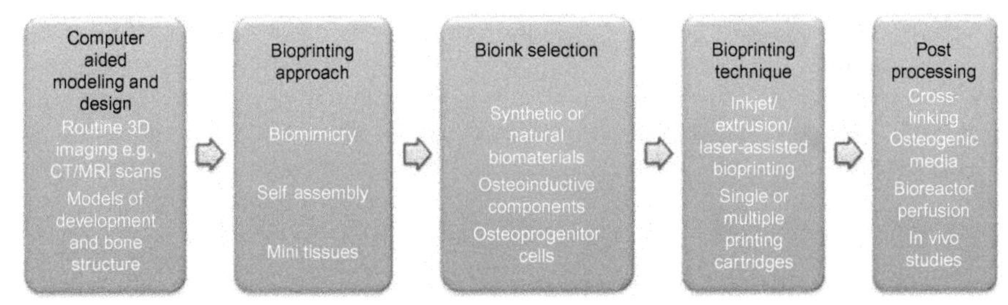

Fig. 13.4 Simplified overview the process required to apply an integrated approach to bioprinting bone tissues. Routine imaging can be used to design anatomically accurate implants while information on bone behavior, structure, and function can be used to model and design functional bone tissues. Bioprinting approach can be to replicate the exact structure, generate self-assembling subunits, or create mini tissues that can then built-up to generate full tissues. In choosing a bioink, biomaterials that can support cell survival and confer osteoinductive properties while permitting high cell density are most desirable. Depending on the approach and bioink selected, bioprinter choice is informed by the complexity and resolution of tissues required, cell type, and density as well as printing time and budgetary constraints. Bioprinted constructs may be further stabilized with cross-linking depending on choice of biomaterials and undergo further culture in osteogenic media or even bioreactor perfusion systems to accelerate bone formation.

require a scaffold but is reliant on detailed understanding of skeletogenesis in the embryo. Although this approach has not yet been demonstrated for bioprinting bone in the literature, recently adipose-derived stem cells under osteogenic culture conditions spontaneously organized to produce self-assembling osseous tissues thus providing the feasibility of applying this to bioprinting [131] (Fig. 13.4).

Bioprinting of mini-tissues offers a compromise between biomimicry and autonomous self-assembly in that it allows the building of bone tissues through bioprinting of biomimetic functional subunits, which can then be assembled either through delivering environmental cues to prompt self-assembly or can be assembled directly [129]. This process enables the generation of higher-resolution complex structures and has been used to bioprint mini-organs such as liver and allows the possibility of creating larger-scale bone constructs and their associated vasculature [132,133].

13.5.1.3 Bioinks

The choice of biomaterials for bone tissue engineering has been discussed earlier in this chapter. 3D bone bioprinting requires aqueous materials as bioinks that can encapsulate cells and other noncellular components. Hence, natural and synthetic hydrogels are at present the most compatible biomaterials for 3D bioprinting [129].

In selecting an appropriate bioink, it is important to consider the optimal cell and scaffold types required for bone tissue engineering and whether the desired materials have the properties required for successful 3D printing. Choice of this biological printing material ("bioink") also depends on the type of bioprinter available that will determine the viscosity, cross-linking/gelation method, and printing approach possible.

Viscosity is one of the most important considerations when selecting a bioink as it affects the sheer stresses on encapsulated cells and increases the risk of clogging where nozzle-head printers are used, which may impair cell viability, affect cell behavior, and reduce uniformity of printed constructs [134–136]. Cross-linking technique should also be considered when choosing an appropriate bioink and bioprinter. In general, bioinks need to maintain a constant viscosity during the printing process to allow uniform deposition of the material. Once printed the bioink does need to, however, provide enough rigidity and stability to maintain its size and shape to allow build-up of layers and in the long-term confer the necessary mechanical strength needed. This can be achieved during the printing process by printing the construct into a cross-linking solution such as for fibrin or alginate, exposure to relevant wavelengths of light for photocurable synthetic polymers, or manipulation of the temperature of the printing environment for materials such as matrigel that undergo thermal gelation [134]. If immediate stability of the printed construct is enough to allow accurate printing, then cross-linking can be employed at the end of the procedure by direct application of the relevant cross-linking agent. Additionally, the cross-linking method should not be cytotoxic or affect cell behavior.

Bioinks can also be designed to incorporate more than one biomaterial or even contain extracellular matrix components or growth factors to better recreate the extracellular matrix, impart greater osteoinductive properties, or improve cell survival [14].

3D bioprinting of bones requires cells that can be rapidly expanded with controllable and predictable proliferation and differentiation profiles. It is thus important to fully characterize chosen cells before assessing the impact of 3D bioprinting and enabling fine tuning. In addition, cells should be evaluated to assess whether they are capable of surviving the stresses associated with bioprinting, which may determine choice of cell type and/or 3D bioprinter. The effect of the bioprinting environment on cell behavior and survival is also important as cell lines should be able to withstand the culture conditions on the bioprinter for the length of time required to print the desired constructs. This is further complicated if more than one cell type is required (for instance, endothelial and osteoprogenitor cells in printing vascularized bone) and again will inform or be limited by choice of printing approach and bioprinter.

13.5.1.4 3D bioprinters

In brief, the main 3D bioprinting technologies commonly investigated in bone bioengineering are inkjet, microextrusion, and laser assisted. The key determinants to the suitability of each technique are the choice of manufacturing printing approach, cells, and biomaterial.

The most commonly used 3D bioprinters are inkjet, which employ thermal or acoustic forces to deposit droplets of bioink onto a substrate in a coordinated fashion. This technique is relatively quick, inexpensive, and multiple nozzle heads can be used to print different bioinks thus allowing for the creation of more complex tissues [129]. Clogging of the nozzle head and impaired cell viability due to thermal and mechanical stress are two key disadvantages posed by this. Additionally, inkjet printing is associated with reduced precision with regard to droplet placement and uniformity of droplets thus limiting the ability to print highly detailed smaller structures. The choice of printing material is also limited to hydrogels with low viscosity and low cell density due to the problems of nozzle head clogging and sheer stress required to expel the droplets.

Microextrusion is another popular bioprinting technology that utilizes temperature to dispense material in the form of continuous beads onto a stage. Either or both the printing head and stage can be robotically moved along the *x–y–z* axes during the printing process and multiple heads can be added to allow serial printing of different materials [129]. A fiberoptic light source can also be added to the stage to enable photoactivation and cross-linking of printed material. These printers allow greater control of the printing environment (e.g., temperature and humidity) and have relatively fast printing speeds thus allowing the production of large-scale and complex tissues in a clinically and biologically acceptable time frame and environment. A wide range of material viscosities and high cell concentrations can be printed at high resolution that allows for producing complex/denser structures and versatility in choice of printing material. The key drawback is that cell viability is the lowest of all three bioprinter choices [137].

Increasingly, laser-assisted bioprinters are proving useful in tissue engineering because of the high resolution and cell viability associated with this technique. Unlike inkjet and microextrusion machines, laser-assisted bioprinters do not possess a dispensing nozzle [134]. Instead, laser pulses are directed onto a laser-absorbing surface that is sandwiched between the laser head and printing material. Laser pulses are absorbed by this surface, which then generates a pressure bubble that propagates the bioink droplet onto a receiving surface to the desired place. This method of bioprinting has the highest cell viability as the process indirectly moves droplets thus reducing the sheer stress on cells. The absence of a printing nozzle also allows for printing higher cell densities, which tend to cause clogging in the other two methods. Importantly, laser-assisted bioprinting allows for high-resolution printing (as low as 10 μm), which is necessary for reproducing detailed ultrastructures in bone as required in biomimicry and mini-tissue approaches. The main limitations to laser-assisted printing are the high cost of the machines and the complexity of the process that makes printing multiple bioinks time consuming.

13.5.2 Review of advances in bone bioprinting to date

To date, the majority of 3D printing studies have focused on generating a biomimetic scaffold for use in supporting osteogenic differentiation of MSC, which are added after printing, or for direct use of the biomaterial as a bone substitute or aid to induce osteoinduction in critical-sized bone defects in vivo [14,138,139]. 3D bioprinting however provides the potential not only to recreate the complex architecture of the extracellular matrix and recreate the different tissue niche but also to dispense the cellular components in densities, distributions skin to native tissue thus recreating the networks, and spatial arrangements necessary to facilitate functional bone engineering. This section provides an overview of the most recent and significant studies that employ bioprinting of cells and biomaterials as well as discusses their limitations and possible future directions.

13.5.3 Case studies of 3D bioprinting in bone

Research into bone bioprinting is still in its infancy with a variety of different approaches and techniques being investigated to recreate the complex structure and biology of functional bone tissues. For simplicity, these studies can be grouped as *fundamental* research that generates rudimentary tissues but provide valuable insight into the optimal bioprinting materials and methods, *transitional* studies are those identified as bioprinting more complex tissues that have some functional activity or address the issue of vascularization or coculture of different cells, and *translational* work that produces full-scale or anatomically accurate bone tissues.

13.5.3.1 Fundamental research

A number of studies have successfully bioprinted cell-laden hydrogels and demonstrated survival of cells, which were then induced to undergo osteoinduction in differentiation media in vitro to produce basic osseous tissues. The work of Campos et al. addressed the issue of maintaining the shape of inkjet bioprinted hydrogels with low viscosity by printing the cellularized constructs directly into a hydrophobic dense fluid that provides support during the printing process. Not only did this produce high-resolution constructs but the encapsulated human osteosarcoma and BMSC survived and proliferated in subsequent in vitro culture and deposited extracellular matrix [140]. In another study, the same group bioprinted MSC-laden hydrogel organoids which when composed of a high-collagen concentration increased stiffness and preferentially directed MSC toward osteogenic differentiation [141]. Bendsten et al. showed that addition of hydroxyapatite to alginate-based hydrogels could be used to produce a printable biomaterial that allows bioprinting of viable mouse osteoblasts in a uniform distribution [142]. The visoelastic properties of this hydrogel constructs exhibited high shape fidelity after extrusion bioprinting that was performed in a calcium bath enabled further cross-linking during the printing process. The value of laser-assisted bioprinting for bone tissue engineering was demonstrated by Gruene et al., who bioprinted undifferentiated and differentiated (chondrogenic and osteogenic) porcine MSC coated onto an alginate-plasma hydrogel and collected onto an alginate and calcium chloride-coated slide [143]. This method allowed bioprinting of high cell densities into predefined patterns (the shape of which was maintained in subsequent culture) without cell damage. Moreover, the undifferentiated MSC maintained their differentiation potential and underwent chondrogenic and osteogenic differentiation when cultured in the relevant differentiation media for 21 days. The osteogenically differentiated MSC, which underwent bioprinting, also retained their osteogenic lineage. Together these findings illustrate that laser-assisted bioprinting can be used to produce grafts of undifferentiated and differentiated MSC, which is potentially useful in recreating the bone microtissue environment that contains a mixture of progenitor and terminally differentiated cells.

The important contribution of cellular distribution was addressed by Carlier et al., who modeled the complex spatial arrangement of cells in regenerating bone and used these results to bioprint agarose-laden with murine embryonic fibroblasts in patterns defined by the model in a variety of cell densities and resolutions [144]. The extrusion bioprinted constructs showed high cell viability produced the expected cell densities. The patterned constructs were then evaluated using the computational model to predict their ability to induce bone formation and achieve complete healing in a large-segmental bone defect. Despite the absence of biological data, this elegant study illustrates how 3D bioprinting can be guided by computer-aided modeling to reproduce the complex cellular geometry in bone and highlights the importance of understanding spatial distribution when attempting to bioengineer clinically relevant bone grafts.

Collectively these studies have helped to identify bioink compositions, patterning, and bioprinting strategies that can support cell survival and direct or preserve osteogenic differentiation. They are however limited by the fact that the bioprinted tissues do not reproduce the complex architecture of bone, were not functional and that the studies were limited to short-term in vitro culture.

13.5.3.2 Transitional research

A key feature of mature bone is the ability to lay down its own extracellular matrix and achieve mineralization through osteoblast activity. Wang et al. showed that this could be achieved using a bioink composed of an alginate/gelatin hydrogel, supplemented with biosilica and polyphosphate, containing human osteosarcoma cells that was bioprinted as cylinders in a calcium-containing dish then cultured in medium containing bioglass nanoparticles [145]. The bioglass significantly increased mineralization even in the absence of additional osteogenic stimuli but short-term differentiation in an osteogenic cocktail resulted in the production of larger crystals. Furthermore, analysis of the crystals showed similar composition to the mineral matrix composition of native bones.

The structural function of bone is determined by its mechanical properties, and thus bioengineered grafts should be able to demonstrate this. The reliance of 3D bioprinting on hydrogel-based bioinks poses a problem, as a major limitation of these biomaterials is low mechanical strength. Sawkins et al. created a bioink using a thermoresponsive copolymer comprising poly(lactic-*co*-glycolic acid) and poly(ethylene glycol), which was mixed with immortalized human BMSC to extrusion bioprint lattice-shaped constructs [146]. This study demonstrated that cellularized constructs with similar compressive strength to cancellous bone could be bioprinted using high cell densities and without any cytotoxic effects. They did not however assess the impact of cell behavior or effect on osteogenic differentiation.

The use of synthetic polymer-based bioinks has also been shown to simultaneously induce osteoinduction and confer mechanical strength to bioprinted constructs. In a study by Gao et al., photoactive poly(ethylene glycol)dimethacrylate was combined with bioglass or hydroxyapatite nanoparticles and used to inket bioprint human BMSC cylindrical constructs [147]. The hydroxyapatite-containing constructs demonstrated relatively high cell viability, increased compressive strength deposited collagen extracellular matrix, and secreted ALP after culture in osteogenic medium for 21 days. These findings are important in that they provide evidence that functional and strong bone-like tissues can be bioprinted successfully by combining biomaterials, which possess mechanical and osteoinductive properties [147].

As previously discussed, vascularization is essential for the survival, growth, and function of bone. The complex organization of native bone was recreated by Ciu et al., who bioprinted growth factor polylactide and cell-laden gelatin methacrylate hydrogels (containing human MSC and human umbilical vein endothelial cells (HUVEC)) in an alternate fashion into a shape representative of the haversian system using a dual printing platform [148]. This process not only preserved viability of both cell types but also the incorporation of the growth factors bone morphogeneic protein 2 and vascular endothelial growth factor into the relevant regions of the construct during the bioprinting process induced osteogenic and angiogenic differentiation of the MSC and HUVEC, respectively, when activated later. Culture of these constructs in a bioreactor that mimics the fluid stress found in native bone grafts resulted in the formation of mature bone-like tissue that lays down its own mineralized extracellular matrix and capillary-like networks. Furthermore, the mechanical strength of bioprinted constructs was comparable with native bone. These mini bone tissues created by this multifaceted study provide proof-of-concept for using bioprinting to recreate vascularized functional bone subunits, which can either be scaled-up or assembled together to create larger vascularized bone tissues.

These studies provide evidence that recreation of the functional and structural complexities of bone could be achieved by combining mechanically strong polymers and bioactive materials to provide strength and induce osteoinduction. In addition, they show that vascularization could be achieved through coculture of MSC with endothelial progenitor cells and patterning of bioprinted constructs to mimic vascular channels. Despite the limitations posed by in vitro studies and that the bioprinted bone constructs did not still replicate the complex architecture of native mature bone, the ability to generate functional and vascularized bone grafts presents the opportunity to scale-up, and generate more complex tissues based on these protocols.

13.5.3.3 Translational research

The end goal of generating autologous, anatomically accurate, and functional bone grafts has been brought closer to clinical translation by a number of studies, which combine modeling of the architecture and biology of native bone as well as biomimetic materials and culture protocols to bioprint more complex bone tissues.

This integrated approach to bone bioengineering can be seen in the work of Kang et al., who combined computer-aided modeling, composite bioinks, and multicartridge extrusion bioprinter to generate anatomically accurate full-sized mandibular and calvarial bone grafts with microchannels that permitted vascularization in vivo [127]. This study showed that the complex microscopic structure of bone could be modeled to extract the key patterns that confer mechanical strength and facilitate osteogenesis and angiogenesis necessary for mature bone bioengineering. They also demonstrated that multiple biomaterials could be used to recreate the different properties of the extracellular matrix by bioprinting PCL and tricalcium phosphate for strength and osteoinduction and cell-laden fibrin hydrogel to deposit accurately osteoprogenitor cells (human amniotic fluid cells) without loss in viability or osteogenic behavior. In coprinting a sacrificial exoskeleton, they were also able to ensure that the constructs maintained their shape and rigidity. In vitro culture in osteogenic media revealed calcium deposition, and in vivo implantation of the calvarial bone in a defect showed further osteogenic differentiation and vascularization of the bioprinted bone constructs.

Another integrated approach was demonstrated by Daly et al., who used the enchondral ossification pathway to bioprint cartilaginous templates containing porcine BMSC which produced a vascularized vertebral body mature bone construct in vivo [149]. This study integrated knowledge of enchondral bone development to design cartilaginous templates based on human vertebral bone models that allowed for incorporation of microchannels that were introduced prior to in vivo implantation. Again, coprinting of multiple bioinks was used to provide the different structural and biological of the biomimetic environment. An alginate containing arginylglycylaspartic acid (RGD) peptides was used to encapsulate and bioprint MSC and enhance chondrogenic differentiation while PCL was simultaneously printed using a second printing cartridge to reinforce the shape and provide mechanical strength to the constructs. After a 4-week chondrogenic culture period, microchannels were introduced into the constructs, which were then subcutaneously implanted in mice. The harvested constructs became mineralized in a pattern similar to trabecular enchondral bone, were vascularized, and even appeared to have developed a bone marrow-like structure.

These studies present a coherent argument for the development of multitool-based protocols to bioprint functional full-sized anatomically accurate bone tissues. They draw and build on the successes and failures of past bone tissue engineering and bioprinting strategies to provide an integrative approach that combines knowledge and understanding of bone biology, and then use this to model and bioprint complex bone tissues using biomimetic patterns and materials. To establish truly the clinical potential of these studies, further long-term data using large animal defect models will be necessary to ascertain safety, survival, functionality, and maintenance of shape.

13.5.4 Limitations

Despite the many promising achievements in the field of bone 3D bioprinting, this area of research is still in its infancy. There remain a number of critical barriers to engineering clinically appropriate tissues because of the complex structure, metabolic requirements, and multifunctional nature of bone.

At a cellular level, none of the studies have so far managed to recreate the multicellular composition of bone, which is necessary for the function, growth, and regeneration of native bone tissues. Although progress has been made with regard to bioprinting vascularized bone grafts, further analysis is required to assess the perfusion patterns and relationship between these vessels and the wider circulation to ensure that they do not disrupt other networks or pose a risk of thromboembolization.

Another major limitation of the studies is that many still use nonhuman MSC and are limited to in vitro culture systems. To accurately assess the value of these constructs as autologous bone tissues, long-term large animal studies are necessary to evaluate the efficacy and safety of these bones in reconstructing critical-sized defects.

13.5.5 *Future directions of research*

3D modeling and bioprinting have the potential to revolutionize the management of bone defects by producing autologous vascularized functional bone tissues. To date studies have provided protocols that allow bioprinting of high-density MSC-laden hydrogels without loss of viability, enhancement of bioinks with bioactive particles, and high-resolution printing of constructs that recreate many of the structural and spatial arrangements in native bone.

Future work should build on this to generate mature bone tissue by printing constructs that replicate the different microniches found in bone as well as the different cell types that inhabit them. The bioprinted samples will need to show that the normal relationship between the osteoprogenitor, osteoblasts, and osteocyte populations is recreated. In addition, constructs should include or encourage the migration of osteoclasts and attempt to establish the normal balance between osteogenesis and remodeling.

Large animal studies using bioprinted bones to reconstruct a defect will be necessary to ascertain the long-term survival and efficacy of these tissues. Moreover, extensive evaluation of the bioprinted bone constructs is also required to assess whether they also provide the metabolic functions required of native bones and when implanted in vivo, how they relate to the surrounding skeleton and systemic signaling.

13.6 Summary

In this chapter, we have provided an introduction into the anatomy and physiology of bone as well as provided an overview of the progress and limitations associated with traditional bone tissue engineering methods. Within this context, we introduced the essential approaches and methods relevant to bone 3D bioprinting and evaluated the most recent studies, which have successfully bioprinted bone tissues.

It is clear that 3D bioprinting provides the ability to place biomaterials and cells in a spatial distribution and structural pattern that can help to promote osteoinduction and permit perfusion. When combined with computer-aided modeling, full-sized bones with anatomically accurate shapes can be bioprinted. Knowledge on bone development can also allow the generation of cellularized templates that can be directed to form mature bone.

Ultimately, the integration of computer-aided modeling and design, bioactive materials, and high resolution multicartridge bioprinting will allow the generation of tissues that can replicate the complex hierarchical architecture and multipurpose nature of native bones. Greater understanding of the developmental pathways and cellular interactions in normal bone is required to enhance current bioprinting bones and generate mature bone tissues. Further in vivo work is required to establish the integration, function, safety, and long-term survival of such bioengineered bone in large animal models. The success of this would then enable clinical studies that can further highlight the feasibility of 3D bioprinted bone for clinical application.

References

[1] Woolf AD. The bone and joint decade 2000–2010. Ann Rheum Dis 2000;59(2):81–2.
[2] Baroli B. From natural bone grafts to tissue engineering therapeutics: brainstorming on pharmaceutical formulative requirements and challenges. J Pharm Sci 2009;98(4):1317–75.
[3] Ou KL, Weng CC, Wu CC, Lin YH, Chiang HJ, Yang TS, et al. Research of stembios cell therapy on dental implants containing nanostructured surfaces: biomechanical behaviors, microstructural characteristics, and clinical trial. Implant Dent 2016;25(1):63–73.
[4] Besinis A, Hadi SD, Le H, Tredwin C, Handy RD. Antibacterial activity and biofilm inhibition by surface modified titanium alloy medical implants following application of silver, titanium dioxide and hydroxyapatite nanocoatings. Nanotoxicology 2017;11(3):1–35.
[5] Rubin JP, Yaremchuk MJ. Complications and toxicities of implantable biomaterials used in facial reconstructive and aesthetic surgery: a comprehensive review of the literature. Plast Reconstr Surg 1997;100(5):1336–53.
[6] Tsukanaka M, Halvorsen V, Nordsletten L, EngesæTer I, EngesæTer LB, Marie Fenstad A, et al. Implant survival and radiographic outcome of total hip replacement in patients less than 20 years old. Acta Orthop 2016;87(5):479–84.
[7] Greenwald AS, Boden SD, Goldberg VM, Khan Y, Laurencin CT, Rosier RN, et al. Bone-graft substitutes: facts, fictions, and applications. J Bone Joint Surg Am 2001;83-A(Suppl. 2 Pt 2):98–103.
[8] Van der Stok J, Van Lieshout EM, El-Massoudi Y, Van Kralingen GH, Patka P. Bone substitutes in the Netherlands—a systematic literature review. Acta Biomater 2011;7(2):739–50.
[9] Finkemeier CG. Bone-grafting and bone-graft substitutes. J Bone Joint Surg Am 2002;84-A(3):454–64.
[10] Brigido SA, Protzman NM, Galli MM, Bleazey ST. The role of demineralized allograft subchondral bone in the treatment of talar cystic OCD lesions that have failed microfracture. Foot Ankle Spec 2014;7(5):377–86.
[11] Zimmermann G, Moghaddam A. Allograft bone matrix versus synthetic bone graft substitutes. Injury 2011;42(Suppl. 2):S16–21.
[12] Lee SH, Yoo CJ, Lee U, Park CW, Lee SG, Kim WK. Resorption of autogenous bone graft in cranioplasty: resorption and reintegration failure. Korean J Neurotrauma 2014;10(1):10–4.
[13] Calori GM, Mazza E, Colombo M, Ripamonti C. The use of bone-graft substitutes in large bone defects: any specific needs? Injury 2011;42(Suppl. 2):S56–63.
[14] Cui H, Zhu W, Holmes B, Zhang LG. Biologically inspired smart release system based on 3D bioprinted perfused scaffold for vascularized tissue regeneration. Adv Sci (Weinh) 2016;3(8):1600058.
[15] Rao RR, Stegemann JP. Cell-based approaches to the engineering of vascularized bone tissue. Cytotherapy 2013;15(11):1309–22.
[16] Lu Z, Roohani-Esfahani SI, Wang G, Zreiqat H. Bone biomimetic microenvironment induces osteogenic differentiation of adipose tissue-derived mesenchymal stem cells. Nanomedicine 2012;8(4):507–15.
[17] Florencio-Silva R, Sasso GR, Sasso-Cerri E, Simões MJ, Cerri PS. Biology of bone tissue: structure, function, and factors that influence bone cells. Biomed Res Int 2015;2015:421746.

[18] Bianco P. Bone and the hematopoietic niche: a tale of two stem cells. Blood 2011;117(20):5281–8.
[19] Chen H, Senda T, Kubo KY. The osteocyte plays multiple roles in bone remodeling and mineral homeostasis. Med Mol Morphol 2015;48(2):61–8.
[20] Paiva KB, Granjeiro JM. Bone tissue remodeling and development: focus on matrix metalloproteinase functions. Arch Biochem Biophys 2014;561:74–87.
[21] Alford AI, Kozloff KM, Hankenson KD. Extracellular matrix networks in bone remodeling. Int J Biochem Cell Biol 2015;65:20–31.
[22] Viguet-Carrin S, Garnero P, Delmas PD. The role of collagen in bone strength. Osteoporos Int 2006;17(3):319–36.
[23] Liu Y, Luo D, Wang T. Hierarchical structures of bone and bioinspired bone tissue engineering. Small 2016;12(34):4611–32.
[24] Anh DJ, Dimai HP, Hall SL, Farley JR. Skeletal alkaline phosphatase activity is primarily released from human osteoblasts in an insoluble form, and the net release is inhibited by calcium and skeletal growth factors. Calcif Tissue Int 1998;62(4):332–40.
[25] Farley JR, Hall SL, Tanner MA, Wergedal JE. Specific activity of skeletal alkaline phosphatase in human osteoblast-line cells regulated by phosphate, phosphate esters, and phosphate analogs and release of alkaline phosphatase activity inversely regulated by calcium. J Bone Miner Res 1994;9(4):497–508.
[26] Clarke B. Normal bone anatomy and physiology. Clin J Am Soc Nephrol 2008;3(Suppl. 3):S131–9.
[27] Auger FA, Gibot L, Lacroix D. The pivotal role of vascularization in tissue engineering. Annu Rev Biomed Eng 2013;15:177–200.
[28] Tomlinson RE, Silva MJ. Skeletal blood flow in bone repair and maintenance. Bone Res 2013;1(4):311–22.
[29] Nicolaije C, van de Peppel J, van Leeuwen JP. Oxygen-induced transcriptional dynamics in human osteoblasts are most prominent at the onset of mineralization. J Cell Physiol 2013;228(9):1863–72.
[30] Ben Shoham A, Rot C, Stern T, Krief S, Akiva A, Dadosh T, et al. Deposition of collagen type I onto skeletal endothelium reveals a new role for blood vessels in regulating bone morphology. Development 2016;143(21):3933–43.
[31] Yao Y, Hua C, Tang X, Wang Y, Zhang F, Xiang Z. Angiogenesis and osteogenesis of non-vascularised autogenous bone graft with arterial pedicle implantation. J Plast Reconstr Aesthet Surg 2010;63(3):467–73.
[32] Santagati F, Rijli FM. Cranial neural crest and the building of the vertebrate head. Nat Rev Neurosci 2003;4(10):806–18.
[33] Berendsen AD, Olsen BR. Bone development. Bone 2015;80:14–8.
[34] Owen TA, Aronow M, Shalhoub V, Barone LM, Wilming L, Tassinari MS, et al. Progressive development of the rat osteoblast phenotype in vitro: reciprocal relationships in expression of genes associated with osteoblast proliferation and differentiation during formation of the bone extracellular matrix. J Cell Physiol 1990;143(3):420–30.
[35] Komori T, Yagi H, Nomura S, Yamaguchi A, Sasaki K, Deguchi K, et al. Targeted disruption of Cbfa1 results in a complete lack of bone formation owing to maturational arrest of osteoblasts. Cell 1997;89(5):755–64.
[36] Ren L, Yang P, Wang Z, Zhang J, Ding C, Shang P. Biomechanical and biophysical environment of bone from the macroscopic to the pericellular and molecular level. J Mech Behav Biomed Mater 2015;50:104–22.
[37] Chaudhary LR, Hofmeister AM, Hruska KA. Differential growth factor control of bone formation through osteoprogenitor differentiation. Bone 2004;34(3):402–11.

[38] Mahmood A, Harkness L, Abdallah BM, Elsafadi M, Al-Nbaheen MS, Aldahmash A, et al. Derivation of stromal (skeletal and mesenchymal) stem-like cells from human embryonic stem cells. Stem Cells Dev 2012;21(17):3114–24.
[39] Si J, Dai J, Zhang J, Liu S, Gu J, Shi J, et al. Comparative investigation of human amniotic epithelial cells and mesenchymal stem cells for application in bone tissue engineering. Stem Cells Int 2015;2015:565732.
[40] de Peppo GM, Marcos-Campos I, Kahler DJ, Alsalman D, Shang L, Vunjak-Novakovic G, et al. Engineering bone tissue substitutes from human induced pluripotent stem cells. Proc Natl Acad Sci U S A 2013;110(21):8680–5.
[41] Levi B, Longaker MT. Concise review: adipose-derived stromal cells for skeletal regenerative medicine. Stem Cells 2011;29(4):576–82.
[42] Guasti L, Prasongchean W, Kleftouris G, Mukherjee S, Thrasher A, Bulstrode N, et al. High plasticity of paediatric adipose tissue-derived stem cells: too much for selective skeletogenic differentiation? Stem Cells Transl Med 2012;1(5):384–95.
[43] Peng W, Gao T, Yang ZL, Zhang SC, Ren ML, Wang ZG, et al. Adipose-derived stem cells induced dendritic cells undergo tolerance and inhibit Th1 polarization. Cell Immunol 2012;278(1-2):152–7.
[44] Gao S, Calcagni M, Welti M, Hemmi S, Hild N, Stark WJ, et al. Proliferation of ASC-derived endothelial cells in a 3D electrospun mesh: impact of bone-biomimetic nanocomposite and co-culture with ASC-derived osteoblasts. Injury 2014;45(6):974–80.
[45] Streckbein P, Jäckel S, Malik CY, Obert M, Kähling C, Wilbrand JF, et al. Reconstruction of critical-size mandibular defects in immunoincompetent rats with human adipose-derived stromal cells. J Craniomaxillofac Surg 2013;41(6):496–503.
[46] Jin Y, Zhang W, Liu Y, Zhang M, Xu L, Wu Q, et al. rhPDGF-BB via ERK pathway osteogenesis and adipogenesis balancing in ADSCs for critical-sized calvarial defect repair. Tissue Eng A 2014;20(23-24):3303–13.
[47] Taylor JA. Bilateral orbitozygomatic reconstruction with tissue-engineered bone. J Craniofac Surg 2010;21(5):1612–4.
[48] Xu S, De Becker A, Van Camp B, Vanderkerken K, Van Riet I. An improved harvest and in vitro expansion protocol for murine bone marrow-derived mesenchymal stem cells. J Biomed Biotechnol 2010;2010:105940.
[49] Shafiee A, Seyedjafari E, Soleimani M, Ahmadbeigi N, Dinarvand P, Ghaemi N. A comparison between osteogenic differentiation of human unrestricted somatic stem cells and mesenchymal stem cells from bone marrow and adipose tissue. Biotechnol Lett 2011;33(6):1257–64.
[50] Pill K, Hofmann S, Redl H, Holnthoner W. Vascularization mediated by mesenchymal stem cells from bone marrow and adipose tissue: a comparison. Cell Regen (Lond) 2015;4:8.
[51] Berner A, Henkel J, Woodruff MA, Steck R, Nerlich M, Schuetz MA, et al. Delayed minimally invasive injection of allogenic bone marrow stromal cell sheets regenerates large bone defects in an ovine preclinical animal model. Stem Cells Transl Med 2015;4(5):503–12.
[52] Rajan A, Eubanks E, Edwards S, Aronovich S, Travan S, Rudek I, et al. Optimized cell survival and seeding efficiency for craniofacial tissue engineering using clinical stem cell therapy. Stem Cells Transl Med 2014;3(12):1495–503.
[53] Mailhot JM, Borke JL. An isolation and in vitro culturing method for human intraoral bone cells derived from dental implant preparation sites. Clin Oral Implants Res 1998;9(1):43–50.
[54] Ishino T, Yajin K, Takeno S, Furukido K, Hirakawa K. Establishment of osteoblast culture from human ethmoidal sinus. Auris Nasus Larynx 2003;30(1):45–51.

[55] Jonsson KB, Frost A, Nilsson O, Ljunghall S, Ljunggren O. Three isolation techniques for primary culture of human osteoblast-like cells: a comparison. Acta Orthop Scand 1999;70(4):365–73.

[56] Wang S, Mundada L, Colomb E, Ohye RG, Si MS. Mesenchymal stem/stromal cells from discarded neonatal sternal tissue: in vitro characterization and angiogenic properties. Stem Cells Int 2016;2016:5098747.

[57] Weisgerber DW, Caliari SR, Harley BA. Mineralized collagen scaffolds induce hMSC osteogenesis and matrix remodeling. Biomater Sci 2015;3(3):533–42.

[58] Machado EG, Issa JP, Figueiredo FA, Santos GR, Galdeano EA, Alves MC, et al. A new heterologous fibrin sealant as scaffold to recombinant human bone morphogenetic protein-2 (rhBMP-2) and natural latex proteins for the repair of tibial bone defects. Acta Histochem 2015;117(3):288–96.

[59] Rossi F, Santoro M, Perale G. Polymeric scaffolds as stem cell carriers in bone repair. J Tissue Eng Regen Med 2015;9(10):1093–119.

[60] Ibrahim A, Bulstrode NW, Whitaker IS, Eastwood DM, Dunaway D, Ferretti P. Nanotechnology for stimulating osteoprogenitor differentiation. Open Orthop J 2016;10:849–61.

[61] Nii M, Lai JH, Keeney M, Han LH, Behn A, Imanbayev G, et al. The effects of interactive mechanical and biochemical niche signaling on osteogenic differentiation of adipose-derived stem cells using combinatorial hydrogels. Acta Biomater 2013;9(3):5475–83.

[62] Minardi S, Corradetti B, Taraballi F, Sandri M, Van Eps J, Cabrera FJ, et al. Evaluation of the osteoinductive potential of a bioinspired scaffold mimicking the osteogenic niche for bone augmentation. Biomaterials 2015;62:128–37.

[63] Terheyden H, Warnke P, Dunsche A, Jepsen S, Brenner W, Palmie S, et al. Mandibular reconstruction with prefabricated vascularized bone grafts using recombinant human osteogenic protein-1: an experimental study in miniature pigs. Part II: transplantation. Int J Oral Maxillofac Surg 2001;30(6):469–78.

[64] Terheyden H, Knak C, Jepsen S, Palmie S, Rueger DR. Mandibular reconstruction with a prefabricated vascularized bone graft using recombinant human osteogenic protein-1: an experimental study in miniature pigs. Part I: prefabrication. Int J Oral Maxillofac Surg 2001;30(5):373–9.

[65] Ceccarelli J, Putnam AJ. Sculpting the blank slate: how fibrin's support of vascularization can inspire biomaterial design. Acta Biomater 2014;10(4):1515–23.

[66] Gasparotto VP, Landim-Alvarenga FC, Oliveira AL, Simões GF, Lima-Neto JF, Barraviera B, et al. A new fibrin sealant as a three-dimensional scaffold candidate for mesenchymal stem cells. Stem Cell Res Ther 2014;5(3):78.

[67] Lohse N, Schulz J, Schliephake H. Effect of fibrin on osteogenic differentiation and VEGF expression of bone marrow stromal cells in mineralised scaffolds: a three-dimensional analysis. Eur Cell Mater 2012;23:413–23. discussion 24.

[68] Giannotti S, Trombi L, Bottai V, Ghilardi M, D'Alessandro D, Danti S, et al. Use of autologous human mesenchymal stromal cell/fibrin clot constructs in upper limb non-unions: long-term assessment. PLoS One 2013;8(8). e73893.

[69] Linsley CS, Wu BM, Tawil B. Mesenchymal stem cell growth on and mechanical properties of fibrin-based biomimetic bone scaffolds. J Biomed Mater Res A 2016;104(12):2945–53.

[70] Chung E, Rytlewski JA, Merchant AG, Dhada KS, Lewis EW, Suggs LJ. Fibrin-based 3D matrices induce angiogenic behavior of adipose-derived stem cells. Acta Biomater 2015;17:78–88.

[71] Maraldi T, Riccio M, Pisciotta A, Zavatti M, Carnevale G, Beretti F, et al. Human amniotic fluid-derived and dental pulp-derived stem cells seeded into collagen scaffold repair critical-size bone defects promoting vascularization. Stem Cell Res Ther 2013;4(3):53.
[72] Matsuura A, Kubo T, Doi K, Hayashi K, Morita K, Yokota R, et al. Bone formation ability of carbonate apatite-collagen scaffolds with different carbonate contents. Dent Mater J 2009;28(2):234–42.
[73] Sader MS, Martins VC, Gomez S, LeGeros RZ, Soares GA. Production and in vitro characterization of 3D porous scaffolds made of magnesium carbonate apatite (MCA)/ anionic collagen using a biomimetic approach. Mater Sci Eng C Mater Biol Appl 2013;33(7):4188–96.
[74] Voss A, McCarthy MB, Hoberman A, Cote MP, Imhoff AB, Mazzocca AD, et al. Extracellular matrix of current biological scaffolds promotes the differentiation potential of mesenchymal stem cells. Arthroscopy 2016;32(11):2381–1.
[75] Antebi B, Zhang Z, Wang Y, Lu Z, Chen XD, Ling J. Stromal-cell-derived extracellular matrix promotes the proliferation and retains the osteogenic differentiation capacity of mesenchymal stem cells on three-dimensional scaffolds. Tissue Eng Part C Methods 2015;21(2):171–81.
[76] Hoch AI, Mittal V, Mitra D, Vollmer N, Zikry CA, Leach JK. Cell-secreted matrices perpetuate the bone-forming phenotype of differentiated mesenchymal stem cells. Biomaterials 2016;74:178–87.
[77] Baino F, Novajra G, Vitale-Brovarone C. Bioceramics and scaffolds: a winning combination for tissue engineering. Front Bioeng Biotechnol 2015;3:202.
[78] Roohani-Esfahani SI, No YJ, Lu Z, Ng PY, Chen Y, Shi J, et al. A bioceramic with enhanced osteogenic properties to regulate the function of osteoblastic and osteocalastic cells for bone tissue regeneration. Biomed Mater 2016;11(3). 035018.
[79] Li H, Xue K, Kong N, Liu K, Chang J. Silicate bioceramics enhanced vascularization and osteogenesis through stimulating interactions between endothelia cells and bone marrow stromal cells. Biomaterials 2014;35(12):3803–18.
[80] Ebrahimi M, Botelho MG, Dorozhkin SV. Biphasic calcium phosphates bioceramics (HA/TCP): concept, physicochemical properties and the impact of standardization of study protocols in biomaterials research. Mater Sci Eng C Mater Biol Appl 2017;71:1293–312.
[81] Mathews S, Gupta PK, Bhonde R, Totey S. Chitosan enhances mineralization during osteoblast differentiation of human bone marrow-derived mesenchymal stem cells, by upregulating the associated genes. Cell Prolif 2011;44(6):537–49.
[82] Yang XB, Bhatnagar RS, Li S, Oreffo RO. Biomimetic collagen scaffolds for human bone cell growth and differentiation. Tissue Eng 2004;10(7-8):1148–59.
[83] Raftery RM, Woods B, Marques AL, Moreira-Silva J, Silva TH, Cryan SA, et al. Multifunctional biomaterials from the sea: assessing the effects of chitosan incorporation into collagen scaffolds on mechanical and biological functionality. Acta Biomater 2016;43:160–9.
[84] Liu X, Zhang G, Hou C, Wang H, Yang Y, Guan G, et al. Vascularized bone tissue formation induced by fiber-reinforced scaffolds cultured with osteoblasts and endothelial cells. Biomed Res Int 2013;2013:854917.
[85] Dan Y, Liu O, Liu Y, Zhang YY, Li S, Feng XB, et al. Development of novel biocomposite scaffold of chitosan-gelatin/nanohydroxyapatite for potential bone tissue engineering applications. Nanoscale Res Lett 2016;11(1):487.

[86] Liu X, Ma PX. Polymeric scaffolds for bone tissue engineering. Ann Biomed Eng 2004;32(3):477–86.
[87] Chen H, Gigli M, Gualandi C, Truckenmüller R, van Blitterswijk C, Lotti N, et al. Tailoring chemical and physical properties of fibrous scaffolds from block copolyesters containing ether and thio-ether linkages for skeletal differentiation of human mesenchymal stromal cells. Biomaterials 2016;76:261–72.
[88] Wang L, Dormer NH, Bonewald LF, Detamore MS. Osteogenic differentiation of human umbilical cord mesenchymal stromal cells in polyglycolic acid scaffolds. Tissue Eng A 2010;16(6):1937–48.
[89] Lee JH, Rhie JW, Oh DY, Ahn ST. Osteogenic differentiation of human adipose tissue-derived stromal cells (hASCs) in a porous three-dimensional scaffold. Biochem Biophys Res Commun 2008;370(3):456–60.
[90] Hao W, Dong J, Jiang M, Wu J, Cui F, Zhou D. Enhanced bone formation in large segmental radial defects by combining adipose-derived stem cells expressing bone morphogenetic protein 2 with nHA/RHLC/PLA scaffold. Int Orthop 2010;34(8):1341–9.
[91] Temple JP, Hutton DL, Hung BP, Huri PY, Cook CA, Kondragunta R, et al. Engineering anatomically shaped vascularized bone grafts with hASCs and 3D-printed PCL scaffolds. J Biomed Mater Res A 2014;102(12):4317–25.
[92] Stevens MM. Biomaterials for bone tissue engineering. Mater Today 2008;11(5):18–25.
[93] Binulal NS, Deepthy M, Selvamurugan N, Shalumon KT, Suja S, Mony U, et al. Role of nanofibrous poly(caprolactone) scaffolds in human mesenchymal stem cell attachment and spreading for in vitro bone tissue engineering—response to osteogenic regulators. Tissue Eng A 2010;16(2):393–404.
[94] Yang M, Zhou G, Shuai Y, Wang J, Zhu L, Mao C. Ca(2+)-induced self-assembly of Bombyx mori silk sericin into a nanofibrous network-like protein matrix for directing controlled nucleation of hydroxylapatite nano-needles. J Mater Chem B Mater Biol Med 2015;3(12):2455–62.
[95] Koroleva A, Deiwick A, Nguyen A, Schlie-Wolter S, Narayan R, Timashev P, et al. Osteogenic differentiation of human mesenchymal stem cells in 3-D Zr-Si organic-inorganic scaffolds produced by two-photon polymerization technique. PLoS One 2015;10(2). e0118164.
[96] Nair M, Nancy D, Krishnan AG, Anjusree GS, Vadukumpully S, Nair SV. Graphene oxide nanoflakes incorporated gelatin-hydroxyapatite scaffolds enhance osteogenic differentiation of human mesenchymal stem cells. Nanotechnology 2015;26(16):161001.
[97] Gandhimathi C, Venugopal JR, Tham AY, Ramakrishna S, Kumar SD. Biomimetic hybrid nanofibrous substrates for mesenchymal stem cells differentiation into osteogenic cells. Mater Sci Eng C Mater Biol Appl 2015;49:776–85.
[98] Yu C, Kornmuller A, Brown C, Hoare T, Flynn LE. Decellularized adipose tissue microcarriers as a dynamic culture platform for human adipose-derived stem/stromal cell expansion. Biomaterials 2017;120:66–80.
[99] Paul A, Manoharan V, Krafft D, Assmann A, Uquillas JA, Shin SR, et al. Nanoengineered biomimetic hydrogels for guiding human stem cell osteogenesis in three dimensional microenvironments. J Mater Chem B Mater Biol Med 2016;4(20):3544–54.
[100] Marinkovic M, Block TJ, Rakian R, Li Q, Wang E, Reilly MA, et al. One size does not fit all: developing a cell-specific niche for in vitro study of cell behavior. Matrix Biol 2016;52-54:426–41.
[101] Lee J, Abdeen AA, Tang X, Saif TA, Kilian KA. Matrix directed adipogenesis and neurogenesis of mesenchymal stem cells derived from adipose tissue and bone marrow. Acta Biomater 2016;42:46–55.

[102] Kittaka M, Kajiya M, Shiba H, Takewaki M, Takeshita K, Khung R, et al. Clumps of a mesenchymal stromal cell/extracellular matrix complex can be a novel tissue engineering therapy for bone regeneration. Cytotherapy 2015;17(7):860–73.
[103] Rakian R, Block TJ, Johnson SM, Marinkovic M, Wu J, Dai Q, et al. Native extracellular matrix preserves mesenchymal stem cell "stemness" and differentiation potential under serum-free culture conditions. Stem Cell Res Ther 2015;6:235.
[104] Kook SH, Heo JS, Lee JC. Crucial roles of canonical Runx2-dependent pathway on Wnt1-induced osteoblastic differentiation of human periodontal ligament fibroblasts. Mol Cell Biochem 2015;402(1-2):213–23.
[105] Siddappa R, Martens A, Doorn J, Leusink A, Olivo C, Licht R, et al. cAMP/PKA pathway activation in human mesenchymal stem cells in vitro results in robust bone formation in vivo. Proc Natl Acad Sci 2008;105(20):7281–6.
[106] Kumar S, Wan C, Ramaswamy G, Clemens TL, Ponnazhagan S. Mesenchymal stem cells expressing osteogenic and angiogenic factors synergistically enhance bone formation in a mouse model of segmental bone defect. Mol Ther 2010;18(5):1026–34.
[107] Meijer GJ, de Bruijn JD, Koole R, van Blitterswijk CA. Cell-based bone tissue engineering. PLoS Med 2007;4. e9.
[108] Langenbach F, Handschel J. Effects of dexamethasone, ascorbic acid and β-glycerophosphate on the osteogenic differentiation of stem cells in vitro. Stem Cell Res Ther 2013;4(5):117.
[109] Lysdahl H, Baatrup A, Foldager CB, Bünger C. Preconditioning human mesenchymal stem cells with a low concentration of BMP2 stimulates proliferation and osteogenic differentiation in vitro. Biores Open Access 2014;3(6):278–85.
[110] An S, Huang X, Gao Y, Ling J, Huang Y, Xiao Y. FGF-2 induces the proliferation of human periodontal ligament cells and modulates their osteoblastic phenotype by affecting Runx2 expression in the presence and absence of osteogenic inducers. Int J Mol Med 2015;36(3):705–11.
[111] Kwan MD, Sellmyer MA, Quarto N, Ho AM, Wandless TJ, Longaker MT. Chemical control of FGF-2 release for promoting calvarial healing with adipose stem cells. J Biol Chem 2011;286(13):11307–13.
[112] Li CJ, Madhu V, Balian G, Dighe AS, Cui Q. Cross-talk between VEGF and BMP-6 pathways accelerates osteogenic differentiation of human adipose-derived stem cells. J Cell Physiol 2015;230(11):2671–82.
[113] Deng Y, Zhou H, Yan C, Wang Y, Xiao C, Gu P, et al. In vitro osteogenic induction of bone marrow stromal cells with encapsulated gene-modified bone marrow stromal cells and in vivo implantation for orbital bone repair. Tissue Eng A 2014;20(13-14):2019–29.
[114] Kim KI, Park S, Im GI. Osteogenic differentiation and angiogenesis with cocultured adipose-derived stromal cells and bone marrow stromal cells. Biomaterials 2014;35(17):4792–804.
[115] Ongaro A, Pellati A, Bagheri L, Fortini C, Setti S, De Mattei M. Pulsed electromagnetic fields stimulate osteogenic differentiation in human bone marrow and adipose tissue derived mesenchymal stem cells. Bioelectromagnetics 2014;35(6):426–36.
[116] Sartori R, Sandri M. BMPs and the muscle-bone connection. Bone 2015;80:37–42.
[117] Duttenhoefer F, Lara de Freitas R, Meury T, Loibl M, Benneker LM, Richards RG, et al. 3D scaffolds co-seeded with human endothelial progenitor and mesenchymal stem cells: evidence of prevascularisation within 7 days. Eur Cell Mater 2013;26:49–64. discussion -5.
[118] David V, Guignandon A, Martin A, Malaval L, Lafage-Proust MH, Rattner A, et al. Ex vivo bone formation in bovine trabecular bone cultured in a dynamic 3D bioreactor is enhanced by compressive mechanical strain. Tissue Eng A 2008;14(1):117–26.

[119] Ding M, Henriksen SS, Wendt D, Overgaard S. An automated perfusion bioreactor for the streamlined production of engineered osteogenic grafts. J Biomed Mater Res B Appl Biomater 2016;104(3):532–7.

[120] Ji J, Sun W, Wang W, Munyombwe T, Yang XB. The effect of mechanical loading on osteogenesis of human dental pulp stromal cells in a novel in vitro model. Cell Tissue Res 2014;358(1):123–33.

[121] Phillippi JA, Miller E, Weiss L, Huard J, Waggoner A, Campbell P. Microenvironments engineered by inkjet bioprinting spatially direct adult stem cells toward muscle- and bone-like subpopulations. Stem Cells 2008;26(1):127–34.

[122] Ker ED, Nain AS, Weiss LE, Wang J, Suhan J, Amon CH, et al. Bioprinting of growth factors onto aligned sub-micron fibrous scaffolds for simultaneous control of cell differentiation and alignment. Biomaterials 2011;32(32):8097–107.

[123] Ferlin KM, Prendergast ME, Miller ML, Kaplan DS, Fisher JP. Influence of 3D printed porous architecture on mesenchymal stem cell enrichment and differentiation. Acta Biomater 2016;32:161–9.

[124] Steinbacher DM. Three-dimensional analysis and surgical planning in craniomaxillofacial surgery. J Oral Maxillofac Surg 2015;73(Suppl. 12):S40–56.

[125] Shi L, Liu W, Yin L, Feng S, Xu S, Zhang ZY. Surgical guide assistant mandibular distraction osteogenesis and sagittal split osteotomy in the treatment of hemifacial microsomia. J Craniofac Surg 2015;26(2):498–500.

[126] Bertol LS, Schabbach R, Dos Santos LA. Dimensional evaluation of patient-specific 3D printing using calcium phosphate cement for craniofacial bone reconstruction. J Biomater Appl 2016;31(6):799–806.

[127] Kang HW, Lee SJ, Ko IK, Kengla C, Yoo JJ, Atala A. A 3D bioprinting system to produce human-scale tissue constructs with structural integrity. Nat Biotechnol 2016;34(3):312–9.

[128] Holmes B, Bulusu K, Plesniak M, Zhang LG. A synergistic approach to the design, fabrication and evaluation of 3D printed micro and nano featured scaffolds for vascularized bone tissue repair. Nanotechnology 2016;27(6). 064001.

[129] Murphy SV, Atala A. 3D bioprinting of tissues and organs. Nat Biotechnol 2014;32(8):773–85.

[130] Costantini M, Idaszek J, Szöke K, Jaroszewicz J, Dentini M, Barbetta A, et al. 3D bioprinting of BM-MSCs-loaded ECM biomimetic hydrogels for in vitro neocartilage formation. Biofabrication 2016;8(3). 035002.

[131] Galbraith T, Clafshenkel WP, Kawecki F, Blanckaert C, Labbé B, Fortin M, et al. A cell-based self-assembly approach for the production of human osseous tissues from adipose-derived stromal/stem cells. Adv Healthc Mater 2017;6(4):1600889.

[132] Zhong C, Xie HY, Zhou L, Xu X, Zheng SS. Human hepatocytes loaded in 3D bioprinting generate mini-liver. Hepatobiliary Pancreat Dis Int 2016;15(5):512–8.

[133] Ozbolat IT. Bioprinting scale-up tissue and organ constructs for transplantation. Trends Biotechnol 2015;33(7):395–400.

[134] Irvine SA, Venkatraman SS. Bioprinting and differentiation of stem cells. Molecules 2016;21(9):1188.

[135] Stolberg S, McCloskey KE. Can shear stress direct stem cell fate? Biotechnol Prog 2009;25(1):10–9.

[136] Kuo YC, Chang TH, Hsu WT, Zhou J, Lee HH, Hui-Chun Ho J, et al. Oscillatory shear stress mediates directional reorganization of actin cytoskeleton and alters differentiation propensity of mesenchymal stem cells. Stem Cells 2015;33(2):429–42.

[137] Choi JW, Kim N. Clinical application of three-dimensional printing technology in craniofacial plastic surgery. Arch Plast Surg 2015;42(3):267–77.

[138] Li C, Jiang C, Deng Y, Li T, Li N, Peng M, et al. RhBMP-2 loaded 3D-printed mesoporous silica/calcium phosphate cement porous scaffolds with enhanced vascularization and osteogenesis properties. Sci Rep 2017;7:41331.

[139] Zhang H, Mao X, Du Z, Jiang W, Han X, Zhao D, et al. Three dimensional printed macroporous polylactic acid/hydroxyapatite composite scaffolds for promoting bone formation in a critical-size rat calvarial defect model. Sci Technol Adv Mater 2016;17(1):136–48.

[140] Duarte Campos DF, Blaeser A, Weber M, Jäkel J, Neuss S, Jahnen-Dechent W, et al. Three-dimensional printing of stem cell-laden hydrogels submerged in a hydrophobic high-density fluid. Biofabrication 2013;5(1). 015003.

[141] Duarte Campos DF, Blaeser A, Buellesbach K, Sen KS, Xun W, Tillmann W, et al. Bioprinting organotypic hydrogels with improved mesenchymal stem cell remodeling and mineralization properties for bone tissue engineering. Adv Healthc Mater 2016;5(11):1336–45.

[142] Bendtsen ST, Quinnell SP, Wei M. Development of a novel alginate-polyvinyl alcohol-hydroxyapatite hydrogel for 3D bioprinting bone tissue engineered scaffolds. J Biomed Mater Res A 2017;105(5):1457–68.

[143] Gruene M, Deiwick A, Koch L, Schlie S, Unger C, Hofmann N, et al. Laser printing of stem cells for biofabrication of scaffold-free autologous grafts. Tissue Eng Part C Methods 2011;17(1):79–87.

[144] Carlier A, Skvortsov GA, Hafezi F, Ferraris E, Patterson J, Koç B, et al. Computational model-informed design and bioprinting of cell-patterned constructs for bone tissue engineering. Biofabrication 2016;8(2). 025009.

[145] Wang X, Tolba E, Schröder HC, Neufurth M, Feng Q, Diehl-Seifert B, et al. Effect of bioglass on growth and biomineralization of SaOS-2 cells in hydrogel after 3D cell bioprinting. PLoS One 2014;9(11).e112497.

[146] Sawkins MJ, Mistry P, Brown BN, Shakesheff KM, Bonassar LJ, Yang J. Cell and protein compatible 3D bioprinting of mechanically strong constructs for bone repair. Biofabrication 2015;7(3). 035004.

[147] Gao G, Schilling AF, Yonezawa T, Wang J, Dai G, Cui X. Bioactive nanoparticles stimulate bone tissue formation in bioprinted three-dimensional scaffold and human mesenchymal stem cells. Biotechnol J 2014;9(10):1304–11.

[148] Cui H, Zhu W, Nowicki M, Zhou X, Khademhosseini A, Zhang LG. Hierarchical fabrication of engineered vascularized bone biphasic constructs via dual 3D bioprinting: integrating regional bioactive factors into architectural design. Adv Healthc Mater 2016;5(17):2174–81.

[149] Daly AC, Cunniffe GM, Sathy BN, Jeon O, Alsberg E, Kelly DJ. 3D bioprinting of developmentally inspired templates for whole bone organ engineering. Adv Healthc Mater 2016;5(18):2353–62.

3D bioprinting cartilage

14

Z.M. Jessop[*,†], N. Gao[‡], S. Manivannan[§], A. Al-Sabah*, I.S. Whitaker*[*,†]*

*Reconstructive Surgery and Regenerative Medicine Research Group, Swansea University Medical School, Swansea, United Kingdom, [†]The Welsh Centre for Burns and Plastic Surgery, Morrison Hospital, Swansea, United Kingdom, [‡]Edinburgh University Medical School, Edinburgh, United Kingdom, [§]Cardiff University Medical School, Cardiff, United Kingdom

14.1 Introduction

Significant facial disfigurements, including nasal and auricular defects following trauma, burns, skin cancer resection, and congenital conditions requiring reconstruction affect 542,000 (or one in 111) people in the United Kingdom (Fig. 14.1). The annual incidence is estimated to be 415,500, some conditions being transient while others life threatening (Changing Faces Prevalence Study 2007). One of the most prevalent muscloskeletal diseases affecting cartilage is osteoarthritis (OA). OA is a multifactorial disease and is characterized by cartilage degeneration, subchondral bone remodeling, osteophyte development, inflammation, and loss of joint function [1]. In the United Kingdom, hospital admissions regarding symptoms of OA, most commonly affecting the hands and larger load-bearing joints such as knees and hips, has incrementally increased in the past decade [2].

Grafting material for repair of cartilaginous defects ranges from alloplastic or synthetic implants, allografts, autologous cells or tissue, and more recently, tissue-engineered cartilage (Table 14.1).

Although the option of synthetic implants for reconstructing cartilaginous facial defects has been explored, it is hampered by the high rates of infection, functional compromise such as nasal blockage, and a sense of inadequacy from the patient's perspective [8–12]. Irradiated homologous cartilage grafts are complicated by high resorption rates (>70%) and risk of disease transmission [13,14]. The gold standard for surgical reconstruction of the face currently involves autologous conchal, nasoseptal, and costal cartilage harvest [15–17] but these are limited by donor-site morbidity [17–20]. Artificial joint replacement, although an extremely successful procedure for large joints such as the knee and hip has had mixed outcomes for smaller joints, such as the carpometacarpal joint. Autologous chondrocyte implantation (ACI) is a cell-based procedure for articular cartilage repair; however, patients are subjected to two surgical procedures, and healing is dependent on the quality and quantity of the patients' autologous cells [21].

Tissue engineering of durable cartilage has the potential to remove the need for a surgical donor (secondary) site and the associated morbidity [22]. The United Kingdom government highlighted regenerative medicine as one of the key areas that could provide a global competitive advantage for the United Kingdom and it was

3D Bioprinting for Reconstructive Surgery. https://doi.org/10.1016/B978-0-08-101103-4.00034-X

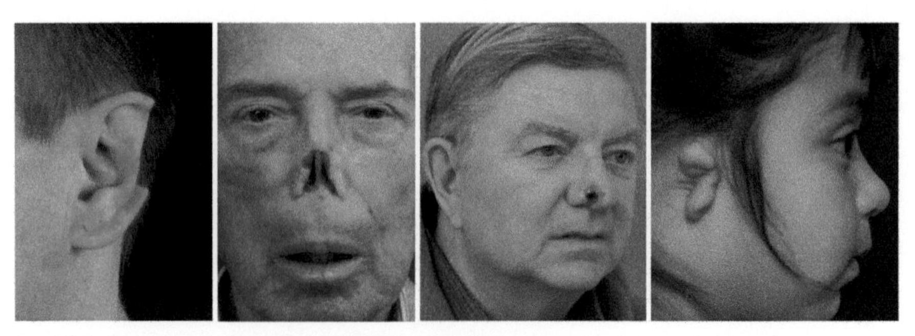

Fig. 14.1 From left to right, facial defects following trauma, cancer, and congenital deformities (patient consent for publication obtained).

Table 14.1 Advantages and disadvantages of reconstructive solutions

Reconstructive solution	Advantages	Disadvantages
Autologous	• No immunological complications [3] • Biologically compatible • Fewer legal restrictions	• Donor-site morbidity • Limited quantity of tissue available • Two separate operative sites, greater risk and cost [4]
Allogeneic	• No donor-site morbidity • Donor cells may have higher viability • Greater quantity of available tissue [5]	• Temporary/high resorption rates • Tissue typing/immunosuppression may be needed [6] • Risk of disease transmission and greater legal hurdles
Synthetic	• Maintain structural integrity • Predictable and reproducible physical and mechanical properties	• Extrusion [7]/infection • Cannot restore tissue function • Do not respond to biological cues/ grow with patient • May provoke inflammatory/fibrotic reaction
Tissue engineered	• Good biofunctionality and biocompatible • Good retention of size and shape • No immunological concerns • Mechanical stability [7]	• Long-term effects and tumorigenic potential unknown • Size often limited by vascularity • Costly

Based on Al-Himdani S, Jessop ZM, Al-Sabah A, Combellack E, Ibrahim A, Doak SH, et al. Tissue-engineered solutions in plastic and reconstructive surgery: principles and practice. Front Surg 2017;4:4.

highlighted as one of the "eight great technologies" worthy of significant investment [23,24]. This chapter outlines the current limitations and considerations of tissue engineering cartilage, in particular, in relation to cartilage development and morphology. It also provides an overview of benefits of bioprinting over other biofabrication techniques and discusses the cell types and bioinks that can be chosen.

14.2 Limitations of current tissue-engineered cartilage

The heterogeneity in approaches for cartilage tissue engineering indicates that we do not yet have a long-lasting durable solution. Current available tissue-engineered auricular and nasal neocartilage, often from unrelated cell sources placed on a variety of scaffolds, has been identified as fragile [25,26] and therefore prone to degradation [27–29], calcification [25,30], and mechanical instability [30–32]. When placed in animal models the constructs are observed to undergo inflammation [27,33], fibrosis [34], and foreign body reaction [35], particularly problematic with synthetic scaffolds, whose degradation products promote antigenicity [36–38] (Fig. 14.2). Contemporary tissue-engineered implants therefore do not produce stable, physiologically relevant cartilage for reconstruction [39].

Lessons from tracheal cartilage tissue engineering indicate that constructs from nonrelated stem cell sources combined with synthetic scaffolds have their limitations [40–42]. This suggests that simply combining the building blocks for tissue engineering is not sufficient for true regeneration. Observations in developmental biology, where different tissue histoarchitecture relies on growth factor gradients and vice versa, have shown that function follows form [43]. This emphasizes the importance of reproducing the anisotropy of native tissue to restore tissue function most effectively [44].

14.3 Cartilage development

Cartilage is a derivative of the embryonic mesenchymal cells, arising from various sources including the neuroectoderm (forming the craniofacial skeleton), paraxial mesoderm (forming the axial skeleton), and lateral plate mesoderm (forming long bones)

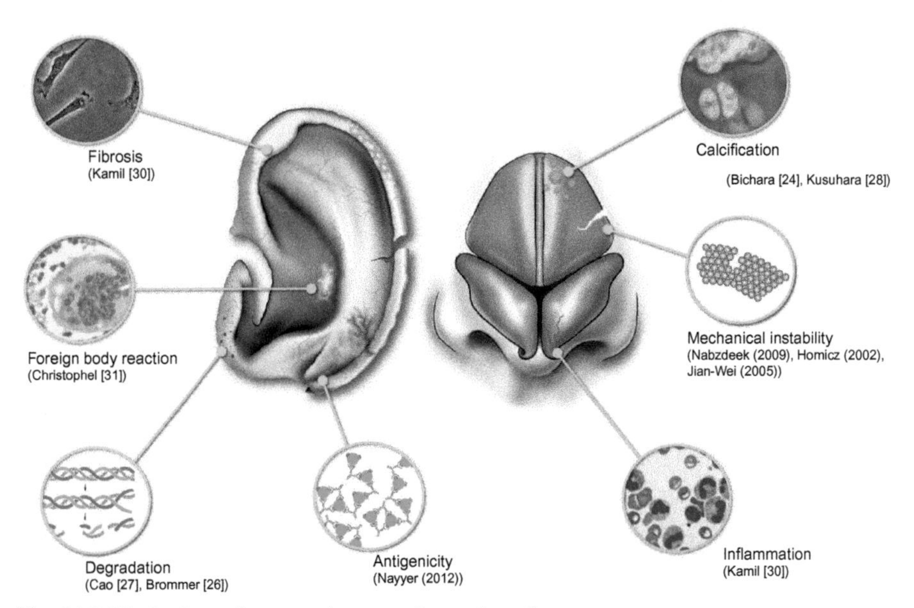

Fig. 14.2 Limitations of current tissue-engineered cartilage. Courtesy of Steve Atherton, medical illustrator ABMU.

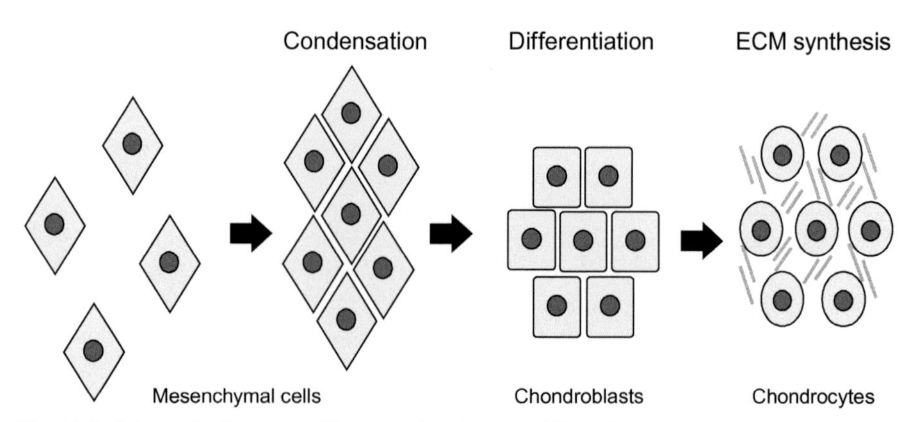

Fig. 14.3 Schematic diagram to illustrate chondrocyte differentiation and cartilage development.

[45]. Lessons from hyaline cartilage development, in the context of limb development, demonstrate that embryonic mesenchymal stem cells (MSCs) condense into clusters (characterized by N-cadherin) before differentiating into chondroblasts (characterized by collagen type II) and subsequently chondrocytes, with concomitant secretion of extracellular matrix (ECM) mediated by cell-to-cell and cell-to-matrix interactions ([46,47]; Fig. 14.3).

It is believed that to be able to tissue engineer durable native cartilage in vitro, it is important to understand and thereby replicate the cartilage developmental pathway, largely controlled by complex FGF-TGFβ-Wnt crosstalk that occurs in vivo [48]. Fibroblast growth factor 2 (FGF2) has been shown to be important for the proliferation of undifferentiated MSCs prior to chondrogenesis and is commonly used for cell expansion in cartilage tissue engineering [47]. Transforming growth factor beta (TGFβ), however, is necessary for the induction of chondrogenesis via the activation of transcription factor SOX9 in the early stages of cartilage formation [49]. The complex interplay of signaling factors is also time sensitive, meaning that exposure of the cells to either FDF2 or TGFβ at the wrong point along the differentiation pathway can lead to cellular hypertrophy, calcification, and fibrosis, which are major limitations of current tissue-engineered cartilage that were highlighted earlier [25,34].

14.4 Understanding cartilage morphology

When designing three-dimensional (3D) tissues and organs it is important to understand the native macro- (i.e., overall shape of the tissue or organ), micro- (composition of the ECM, pore shape and size vs cell shape and size), and nanoarchitecture (nanotopography and biomolecule attachments of ECM for optimal cell adhesion and proliferation) [50] to create true "like for like" tissue for reconstructive surgery. Ultimately, cartilage consists of isolated chondrocytes within lacunae amidst ECM containing type II collagen, proteogylcans, elastic fibers, and other proteins that satisfy its structural and functional role. However, it is the avascular, aneural, and immune-privileged nature of cartilage that means it has limited capacity for spontaneous repair and regeneration [51–53] but also makes it an ideal target for tissue engineering [54].

Articular cartilage has a well-defined zonal organization [55] and this varying organization of the cartilage ECM and cells with depth has been shown to affect its physical properties [56–58]. Until recently, nasoseptal cartilage has been assumed to be isotropic [59–63]. However, malrotation of cartilage grafts, in several studies, was suggested to lead to collagen misalignment between graft and host and therefore to lead to graft absorption using both articular [64] as well as nasoseptal samples [65,66]. This indicates that orientation affects the strength of grafts, suggesting potential for anisotropy, which was confirmed by biomechanical studies of human and bovine nasal septal cartilage [67,68]. However, there have been no further investigations elucidating the exact morphology of nasoseptal cartilage to explain these biomechanical differences. Moreover, although aging has been shown to affect mechanical and biochemical properties of human nasal septal cartilage [69], no studies have investigated the effect of growth and development, in particular, the time point at which anisotropy develops, that is, prenatal or postnatal. Since anisotropy has been shown to affect the biomechanical strength of cartilage, understanding this process in native cartilage may inform strategies for tissue engineering durable constructs.

14.5 Combining strategies to tissue-engineer durable cartilage

Strategies for cartilage tissue engineering require three main components, which include cells, scaffolds, and growth factors [3]. It is suggested that combining regenerative medicine strategies such as tissue-specific stem cells and novel biofabrication techniques such as 3D bioprinting may allow us to replicate better the native tissue anisotropy discussed previously and engineer durable cartilage for surgical reconstruction [70].

14.5.1 Cell source

A variety of cell sources have been implicated for cartilage tissue engineering but widespread clinical use remains elusive. The choice of cell type can be broadly divided into fully differentiated chondrocytes or stem cells, which can be embryonic or adult in origin (Table 14.2).

Although cultured adult chondrocytes have been described for facial cartilaginous tissue engineering [79,80], the harvest of a sufficient quantity of chondrocytes to repair any given defect is limited by the previously mentioned burden on the donor site. Also, the dedifferentiation phenomenon observed with chondrocytes may explain the preferential synthesis of fibrocartilage rather than functional hyaline cartilage as well as poor long-term tissue maintenance [81–83] and this means that a cell source is required that can provide an increased cell yield along with maintenance of chondrogenic potential.

Table 14.2 Summary of advantages and disadvantages of different cell sources for cartilage tissue engineering

Cell type	Advantages	Disadvantages
Chondrocytes	Variety of sources	Donor-site morbidity Dedifferentiation phenomenon
Stem cells		
BMSC	Easily harvested, e.g., bone marrow aspirate Multipotency	Concerns over osteogenic lineage preference [71,72]
ADSC	Easily harvested, e.g., during liposuction Multipotency	Limited chondrogenic potential compared to other MSCs [73,74]
SSC	Superior ECM secretion to other MSCs [75] Multipotency	Compromised proliferative/ chondrogenic capability in OA patients [76,77]
ESC	Pluripotency	Ethical considerations
iPSCs	Pluripotency	Teratoma formation [78]

ADSC—adipose-derived stem cell, BMSC—bone marrow-derived stem cell, ESC—embryonic stem cell, iPSC—induced pluripotent stem cell, SSC—synovium-derived stem cell.

MSCs, found in multiple tissue types (Table 14.2), are defined by specific cell markers, adherence to plastic, and multipotency [84]. MSCs have the capability of continuous self-renewal and hence predicted to be a useful cell type for tissue engineering [85]. Despite this MSCs are reported to have a transient/low chondrogenic potential, insufficient cartilaginous matrix synthesis, followed by terminal chondrocyte differentiation and mineralization [75,86–91], in contrast to transplanted chondrocytes which tend to mature into cartilage [92], and which hinders their application in cartilage tissue engineering. More recently, the healing effects of MSCs, through their interaction with immune cells, have been exploited to modulate inflammatory conditions such as OA by combining them with autologous chondrons [93]. Since cartilage is immune-privileged, in theory it should be possible to use heterologous cells in grafting, such as human-derived induced pluripotent stem cells (iPSCs) [94], but concerns remain about genetic stability and potential for teratoma formation [78].

The latest additions to the potential cell repertoire for cartilage tissue engineering are chondroprogenitor cells, initially identified in articular cartilage [95] and they are gaining interest in their potential for tissue engineering [95–97]. Initial work on articular cartilage identified these progenitor cell populations using colony-forming assays (i.e., clonally derived, able to proliferate to >50 population doublings while maintaining phenotype and trilineage potential) [95,98,99] and MSC marker expression [100,101]. There is no consensus in the literature regarding whether these cells are chondroprogenitors or true MSCs (Table 14.3). The fact that these stem cells exist in the very tissue [103] that requires replacement means that they could provide a promising approach to tissue engineering functional cartilage.

Table 14.3 **Table comparing characteristics of cartilage-derived stem cells versus progenitor cells**

Cartilage-derived stem cell (CDSC)	Cartilage progenitor cell (CPC)
Multipotency [99]	Limited adipogenecity [102]
CD105$^+$, CD166$^+$, Notch-1, STRO-1 and other stem cell-related markers	Overexpression of chondrocyte markers COL2A1, ACAN, PRG4, S100A1, and S100B compared to MSC [103]
Chemotactic/migratory activity [100]	
Clonogenicity	Clonogenicity [103]
Fibronectin adhesion ($\alpha5\beta1$ integrin expression)	Fibronectin adhesion [95]

14.5.2 Evolution of biofabrication into 3D bioprinting

Tissue engineering pioneer W.T. Green laid out important theoretical and practical concepts for cartilage tissue engineering [104] and his experiments in 1977, although not entirely successful, attempted to produce new cartilage using a decalcified bone scaffold seeded with chondrocytes. Research was directed toward the development of scaffolds and microenvironments capable of stimulating cells toward chondrogenic lineage and the production of ECM. Improving cell isolation and culture techniques in the 1990s further pushed the research forward and was even brought forward to the mainstream through media coverage of J. Vacanti's "*ear-mouse*" in 1997 [29]. Vacanti's experiments demonstrated that bovine chondrocytes seeded on an injection molded, ear-shaped scaffold could survive and even produce ECM when implanted in vivo, although long-term tissue maintenance was uncertain. The challenge was to find a biofabrication method that allows control of both micro- and macroarchitecture simultaneously. 3D bioprinting, an automated, computer-aided deposition of cells, biomaterials, and biomolecules [105–108] provides this capability. The high spatial resolution of 3D bioprinting has the potential to replicate complex native-like tissue architecture more faithfully in the laboratory than traditional tissue engineering methods of assembly, consisting of nonspecific cell seeding of scaffolds [108]. This emerging enabling technology represents a developmental biology inspired alternative to classic biodegradable solid scaffold-based approaches in tissue engineering and has the ability to assemble biological components in prescribed 3D structure with the potential of restoring biological function as well as patient customization ([109]; Fig. 14.4).

Most tissue engineering as well as 3D bioprinting research to date has focused on cartilage as it is relatively acellular, avascular, and aneural; so theoretically it should be a relatively simple tissue type to replicate. There have been a number of heterogenous studies investigating cartilage 3D bioprinting, many using a variety of cell and scaffold types but the majority, as with many other connective tissues, involve extrusion-based techniques (Table 14.4). Most focus on cell postprinting viability, which tends to vary between 69% and 99%, as well as evidence of cartilage ECM formation at the gene and protein level. The key to the clinical translation of these constructs will be consensus on the parameters that represent structural integrity, such

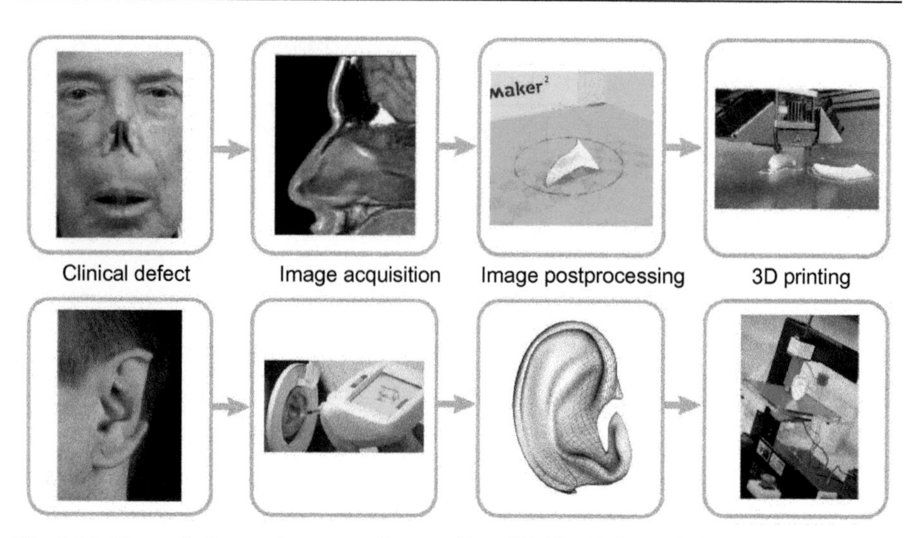

Fig. 14.4 Theoretical stepwise approach to cartilage 3D bioprinting. (1) Clinical defect and patient expectations. (2) Image acquisition (CT or 3D scanning). (3) Image postprocessing (3D model converted to STL; standard tessellation language, and processed with "slicer" software). (4) 3D bioprinting for surgical reconstruction.

as ability to withstand compressive and tensile loads at the physiological level, and hence allow surgical handling and implantation.

14.5.3 Cartilage bioinks

One of the major current challenges still faced in cartilage bioprinting is the fabrication of a scaffold which is printable, capable of holding its structure, maintaining close biomechanical and surface properties to that of native cartilage, and providing an environment in which cells can thrive, differentiate, and produce extra-cellular matrix [142]. The increased accessibility of 3D printers at the turn of the millennium alongside developments in the fields of polymers and biomaterials, rheology, and fluid dynamics has driven advancements in scaffold design and naturally the research is shifting toward the development of printable bioinks, which already contain cells [143]. Below we discuss the most commonly used bioinks for cartilage bioprinting and their potential benefits and drawbacks.

14.5.3.1 Alginate

Alginate is a polysaccharide commonly found in the cell wall of brown algae which has been used widely in medical application due to its favorable biological properties. Alginate bioinks are cross-linked chemically using divalent salts (Zn^{2+}, Ca^{2+}, Na^{2+}, etc.) [144], which allows maintenance of structural integrity for long-term culture, but the speed of cross-linking reduces adhesion between the layers during bioprinting, which can compromise the overall shape. To reduce the rate of cross-linking, alginate bioinks tend to be relatively low-weight percentage gels [113] but this in turn reduces resolution and viscosity, enabling the printing of simple structures prone to flattening. Furthermore, even after cross-linking, alginate hydrogels do not match the mechanical characteristics of native cartilage, which limits their utility for cartilage 3D bioprinting. Alginate has been used widely to encapsulate cells for 3D culture and has shown potential in promoting chondrogenicity. A study investigating the potential of different compositions of hydrogels stimulated with TGF-β3 in promoting hyaline or fibrocartilage formation in vitro has demonstrated that MSCs embedded in alginate were more inclined to form hyaline cartilage [116]. Alginate bioninks have been additionally used for printing heterogeneous constructs to study their potential use for osteochondral defects. Alginate powder was mixed at 10% w/v with MSC expansion, chondrogenic, or osteogenic media and each variant mixture was printed as a porous scaffold with different designs in terms of fiber spacing and angle with human chondrogenic and osteogenic progenitors. After 3 weeks culture in vitro, and 6 weeks culture in vivo the constructs demonstrated tissue formation at different areas of the construct corresponding to the alginate variant mixture indicating that simple constructs could be made more complex by altering alginate mixtures for specific cell differentiation [117].

Hybrid alginate bioinks have shown to be a promising approach in improving structural fidelity by using additives such as methylcellulose (MC) and polycaprolactone (PCL). MC is a biocompatible chemical compound derived from cellulose, which is used to enhance the viscosity of alginate mixture enabling the printing of complex structures with higher precision, while allowing the retention of the positive attributes of the alginate bioink in terms of cell differentiation potential [135]. The other additive, PCL, is a biodegradable synthetic polymer that has been used to enhance the viscosity of alginate and especially useful for 3D extrusion bioprinting due to its low melting point of 60°C [145]. PCL/alginate bioink has been shown to improve the mechanical strength of constructs as well as increasing cell proliferation, viability, and chondrogenic differentiation in vitro and in vivo as a possible suitable approach for cartilage tissue engineering [146,147]. PCL and poly-ethylene glycol (PEG), a synthetic polymer, are additionally used to provide a porous scaffold and act as a sacrificial layer to permit the printing of complex structures such as the ear. PEG was used as a mold to hold the main construct body while PCL was added in the structure as a temporary framework for the alginate construct which is then removed postprinting. This method was utilized to print alginate with chondrocytes and adipocytes differentiated from adipose-derived stromal cells to determine the feasibility of printing 3D ear regeneration. Both cell types have been printed in specific regions (chondrocytes predominantly and adipocytes localized to the ear lobe region) of the framework and have shown tissue formation, with the bioink favoring a chondrogenic phenotype [129].

Synthetic polymers such as PCL and PEG may provide excellent structural stability of fabricated hydrogels, but these polymers impose a risk due to their slow degradation that may elicit an immune reaction, which may damage the implanted construct [148]. Furthermore, synthetic polymers are less biologically active than naturally derived biomaterials [149].

Table 14.4 Summary of cartilage 3D bioprinting studies

Printing technique	Bioink/scaffold material	Cell type	Polymerization	In vivo/ in vitro
Coaxial extrusion	Alginate + GelMA/ CS-AEMA/HAMA	hMSC	$CaCl_2$	In vitro
Coaxial extrusion ($CaCl_2$ core + alg sheath)	Alginate	Bovine cartilage progenitor cells	$CaCl_2$	In vitro
Coaxial extrusion – handheld	HA-GelMA	Sheep MSCs	UV	In vivo (sheep)
Extrusion	Alginate/Pluronic F127	hMSC	$CaCl_2$	In vitro
Extrusion	Collagen-coated PCL (nonbiological/temp)	Porcine articular chondrocytes	–	In vitro
Extrusion	PEGDMA in osteochondral plug	Human articular chondrocytes	UV	In vitro
Extrusion	Agarose/Alginate/ GelMA/BioINK	Porcine BM-MSCs	$CaCl_2$/UV	In vitro
Extrusion	Alginate	hMSC	$CaCl_2$	In vivo (mice)
Extrusion	PEGDMA + Gel MA	hMSC	–	In vitro
Extrusion	Alginate bioink in PCL canals	Primary chondrocytes (chick embryos)	$CaCl_2$	In vitro
Extrusion	Alginate PCL hybrid construct	Human nasoseptal chondrocytes	$CaCl_2$	In vitro + in vivo (mice)
Extrusion	Nanocellulose + alginate	Human iPSC	$CaCl_2$/H_2O_2	In vitro

Cell density (million/ml)	Print resolution (mm)	Structural integrity	Cell viability	Reference
6.6	0.15	Individual fibers maintain rod-like shape	85%–95% 3 h postfabrication, slight decrease after 1–3 weeks. Cells not affected by shear stress (3 h postfabrication)	[110]
–	0.445	Rupture if high flow rates through channels used	97% day 1, 95% day 7	[111]
2.5	–	Compressive modulus 0.5 mPa	–	[112]
10	0.19	45 kPa	83% at 7 days	[113]
10	–	Retained intact scaffold structure and uniform surface	–	[114]
8	4	Compressive modulus 36.9 kPa comparable to human articular cartilage	85% after printing	[115]
20	0.25	Compressive modulus below human articular cartilage	>70% even with PCL	[116]
10	0.8	–	88% (compared to 89% control)	[117]
6	–	30–50 kPa following chondrogenic differentiation after 3 weeks	80%–90% shown in bar graph	[118]
5.6	–	Hybrid constructs maintained their structural integrity	Both fibroblastic cells viability >80% after 2 weeks. Initial drop to 76% for fibroblastic cells at 1 week. PCL temp drops within 10 s at 60–75°C	[119]
1	0.2	SEM images confirming structural integrity	85% in alginate/PCL scaffolds, 95%–97% in single alginate layer scaffolds, decreased when multiple layers	[120]
–	–	60/40 poor at holding shape, 80/20 good but reduces cell viability	Poor cell viability in NFC HA, large growth of round cells in NFC A 60/40—although did not hold shape	[94]

(Continued)

Table 14.4 **Continued**

Printing technique	Bioink/scaffold material	Cell type	Polymerization	In vivo/ in vitro
Extrusion	HA/Col-1	hMSC	$CaCl_2$	In vitro
Extrusion	High-density collagen hydrogel	Bovine meniscal fibrochondrocytes	Temperature	In vitro
Extrusion	Atelocollagen/CB-DAH interlaced in PCL scaffold	Turbinate derived hMSCs	–	In vitro + In vivo (rabbit)
Extrusion	Alginate (interlaced in PCL scaffold)	hMSCs	$CaCl_2$	In vitro + in vivo (mice)
Extrusion	GelMA	Equine articular chondrocytes	UV	In vitro
Extrusion (valve based droplet ejector)	GelMA	hMSC	UV	In vitro
Extrusion + microinjection	Alginate (microinjection of cell pellet)	Bovine articular chondrocytes	$CaCl_2$	In vitro + in vivo (bovine)
Heated extrusion 37°C – fluorocarbon substrate	Agarose	hMSC/hMG-63 from osteosarcoma	–	In vitro
Plunger/pneumatic extrusion	Alginate (on PCL scaffold)	Human adipose-derived stem cells	$CaCl_2$	In vitro
Plunger/pneumatic extrusion	Alginate + dECM (porcine auricular) on PCL scaffold	Human auricular chondrocytes	$CaCl_2$	In vitro
Plunger/pneumatic extrusion	Nanofibrillated cellulose + alginate (support structures in PCL)	Human nasoseptal chondrocytes	$CaCl_2$	In vitro
Pneumatic extrusion	Nanofibrillated cellulose + alginate	BM hMSC	$CaCl_2$	In vivo (mice)
Pneumatic extrusion	Nanocellulose + alginate sulfate	Bovine articular chondrocytes	$CaCl_2$	In vitro

Cell density (million/ ml)	Print resolution (mm)	Structural integrity	Cell viability	Reference
2	–	–	>90%	[121]
10	–	Compressive modulus 30 kPa	90% immediately postfabrication—no variability 10 days postculture	[122]
1	–	–	93% atelocollagen, 86% CB-DHA	[123]
10	–	–	–	[124]
5	–	36 kPa at 10% GelMA, 199 ka at 20% GelMA	82% with HA at day 3 (increase compared to day 1 and better than without HA)	[125]
–	0.3 droplets	–	–	[126]
–	0.7	Young's modulus 1 mPa	75% day 1, 87% day 2	[127]
10	–	Compressive modulus 15 kPa	>90% at 21 days	[128]
1	0.2	Tensile modulus 14–16 MPa	95% day 1	[129]
5	0.317	–	99% at 8 h in 6, 12, 18°C	[130]
24	–	Excellent size and shape fidelity after 28 days in culture	68.5%, increased in dead cell numbers after day 28	[131]
10	–	Increased stability over 60 days of implantation	–	[132]
6	–	–	>85%, initially lower with addition of nanocellulose but proliferation increased cell viability numbers	[133]

(Continued)

Table 14.4 Continued

Printing technique	Bioink/scaffold material	Cell type	Polymerization	In vivo/ in vitro
Pneumatic extrusion	Alginate (in PCL scaffold—cell seeded)	Rabbit auricular chondrocyte	$CaCl_2$	In vitro + in vivo (rabbit)
Pneumatic extrusion	Alginate + methylcellulose	hMSC	$CaCl_2$	In vitro
Pressure extrusion	Alginate hydrogel on polyethelinimine scaffold	ATDC5 mouse cell line (teratocarcinoma)	$CaCl_2$	In vitro
Pressure extrusion	GelMA + Poloxamer reinforcements	Equine chondrocytes	UV	In vitro
Pressure extrusion	PEG/Partially methacrylated poly(*N*-(2-hydroxypropyl) methacrylamide mono/ dilactate)	Equine primary chondrocytes	UV	In vitro
Piezoelectric inkjet	Nanofibrillated cellulose + alginate	Human nasoseptal chondrocytes	$CaCl_2$	In vitro
Thermal inkjet	PEGDMA	hMSC	UV	In vitro
Electrospin PCL + injected hydrogel	GelMA/alginate on PCL microfiber scaffold (electrospun)	Equine chondrocytes	Temperature	In vitro
Electrospun + inkjet deposition of cells	PCL	Rabbit auricular chondrocytes	–	In vitro + in vivo (mice)

BM—bone marrow, $CaCl_2$—calcium chloride, CB-DAH, Col-1—collagen type 1, CS-AEMA—chondroitin sulfate amino ethyl methacrylate, dECM—decellularized extracellular matrix, GelMA—gelatin methacrylamide, H_2O_2—hydrogen peroxide, HA—hyaluronic acid, HAMA—hyaluronic acid methacrylate, hMSC—human mesenchymal stem cell, iPSC—induced pluripotent stem cell, PCL—polycaprolactone, PEG—poly-ethylene glycol, PEGDMA—Polyethylene glycol dimethacrylate, UV—ultraviolet.

Cell density (million/ml)	Print resolution (mm)	Structural integrity	Cell viability	Reference
3	0.2	Elastic modulus and tensile modulus similar to native auricular cartilage	Good' no values given	[134]
5	–	Compressive modulus at 77 kPa after 3 weeks in culture (initially 100–150 kPa)	Good' no values given	[135]
5	–	Compressive modulus decreased with time due to swelling (40–70 to 20–50 kPa)	76% day 1, increased to 84% day 14	[136]
1	–	–	90% day 14	[137]
20	–	–	85%–95% at 7 days	[138]
15	–	Good shape fidelity, shapes deform and restore on squeezing	85.7% at 7 days	[139]
6	–	–	High' no values given	[118]
–	–	Stress-strain curve similar to articular cartilage	–	[140]
3–4	0.001	Young's modulus 1.76 MPa (<native cartilage)	81%	[141]

14.5.3.2 GelMA

An alternative additive which is additionally used on its own is gelatin methacrylamide (GelMA). GelMA is a chemically modified gelatin which enables it to be photopolymerizable by ultraviolet (UV) exposure [150]. As gelatin is additionally present in the ECM it is an attractive material to use due to its biological potential. Gelatin retains its cell-binding motifs and integral enzymatic degradation sites, which contributes to its use as a biomimetic hydrogel [150]. Culture of printed samples of bone-derived human MSCs in GelMA with alginate have demonstrated an enhanced expression of chondrogenic markers after 3 weeks in culture with chondrogenic media [110]. GelMA with alginate bioink was shown to improve the accuracy of deposition and printing of intricate structures [116]. As GelMA is known to enhance chondrogenic potential, studies have used it as an additive to different bioink mixtures. A study investigated the potential of using GelMA with hyaluronic methacrylate (HAMA) to regenerate cartilage in vivo by directly 3D bioprinting into the cartilage defect. Sheep models were used to create full-thickness chondral defect in the load-bearing region of the femur to investigate the feasibility of 3D surgical bioprinting. A Biopen was used containing two cartridges, one containing the bioink and the other containing allogenic adipose-derived MSCs in which the bioink is printed on the exterior and cells are interior. After 8 weeks, neocartilage formation was shown without any postoperative complications [112]. As cartilage tissue requires high mechanical integrity, to recapitulate that in engineered cartilage tissue requires further synthetic additives such as PEG. A study that combined PEG with GelMA with MSCs to bioprint scaffolds for chondrogenic differentiation exhibited uniform cell distribution within the construct and increased mechanical strength [118]. In addition, the chondrogenic marker expression was highly elevated in comparison to printing with PEG alone. Success in printing homogenous layers of single type cells with a specific bioink has further evolved into creating constructs mimicking anisotropic tissues. Using GelMA with MSCs, it was possible to 3D print anisotropic single layers which are photocured to create a multiphasic construct. Within each layer, nanoliter droplets of GelMA with cells containing either TGF-β1 or BMP-2 were deposited creating a biochemical gradient which has driven chondrogenic and osteogenic differentiation, respectively. The overlapping region where both factors were present resulted in fibrocartilage formation, which is representative of full-depth cartilage [126]. Although GelMA has been shown to support chondrogenesis and provide mechanical stability depending on several cross-linking parameters, its low viscosity at 37°C results in it not being compatible with most printing methods unless it is incorporated with other additives as mentioned previously [125].

14.5.3.3 Nanocellulose

A natural novel nanocellulose bioink, which can be harnessed from both bacterial- and plant-based sources and is available in fibril and crystalline forms with favorable rheological properties, was recently introduced into the world of 3D bioprinting [143]. As mentioned previously, the low viscosity of natural biomaterials results in the compromise of structural integrity and consequently synthetic components have to be incorporated to enhance structural fidelity. However, nanocellulose combined with alginate was shown to retain structure during the printing process through ionic cross-linking of the alginate in the mixture. Altering concentration of the alginate in the nanocellulose bioink tunes the mechanical properties of the construct. This bioink mixture was used to print complex structures such as the ear with human nasal chondrocytes while retaining excellent shape fidelity and high cell viability [139]. Another study demonstrated cell spreading, proliferation, and increased chondrogenic marker expression of bovine articular chondrocytes within a nanocellulose/alginate bioink as long as extrusion pressure and shear stress were minimized [133]. The chondrogenic potential of the bioink has been further investigated in vivo [132]. Nude mice were implanted subcutaneously with a nanocellulose/alginate construct containing either human nasal chondrocytes, bone marrow-derived MSCs, or coculture of both cell types. The constructs retained their structure after 60 days indicating the robustness of the biomaterial. Constructs with human nasal chondrocytes and coculture displayed increased synthesis of ECM including glycosaminoglycans, collagen type II, and cell proliferation, demonstrating the potential use of this bioink for cartilage 3D bioprinting in a clinical setting [132]. In addition to chondrocytes and MSCs, IPSCs and irradiated chondrocytes have been incorporated with nanocellulose/alginate bioink as a potential therapeutic option for treating cartilage lesions. Two bioink combinations were utilized, either nanocellulose/alginate or nanocellulose/hyaluronic acid both with IPS and irradiated chondrocytes. Nanocellulose-/alginate-bioprinted structures have shown pluripotency to be maintained initially; however, 5 weeks postprinting there were indication of neocartilage formation as shown by enhanced collagen type II expression and cellular proliferation. In contrast, a nanocellulose/hyaluronic acid mixture resulted in the cells expressing tumorigenic factors, demonstrating that the nanocellulose/alginate mixture is more suitable for cartilage bioprinting [94]. One of the few disadvantages with nanocellulose is that the human body is not capable of degrading it; however, it is biologically inert and theoretically should not interfere with homeostatic processes such as matrix remodeling [133].

14.5.3.4 Agarose

Agarose is a natural polymer which has been used as a bioink due to its low gelling temperature of 32°C, biocompatibility, and mechanical strength. A comparative study of various bioinks available has identified alginate and agarose to provide the best support for chondrogenic differentiation of MSCs in which the hyaline cartilage phenotype was observed [116]. Furthermore, the printing of complex structures is feasible with agarose due to the unique gelling properties of agarose which does not require a cross-linker. A study printed complex structures using agarose embedded with human MSCs and MG-63 cells, which were viable after 3 weeks in culture and maintained shape fidelity for over 6 months [128].

14.5.3.5 Collagen

Collagens, being a major component of the ECM, can enhance cell adherence, migration, and proliferation and have been used extensively in tissue engineering, especially in relation to cartilage. There are some limitations with collagen bioinks, which are similar to alginate bioinks, mainly poor mechanical strength and failure to retain structure fidelity. High-density collagen bioinks were shown to improve shape retention with the application of heat during the printing process to prevent structural collapse. A study has seeded fibrochondrocytes into high-density collagen bioinks and they have displayed enhanced mechanical stability and cell viability [122]. However, collagen-based bioinks have been known to decrease in volume up to 20%–30% due to cellular contractile remodeling, which is a major issue when printing anatomically accurate constructs [148].

14.6 Summary and future perspectives

3D bioprinting remains a promising approach to regenerating damaged cartilage. With increased accessibility to advancing technology, tissue engineering research is beginning to realize fully the potential of 3D bioprinting to produce accurately complex 3D shapes. The huge variety of cartilage bioprinting techniques, cell types, and bioinks studied represents the complexity of this field. Despite cartilage being one of the simpler tissues to replicate there is not yet one optimum process.

In vitro studies demonstrate that printed cells are viable for at least 6 weeks in culture and can be stimulated to produce ECM. Studies of implanted constructs have also demonstrated neocartilage formation with compressive and tensile moduli approaching native levels. However, most studies have used small-scale constructs, short-time frames with few large animal studies giving little indication of the effects of long-term in vivo implantation of more complex constructs in humans.

Many challenges still exist before this technology can be utilized in a clinical setting. Namely demonstrating cell survival, safe proliferation, and differentiation under long-term physiological conditions, where complex constructs are subject to loading forces experienced by native cartilage. More research is also needed to elucidate scaffold degradation times and whether they allow adequate amounts of neocartilage to be formed in vivo. Ultimately, while there is positive evidence that 3D bioprinting may provide a useful platform technology for tissue assembly, further research and optimization to obtain reproducible results is required before this technology can be translated into clinical practice. Progress in this field will most likely be made when cartilage development and morphology are better understood and in order to be replicated.

References

[1] Heinegård D, Saxne T. The role of the cartilage matrix in osteoarthritis. Nat Rev Rheumatol 2011;7(1):50–6.
[2] Chen A, et al. The global economic cost of osteoarthritis: how the UK compares. Arthritis 2012;2012:. 698709. PMC Web 29 July 2017.
[3] Lanza R, Langer R, Vacanti JP. Principles of tissue engineering. Elsevier; 2013.

[4] Fodor WL. Tissue engineering and cell based therapies, from the bench to the clinic: the potential to replace, repair and regenerate. Reprod Biol Endocrinol 2003;1:102.
[5] Meijer GJ, de Bruijn JD, Koole R, van Blitterswijk CA. Cell-based bone tissue engineering. PLoS Med 2007;4:e9.
[6] Nerem RM. Tissue engineering: the hope, the hype, and the future. Tissue Eng 2006;12:1143–50.
[7] Ciorba A, Martini A. Tissue engineering and cartilage regeneration for auricular reconstruction. Int J Pediatr Otorhinolaryngol 2006;70:1507–15.
[8] Andretto AC. The central role of the nose in the face and the psyche: review of the nose and the psyche. Aesthetic Plast Surg 2007;31(4):406–10.
[9] Lovice DB, Mingrone MD, Toriumi DM. Grafts and implants in rhinoplasty and nasal reconstruction. Otolaryngol Clin North Am 1999;32(1):113–41.
[10] Matarasso A, Elias AC, Elias RL. Labial incompetence: a marker for progressive bone resorption in silastic chin augmentation. Plast Reconstr Surg 1996;98:1007–14. discussion 1015.
[11] NaBadalung DP. Prosthetic rehabilitation of a total rhinectomy patient resulting from squamous cell carcinoma of the nasal septum: a clinical report. J Prosthet Dent 2003;89:234–8.
[12] Zeng Y, Wu W, Yu H, Yang J, Chen G. Silicone implants in augmentation rhinoplasty. Aesthetic Plast Surg 2002;26:85–8.
[13] Donald PJ. Cartilage grafting in facial reconstruction with special consideration of irradiated grafts. Laryngoscope 1986;96:786–807. PubMed:3523085.
[14] Welling D, Maves MD, Schuller DE, Bardach J. Irradiated homologous cartilage grafts: long-term results. Arch Otolaryngol Head Neck Surg 1988;114(3):291–5.
[15] Cochran CS, DeFatta RJ. Tragal cartilage grafts in rhinoplasty: a viable alternative in the graft-depleted patient. Otolaryngol Head Neck Surg 2008;138(2):166–9.
[16] Sancho BV, Molina AR. Use of septal cartilage homografts in rhinoplasty. Aesthetic Plast Surg 2000;24(5):357–63.
[17] Uppal RS, Sabbagh W, Chana J, Gault DT. Donor-site morbidity after autologous costal cartilage harvest in ear reconstruction and approaches to reducing donor-site contour deformity. Plast Reconstr Surg 2008;121(6):1949–55.
[18] Osorno G. A 20-year experience with the Brent technique of auricular reconstruction: pearls and pitfalls. Plast Reconstr Surg 2007;119(5):1447–63.
[19] Skouteris CA, Sotereanos GC. Donor site morbidity following harvesting of autogenous rib grafts. J Oral Maxillofac Surg 1989;47:808–12.
[20] Wallace CG, Mao HY, Wang CJ, et al. Three-dimensional computed tomography reveals different donor-site deformities in adult and growing microtia patients despite total subperichondrial costal cartilage harvest and donor-site reconstruction. Plast Reconstr Surg 2014;133(3):640.
[21] Brittberg M, et al. Treatment of deep cartilage defects in the knee with autologous chondrocyte transplantation. N Engl J Med 1994;331:889–95.
[22] Jessop ZM, Al-Himdani S, Clement M, Whitaker IS. The challenge for reconstructive surgeons in the twenty-first century: manufacturing tissue-engineered solutions. Front Surg 2015;2:52. https://doi.org/10.3389/fsurg.2015.00052.
[23] House of Lords Science and Technology Committee. 28 Jun 2013.
[24] O'Dowd A. Peers call for UK to harness "enormous" potential of regenerative medicine. BMJ 2013;347:f4248.
[25] Bichara DA, O'Sullivan NA, Pomerantseva I, et al. The tissue-engineered auricle: past, present, and future. Tissue Eng Part B Rev 2012;18:51–61.

[26] Yanaga H, Imai K, Koga M, Yanaga K. Cell-engineered human elastic chondrocytes regenerate natural scaffold in vitro and neocartilage with neoperichondrium in the human body post-transplantation. Tissue Eng Part A 2012;18(19–20):2020–9.

[27] Britt C, Park SS. Autogenous tissue-engineered cartilage: evaluation as an implant material. Arch Otolaryngol Head Neck Surg 1998;124:671–7.

[28] Brommer H, Brama PAJ, Laasanen MS, et al. Functional adaptation of articular cartilage from birth to maturity under the influence of loading: a biomechanical analysis. Equine Vet J 2005;37(2):148–54.

[29] Cao Y, Vacanti JP, Paige KT, et al. Transplantation of chondrocytes utilizing a polymer-cell construct to produce tissue-engineered cartilage in the shape of a human ear. Plast Reconstr Surg 1997;100:297–302.

[30] Kusuhara H, Isogai N, Enjo M, et al. Tissue engineering a model for the human ear: assessment of size, shape, morphology, and gene expression following seeding of different chondrocytes. Wound Repair Regen 2009;17(1):136–46.

[31] Homicz MR, Schumacher BL, Sah RL, Watson D. Effects of serial expansion of septal chondrocytes on tissue-engineered neocartilage composition. Otolaryngol Head Neck Surg 2002;127:398–408.

[32] Nabzdyk C, Pradhan L, Molina J, Perin E, Paniagua D, Rosenstrauch D. Auricular chondrocytes—from benchwork to clinical applications. In Vivo 2009;23(3):369–80.

[33] Kamil SH, Kojima K, Vacanti MP, Bonassar LJ, Vacanti CA, Eavey RD. In vitro tissue engineering to generate a human-sized auricle and nasal tip. Laryngoscope 2003;113(1):90–4.

[34] Kamil SH, Vacanti MP, Aminuddin BS, et al. Tissue engineering of a human sized and shaped auricle using a mold. Laryngoscope 2004;114:867.

[35] Christophel JJ, Chang JS, Park SS. Transplanted tissue-engineered cartilage. Arch Facial Plast Surg 2006;8(2):117–22.

[36] Puelacher WC, Vacanti JP, Kim SW, Upton J, Vacanti CA. Fabrication of nasal implants using human shape-specific polymer scaffolds seeded with chondrocytes. Surg Forum 1993;44:678–80.

[37] Peulacher WC, Mooney D, Langer R, Upton J, Vacanti JP, Vacanti CA. Design of nasoseptal cartilage replacements synthesized from biodegradable polymers and chondrocytes. Biomaterials 1994;15:774–8.

[38] Nayyer L, Patel KH, Esmaelli A. Tissue engineering: revolution and challenge in auricular cartilage reconstruction. Plast Reconstr Surg 2012;129(5):1123–37.

[39] Abbott RD, Kaplan DL. Strategies for improving the physiological relevance of human engineered tissues. Trends Biotechnol 2015;33:401–7.

[40] Delaere P, Vranckx J, Verleden G, et al. Tracheal transplant group. Tracheal allotransplantation after withdrawal of immunosuppressive therapy. N Engl J Med 2010;362:138–45.

[41] Gonfiotti A, Jaus MO, Barale S, et al. The first tissue-engineered airway transplantation: 5-year follow-up results. Lancet 2014;383:238–44.

[42] Vogel G. Trachea transplants test the limits. Science 2013;340:266–8.

[43] Ingber DE. Mechanical control of tissue growth: function follows form. Proc Natl Acad Sci U S A 2005;102:11571–2.

[44] Guillotin B, Guillemot F. Cell patterning technologies for organotypic tissue fabrication. Trends Biotechnol 2011;29(4):183–90.

[45] Olsen BR, Reginato AM, Wang W. Bone development. Annu Rev Cell Dev Biol 2000;16:191.

[46] Archer CW, Dowthwaite GP, Francis-West P. Development of synovial joints. Birth Defects Res C Embryo Today 2003;69(2):144–55.

[47] Hellingman CA, Koevoet W, Kops N, et al. Fibroblast growth factor receptors in in vitro and in vivo chondrogenesis: relating tissue engineering using adult mesenchymal stem cells to embryonic development. Tissue Eng Part A 2010;16:545–56.
[48] Cleary MA, van Osch GJ, Brama PA, Hellingman CA, Narcisi R. FGF, TGFβ and Wnt crosstalk: embryonic to in vitro cartilage development from mesenchymal stem cells. J Tissue Eng Regen Med 2015;9(4):332–42.
[49] Tuli R, Tuli S, Nandi S, et al. Transforming growth factor-b-mediated chondrogenesis of human mesenchymal progenitor cells involves N-cadherin and mitogen-activated protein kinase and Wnt signaling crosstalk. J Biol Chem 2003;278:41227–36.
[50] Chia HN, Wu BM. Recent advances in 3D printing of biomaterials. J Biol Eng 2015;9:4. https://doi.org/10.1186/s13036-015-0001-4. eCollection 2015.
[51] Cancedda R, Dozin B, Giannoni P, Quarto R. Tissue engineering and cell therapy of cartilage and bone. Matrix Biol 2003;22(1):81–91.
[52] Detterline AJ, Goldberg S, Bach Jr BR, Cole BJ. Treatment options for articular cartilage defects of the knee. Orthop Nurs 2005;24(5):361–6.
[53] Vinatier C, Bouffi C, Merceron C, Gordeladze J, Brondello JM, Jorgensen C, et al. Cartilage tissue engineering: towards a biomaterial-assisted mesenchymal stem cell therapy. Curr Stem Cell Res Ther 2009;4(4):318–29.
[54] Oseni AO, Butler PE, Seifalian AM. Optimization of chondrocyte isolation and characterization for large-scale cartilage tissue engineering. J Surg Res 2013;181(1):41–8.
[55] Eyre D. Collagen of articular cartilage. Arthritis Res 2002;4(1):30–5. Epub 2001 Oct.
[56] Arokoski J, Kiviranta I, Jurvelin J, Tammi M, Helminen HJ. Long-distance running causes site- dependent decrease of cartilage glycosaminoglycan content in the knee joints of beagle dogs. Arthritis Rheum 1993;36:1451–9. PubMed:7692860.
[57] Jeffery AK, Blunn GW, Archer CW, Bentley G. Three-dimensional collagen architecture in bovine articular cartilage. J Bone Joint Surg Br 1991;73(5):795–801.
[58] Muir H, Bullough P, Maroudas A. The distribution of collagen in human articular cartilage with some of its physiological implications. J Bone Joint Surg 1970;52(B):554–63.
[59] Jelicks LA, Paul PK, O'Byrne E, Gupta RK. Hydrogen-1, Sodium-23, and Carbon-13 MR spectroscopy of cartilage degradation in vitro. J Magn Reson Imaging 1993;3:565–8. PubMed:8347947.
[60] Reiter DA, Lin PC, Fishbein KW, Spencer RG. Multicomponent T(2) relaxation analysis in cartilage. Magn Reson Med 2009;61:803–9. PubMed:19189393.
[61] Zheng S, Xia Y. Effect of phosphate electrolyte buffer on the dynamics of water in tendon and cartilage. NMR Biomed 2009;22:158–64. PubMed:18720450.
[62] Zheng S, Xia Y. Multi-components of T2 relaxation in ex vivo cartilage and tendon. J Magn Reson 2009;198:188–96. PubMed:19269868.
[63] Zheng S, Xia Y. On the measurement of multi-component T(2) relaxation in cartilage by MR spectroscopy and imaging. Magn Reson Imaging 2010;28:537–45. PubMed:20061115.
[64] Bisson L, Brahmabhatt V, Marzo J. Split-line orientation of the talar dome articular cartilage. Arthroscopy 2005;21:570–3. PubMed: 15891723.
[65] Grellmann W, Berghaus A, Haberland EJ, Jamali Y, Holweg K, Reincke K, et al. Determination of strength and deformation behavior of human cartilage for the definition of significant parameters. J Biomed Mater Res A 2006;78:168–74. PubMed:16628548.
[66] Maas CS, Monhian N, Shah SB. Implants in rhinoplasty. Facial Plast Surg 1997;13(4):279–90. PubMed:9656882.
[67] Richmon JD, Sage A, Van Wong W, Chen AC, Sah RL, Watson D. Compressive biomechanical properties of human nasal septal cartilage. Am J Rhinol 2006;20:496–501. PubMed: 1706374.

[68] Xia Y, Zheng S, Szarko M, Lee J. Anisotropic properties of bovine nasal cartilage. Microsc Res Tech 2012;75(3):300–6.
[69] Rotter N, Tobias G, Lebl M, Roy AK, Hansen MC, Vacanti CA, et al. Age-related changes in the composition and mechanical properties of human nasal cartilage. Arch Biochem Biophys 2002;403:132–40.
[70] Jessop ZM, Javed M, Otto IA, et al. Combining regenerative medicine strategies to provide durable reconstructive options: auricular cartilage tissue engineering. Stem Cell Res Ther 2016;7:19. https://doi.org/10.1186/s13287-015-0273-0.
[71] Djouad F, Bony C, Häupl T, Uzé G, Lahlou N, Louis-Plence P, et al. Transcriptional profiles discriminate bone marrow-derived and synovium-derived mesenchymal stem cells. Arthritis Res Ther 2005;7:R1304–15.
[72] Segawa Y, Muneta T, Makino H, Nimura A, Mochizuki T, Ju YJ, et al. Mesenchymal stem cells derived from synovium, meniscus, anterior cruciate ligament, and articular chondrocytes share similar gene expression profiles. J Orthop Res 2009;27:435–41.
[73] Diekman BO, Rowland CR, Lennon DP, Caplan AI, Guilak F. Chondrogenesis of adult stem cells from adipose tissue and bone marrow: induction by growth factors and cartilage-derived matrix. Tissue Eng Part A 2010;16(2):523–33.
[74] Diekman BO, Christoforou N, Willard VP, Sun H, Sanchez-Adams J, Leong KW, et al. Cartilage tissue engineering using differentiated and purified induced pluripotent stem cells. Proc Natl Acad Sci U S A 2012;109(47):19172–7.
[75] Sakaguchi Y, Sekiya I, Yagishita K, Muneta T. Comparison of human stem cells derived from various mesenchymal tissues: superiority of synovium as a cell source. Arthritis Rheum 2005;52:2521–9.
[76] Jones E, Churchman SM, English A, Buch MH, Horner EA, Burgoyne CH, et al. Mesenchymal stem cells in rheumatoid synovium: enumeration and functional assessment in relation to synovial inflammation level. Ann Rheum Dis 2010;69:450–7.
[77] Murphy JM, Dixon K, Beck S, Fabian D, Feldman A, Barry F. Reduced chondrogenic and adipogenic activity of mesenchymal stem cells from patients with advanced osteoarthritis. Arthritis Rheum 2002;46:704–13.
[78] Uto S, Nishizawa S, Takasawa Y, Asawa Y, Fujihara Y, Takato T, et al. Bone and cartilage repair by transplantation of induced pluripotent stem cells in murine joint defect model. Biomed Res 2013;34(6):281–8.
[79] Terada S, Fuchs JR, Yoshimoto H, Fauza DO, Vacanti JP. In vitro cartilage regeneration from proliferated adult elastic chondrocytes. Ann Plast Surg 2005;55(2):196–201.
[80] Yanaga H, Yanaga K, Imai K, Koga M, Soejima C, Ohmori K. Clinical application of cultured autologous human auricular chondrocytes with autologous serum for craniofacial or nasal augmentation and repair. Plast Reconstr Surg 2006;117(6):2019–30.
[81] Benya PD, Shaffer JD. Dedifferentiated chondrocytes reexpress the differentiated collagen phenotype when cultured in agarose gels. Cell 1982;30:215e24.
[82] Bonaventure J, Kadhom N, Cohen-Solal L, Ng KH, Bourguignon J, Lasselin C, et al. Reexpression of cartilage-specific genes by dedifferentiated human articular chondrocytes cultured in alginate beads. Exp Cell Res 1994;212:97e104.
[83] Dell'Accio F, De Bari C, Luyten FP. Molecular markers predictive of the capacity of expanded human articular chondrocytes to form stable cartilage in vivo. Arthritis Rheum 2001;44(7):1608–19.
[84] Dominici M, Le Blanc K, Mueller I, Slaper-Cortenbach I, Marini F, Krause D, et al. Minimal criteria for defining multipotent mesenchymal stromal cells. The International Society for Cellular Therapy position statement. Cytotherapy 2006;8(4):315–7.
[85] Bianco P, Robey PG. Stem cells in tissue engineering. Nature 2001;414:118–21.

[86] Afizah H, Yang Z, Hui JH, Ouyang HW, Lee EH. A comparison between the chondrogenic potential of human bone marrow stem cells (BMSCs) and adipose-derived stem cells (ADSCs) taken from the same donors. Tissue Eng 2007;13:659–66.
[87] Cui JH, Park SR, Park K, Choi BH, Min BH. Preconditioning of mesenchymal stem cells with low-intensity ultrasound for cartilage formation in vivo. Tissue Eng 2007;13(2):351–60.
[88] Dickhut A, et al. Calcification or dedifferentiation: requirement to lock mesenchymal stem cells in a desired differentiation stage. J Cell Physiol 2009;219:219–26.
[89] Gelse K, Ekici AB, Cipa F, Swoboda B, Carl HD, Olk A, et al. Molecular differentiation between osteophytic and articular cartilage- clues for a transient and permanent chondrocyte phenotype. Osteoarthritis Cartilage 2012;20(2):162–71.
[90] Koga H, et al. Comparison of mesenchymal tissues-derived stem cells for in vivo chondrogenesis: suitable conditions for cell therapy of cartilage defects in rabbit. Cell Tissue Res 2008;333:207–15.
[91] Shieh SJ, Terada S, Vacanti JP. Tissue engineering auricular reconstruction: in vitro and in vivo studies. Biomaterials 2004;25:1545–57.
[92] Tallheden T, et al. Phenotypic plasticity of human articular chondrocytes. J Bone Joint Surg Am 2003;85-A(Suppl 2):93–100.
[93] de Windt TS, et al. Allogeneic mesenchymal stem cells stimulate cartilage regeneration and are safe for single-stage cartilage repair in humans upon mixture with recycled autologous chondrons. Stem Cells 2017;35:256–64.
[94] Nguyen D, Hägg DA, Forsman A, Ekholm J, Nimkingratana P, Brantsing C, et al. Cartilage tissue engineering by the 3D bioprinting of iPS cells in a nanocellulose/alginate bioink. Sci Rep 2017;7(1):658.
[95] Dowthwaite GP, Bishop JC, Redman SN, et al. The surface of articular cartilage contains a progenitor cell population. J Cell Sci 2004;117:889–97.
[96] Kobayashi S, Takebe T, Zheng Y-W, Mizuno M, Yabuki Y, Maegawa J, et al. Presence of cartilage stem/progenitor cells in adult mice auricular perichondrium. PLoS One 2011;6(10):e26393.
[97] Derks M, Sturm T, Haverich A, Hilfiker A. Isolation and chondrogenic differentiation of porcine perichondrial progenitor cells for the purpose of cartilage tissue engineering. Cells Tissues Organs 2013;198:179–89.
[98] Fickert S, Fiedler J, Brenner RE. Identification of subpopulations with characteristics of mesenchymal progenitor cells from human osteoarthritic cartilage using triple staining for cell surface markers. Arthritis Res Ther 2004;6:422–32.
[99] Alsalameh S, Amin R, Gemba T, Lotz M. Identification of mesenchymal progenitor cells in normal and osteoarthritic human articular cartilage. Arthritis Rheum 2004;50:1522–32.
[100] Koelling S, Kruegel J, Imer M, Path JR, Sadowski B, Miro X, et al. Migratory chondrogenic progenitor cells from repair tissue during the later stages of human osteoarthritis. Cell Stem Cell 2009;4(4):324–35.
[101] Williams R, Khan IM, Richardson K, et al. Identification and clonal characterisation of a progenitor cell sub- population in normal human articular cartilage. PLoS One 2010;5(10). e13246.
[102] Grogan SP, Miyaki S, Asahara H, D'Lima DD, Lotz MK. Mesenchymal progenitor cell markers in human articular cartilage: normal distribution and changes in osteoarthritis. Arthritis Res Ther 2009;11(3):R85.
[103] Seol D, McCabe DJ, Choe H, Zheng H, Yu Y, Jang K, et al. Chondrogenic progenitor cells respond to cartilage injury. Arthritis Rheum 2012;64(11):3626–37.

[104] Meyer U. The history of tissue engineering and regenerative medicine in perspective. In: Fundamentals of tissue engineering and regenerative medicine. Berlin, Heidelberg: Springer; 2009. p. 5–12.
[105] Langer R, Vacanti JP. Tissue engineering. Science 1993;260:920–6.
[106] Mironov V, Boland T, Trusk T, Forgacs G, Markwald RR. Organ printing:computer-aided jet-based 3d tissue engineering. Trends Biotechnol 2003;21:157e61.
[107] Mironov V. The Second International Workshop on Bioprinting, Biopatterning and Bioassembly. Expert Opin Biol Ther 2005;5(8):1111–5.
[108] Jessop ZM, Al-Sabah A, Gardiner M, Combellack E, Hawkins K, Whitaker IS. 3D bioprinting for reconstructive surgery: principles, applications and challenges. J Plast Reconstr Aesthet Surg 2017;70(9):1155–70.
[109] Mironov V. Toward human organ printing: Charleston bioprinting symposium. ASAIO J 2006;52(6):e27–30.
[110] Costantini M, et al. 3D bioprinting of BM-MSCs-loaded ECM biomimetic hydrogels for in vitro neocartilage formation. Biofabrication 2016;8(3). 035002.
[111] Zhang Y, Yu Y, Chen H, Ozbolat IT. Characterization of printable cellular micro-fluidic channels for tissue engineering. Biofabrication 2013;5(2):025004.
[112] Di Bella C, et al. In-situ handheld 3D bioprinting for cartilage regeneration. J Tissue Eng Regen Med 2017;https://doi.org/10.1002/term.2476.
[113] Armstrong JPK, et al. 3D bioprinting using a templated porous bioink. Adv Healthc Mater 2016;5(14):1724–30.
[114] Chen CH, Lee MY, Shyu VB, Chen YC, Chen CT, Chen JP. Surface modification of polycaprolactone scaffolds fabricated via selective laser sintering for cartilage tissue engineering. Mater Sci Eng C Mater Biol Appl 2014;40:389–97.
[115] Cui X, Breitenkamp K, Finn MG, Lotz M, D'Lima DD. Direct human cartilage repair using three-dimensional bioprinting technology. Tissue Eng Part A 2012;18(11–12):1304–12.
[116] Daly AC, et al. A comparison of different bioinks for 3D bioprinting of fibrocartilage and hyaline cartilage. Biofabrication 2016;8(4). 045002.
[117] Fedorovich NE, Schuurman W, Wijnberg HM, Prins HJ, van Weeren PR, Malda J, et al. Biofabrication of osteochondral tissue equivalents by printing topologically defined, cell-laden hydrogel scaffolds. Tissue Eng Part C Methods 2012;18(1):33–44.
[118] Gao G, et al. Improved properties of bone and cartilage tissue from 3D inkjet-bioprinted human mesenchymal stem cells by simultaneous deposition and photocrosslinking in PEG-GelMA. Biotechnol Lett 2015;37(11):2349–55.
[119] Izadifar Z, Chang T, Kulyk W, Chen X, Eames BF. Analyzing biological performance of 3D-printed, cell-impregnated hybrid constructs for cartilage tissue engineering. Tissue Eng Part C Methods 2016;22(3):173–88.
[120] Kundu J, et al. An additive manufacturing-based PCL–alginate–chondrocyte bioprinted scaffold for cartilage tissue engineering. J Tissue Eng Regen Med 2015;9(11):1286–97.
[121] Park JY, Choi JC, Shim JH, Lee JS, Park H, Kim SW, et al. A comparative study on collagen type I and hyaluronic acid dependent cell behavior for osteochondral tissue bioprinting. Biofabrication 2014;6(3):035004.
[122] Rhee S, et al. 3D bioprinting of spatially heterogeneous collagen constructs for cartilage tissue engineering. ACS Biomater Sci Eng 2016;2(10):1800–5.
[123] Shim JH, Jang KM, Hahn SK, Park JY, Jung H, Oh K, et al. Three-dimensional bioprinting of multilayered constructs containing human mesenchymal stromal cells for osteochondral tissue regeneration in the rabbit knee joint. Biofabrication 2016;8(1):014102.

[124] Yi H, Park M, Kang K, Hong J, Jang J, Cho D. Pretreatment of electromagnetic field to 3D printed chondrocytes for the enhancement of cartilage regeneration. Tissue Eng Part A 2014;20:S116–7.

[125] Schuurman W, et al. Gelatin-methacrylamide hydrogels as potential biomaterials for fabrication of tissue-engineered cartilage constructs. Macromol Biosci 2013;13(5):551–61.

[126] Gurkan UA, et al. Engineering anisotropic biomimetic fibrocartilage microenvironment by bioprinting mesenchymal stem cells in nanoliter gel droplets. Mol Pharm 2014;11(7):2151–9.

[127] Yu Y, Moncal KK, Li J, Peng W, Rivero I, Martin JA, et al. Three-dimensional bioprinting using self-assembling scalable scaffold-free "tissue strands" as a new bioink. Sci Rep 2016;6:28714.

[128] Duarte Campos DF, et al. Three-dimensional printing of stem cell-laden hydrogels submerged in a hydrophobic high-density fluid. Biofabrication 2012;5(1). 015003.

[129] Lee JS, et al. 3D printing of composite tissue with complex shape applied to ear regeneration. Biofabrication 2014;6(2). 024103.

[130] Lee JS, Kim BS, Seo D, Park JH, Cho DW. Three-dimensional cell printing of large-volume tissues: application to ear regeneration. Tissue Eng Part C Methods 2017;23(3):136–45.

[131] Ávila HM, Schwarz S, Rotter N, Gatenholm P. 3D bioprinting of human chondrocyte-laden nanocellulose hydrogels for patient-specific auricular cartilage regeneration. Bioprinting 2016;1:22–35.

[132] Möller T, et al. In vivo chondrogenesis in 3D bioprinted human cell-laden hydrogel constructs. Plast Reconstr Surg Glob Open 2017;5(2).

[133] Müller M, Öztürk E, Arlov Ø, Gatenholm P, Zenobi-Wong M. Alginate sulfate-nanocellulose bioinks for cartilage bioprinting applications. Ann Biomed Eng 2017;45(1):210–23.

[134] Park JY, Choi YJ, Shim JH, Park JH, Cho DW. Development of a 3D cell printed structure as an alternative to autologs cartilage for auricular reconstruction. J Biomed Mater Res B Appl Biomater 2017;105(5):1016–28.

[135] Schütz K, Placht AM, Paul B, Brüggemeier S, Gelinsky M, Lode A. Three-dimensional plotting of a cell-laden alginate/methylcellulose blend: towards biofabrication of tissue engineering constructs with clinically relevant dimensions. J Tissue Eng Regen Med 2017;11(5):1574–87.

[136] You F, Wu X, Zhu N, Lei M, Eames BF, Chen X. 3D printing of porous cell-laden hydrogel constructs for potential applications in cartilage tissue engineering. ACS Biomater Sci Eng 2016;2(7):1200–10.

[137] Melchels FP, Blokzijl MM, Levato R, Peiffer QC, Ruijter MD, Hennink WE, et al. Hydrogel-based reinforcement of 3D bioprinted constructs. Biofabrication 2016;8(3):035004.

[138] Abbadessa A, Mouser VH, Blokzijl MM, Gawlitta D, Dhert WJ, Hennink WE, et al. A synthetic thermosensitive hydrogel for cartilage bioprinting and its biofunctionalization with polysaccharides. Biomacromolecules 2016;17(6):2137–47.

[139] Markstedt K, et al. 3D bioprinting human chondrocytes with nanocellulose–alginate bioink for cartilage tissue engineering applications. Biomacromolecules 2015;16(5):1489–96.

[140] Visser J, Melchels FP, Jeon JE, van Bussel EM, Kimpton LS, Byrne HM, et al. Reinforcement of hydrogels using three-dimensionally printed microfibres. Nat Commun 2015;6:6933.

[141] Xu T, Binder KW, Albanna MZ, Dice D, Zhao W, Yoo JJ, et al. Hybrid printing of mechanically and biologically improved constructs for cartilage tissue engineering applications. Biofabrication 2013;5(1):015001.
[142] Hutmacher DW. Scaffolds in tissue engineering bone and cartilage. Biomaterials 2000;21(24):2529–43.
[143] Kyle S, Jessop ZM, Al-Sabah A, Whitaker IS. 'Printability' of candidate biomaterials for extrusion based 3D printing: state-of-the-art. Adv Healthc Mater 2017;6(16).
[144] Kuo CK, Ma PX. Ionically crosslinked alginate hydrogels as scaffolds for tissue engineering: part 1. Structure, gelation rate and mechanical properties. Biomaterials 2001;22(6):511–21.
[145] Kim BS, et al. Three-dimensional bioprinting of cell-laden constructs with polycaprolactone protective layers for using various thermoplastic polymers. Biofabrication 2016;8(3). 035013.
[146] Izadifar Z, Chang T, Kulyk W, Chen X, Eames BF. Analyzing biological performance of 3D-printed, cell-impregnated hybrid constructs for cartilage tissue engineering. Tissue Eng Part C Methods 2016;22(3):173–88.
[147] Kundu J, Shim JH, Jang J, Kim SW, Cho DW. An additive manufacturing-based PCL-alginate-chondrocyte bioprinted scaffold for cartilage tissue engineering. J Tissue Eng Regen Med 2015;9(11):1286–97.
[148] Visscher DO, et al. Cartilage tissue engineering: preventing tissue scaffold contraction using a 3D-printed polymeric cage. Tissue Eng Part C Methods 2016;22(6):573–84.
[149] Do AV, et al. 3D printing of scaffolds for tissue regeneration applications. Adv Healthc Mater 2015;4(12):1742–62. PMC. Web. 29 July 2017.
[150] Nichol JW, et al. Cell-laden microengineered gelatin methacrylate hydrogels. Biomaterials 2010;31(21):5536–44.

Further reading

[1] Ahrens MJ, Dudley AT. Chemical pretreatment of growth plate cartilage increases immunofluorescence sensitivity. J Histochem Cytochem 2011;59(4):408–18.
[2] Aruffo A, Stamenkovic I, Melnick M, Underhill CB, Seed B. CD44 is the principal cell surface receptor for hyaluronate. Cell 1990;61:1303–13.
[3] Azandeh S, Orazizadeh M, Hashemitabar M, Khodadadi A, Shayesteh A, Nejad D, et al. Mixed enzymatic-explant protocol for isolation of mesenchymal stem cells from Wharton's jelly and encapsulation in 3D culture system. J Biomed Sci Eng 2012;5:580–6.
[4] Barker N. Adult intestinal stem cells: critical drivers of epithelial homeostasis and regeneration. Nat Rev Mol Cell Biol 2014;15:19–33.
[5] Berry L, Grant ME, McClure J, Rooney P. Bone-marrow-derived chondrogenesis in vitro. J Cell Sci 1992;101:333–42.
[6] Brack AS, Rando TA. Tissue-specific stem cells: lessons from the skeletal muscle satellite cell. Cell Stem Cell 2012;10:504–14.
[7] https://www.changingfaces.org.uk/Health-Care-Professionals/Introduction-to-patient- needs/Statistics.
[8] Chow G, Knudson CB, Homandberg G, Knudson W. Increased expression of CD44 in bovine articular chondrocytes by catabolic cellular mediators. J Biol Chem 1995;270:27734–41.
[9] Derby B. Printing and prototyping of tissues and scaffolds. Science 2012;338:921–6.

[10] Eyre DR, Dickson IR, Van Ness KP. Collagen cross-linking in human bone and articular cartilage. Age-related changes in the content of mature hydroxypyridinium residues. Biochem J 1988;252:495–500.

[11] Golas AR, Hernandez KA, Spector JA. Tissue engineering for plastic surgeons: a primer. Aesthetic Plast Surg 2014;38:207–21. https://doi.org/10.1007/s00266-013-0255-5.

[12] Green E, Ellis R, Winlove P. The molecular structure and physical properties of elastin fibers as revealed by Raman microspectroscopy. Biopolymers 2008;89(11):931–40.

[13] Ikada Y. Challenges in tissue engineering. J R Soc Interface 2006;3:589–601.

[14] Jakab K, Damon B, Neagu A, Kachurin A, Forgacs G. Three-dimensional tissue constructs built by bioprinting. Biorheology 2006;43(3-4):509–13.

[15] Jako M, Démarteau O, Schäfer D, Hintermann B, Dick W, Heberer M, et al. Specific growth factors during the expansion and redifferentiation of adult human articular chondrocytes enhance chondrogenesis and cartilaginous tissue formation in vitro. J Cell Biochem 2001;81:368–77.

[16] Jang KW, Ding L, Seol D, Lim TH, Buckwalter JA, Martin JA. Low-intensity pulsed ultrasound promotes chondrogenic progenitor cell migration via focal adhesion kinase pathway. Ultrasound Med Biol 2014;40(6):1177–86.

[17] Jones PH, Watt F. Separation of epidermal stem cells from transit amplifying cells on the basis of differences in integrin function and expression. Cell 1993;73:713–24.

[18] Kafienah W, Mistry S, Williams C, Hollander AP. Nucleostemin is a marker of proliferating stromal stem cells in adult human bone marrow. Stem Cells 2006;24(4):1113–20.

[19] Kafienah W, Mistry S, Dickinson S, Sims TJ, Learmonth I, Hollander AP. Three-dimensional cartilage tissue engineering using adult stem cells from osteoarthritis patients. Arthritis Rheum 2007;56(1):177–87.

[20] Knudson W, Loeser RF. CD44 and integrin matrix receptors participate in cartilage homeostasis. Cell Mol Life Sci 2002;59:36–44.

[21] Lee JW, Kim YH, Park KD, Jee KS, Shin JW, Hahn SB. Importance of integrin beta1-mediated cell adhesion on biodegradable polymers under serum depletion in mesenchymal stem cells and chondrocytes. Biomaterials 2004;25(10):1901–9.

[22] Mathur D, Pereira WC, Anand A. Emergence of chondrogenic progenitor stem cells in transplantation biology—prospects and drawbacks. J Cell Biochem 2012;113:397–403. https://doi.org/10.1002/jcb.23367.

[23] Morrison SJ, Shah NM, Anderson DJ. Regulatory mechanisms in stem cell biology. Cell 1997;88:287–98.

[24] Naderi H, Matin MM, Bahrami AR. Review paper: critical issues in tissue engineering: biomaterials, cell sources, angiogenesis, and drug delivery systems. J Biomater Appl 2011;26:383–417.

[25] Nodzo SR, Hohman DW, Chakravarthy K. Nanotechnology: why should we care? Am J Orthop 2015;44:E87–8.

[26] Nakamura H, Kato R, Hirata A, Inoue M, Yamamoto T. Localization of CD44 (hyaluronan receptor) and hyaluronan in rat mandibular condyle. J Histochem Cytochem 2005;53:113–20.

[27] Ogata Y, Mabuchi Y, Yoshida M, Suto EG, Suzuki N, Muneta T, et al. Purified human synovium mesenchymal stem cells as a good resource for cartilage regeneration. PLoS One 2015;10(6):e0129096.

[28] Osyczka AM, Noth U, O'Connor J, et al. Multilineage differentiation of adult human bone marrow progenitor cells transduced with human papilloma virus type 16 E6/E7 genes. Calcif Tissue Int 2002;71:447–58.

[29] Pietila K, Kantomaa T, Pirttiniemi P, Poikela A. Comparison of amounts and properties of collagen and proteoglycans in condylar, costal and nasal cartilages. Cells Tissues Organs 1999;164:30–6. PubMed:10940671.

[30] Pittenger MF, Mackay AM, Beck SC, et al. Multilineage potential of adult human mesenchymal stem cells. Science 1999;284:143–7.

[31] Rees A, Powell LC, Chinga-Carrasco G, et al. 3D bioprinting of carboxymethylated-periodate oxidized nanocellulose constructs for wound dressing applications. Biomed Res Int 2015;2015:925757. 7 p..

[32] Takebe T, Kobayashi S, Suzuki H, Mizuno M, Chang YM, Yoshizawa E, et al. Transient vascularization of transplanted human adult-derived progenitors promotes self-organized cartilage. J Clin Invest 2014;124(10):4325–34.

[33] Zhang Y, Li J, Davis ME, Pei M. Delineation of in vitro chondrogenesis of human synovial stem cells following preconditioning using decellularized matrix. Acta Biomater 2015;20:39–50.

[34] Zuk PA, Zhu M, Mizuno H, Huang J, Futrell JW, et al. Multilineage cells from human adipose tissue: implications for cell-based therapies. Tissue Eng 2001;7:211–28.

3D bioprinting adipose tissue for breast reconstruction

15

M.P. Chae*[,†], D.J. Hunter-Smith*[,†], S.V. Murphy[‡], M.W. Findlay[†,§,¶]
*Monash University, Clayton, VIC, Australia, [†]Peninsula Health, Frankston, VIC, Australia, [‡]Wake Forest University School of Medicine, Winston-Salem, NC, United States, [§]University of Melbourne, Parkville, VIC, Australia, [¶]The Peter MacCallum Cancer Center, Parkville, VIC, Australia

15.1 Introduction

Three-dimensional (3D) bioprinting has garnered immense interest over the past decade based on its potential to provide a means to rapidly manufacture replacement body parts that replace like with like and are immediately biocompatible. Despite promising advances in bioprinting, it is still immensely difficult to reproduce the delicate structure-function relationships of complex tissues and organs using this approach. In this way, the bioprinting of functional autologous solid organs (e.g., kidney or heart) remains an aspirational goal. However, the formation of more simple tissues such as adipose tissue for breast reconstruction, represents a lower-hanging fruit in translational bioprinting research. The ability to bioprint autologous fat tissue would be transformational in the management of breast cancer patients. This centers around the combination of great need for breast reconstruction in today's society and a lack of an ideal form of breast reconstruction characterized by simplicity along with a low complication profile.

Postmastectomy breast reconstruction has become an important component of breast cancer treatment due to its high prevalence and incidence in our aging populations and the significant improvements in patient survival over the past few decades [1]. One in eight women in the United States are likely to develop breast cancer in their lifetime [2] and it accounts for 14.6% of all new cancers affecting women [3]. In addition to breast reconstruction following mastectomy for established breast cancer, an increasing number of women (25%) are opting for it early in their lives as a prophylactic measure for gene mutations (BRCA mutations) or as definitive management of early stages of breast cancers, where there is a strong family history [4]. The 5-year survival rate for patients following therapy has improved significantly in the last decade (89.7%) [3]. To this effect, numerous studies have demonstrated that postmastectomy breast reconstruction improves the psychosexual well-being of these women and the above factors have significantly increased the demand for breast reconstruction [1,5–8].

Breast reconstruction can be largely classified into two groups: implant-based or autologous. Implant-based reconstruction can be performed in a single stage (i.e., direct- to-implant), if there is adequate mastectomy skin flap or in two stages using tissue expanders [9,10]. Despite their shorter operative time and faster recovery, breast

3D Bioprinting for Reconstructive Surgery. https://doi.org/10.1016/B978-0-08-101103-4.00028-4

implants are not considered the gold-standard in breast reconstruction due to their significant risk of long-term complications such as infection, foreign body reaction, capsular contracture, and anaplastic large-cell lymphoma [11–15].

In contrast, autologous breast reconstruction, most commonly using the abdominal wall-based free deep inferior epigastric artery perforator (DIEP) flaps or muscle-sparing transverse rectus abdominus myocutaneous (TRAM) flaps, can produce more esthetically pleasing, natural-looking breasts with fewer long-term complications [16]. Microsurgical breast reconstruction has evolved into a safe, reliable procedure due to recent advancements in microsurgical techniques [17–21]. To improve esthetic outcome of free flaps, numerous surgical techniques have been described, such as Blondeel's "three-suture" technique [22], Nahabedian's technique of flap inset [23], Wang's conical folding technique [24], St Andrew's coning sutures [25], and breast esthetic unit-based flap inset [26]. One of the major disadvantages of free flaps is related to donor site morbidity. Overall rate of abdominal wall hernia repair after a DIEP or transverse rectus abdominis muscle (TRAM) flap reconstruction is low (2.45% of 7929 cases) [27]. However, this is still greater than the age-matched population risk (0.28% of 15,679 women) [27]. Furthermore, in some women, abdominal pannus may not a suitable donor site due to lack of adequate volume [28–30].

First described by van der Meulen in 1889 and Neuber in 1893 [31], fat grafting for breast reconstruction in the last decade has returned to popularity following the improvement of liposuction techniques [32]. Fat grafting offers numerous advantages for use in breast reconstruction as it provides (usually) abundant autologous donor tissue, low donor site morbidity, and relative ease of harvest [33,34]. However, the need for engraftment and variable survival of fat following fat grafting make this more suited as an adjunct to conventional breast reconstruction techniques with injections of smaller volumes, spread over multiple stages [35–40]. Investigators have attempted to improve the adipogenic potential of fat grafts by enriching them with additional adipose-derived stem cells (ADSCs), platelet-rich plasma (PRP) or growth factors, however, results remain mixed [34,41–51]. Despite these advances, the current consensus recommends that fat grafting is most appropriately reserved to supplement the conventional breast reconstructive techniques during secondary contouring procedures [52,53].

3D bioprinting provides a novel means to combine cells, scaffolds, and growth factors into a carefully designed 3D structure that includes an intrinsic circulation (either de novo or ex vivo) [54] and builds on decades of experience and expertise in autologous fat grafting, ADSC biology and tissue engineering. Today's highly flexible 3D printing technology provides a potential solution for unmet clinical needs in breast reconstruction. Here, we discuss the advances in 3D bioprinting technology with regard to tissue engineering and its use and potential in breast reconstruction.

15.2 3D bioprinting

15.2.1 Background

3D bioprinting describes a method of creating individualized 3D tissue constructs by incorporating novel 3D printing technology in traditional tissue engineering [55].

15.2.1.1 Tissue engineering

Tissue engineering utilizes cells, biomaterials, and biologically active growth factors to produce tissues that mature into functional, vascularized structures [56]. Cells are usually derived from a sample of the desired tissue and are enriched through selection strategies and/or expanded ex vivo. [57–62]. Biomaterials and growth factors must provide a controlled gradient in mechanical properties and cellular signaling for optimal cell growth in vitro [63]. As tissues and organs grow, it is essential that they develop an intrinsic vascular network. Given that the maximum nutrient diffusion distance for cells for survival is only 100–200 μm [64], manufacturing complex, well-vascularized tissues of clinically relevant size has proved challenging using conventional laboratory techniques [56,65]. To this effect, investigators have studied 3D culture systems such as cell suspension culture [66] and ceiling culture [67], which can regulate cell tension [68] and enhance adipogenic differentiation [69]. However, they fail to reproduce accurately the in vivo microenvironment [70,71]. Furthermore, adipocytes in cell suspension cultures are not exposed to nutrition equally leading to progressive cell lysis within 72 h of incubation [72]. In ceiling cultures, preadipocytes proliferate and differentiate; however, they display spindle, fibroblast-like morphology, rather than the round unilocular phenotype typical of mature adipocytes [67,73,74]. Addition of angiogenic growth factors and endothelial precursor cells may address these issues; however, conditions for their culture and expansion are not readily compatible with adipocytes, which poses practical issues that will be discussed later in this review [63,75,76].

Additional approaches have included the use of existing tissues to try to generate new tissues and organs through decellularization. In this technique, cells are carefully removed from a donor tissue while its extracellular matrix (ECM) and mechanical properties are preserved [77]. These constructs elicit minimal host immune response [78,79] and the donor ECM is gradually replaced by ECM excreted by repopulated cells [80]. Unfortunately, allogeneic donor tissues are rare and autologous options pose donor site morbidity. Various synthetic scaffolds have been described that seek to encourage neovascularization, tissue ingrowth, and development, such as 3D fiber deposition [81], particulate leaching technique [82], and electrospinning [83]. The microarchitecture of these can be very difficult to control across the entire construct. In comparison, 3D-printed scaffolds can be generated with customizable form and interconnected pores [84–86] to facilitate neovascularization and nutrient flow and the control offered by this technology is preferable to the existing techniques [55,87–94].

15.2.1.2 3D printing

3D printing, also known as rapid prototyping or additive manufacturing, describes a process by which a 3D construct derived from computer-aided design (CAD) is built in a layer-by-layer fashion [95–98]. One of the major advantages of 3D printing is the ability to produce custom designs with complex internal details [99–101]. 3D printing has been utilized in industrial design for decades, however, it has only been adopted for medical application in the last decade [102]. Imaging data from routine computed tomography (CT) or magnetic resonance imaging (MRI) scans is converted into a CAD file using 3D printing softwares such as 3D Slicer (Surgical Planning Laboratory; Boston, MA, United States). The file is used by a 3D printer to fabricate the final model. Aided by expiration of key patents in the last decade, 3D printers have become affordable to lay consumers and has been adopted for 3D bioprinting.

15.2.2 3D bioprinting composition

A 3D bioprinted construct is typically composed of a scaffold and cells or precursor cells of the desired tissue.

15.2.2.1 Scaffold

The scaffold provides necessary 3D structural integrity, mimicking local microenvironment, contains appropriate cell-specific signalling cues, and must possess negligible cytotoxicity [103,104]. The external shape of a scaffold can be designed using CT/MRI scans of the desired organ and the internal porous architecture must be optimized for vascular growth and nutrient diffusion [105]. There is a distinct absence of an ideal scaffold material for 3D bioprinting. A variety of biological and synthetic materials have been used to build scaffolds such as alginate [86,106–109], fibrin [86,106–108,110], gelatin [110], hyaluronic acid [110], glycerol [110], and Pluronic F-127 Derby, 2012 #2409} [106–108]. Despite their superior biocompatibility and cytocompatibility, biological scaffolds often lack mechanical strength. Likewise, synthetic polymers can often lack biocompatibility but this can be reversed by incorporating biologically active domains [60,111–113] such as cell-adhesion peptides [114], silk functionalized with titanium-binding peptides [115], and collagen [116].

One of the most commonly used polymers in 3D bioprinted scaffolds is polycaprolactone (PCL) [117,118]. PCL has a low melting temperature (60°C) and cools rapidly on deposition, making it cytocompatible [117]. It is durable with a long degradation period (1.5–2 years), and is completely excreted by the body [118]. However, it is elastic and therefore one of its main disadvantages is its inability to provide mechanical strength. In contrast, polymers such as Pluronic F-127 (BASF SE; Ludwigshafen, Germany) composed of hydrophobic polypropylene glycol and hydrophilic polyethylene glycol (PEG), can provide strength and is easy to use [119], but lacks good cytocompatibility [120].

15.2.2.2 Cells

Stem cells such as mesenchymal stem cells (MSC) are the most commonly used cells in 3D bioprinting. The microenvironment of these pluripotent or multipotent cells should be tightly regulated during tissue growth so that the necessary tissue is produced, rather than any of the other tissues that the stem cell is capable of producing (or some mixture of these). As a result, the final construct is most commonly matured ex vivo or inside an in vivo bioreactor before implantation [121–125]. Autologous or allogeneic cells are required. Despite their obvious advantages, autologous stem cells can be difficult to culture and expand to a sufficient number for clinical application without a significant loss of proliferative function, phenotype, and additional regulatory hurdles [126,127]. Encouragingly, this can be partly addressed by coculture with precursor, progenitor, and supportive cells as well as seen with the coculture of preadipocytes and endothelial cells for adipose tissue engineering [128]. In contrast, allogeneic stem cells can be stored and accessed when needed. However, they pose a risk of immune rejection.

15.2.3 3D bioprinting techniques

3D bioprinting techniques can be broadly classified by their mechanism of cell deposition: inkjet [93,129–131], microextrusion [132–134], or laser [135–137]. Integrated tissue organ printer (ITOP) is a novel 3D bioprinting technique that simultaneously deposits cell-laden hydrogel with synthetic polymer by a pneumatic microextrusion controller [110] (Table 15.1).

15.2.3.1 Inkjet bioprinting

Inkjet bioprinting is the earliest described technique where either thermal [131] or acoustic [146–148] forces are used to eject drops of liquid on to a scaffold. Electrically heated thermal print-heads can produce a localized temperature increase to 200–300°C for a short duration (2 ms) but produce only 4–10°C rise in the overall temperature [149]. Despite some studies demonstrating minimal impact on the stability of biological molecules such as DNA [146,147], there still remains a potential risk using thermal inkjet bioprinters by exposing cells and the tissue construct to heat and mechanical stress. Acoustically based printing uses acoustic waves created by a piezoelectric crystal to break liquid into regular droplets [150]. Pulse, duration, and amplitude of the sound waves can be adjusted to alter the size of droplets and rate of ejection. The major disadvantage of using acoustic forces lies with the potential risk of cell damage and lysis from 15 to 25 kHz frequencies emitted by the piezoelectric crystals [151]. In summary, despite their low cost, high resolution, high speed, and biocompatibility, both thermal and acoustic inkjet bioprinting is limited by its requirement of the biological material to be in liquid form. This limitation can be potentially addressed by immediately curing the material with chemical, pH, or ultraviolet [152,153]. However, this increases the printing time significantly and introduces chemical modifications leading to cell damage. To date, inkjet 3D bioprinters have been utilized to fabricate functional skin [138], cartilage [139], and bone [140] in preclinical models, but not adipose tissue.

15.2.3.2 Microextrusion bioprinting

Microextrusion bioprinters are the most common and affordable bioprinters used in research [105]. In comparison with an inkjet bioprinter that extrudes liquid droplets, a microextrusion bioprinter ejects microbeads of a material such as hydrogel, biocompatible copolymers, and cell spheroids, using pneumatic [119,153–155] or mechanical [156,157] dispensing systems. Pneumatic printers are built with simpler components, but mechanical dispensers provide a greater spatial control. Major advantages of microextrusion printers include their compatibility with materials with a wide range of fluid

Table 15.1 **Summary of 3D bioprinting techniques**

Bioprinting techniques	Mechanism	Advantages	Disadvantages	Clinical application
Inkjet	Thermal	High resolution Low cost High speed Biocompatibility	Exposure to high heat (300°C) Absolute requirement for biological material to be liquid	Skin [138] Cartilage [139] Bone [140]
	Acoustic	High resolution Low cost High speed Biocompatibility	Cell lysis at 15–25 kHz Absolute requirement for biological material to be liquid	
Microextrusion	Pneumatic	Affordable Simpler components Viscous biologic material	Lower spatial control Low cell viability Low resolution Slow speed	Aortic valve [141] Blood vessels [142] Ovarian cancer model [143]
	Mechanical	Affordable Superior spatial control Viscous biologic material	Sophisticated components Low cell viability Low resolution Slow speed	
Laser-assisted	LIFT	Compatibility with range of viscosity, resolution, and speed	Slow print speed	Skin [144] Skull defect [145]
ITOP	Pneumatic	Microchannel formation Microscale nozzle Produce well-vascularized, human-scale tissue construct	Limited accessibility	Mandibular bony defect [110] Ear cartilage [110] Skeletal muscle [110]

LIFT, laser-induced forward transfer; *ITOP*, integrated tissue organ printer.
Reproduced with permission from Chae MP, Hunter-Smith DJ, Murphy SV, Atala A, Rozen WM. 3D bioprinting in Nipple-areolar complex reconstruction. In: Shiffman MA, editors. Nipple-Areolar Complex reconstruction: principles and clinical techniques. Heidelberg, Germany: Springer; 2016.

properties such as viscosity [91], and the ability to deposit very high cell densities such as tissue spheroids that can self-assemble directly into complex structures [158,159]. One of the major disadvantages of microextrusion as a technique is its relatively low cell viability rate (40%–86%) due to shear stress [155,160], low print resolution, and speed [140]. To date, microextrusion technology has been used to fabricate aortic valves [141], blood vessels [142], and in vitro ovarian cancer model [143] in preclinical studies.

15.2.3.3 Laser-assisted bioprinting

Laser-assisted bioprinting (LAB) is the least commonly used technique and relies on the principle of laser-induced forward transfer (LIFT) [161,162]. In a LIFT system, a pulsed laser beam is directed on to the laser-energy-absorbing layer (e.g., gold or titanium) over a "ribbon" containing the donor transport system. The laser induces the formation of a high-pressure bubble that propels biological material containing cells forward toward a scaffold. Microscale resolution of LAB means that, in addition to cells [163], it can be used to deposit peptides [164] and DNA [165]. One of the main advantages of LAB technology is its flexibility, as it is compatible with a wide range of viscosity, resolution (i.e., single cell per drop to 10^8 cell per mL), and speed (i.e., 5 kHz to 1600 mm/s) [137]. Moreover, this technique has a negligible effect on cell viability and function [166–168]. A major drawback is its slow print speed due to the requirement of rapid gelation of the deposited material owing to its high resolution [169]. In preclinical studies, LAB has been used to create a small cellularized skin construct [144] and a skull defect [145].

15.2.3.4 Integrated tissue organ printer

Integrated tissue organ printer (ITOP) is an innovative bioprinting technique, developed by Kang et al., that consists of a multimaterial-dispensing printer system controlled by a custom-designed microscale nozzle motion program enabling simultaneous deposition of both cell-laden hydrogel and synthetic biodegradable polymer to deliver a human-scale tissue construct [110]. Despite its relatively high resolution, inkjet bioprinting is limited by its requirement of liquid hydrogel that results in low structural integrity and mechanical strength. Microextrusion method utilizes viscous fluid and can produce more stable 3D constructs, but the generated shear stress reduces cell viability, printing resolution, speed, and size. LAB technique requires rapid gelation of hydrogels to achieve its very high resolution, leading to low flow rates.

ITOP deposits PCL-based scaffolds in various designs that provide mechanical strength to the construct and forms networks of microchannels that facilitate cell nutrient and oxygen diffusion. However, the bulk of mechanical stability is provided by Pluronic F-127 hydrogel extruded from a separate nozzle that acts as an outer sacrificial support layer. The composite hydrogel system in ITOP consists of fibrinogen, gelatin, hyaluronic acid, and glycerol in disparate concentrations optimized for each target tissue. The nozzle motion program is customized based on the printing pattern and the fabrication condition (i.e., scan speed, temperature, material information, and air pressure). Once the printing is completed, thrombin is added to crosslink fibrinogen into stable fibrin, whereas the other hydrogels, including Pluronic F-127, are washed out. Using 3T3 fibroblast cell model, authors demonstrate ≥95% cell viability at 6 days and persistent tissue growth at 15 days.

As a result, ITOP can manufacture well-vascularized, human-scale, complex shape, structurally stable tissue constructs. ITOP has demonstrated proof of principle in the use of human amniotic fluid-derived stem cells to form a construct for a human mandibular bony defect in vitro, and ITOP-printed constructs have been successfully implanted to treat skull defects in rodents. Rabbit chondrocytes have been used with ITOP to form a human ear-shaped cartilage with demonstrated viability at 1 month after implantation in preclinical models. Furthermore, functional 15×5×1 mm rodent skeletal muscle tissue has been formed.

15.3 3D bioprinting adipose tissue for breast reconstruction

15.3.1 Challenges facing adipose tissue engineering

For decades, researchers have investigated adipose tissue engineering for postmastectomy breast reconstruction [170]. In contrast to other organs and tissues such as bladder [57] and mandible [171,172] that have already been tissue engineered and successfully implanted in patients, building stable, large-volume adipose tissue by conventional tissue engineering presents numerous challenges. Standard subcutaneous adipose tissue consists of fully differentiated adipocytes that constitutes 90% of the total volume and 15% of total cells [173]. All adipocytes are found in close proximity to a capillary network (within 200 μm) [174] and are mechanically supported by a thin layer of basement membrane and the stromal ECM [175]. As a result, they are highly sensitive to hypoxia and physically fragile. Moreover, adipogenesis is closely accompanied by angiogenesis in physiological circumstances [174]. Therefore, adipose tissue engineering cannot be achieved without concomitant neovascularization, posing further challenges [176–180]. To this effect, the ability of 3D bioprinters that can deposit multiple types of cells and materials, and build composite structures appears promising (e.g., ITOP).

15.3.2 Composition of 3D-bioprinted adipose tissue

The two major components in the formation of a 3D-bioprinted adipose tissue construct for breast reconstruction are a scaffold and cells.

15.3.2.1 Scaffolds for 3D-bioprinted adipose tissue

Role of a scaffold

In addition to providing form and encasing biological materials, the mechanical properties of a scaffold are important in determining the differentiation fate of stem cell differentiation lineages [181]. Cytoskeletal tension transmitted from actomyosin contractility and reaction forces generated from surrounding ECM can influence gene expression, cell shape, and differentiation [182]. Extracellularly, mechanical forces on ECM, such as tissue stretch, compression or shear stress, is transmitted to cells via integrin-regulated adhesions [183] into cytoskeleton and cell nucleus [184], leading to the activation of various signaling pathways [185–189] and mechanosensitive ion channels [190]. As a result, a scaffold that mimics adipose tissue stiffness (i.e., Young's elastic modulus of 2–4 kPa) promotes adipogenic differentiation even in the absence of exogenous adipogenic growth factors [181,191–196]. On the contrary, if the stiffness is increased, ADSCs lose their typical rounded morphology and fail to upregulate adipogenic markers [181]. Likewise, the pattern of microstructure (i.e., square vs rectangle) [197] and the surface nanotopography (i.e., round nanogroove vs straight grooves and grids) [187] of a scaffold can all influence adipogenic differentiation.

Scaffold design

To support concomitant angiogenesis, the design of a scaffold must incorporate microchannels that enable vascular infiltration and growth, facilitating subsequent oxygen and nutrient perfusion [198,199]. Furthermore, the porosity of interconnect cell structure is crucial for stem cell migration, proliferation, and differentiation [200]. The pore size must be able to accommodate coexistence of ADSCs, differentiated adipocytes, and mature adipose tissue lobules (i.e., 10 vs 100 vs 300–500 μm, respectively) [192,201,202]. In addition, surface-coating scaffolds with silica nanoparticles can increase ADSCs proliferation by activating downstream ERK 1/2 pathways and improves mechanical strength [203].

Scaffold building techniques

Numerous techniques for constructing 3D scaffolds have been described in the past, such as cell patterning [204], particulate leaching [82], electrospinning [83,205], lithography [206], and microfabrication [206]. In comparison, 3D bioprinting provides unique advantages of being able to create 3D spheroid constructs rapidly with complex internal structures like microchannels, and perform controlled material extrusion to achieve the desired biomechanical properties [119,160,164,207–209]. Moreover, 3D bioprinters can safely handle delicate stem cells like ADSCs [151,210].

Scaffold materials

Materials used to construct 3D scaffolds for tissue engineering and 3D bioprinting of adipose tissue can be broadly classified into biological materials, biodegradable polymers, and composite scaffolds.

Biological materials

Biological materials used to build 3D scaffolds consist of naturally derived substances that resemble human ECM [211,212]. Their major advantages are biocompatibility, biodegradability, and inherent biological functions that make them suitable for tissue engineering [213–216]. Their potential limitations such as immunogenicity stemming from allogeneic or xenogenic origin or contamination from endotoxins can be controlled by careful preparation and routine quality control measures [217]. To date, the following biological scaffold materials have been used for adipose tissue regeneration: collagen, hyaluronic acid, silk, gelatin sponge, fibrin, alginate, agarose, chitosan, calcium phosphate, Matrigel, and decellularized adipose tissue (DAT) (Table 15.2).

Table 15.2 Summary of 3D scaffold materials used in adipose tissue regeneration

	Description	Advantages	Limitations	Adipose tissue regeneration in vivo
Biological materials				
Collagen	Natural protein found abundantly in ECM	• Biocompatibility • Nontoxic • Excellent for ADSC adhesion, differentiation, and proliferation • FDA approved for clinical use	• Rapid degradation • Crosslinking, required for durability, compromises downstream cell signaling	[218] [219] [220] [221] [222]
Gelatin	Hydrolyzed, water-soluble derivative of collagen	• Mimic natural ECM	• Relatively unstable at room temperature	[223]
HA	Natural polymer that is a component of ECM, of which the insoluble (derivatized or crosslinked) form is used in tissue engineering	• Biocompatibility • Nonimmunogenicity • Intrinsic porosity • High hygroscopicity • Degrade into safe byproducts	• Too fragile	[224] [225]
Silk	Proteinaceous substance (fibroin and sericin) extracted from *Bombyx mori* silkworm cocoons	• Good cytocompatibility • Low immunogenicity • Intrinsic porosity • No requirement for further chemical or photo-crosslinking for stability • Slow degradation in vivo • Modifiable to express various growth factors	• Available in aqueous form for use as an injectable	[196]

Fibrin	Natural polymer of fibrous protein found in blood clots	• Clinically used as a surgical sealant for hemostasis • More angiogenic than collagen	• Relatively weak mechanical properties • Rapid degradation in vivo	[226]
Alginate	Seaweed-derived anionic polysaccharide	• Biocompatibility • Semipermeability • High malleability • Available in gel, foam, nanoparticles, beads	• Negative overall change, which prevents mammalian cell attachment	[227]
DAT	Technique where cells are removed from a tissue but its native ultrastructure and ECM are preserved	• Retention of mechanical properties of native tissue • Preservation of essential ECM components for adipogenesis • Low immunogenicity due to allogeneicity	• Laborious tissue preparation and manual handling • Availability of allogeneic tissue	[228] [229] [194] [211] [195] [230]
Matrigel	Natural polymers secreted by EHS mouse sarcoma cells	• Commonly used cell culture matrix • Sensitive to cell culture conditions • Contain growth factors and cytokines for cells	• Derived from mouse sarcoma cell lines	[231] [232] [178]
Biodegradable polymers				
PCL	Most commonly used filament in 3D bioprinting	• Readily available • Used in FDA-approved suture (Monocryl) • Completely excreted by body	• Long degradation (1.5–2 years) • Stiff mechanical properties favoring osteogenic differentiation of ADSC	N/A

Continued

Table 15.2 Continued

	Description	Advantages	Limitations	Adipose tissue regeneration in vivo
PLA	Most commonly used filament in desktop 3D printers	• Readily available • Used in FDA-approved suture (Vicryl) • Completely excreted by body	• Long degradation (up to 2 years) • Stiff mechanical properties favoring osteogenic differentiation of ADSC	N/A
PLGA	Copolymer synthesized by combining monomers, glycolic acid and lactic acid	• Adjustable mechanical properties • Safe by-products	• Poor cell adhesion and differentiation • Relatively rapid degradation	[233] [234]
PEG	Polymer of ethylene oxide	• Water-soluble • Elastic • Adequate degradation time • Minimal inflammatory reaction • Chemical versatility	• Toxic degradation products	[235] [198]
OPAA	Created by free radical polymerization of agamantine-containing, cross-linked PAA oligomers and RGD tripeptide	• Macroporous foam of OPAA has the same mechanical properties as adipose tissue • Macroporous structure supports nutrient supply and complex tissue growth	• Expensive to prepare	[236]
Composite scaffolds				
Collagen-HA	Crosslinked 3D porous scaffold	• Enhanced efficiency of collagen crosslinkage • More robust collagen scaffold	• Requires manually intensive preparation	N/A

Collagen-gelatin	3D porous scaffold	• Cytocompatibility • Improved scaffold stability at room temperature • Improved scaffold strength	• Relatively weak scaffold still	N/A
Alginate-gelatin	Microspheres	• Flexible mechanical properties • Improved cell adhesion	• Large-volume tissue production may be limited	N/A
PGS-PLLA	Porous scaffold	• Alter mechanical properties to closely mimic adipose tissue • Hydrophilic • Large-volume tissue production	• Requires manually intensive preparation	N/A
PDM-XLHA	3D scaffold	• Improved angiogenesis and adipogenesis	• Relative lack of availability of PDM • Potential ethical issues	[237]

ADSC, adipose-derived stem cells; *DAT*, decellularized adipose tissue; *ECM*, extracellular matrix; *EHS*, Engelbreth-Holm-Swarm; *HA*, hyaluronic acid; *N/A*, not applicable; *OPAA*, poly(amidoamine) oligomer; *PAA*, poly(amidoamine); *PCL*, polycaprolactone; *PDM*, placental decellular matrix; *PEG*, polyethylene glycol; *PGS*, poly(glycerol sebacate); *PLA*, polylactic acid; *PLGA*, poly(lactic-*co*-glycolic acid); *PLLA*, poly(L-lactic acid); *XLHA*, cross-linked hyaluronic acid.

Collagen Collagen is a natural protein that provides structural support and is found abundantly in extracellular spaces across the human body [238]. It is nontoxic, biocompatible and, via other clinical indications, already approved by the Food and Drug Administration (FDA) for use in humans. Moreover, it provides an ideal microenvironment for ADSCs adhesion, differentiation, and proliferation both in vitro and in vivo [194,218,229,239–242], which is superior to other biological substances such as silk, and synthetic polymers such as polylactic acid (PLA) [243]. In animal models, ADSCs cultured on 3D collagen scaffolds have increased ECM production and subsequently yield a well-vascularized adipose tissue construct [219–221]. In an interesting study, Xu et al. reported that adding ginsenoside Rg1, an active component of ginseng, and platelet-rich fibrin (PRF) promotes adipogenesis of ADSCs in collagen sponges [222]. Nonetheless, one of the major limitations of collagen scaffolds is its rapid degradation on implantation [243]. Crosslinking of collagen is required to improve its durability, but can be cytotoxic and compromise cell signaling [244].

Gelatin Gelatin is a hydrolyzed, water-soluble derivative of collagen. It closely mimics the natural ECM and when a gelatin sponge is populated with bone marrow-derived mesenchymal stem cells (BM-MSCs) in adipogenic medium, lipid droplets accumulate in vitro [245]. Without any cells, Vashi et al. reported that simply implanting a dome tissue engineering chamber (TEC) filled with gelatin microspheres impregnated with slow-release basic fibroblast growth factor-2 (bFGF-2) suspended in collagen gel leads to successful adipogenesis in mice, similar to the popular cell culture matrix, Matrigel [223]. One of the major limitations of gelatin sponges as scaffolds is their relative instability at room temperature. This can be improved by using a stable, photo-cross-linked methacrylated gelatin (GM), instead.

Hyaluronic acid Hyaluronic acid is a natural component of ECM, of which its insoluble form, via derivatization or crosslinkage, is utilized in tissue engineering. Its advantages include excellent cytocompatibility, biocompatibility, nonimmunogenicity, high hygroscopicity, intrinsic porosity for ideal cell proliferation, and its ability to degrade into safe by-products [246–250]. In 2001, von Heimburg et al. have reported that preadipocytes embedded in esterified hyaluronic acid (or HYAFF 11) sponges lead to more vascularized, higher cell-density adipose tissue in nude mice [224]. In a later study, the same group show that coating HYAFF 11 sponges with ECM glycosaminoglycan hyaluronic acid results in superior cell penetration and tissue vascularization [225]. However, the authors concede that the volume of fully differentiated adipose tissue from this method is still inadequate for clinical application [225]. One of the major limitations of hyaluronic acid scaffolds is related to its high hygroscopicity, which results in a fragile structure [247]. As a result, hyaluronic acid is increasingly being used as a cytocompatible cell carrier, rather than as a solid scaffold.

Silk Silk is a naturally occurring substance extracted from *Bombyx mori* silkworm cocoons, consisting of a core filament protein, fibroin, coated by a glue-like sericin protein [251,252]. It is a well-known, clinically accepted, biocompatible material that is already being used to manufacture FDA-approved surgical sutures. The major advantages of silk are its good cytocompatibility, low immunogenicity, intrinsic porosity, no requirement for toxic chemical or photo-crosslinking for stability and function [253], slow-degradation rate [243], and modifiability to express various growth factors [251]. Numerous in vitro experiments demonstrate that silk scaffolds can support cell adhesion and viability for long-term coculture (i.e., up to 6 months) [254] between undifferentiated [255–257] and differentiated [254,257,258] adipocytes with endothelial cells. One of the significant limitations of silk is its existence in mainly aqueous form, which deems it more suitable in an injectable form. Recently, Bellas et al. have injected silk foams embedded with ADSCs subcutaneously in rats showing that the neoadipose tissue integrates well with the host tissue [196].

Fibrin Fibrin is a natural polymer of fibrous proteins found in blood clots during hemostasis [259]. It is already being used routinely as a surgical sealant to protect wounds from infection and allow cellular repair [259–261]. Ironically, it is more angiogenic than collagen [262]. Experiments using fibrin gel demonstrate that it can provide a suitable microenvironment for adipogenesis in adipocyte/endothelial cell coculture in vitro [263,264]. Wittmann et al. reported that implanting cultivated adipocytes in stable fibrin gel into mice leads to the formation of well-vascularized adipose tissue in vivo [226]. However, fibrin has weak mechanical properties and degrades rapidly in vivo. To this effect, Chung et al. reported the use of PEGylated fibrin (P-fibrin), where fibrin is covalently modified with amine-reactive PEG, which improves its strength and durability, but also maintains its adipogenic and angiogenic effect on ADSCs [262].

Alginate Alginate is a seaweed-derived anionic polysaccharide that is available in several forms for tissue engineering: soft gel, foam, nanoparticles, and spherical beads [265]. Its biocompatibility, semipermeability, and high malleability make it an attractive scaffold material for tissue engineering and in drug delivery. One of the major limitations of alginate is its overall negative charge preventing mammalian cell binding [266,267]. This can be overcome by structural modification with laminin, which enables cell adhesion [268], facilitates physiological adipogenesis by binding to integrin [269,270], and supports angiogenesis [271]. A number of studies support the use of alginate-based scaffolds to facilitate adequate proliferation and differentiation of ADSCs to yield mature adipose tissue both in vitro [272,273] and in vivo [227].

Scaffolds from decellularized tissues Flynn et al. first described DAT in 2010, where cells are removed but its native ultrastructure and ECM are preserved [274], and Wang et al. have first utilized DAT for adipose tissue engineering [211]. The main advantages of DAT are retention of mechanical properties, preservation of essential ECM components for adipogenesis, such as collagen, glycosaminoglycan, and growth factors [77,275–277], and low immunogenicity [78,79]. Interestingly, ECM of DAT undergoes host integration and is replaced by the ECM produced by the seeded cells [80,230]. Encouragingly, DAT produced using adipose tissue from abdomen, flank, and omentum, all have similar mechanical properties to ex vivo breast adipose tissue [278,279]. Numerous animal studies have demonstrated that ADSC-laden DAT can facilitate adipogenesis and concomitant angiogenesis in vivo up to 8 weeks [194,195,211,228,229]. Current limitations that prevent widespread application of DAT are laborious manual handling required for safe tissue preparation and relative unavailability of allogeneic tissues.

Decellularized muscle tissue Muscle tissue was first used more than a decade ago in an unprocessed form within TECs to promote de novo adipogenesis around an arteriovenous loop based on the propensity for denervated muscle to undergo fatty changes [280]. The same group then derived an adipogenic matrix from muscle tissue with successful adipogenesis demonstrated from the matrix derived from various species, but with significant batch to batch variation. Myogel, a novel, basement membrane-rich ECM derived from skeletal muscles, is highly adipogenic in vivo and in vitro [281]. The composite nature of the matrix make it suitable for injection but direct application, codelivered with cells in 3D bioprinting would be challenging.

Matrigel Matrigel is a commonly used cell culture matrix, consisting of natural polymers secreted by Engelbreth-Holm-Swarm (EHS) mouse sarcoma cells. It contains critical growth factors and cytokines for cell growth [282,283]. Kawaguchi et al. reported that autografting preadipocytes reconstituted in Matrigel and bFGF in mice leads to successful growth of the fat pad for 3 weeks and is maintained over 10 weeks [231]. Interestingly, implanting a silicone dome filled with Matrigel and bFGF without cells also leads to adipose tissue growth due to infiltrative host preadipocytes in rats [232]. Subsequently, Findlay et al. have shown that implanting a silicone chamber encasing an epigastric vascular pedicle, filled with Matrigel, facilitates early angiogenesis and adipogenesis in mice with long-term stability of the adipose tissue through to 18 months [284]. Despite its excellent biocompatibility and cytocompatibility, its derivation from murine sarcoma cell lines makes Matrigel unsuitable for clinical application.

Other biological materials

Other biocompatible and cytocompatible materials such as agarose, seaweed-derived polysaccharide [285], and chitosan, crustacean shell-derived polysaccharide [286] have shown to promote adipogenesis in vitro. However, further studies are required to support their adipogenic utility in vivo.

Biodegradable polymers

A number of synthetic biodegradable polymers have been utilized in adipose tissue engineering: PCL, PLA, poly(lactic-*co*-glycolic acid) (PLGA), PEG, and poly(amidoamine) oligomer (OPAA). Major advantages of synthetic polymers are their versatility; researchers can achieve optimal porosity, surface characteristics, and degradation rate by altering their chemical and physical characteristics [287–290].

Moreover, there is low batch-to-batch variability leading to consistent outcomes [291]. However, their main limitation is poor cytocompatibility, which requires expensive surface design or time-consuming chemical modification to promote cell adhesion [292–297].

PCL PCL is the most commonly used plastic filament in 3D bioprinting. It is readily available, biodegradable, and has already been used as an FDA-approved suture material found in Monocryl (Ethicon Inc, Somerville, NJ, United States). However, PCL prior to any chemical modification may be unsuitable for adipose tissue engineering as ADSC cultured on PCL-based scaffolds preferentially undergoes osteogenic differentiation due to its mechanical properties [267].

PLA Similar to PCL, PLA is the most commonly used plastic filament in desktop 3D printers [97]; hence, it is readily available and is already being used in FDA-approved sutures like Vicryl (Ethicon Inc). In vitro experiments demonstrate that ADSCs attach well to PLA-based scaffolds and undergo adipogenic differentiation under appropriate culture medium [298]. However, their long degradation rate in vivo may be undesirable for adipose tissue engineering.

PLGA PLGA is a copolymer synthesized by combining monomers, glycolic acid, and lactic acid. It is a versatile material with adjustable mechanical properties achieved by altering the ratio of monomers. Furthermore, its by-products, glycolic and lactic acid, are also physiological metabolites and, thus, are safe. Under favorable culture conditions, preadipocytes and BM-MSCs undergo adipogenic differentiation and produce functional secretory adipose tissue in vitro and in animal models [233,234]. However, its utility in adipose tissue engineering has been limited due to its rapid degradation in vivo and its hydrophobic surface, which prevents strong cell adhesion and differentiation [299]. This is overcome by combining PLGA with cytocompatible silk fibroin and hydroxyapatite nanoparticles creating a hybrid scaffold [299].

PEG PEG is a polymer of ethylene oxide that is used widely in several medical applications already such as laxatives and for bowel preparation before colonoscopy [300]. It is water-soluble, elastic, has a degradation time of 16 weeks, and exerts minimal inflammatory response in vivo [301]. Furthermore, its chemical versatility enables various modifications to enhance cytocompatibility [302–305]. Patel et al. demonstrated that PEG hydrogel containing collagen-like peptide (LGPA), which is degraded by ADSC-secreted collagenase, and laminin-binding peptide (YIGSR), which promotes cell adhesion, can promote adipogenic differentiation of preadipocytes in vitro [306]. More recently, Clevenger et al. reported that adding RGD tripeptide (Arg-Gly-Asp) enhances its biostability by increasing the degradation time by a month [307]. Interestingly, when the authors added another peptide (MMPc), which contains matrix metalloproteinase (MMP) cleavage sites for MMP secreted by differentiating ADSC [308], this provides more space for ADSC to deposit ECM in its microenvironment and promote angiogenesis [307]. In animal models, Alhadlaq et al. presented that BM-MSCs in PEG hydrogel must be differentiated in vitro prior to in vivo implantation for adipogenesis [235]. Similarly, Stosich et al. demonstrated that microchanneled

PEG cylinders seeded with BM-MSCs and bFGF successfully lead to adipose tissue formation [198]. One of the major limitations of PEG-based scaffold is its potentially toxic degradation products including ethylene oxide.

OPAA OPAA is created by free radical polymerization of agamantine-containing, cross-linked poly(amidoamine) (PAA) oligomers and RGD tripeptide [288,289]. Cell adhesion and cytocompatibility of PAA is improved by the addition of agamantine [309] and RGD tripeptide [294]. Recently, Rossi et al. have created a microporous foam of OPAA (OPAAF), using a gas foaming technique [236]. OPAAF has a complex, porous 3D architecture that resembles the mechanical properties of native adipose tissue (Young's modulus of elasticity of 3.2–4.4 kPa). It enhances cell infiltration and nutrient delivery for successful adipogenesis both in vitro and in vivo [236]. Building on from this finding, the authors need to upscale the tissue volume generated for clinical translation.

Composite scaffold

To overcome deficiencies of individual scaffold materials, researchers have developed composite scaffolds that complement the benefits of each material. Davidenko et al. described a cross-linked 3D porous collagen-hyaluronic acid composite scaffold for culturing mouse preadipocytes in vitro [310]. Despite its excellent biocompatibility and cytocompatibility, collagen undergoes rapid degradation in vivo, which can be improved by crosslinking with 1-ethyl-3-(3-dimethylaminopropyl)-carbodiimide hydrochloride. The presence of hyaluronic acid enhances the efficiency of the crosslinkage and robustness of the overall scaffold, and facilitates the adipogenic differentiation of preadipocytes. Similarly, Lin et al. reported their 3D porous collagen-gelatin composite scaffold supports the growth of viable adipocytes for 28 days from lipoaspirate-derived stromal volume fraction (SVF) cells in vitro [311]. Weak mechanical properties of gelatin and its instability at room temperature are ameliorated by the addition of collagen matrix. Yao et al. described alginate-gelatin composite microspheres, which enables flexible control of the scaffold's mechanical properties in vitro [273]. Alginate is malleable, which allows researchers to alter its porosity and swelling behavior to suit cultured cells, but it has poor cell adhesion. This is amended by the addition of cytocompatible gelatin. However, it is not yet clear if a large-volume adipose tissue with a predefined shape can be manufactured using microsphere-based scaffolds. Frydrych et al. reported that adding poly(glycerol sebacate) (PGS) to stiff poly(L-lactic acid) (PLLA) enables the manipulation of scaffold strength to mimic closely adipose tissue while providing robustness (tensile Young's modulus of 30 kPa and tensile strength of 7 kPa) for large-volume adipose tissue generation in vitro [312]. In an interesting study, Flynn et al. demonstrated that placental decellularized matrix (PDM) combined with cross-linked hyaluronic acid (XLHA) can improve angiogenesis and adipogenesis in vivo, compared with PDM alone [237]. Lack of availability and potential ethical issues surrounding the usage of placental tissues may curtail the future application of this scaffold.

Cells

Without exogenous cells, an empty 3D-bioprinted scaffold elicits host inflammatory response after implantation resulting in fibrotic scarring and dysregulated tissue regeneration [313]. Various stem cells have been utilized for adipose tissue engineering. However, current evidence suggests that using ADSC seeded with appropriate growth factors and endothelial precursor cells in coculture appears to be the most successful.

Cell types

In animal models, mouse-derived preadipocyte cell lines such as 3T3L1 [314–318] and 3T3-F442 A [319,320] have been well-characterized to generate adipose tissue both in vitro and in vivo. However, their fully differentiated state cannot reproduce the full endocrine function of a mature human adipose tissue [314,321]. Furthermore, their xenogeneic origin precludes direct translation into clinical application. To this effect, a variety of human stem cells can be harvested and induced for adipogenesis: embryonic [322,323], foetal [324], adult stem cells, and induced pluripotent stem cells (iPSC) [325]. Despite their excellent pluripotency and nonimmunogenicity, embryonic and foetal stem cells pose significant ethical issues and lack of large sources for practical application. First discovered in 2007 [325], iPSCs are derived from fully differentiated adult somatic cells, such as skin fibroblasts, and are advantageous as they avoid ethical issues and are potentially available in abundance. With the current technology, iPSCs are relatively difficult to induce, limiting its widespread application. In comparison, adult stem cells such as BM-MSCs and ADSCs, are more readily accessible and available. BM-MSC is harvested most commonly from the iliac crest and is used extensively in the treatment of hematooncological disorders [326]. However, BM-MSCs are relatively difficult and painful to extract, poses significant donor site morbidity, and has a low cell yield rate for tissue engineering [327]. In contrast, ADSCs are abundantly available [328], easy to harvest [307,329–332], generates a high cell yield [333], is multipotent [334], and safe [335].

ADSCs

First described by Zuk et al. in 2001, ADSCs represent a host of mesenchymal stem cells isolated from the SVF of adipose tissue [336]. They are found perivascularly along capillaries between adipocytes [173,337] and are the main cell population in adipocyte regeneration [334]. ADSCs have excellent self-renewal properties and demonstrate multipotent differentiation into osteoblasts [338–343], chondrocytes [344], myocytes [345], neurocytes [346], vascular endothelial cells [347], and adipocytes [338]. Currently, the immune-modulatory and regenerative capacity of ADSCs are already being exploited for therapeutic application in multiple disorders: graft-versus-host disease [348], autoimmune diseases [349,350], multiple sclerosis [351], type 1 diabetes mellitus [352], tracheomediastinal fistula [353], and wound healing [354–356].

ADSCs harvest ADSCs are most commonly harvested from the abdomen, thighs, flanks, and axilla [357,358] using liposuction [359] or direct excision technique [329]. Compared with the flank and axilla, more SVF cells can be obtained from the abdomen [358]. Moreover, the latest Coleman liposuction technique [32] yields more ADSCs than the conventional liposuction technique [358]. In contrast, the difference in ADSCs yield rate between Coleman method and direction excision appears unclear [359,360]. Nevertheless, if a large number of cells are required, ADSCs can be expanded in vitro or pooled from multiple donor sites. Interestingly, the number of mature adipocytes per volume remains constant throughout adulthood [361]. However, proliferative activity and differentiation potential of ADSCs are reduced in older adults (age of 50–70) compared with that in young adults (>20) [362].

ADSCs isolation Once they are harvested, ADSCs are isolated by enzymatic digestion, cell culture, and surface marker antibodies. Collagenase is the most commonly used digestive enzyme before the tissue is centrifuged to separate the SVF [332,363–366]. Alternatively, researchers have used trypsin and red blood cell lysis buffer solution to lyse adipose tissue blocks with varying results [367]. Subsequently, the SVF is set in a cell culture, from which multipotent cells can be segregated, as they will float whereas other cells adhere strongly to the plastic surface [336,368–372]. Among the multipotent cells, ADSC can be separated using fluorochrome-conjugated or magnetic bead-attached antibodies that attach to specific cell surface markers, known as fluorescence-activated cell sorting (FACS) [373,374] and magnetic-activated cell sorting (MACS) [370,375–377], respectively. FACS is used widely in basic science experiments and diagnostic tests, but is not suitable for tissue engineering due to its cytotoxicity and poor efficacy. In comparison, MACS is safe, affordable, and readily accessible. Using surface antibodies, ADSC is defined as $CD73^+CD90^+CD34^+CD45^-CD3^+CD13/CD105^+$ according to the International Federation for Adipose Therapeutics (IFATS) and the International Society for Cellular Therapy (ISCT) [378]. In a recent study, Lauvrud et al. reported that a subgroup of ADSC expressing CD146 ($CD146^+$ ADSC) has superior adipogenic and angiogenic potential through increased gene expression of *VEGF-A*, *FGF-1*, and *angiopoietin-1*, differentiates more efficiently, and proliferates faster [379].

ADSCs adjuncts ADSC alone may not consistently produce mature adipose tissue and require addition of growth factors, endothelial precursor cell coculture, and interaction with immune cells.

Growth factors Growth factors are important for activating adipogenesis and promoting angiogenesis from ADSCs [380]. Researchers have reported the use of numerous adipogenic growth factors such as FGF-2 [381], transforming growth factor-beta 1 (TGF-β1) [382], and PDGF-BB [383]; angiogenic growth factors such as VEGF [384], PDGF-BB [385], bFGF [223,386]; and epidermal growth factor (EGF) [387,388]. Notably, TGF-β1 may be associated with tumorigenesis due to their action via MAPK cell signaling pathway and, hence, must be used with extreme caution [389–391] (Table 15.3).

Coculture with endothelial precursor cells Unsurprisingly, adding endothelial progenitor cells appear to facilitate simultaneous neovascularization and complement adipogenesis both in vitro and in vivo [254–257,263,264,396–399]. However, practically this is difficult to achieve due to differing culture conditions preferred by either cell type. Endothelial cell culture typically contains: VEGF and FGF that are known

Table 15.3 List of adipogenic and angiogenic growth factors used to support 3D bioprinting adipose tissue

Growth factors	References
Adipogenic	
FGF-2	[381]
TGF-β1	[382]
PDGF-BB	[383]
Angiogenic	
VEGF	[384]
PDGF-BB	[385]
bFGF	[389]
	[386]
EGF	[387]
	[388]
IGF-1	[392]
	[393]
Ascorbic acid	[394]
Heparin	[395]

bFGF, basic fibroblast growth factor; *EGF*, epidermal growth factor; *FGF-2*, fibroblast growth factor-2; *IGF-1*, insulin-like growth factor; *PDGF-BB*, platelet-derived growth factor receptor B; *TGF-β1*, transforming growth factor-beta 1; *VEGF*, vascular endothelial growth factor.

mitogenic reagents [400–402]; EGF has controversial effect on adipocytes as some reports suggest it inhibits preadipocyte differentiation [403–405], whereas others suggest it improves lipogenesis [406–408]; and hydrocortisone, at high level exclusively, induces dedifferentiation of mature adipocytes leading to lipolysis [409–411]. As a result, other angiogenic growth factors with no lipolytic effect, albeit weaker angiogenic effect, may have to be preferred such as insulin-like growth factor-1 (IGF-1) [392,393], ascorbic acid [394], and heparin [395].

Role of immune system In recent times, investigators have highlighted the potential importance of the immune cells in adipose tissue engineering. For example, Chazaud et al, acknowledges that different macrophage phenotypes affect adipose tissue expansion, metabolism, and remodeling [412]. Similar to the physiological tissue repair process, called M1-to-M2 macrophage shift, where proinflammatory M1 macrophages that initially infiltrate a wound are eventually replaced by antiinflammatory M2 macrophages [413–416], Li et al. note similar changes in infiltrating macrophage profiles at day 3 after implanting a synthetic TEC with a vascular pedicle in rats [417]. Furthermore, Rophael et al. and Debels et al. demonstrate that depleting macrophages in host mouse significantly impairs adipogenesis and angiogenesis [178,418].

ADSCs limitations The main limitations of ADSCs are their potential role in tumorigenesis and their preparation using xenogenic substances. Current literature presents contrasting views of ADSCs in carcinogenesis. Chandler et al. [419] and

Koellensperger et al. [420] reported that coculturing ADSCs with glioma or squamous cell carcinoma cells increases tumor cell viability, invasiveness, and induce apoptosis of the surrounding normal cells. In contrast, Cousin et al. [421] and Zhao et al. [422] reported that ADSC can inhibit proliferation and induce apoptosis of pancreatic tumor and hepatic cell carcinoma cells. Importantly, for adipose tissue engineering in breast reconstruction, when ADSCs are cocultured with breast cancer cells, they undergo myofibroblastic differentiation behaving like breast cancer cells [419,423]. This has not been found in vivo to date, and there is currently no compelling evidence to suggest that autologous fat transfer, the most common form of ADSCs use in clinical practice, has any association with increased recurrence of breast cancer in humans. Furthermore, several key reagents used for ADSCs preparation are animal-derived: cell culture media from foetal bovine serum, collagen scaffold from rat tail, and gelatin from porcine skin [424,425]. Xenogenic substances can potentially elicit acute graft rejection or chronic inflammation and may lead to host antibody production resulting in severe autoimmune reactions. Currently, serum-free or xeno-free culture media such as human serum-based media and PRP are not readily available and have not been proved for safety and efficacy in vivo [425–428].

15.3.3 3D-bioprinted adipose tissue in vitro

A number of in vitro studies have demonstrated that 3D bioprinting is not cytotoxic, preserves proliferative and adipogenic differentiation capabilities of ADSCs, and produces lasting mature lipid droplets, while enabling flexible design of the final adipose tissue [272,429–431].

15.3.3.1 3D bioprinting ADSCs

Concerned about potential cytotoxicity of laser, Ovsianikov et al. seeded ADSCs on to the photopolymerized, methacrylamide-modified gelatin scaffold after it had been laser 3D-printed [429]. Modified gelatin maintains its biodegradability while supporting ADSCs adhesion, proliferation, and adipogenic differentiation. In 2010, Koch et al. have described a laser bioprinting technique using two coplanar glass slides in close proximity to each other (500 μm) so that direct exposure of laser to cells could be avoided [168]. When laser is directed on to the gold-coated upper slide, it generates a jet dynamic that pushes the cell-laden hydrogel forward and drops on to the lower slide while the laser energy is absorbed by gold. Using this method, Gruene et al. have successfully 3D-bioprinted an alginate-blood plasma product-based composite scaffold, laden with ADSCs that had been cultured for seven days prior in an adipogenic medium [272]. Importantly, the authors demonstrated that laser bioprinting does not compromise ADSCs proliferation or differentiation. Similarly, Williams et al. show that a pressure-based microextrusion 3D bioprinting technique [119,160,207] is safe for depositing human adipose-derived SVF spheroids in alginate-based scaffolds [430].

15.3.3.2 3D bioprinting mature adipocytes

In an interesting proof-of-concept study, Huber et al. demonstrated that laser 3D bioprinting of photocurable GM scaffold laden with mature adipocytes can produce a larger (7 mm in height), functional adipose tissue that can last up to 14 days in vitro [431]. The authors cited the benefits of using mature adipocytes including improved efficiency as only 42% of ADSCs undergo adipogenic differentiation physiologically [336,432], potentially less immunogenicity of mature cells, ability to forego extensive laboratory processes to isolate ADSCs, time spent inducing adipogenic differentiation ex vivo prior to implantation, and the ability to produce immediately functional, long-lasting mature adipose tissue [361]. One of the main limitations of using mature adipocytes are their fragile univacuolar morphology and reduced renewal capacity due to their terminal differentiation and low adipocyte annual turnover rate [237,433].

15.3.4 3D-bioprinted adipose tissue in vivo

3D bioprinting adipose tissue for breast reconstruction is such a novel technology that only a small number of studies have explored its in vivo application to date.

15.3.4.1 TEC

TEC has revolutionized adipose tissue engineering in the last decade, enabling researchers to produce a large amount of tissue (up to 78.5 mL) without exogenous scaffold implantation, growth factors, and ex vivo cell preparation [434]. Using this method, investigators surgically prepare a vascular pedicle, usually from superficial inferior epigastric vessels [435]. This arteriovenous (AV) loop is placed inside a larger, perforated, dome-shaped TEC, which can be manufactured using a range of biocompatible synthetic polymers: polycarbonate [284,434,436–439], PLGA [435,440–442], and silicone [223,232,386,443,444]. Most studies have relied on standard manufactured TEC; however, as 3D printing has become more affordable and convenient to use in the past few years [96–98], future investigators have the potential to 3D print patient-specific TECs. Encouragingly, Morrison et al. have recently reported their use of customized acrylic-based TEC that was produced by a well-known medical 3D printing company called Anatomics (Melbourne, VIC, Australia), however, using conventional manufacturing technology [445].

15.3.4.2 TEC mechanism of action

The exact mechanism of tissue growth within TEC is still unclear, be it mechanotransduction [446,447], intrachamber edema-induced adipogenesis [448], or immune-mediated response [437]. It is most likely that once a TEC is implanted subcutaneously, it induces ischemia and apoptosis of the separated adipocytes [443] stimulating aseptic inflammatory response [438], generating mitogenic stimuli that promote angiogenesis and adipogenesis [449], leading to infiltration by a host of endogenous cells such as immune cells like macrophages [417], preadipocytes [438], fibroblasts that differentiate into myofibroblasts on entry and deposit collagen [450,451], and endothelial precursor cells that develop into a complicated vascular network [386,435,452,453]. In addition, TEC protects the developing adipose tissue from deforming external mechanical forces [436]. Despite lacking exogenous scaffolds for cell adhesion, proliferation, and differentiation, proinflammatory environment created by the TEC induces self-synthesis of ECM, on to which adipogenesis and neovascularization occur [443].

15.3.4.3 TEC in human patients

In contrast to other animal studies where TEC could only produce a small volume of adipose tissue (0.44–30.7 mL) [284,435,439,440,442–444,454], Findlay et al., in a landmark paper, have successfully regenerated up to 78.5 mL of well-vascularized adipose tissue in eight pigs using polycarbonate-based TEC encasing a small pedicled superficial circumflex iliac artery flap (<5 mL) [434]. Encouraged by this finding, the research group, in a proof-of-concept study, implanted acrylic-based perforated TEC that was size-matched to the contralateral breast, encasing thoracodorsal artery perforator flap, into five patients undergoing unilateral postmastectomy reconstruction [445]. Although well-tolerated, there was no tissue growth beyond the original flap in three out of five patients. In one case, TEC had to be explanted earlier than planned due to pain. Ironically, in the fifth patient with diabetes, adipose tissue growth of 210 mL was demonstrated at TEC removal in 12 months. However, this tissue did not esthetically match the contralateral breast and, hence, required further surgical intervention.

15.3.4.4 Patient-specific scaffold prevascularization and delayed lipoinjection technique

Recently, Chhaya et al. reported an innovative proof-of-concept study using prevascularized patient-specific scaffold and delayed lipoaspirate injection in minipigs [455]. The authors subcutaneously implanted 3D-printed PCL-based porous scaffold (volume of 75 mm^3) with immediate or delayed (2 weeks) fat graft injection (4 cm^3). In the delayed group, it was hypothesized that the scaffold would undergo host integration and neovascularization, preparing a rich vascular bed for adipocyte adhesion and proliferation. After 6 months of incubation, angiogenesis and adipogenesis were noted in all groups. However, the relative adipose tissue content in the scaffold is greatest in the delayed lipoinjection group, compared with that in the immediate group (47.32% vs 39.67%). Encouragingly, this result is similar to the ratio of adiposity in native porcine adipose tissue (44.97%). This technique has not yet been translated in humans.

15.3.5 Current limitations

Currently, researchers are faced with difficulty vascularizing large-volume adipose tissue using the latest 3D bioprinters. Furthermore, a wide-scale production of clinically suitable 3D-bioprinted adipose tissue is hindered by the exorbitant cost associated with scaling up to complex Good Manufacturing Practice (GMP)-certified laboratory practices [456–460]. Recent introduction of novel ITOP 3D bioprinters that can deposit multiple polymeric and cellular materials to produce large vascularized tissue appears promising [110].

15.4 Conclusions

Researchers have faced difficulty in producing adipose tissue in clinically relevant large volumes using standard tissue engineering practices. Most studies have used

ADSCs with varying success. However, it requires laborious ex vivo preparation, long-term tissue integration, and volume maintenance is uncertain and its potential to support tumorigenesis needs to be thoroughly addressed prior to therapeutic application. Using mature adipocytes may bypass such risks. Nevertheless, adipogenesis may need to be supported by inclusion of growth factors, endothelial precursor cells, and immune cells. Unfortunately, adipocytes and endothelial cells have disparate preferred culture conditions, requiring compromised solutions. With the advent of 3D printing and subsequent 3D bioprinting technology, investigators are now able to fabricate custom-shaped, complex internal porous design scaffolds with multiple cell types. Building on promising in vitro and animal studies, 3D bioprinting of adipose tissue is poised for clinical translation.

Conflict of interest

The authors declare no conflict of interests.

Declarations

The content of this chapter has not been submitted or published elsewhere. There was no source of funding for the chapter. The authors declare that there is no source of financial or other support, or any financial or professional relationships, which may pose a competing interest. All the authors contributed to the preparation of this manuscript. The manuscript has been seen and approved by all the authors. No color reproduction is required in this publication.

References

[1] Ng SK, Hare RM, Kuang RJ, Smith KM, Brown BJ, Hunter-Smith DJ. Breast reconstruction post mastectomy: patient satisfaction and decision making. Ann Plast Surg 2016;76(6):640–4.
[2] DeSantis C, Ma J, Bryan L, Jemal A. Breast cancer statistics, 2013. CA Cancer J Clin 2014;64:52–62.
[3] Howlader N, Noone AM, Krapcho M, et al. SEER cancer statistics review, 1975–2013. Bethesda, MD: National Cancer Institute; 2015.
[4] Shons AR, Mosiello G. Postmastectomy breast reconstruction: current techniques. Cancer Control 2001;8:419–26.
[5] Al-Ghazal SK, Fallowfield L, Blamey RW. Comparison of psychological aspects and patient satisfaction following breast conserving surgery, simple mastectomy and breast reconstruction. Eur J Cancer 2000;36:1938–43.
[6] Wilkins EG, Cederna PS, Lowery JC, et al. Prospective analysis of psychosocial outcomes in breast reconstruction: one-year postoperative results from the Michigan Breast Reconstruction Outcome Study. Plast Reconstr Surg 2000;106:1014–25. discussion 1026–1017.
[7] Nano MT, Gill PG, Kollias J, Bochner MA, Malycha P, Winefield HR. Psychological impact and cosmetic outcome of surgical breast cancer strategies. ANZ J Surg 2005;75:940–7.

[8] Neto MS, de Aguiar Menezes MV, Moreira JR, Garcia EB, Abla LE, Ferreira LM. Sexuality after breast reconstruction post mastectomy. Aesthetic Plast Surg 2013;37:643–7.

[9] Kalus R, Dixon Swartz J, Metzger SC. Optimizing safety, predictability, and aesthetics in direct to implant immediate breast reconstruction: evolution of surgical technique. Ann Plast Surg 2016;76(Suppl 4):S320–7.

[10] Ascherman JA, Zeidler K, Morrison KA, et al. Carbon dioxide-based versus saline tissue expansion for breast reconstruction: results of the XPAND prospective, randomized clinical trial. Plast Reconstr Surg 2016;138:1161–70.

[11] Henriksen TF, Holmich LR, Fryzek JP, et al. Incidence and severity of short-term complications after breast augmentation: results from a nationwide breast implant registry. Ann Plast Surg 2003;51:531–9.

[12] Siggelkow W, Klosterhalfen B, Klinge U, Rath W, Faridi A. Analysis of local complications following explantation of silicone breast implants. Breast 2004;13:122–8.

[13] Handel N, Garcia ME, Wixtrom R. Breast implant rupture: causes, incidence, clinical impact, and management. Plast Reconstr Surg 2013;132:1128–37.

[14] Spear SL, Onyewu C. Staged breast reconstruction with saline-filled implants in the irradiated breast: recent trends and therapeutic implications. Plast Reconstr Surg 2000;105:930–42.

[15] Doren EL, Miranda RN, Selber JC, et al. United States Epidemiology of Breast Implant-Associated Anaplastic Large Cell Lymphoma. Plast Reconstr Surg 2017;139(5):1042–50.

[16] Kroll SS. Why autologous tissue? Clin Plast Surg 1998;25:135–43.

[17] Masia J, Clavero JA, Larranaga JR, Alomar X, Pons G, Serret P. Multidetector-row computed tomography in the planning of abdominal perforator flaps. J Plast Reconstr Aesthet Surg 2006;59:594–9.

[18] Rozen WM, Phillips TJ, Stella DL, Ashton MW. Preoperative CT angiography for DIEP flaps: 'must-have' lessons for the radiologist. J Plast Reconstr Aesthet Surg 2009;62:e650–1.

[19] Hamdi M, Khuthaila DK, Van Landuyt K, Roche N, Monstrey S. Double-pedicle abdominal perforator free flaps for unilateral breast reconstruction: new horizons in microsurgical tissue transfer to the breast. J Plast Reconstr Aesthet Surg 2007;60:904–12. discussion 913-904.

[20] Blondeel PN. One hundred free DIEP flap breast reconstructions: a personal experience. Br J Plast Surg 1999;52:104–11.

[21] Healy C, Allen Sr. RJ. The evolution of perforator flap breast reconstruction: twenty years after the first DIEP flap. J Reconstr Microsurg 2014;30:121–5.

[22] Blondeel PN, Hijjawi J, Depypere H, Roche N, Van Landuyt K. Shaping the breast in aesthetic and reconstructive breast surgery: an easy three-step principle. Plast Reconstr Surg 2009;123:455–62.

[23] Nahabedian MY. Achieving ideal breast aesthetics with autologous reconstruction. Gland Surg 2015;4:134–44.

[24] Wang T, He J, Xu H, Ma S, Dong J. Achieving Symmetry in Unilateral DIEP Flap Breast Reconstruction: An Analysis of 126 Cases over 3 Years. Aesthetic Plast Surg 2015;39:63–8.

[25] Chae MP, Rozen WM, Hunter-Smith DJ, Ramakrishnan V. Enhancing breast projection in autologous reconstruction using St Andrew's coning technique and 3D photography: case series. Gland Surg 2016;5(2):99–106.

[26] Gravvanis A, Smith R. W. Shaping the breast in secondary microsurgical breast reconstruction: single- vs. two-esthetic unit reconstruction. Microsurgery 2010;30:509–16.

[27] Mennie JC, Mohanna PN, O'Donoghue JM, Rainsbury R, Cromwell DA. Donor-Site Hernia Repair in Abdominal Flap Breast Reconstruction: A Population-Based Cohort Study of 7929 Patients. Plast Reconstr Surg 2015;136:1–9.
[28] Serletti JM, Moran SL. Microvascular reconstruction of the breast. Semin Surg Oncol 2000;19:264–71.
[29] Wu LC, Bajaj A, Chang DW, Chevray PM. Comparison of donor-site morbidity of SIEA, DIEP, and muscle-sparing TRAM flaps for breast reconstruction. Plast Reconstr Surg 2008;122:702–9.
[30] Arnez ZM, Khan U, Pogorelec D, Planinsek F. Rational selection of flaps from the abdomen in breast reconstruction to reduce donor site morbidity. Br J Plast Surg 1999;52:351–4.
[31] Neuber F. Fettransplantation. Chir Kongr Verhandl Dtsch Ges Chir 1893;22:66.
[32] Coleman SR. Structural fat grafting: more than a permanent filler. Plast Reconstr Surg 2006;118:108S–20S.
[33] Lafontan M. Fat cells: afferent and efferent messages define new approaches to treat obesity. Annu Rev Pharmacol Toxicol 2005;45:119–46.
[34] Kolle SF, Fischer-Nielsen A, Mathiasen AB, et al. Enrichment of autologous fat grafts with ex-vivo expanded adipose tissue-derived stem cells for graft survival: a randomised placebo-controlled trial. Lancet 2013;382:1113–20.
[35] Ersek RA. Transplantation of purified autologous fat: a 3-year follow-up is disappointing. Plast Reconstr Surg 1991;87:219–27. discussion 228.
[36] Largo RD, Tchang LA, Mele V, et al. Efficacy, safety and complications of autologous fat grafting to healthy breast tissue: a systematic review. J Plast Reconstr Aesthet Surg 2014;67:437–48.
[37] Trojahn Kolle S, Oliveri R, Glovinski P, Elberg J, Fischer-Nielsen A, Drzewiecki K. Importance of mesenchymal stem cells in autologous fat grafting: a systematic review of existing studies. J Plast Surg Hand Surg 2012;42:59–68.
[38] Choi M, Small K, Levovitz C, Lee C, Fadl A, Karp NS. The volumetric analysis of fat graft survival in breast reconstruction. Plast Reconstr Surg 2013;131:185–91.
[39] Chung MT, Hyun JS, Lo DD, et al. Micro-computed tomography evaluation of human fat grafts in nude mice. Tissue Eng Part C Methods 2013;19:227–32.
[40] Saint-Cyr M, Rojas K, Colohan S, Brown S. The role of fat grafting in reconstructive and cosmetic breast surgery: a review of the literature. J Reconstr Microsurg 2012;28:99–110.
[41] Matsumoto D, Sato K, Gonda K, et al. Cell-assisted lipotransfer: supportive use of human adipose-derived cells for soft tissue augmentation with lipoinjection. Tissue Eng 2006;12:3375–82.
[42] Moseley TA, Zhu M, Hedrick MH. Adipose-derived stem and progenitor cells as fillers in plastic and reconstructive surgery. Plast Reconstr Surg 2006;118:121S–8S.
[43] Toyserkani NM, Quaade ML, Sorensen JA. Cell-assisted lipotransfer: a systematic review of its efficacy. Aesthetic Plast Surg 2016;40:309–18.
[44] Zhou Y, Wang J, Li H, et al. Efficacy and safety of cell-assisted lipotransfer: a systematic review and meta-analysis. Plast Reconstr Surg 2016;137:44e–57e.
[45] Conde-Green A, Wu I, Graham I, et al. Comparison of 3 techniques of fat grafting and cell-supplemented lipotransfer in athymic rats: a pilot study. Aesthet Surg J 2013;33:713–21.
[46] Gentile P, Orlandi A, Scioli MG, et al. A comparative translational study: the combined use of enhanced stromal vascular fraction and platelet-rich plasma improves fat grafting maintenance in breast reconstruction. Stem Cells Transl Med 2012;1:341–51.

[47] Perez-Cano R, Vranckx JJ, Lasso JM, et al. Prospective trial of adipose-derived regenerative cell (ADRC)-enriched fat grafting for partial mastectomy defects: the RESTORE-2 trial. Eur J Surg Oncol 2012;38:382–9.
[48] Yoshimura K, Asano Y, Aoi N, et al. Progenitor-enriched adipose tissue transplantation as rescue for breast implant complications. Breast J 2010;16:169–75.
[49] Peltoniemi HH, Salmi A, Miettinen S, et al. Stem cell enrichment does not warrant a higher graft survival in lipofilling of the breast: a prospective comparative study. J Plast Reconstr Aesthet Surg 2013;66:1494–503.
[50] Missana MC, Laurent I, Barreau L, Balleyguier C. Autologous fat transfer in reconstructive breast surgery: indications, technique and results. Eur J Surg Oncol 2007;33:685–90.
[51] Weichman KE, Broer PN, Tanna N, et al. The role of autologous fat grafting in secondary microsurgical breast reconstruction. Ann Plast Surg 2013;71:24–30.
[52] Nelissen X, Lhoest F, Preud'Homme L. Refined method of lipofilling following DIEP breast reconstruction: 3D analysis of graft survival. Plast Reconstr Surg Glob Open 2015;3: e526.
[53] Kim HY, Jung BK, Lew DH, Lee DW. Autologous fat graft in the reconstructed breast: fat absorption rate and safety based on sonographic identification. Arch Plast Surg 2014;41:740–7.
[54] Chang EI, Bonillas RG, El-ftesi S, et al. Tissue engineering using autologous microcirculatory beds as vascularized bioscaffolds. FASEB J 2009;23:906–15.
[55] Murphy SV, Atala A. 3D bioprinting of tissues and organs. Nat Biotechnol 2014;32:773–85.
[56] Atala A, Kasper FK, Mikos AG. Engineering complex tissues. Sci Transl Med 2012;4: 160rv112.
[57] Atala A, Bauer SB, Soker S, Yoo JJ, Retik AB. Tissue-engineered autologous bladders for patients needing cystoplasty. Lancet 2006;367:1241–6.
[58] Raya-Rivera A, Esquiliano DR, Yoo JJ, Lopez-Bayghen E, Soker S, Atala A. Tissue-engineered autologous urethras for patients who need reconstruction: an observational study. Lancet 2011;377:1175–82.
[59] Raya-Rivera AM, Esquiliano D, Fierro-Pastrana R, et al. Tissue-engineered autologous vaginal organs in patients: a pilot cohort study. Lancet 2014;384:329–36.
[60] Amini AR, Laurencin CT, Nukavarapu SP. Bone tissue engineering: recent advances and challenges. Crit Rev Biomed Eng 2012;40:363–408.
[61] Bichara DA, O'Sullivan NA, Pomerantseva I, et al. The tissue-engineered auricle: past, present, and future. Tissue Eng Part B Rev 2012;18:51–61.
[62] Ostrovidov S, Hosseini V, Ahadian S, et al. Skeletal muscle tissue engineering: methods to form skeletal myotubes and their applications. Tissue Eng Part B Rev 2014;20:403–36.
[63] Lee YB, Polio S, Lee W, et al. Bio-printing of collagen and VEGF-releasing fibrin gel scaffolds for neural stem cell culture. Exp Neurol 2010;223:645–52.
[64] Jain RK, Au P, Tam J, Duda DG, Fukumura D. Engineering vascularized tissue. Nat Biotechnol 2005;23:821–3.
[65] Mikos AG, Herring SW, Ochareon P, et al. Engineering complex tissues. Tissue Eng 2006;12:3307–39.
[66] Wang YH, Wu JY, Chou PJ, et al. Characterization and evaluation of the differentiation ability of human adipose-derived stem cells growing in scaffold-free suspension culture. Cytotherapy 2014;16:485–95.
[67] Sugihara H, Yonemitsu N, Miyabara S, Yun K. Primary cultures of unilocular fat cells: characteristics of growth in vitro and changes in differentiation properties. Differentiation 1986;31:42–9.

[68] Kural MH, Billiar KL. Regulating tension in three-dimensional culture environments. Exp Cell Res 2013;319:2447–59.
[69] Wang W, Itaka K, Ohba S, et al. 3D spheroid culture system on micropatterned substrates for improved differentiation efficiency of multipotent mesenchymal stem cells. Biomaterials 2009;30:2705–15.
[70] Cukierman E, Pankov R, Yamada KM. Cell interactions with three-dimensional matrices. Curr Opin Cell Biol 2002;14:633–9.
[71] Edelman DB, Keefer EW. A cultural renaissance: in vitro cell biology embraces three-dimensional context. Exp Neurol 2005;192:1–6.
[72] Zhang HH, Kumar S, Barnett AH, Eggo MC. Ceiling culture of mature human adipocytes: use in studies of adipocyte functions. J Endocrinol 2000;164:119–28.
[73] Sugihara H, Yonemitsu N, Miyabara S, Toda S. Proliferation of unilocular fat cells in the primary culture. J Lipid Res 1987;28:1038–45.
[74] Shen JF, Sugawara A, Yamashita J, Ogura H, Sato S. Dedifferentiated fat cells: an alternative source of adult multipotent cells from the adipose tissues. Int J Oral Sci 2011;3:117–24.
[75] Lee W, Debasitis JC, Lee VK, et al. Multi-layered culture of human skin fibroblasts and keratinocytes through three-dimensional freeform fabrication. Biomaterials 2009;30:1587–95.
[76] Boland T, Xu T, Damon B, Cui X. Application of inkjet printing to tissue engineering. Biotechnol J 2006;1:910–7.
[77] Dahms SE, Piechota HJ, Dahiya R, Lue TF, Tanagho EA. Composition and biomechanical properties of the bladder acellular matrix graft: comparative analysis in rat, pig and human. Br J Urol 1998;82:411–9.
[78] Chen F, Yoo JJ, Atala A. Acellular collagen matrix as a possible "off the shelf" biomaterial for urethral repair. Urology 1999;54:407–10.
[79] Probst M, Dahiya R, Carrier S, Tanagho EA. Reproduction of functional smooth muscle tissue and partial bladder replacement. Br J Urol 1997;79:505–15.
[80] Brown BN, Valentin JE, Stewart-Akers AM, McCabe GP, Badylak SF. Macrophage phenotype and remodeling outcomes in response to biologic scaffolds with and without a cellular component. Biomaterials 2009;30:1482–91.
[81] Moroni L, de Wijn JR, van Blitterswijk CA. 3D fiber-deposited scaffolds for tissue engineering: influence of pores geometry and architecture on dynamic mechanical properties. Biomaterials 2006;27:974–85.
[82] Kretlow JD, Mikos AG. Founder's award to Antonios G. Mikos, Ph.D., 2011 Society for Biomaterials annual meeting and exposition, Orlando, Florida, April 13-16, 2011: Bones to biomaterials and back again--20 years of taking cues from nature to engineer synthetic polymer scaffolds. J Biomed Mater Res A 2011;98:323–31.
[83] Tan JY, Chua CK, Leong KF. Fabrication of channeled scaffolds with ordered array of micro-pores through microsphere leaching and indirect Rapid Prototyping technique. Biomed Microdevices 2013;15:83–96.
[84] Hutmacher DW. Scaffolds in tissue engineering bone and cartilage. Biomaterials 2000;21:2529–43.
[85] Hollister SJ. Porous scaffold design for tissue engineering. Nat Mater 2005;4:518–24.
[86] Derby B. Printing and prototyping of tissues and scaffolds. Science 2012;338:921–6.
[87] Stachowiak AN, Bershteyn A, Tzatzalos E, Irvine DJ. Bioactive hydrogels with an ordered cellular structure combine interconnected macroporosity and robust mechanical properties. Adv Mater 2005;17:399–403.
[88] Cabodi M, Choi NW, Gleghorn JP, Lee CS, Bonassar LJ, Stroock AD. A microfluidic biomaterial. J Am Chem Soc 2005;127:13788–9.

[89] Ling Y, Rubin J, Deng Y, et al. A cell-laden microfluidic hydrogel. Lab Chip 2007;7:756–62.
[90] Mironov V, Visconti RP, Kasyanov V, Forgacs G, Drake CJ, Markwald RR. Organ printing: tissue spheroids as building blocks. Biomaterials 2009;30:2164–74.
[91] Jones N. Science in three dimensions: the print revolution. Nature 2012;487:22–3.
[92] Ferris CJ, Gilmore KG, Wallace GG, het Panhuis M. Biofabrication: an overview of the approaches used for printing of living cells. Appl Microbiol Biotechnol 2013;97:4243–58.
[93] Xu T, Zhao W, Zhu JM, Albanna MZ, Yoo JJ, Atala A. Complex heterogeneous tissue constructs containing multiple cell types prepared by inkjet printing technology. Biomaterials 2013;34:130–9.
[94] Durmus NG, Tasoglu S, Demirci U. Bioprinting: functional droplet networks. Nat Mater 2013;12:478–9.
[95] Chae MP, Hunter-Smith DJ, Rozen WM. Image-guided 3D-printing and haptic modeling in plastic surgery. In: Saba L, Rozen WM, Alonso-Burgos A, Ribuffo D, editors. Imaging in plastic surgery. London: CRC Taylor and Francis Press; 2014.
[96] Gerstle TL, Ibrahim AM, Kim PS, Lee BT, Lin SJ. A plastic surgery application in evolution: three-dimensional printing. Plast Reconstr Surg 2014;133:446–51.
[97] Chae MP, Rozen WM, McMenamin PG, Findlay MW, Spychal RT, Hunter-Smith DJ. Emerging applications of bedside 3D printing in plastic surgery. Front Surg 2015;2:25.
[98] Kamali P, Dean D, Skoracki R, et al. The current role of three-dimensional (3D) printing in plastic surgery. Plast Reconstr Surg 2016;137(3):1045–55.
[99] Levy GN, Schindel R, Kruth JP. Rapid manufacturing and rapid tooling with layer manufacturing (LM) technologies, state of the art and future perspectives. CIRP Ann Manuf Technol 2003;52:589–609.
[100] Sealy W. Additive manufacturing as a disruptive technology: how to avoid the pitfall. Am J Eng Technol Res 2011;11:86–93.
[101] Hoy MB. 3D printing: making things at the library. Med Ref Serv Q 2013;32:93–9.
[102] Klein GT, Lu Y, Wang MY. 3D printing and neurosurgery—ready for prime time? World Neurosurg 2013;80:233–5.
[103] Schuurman W, Khristov V, Pot MW, van Weeren PR, Dhert WJ, Malda J. Bioprinting of hybrid tissue constructs with tailorable mechanical properties. Biofabrication 2011;3: 021001.
[104] Shim JH, Lee JS, Kim JY, Cho DW. Bioprinting of a mechanically enhanced three-dimensional dual cell-laden construct for osteochondral tissue engineering using a multi-head tissue/organ building system. J Micromech Microeng 2012;22: 085014.
[105] Peltola SM, Melchels FP, Grijpma DW, Kellomaki M. A review of rapid prototyping techniques for tissue engineering purposes. Ann Med 2008;40:268–80.
[106] Fedorovich NE, De Wijn JR, Verbout AJ, Alblas J, Dhert WJ. Three-dimensional fiber deposition of cell-laden, viable, patterned constructs for bone tissue printing. Tissue Eng Part A 2008;14:127–33.
[107] Jakab K, Neagu A, Mironov V, Markwald RR, Forgacs G. Engineering biological structures of prescribed shape using self-assembling multicellular systems. Proc Natl Acad Sci U S A 2004;101:2864–9.
[108] Landers R, Hubner U, Schmelzeisen R, Mulhaupt R. Rapid prototyping of scaffolds derived from thermoreversible hydrogels and tailored for applications in tissue engineering. Biomaterials 2002;23:4437–47.
[109] Joddar B, Garcia E, Casas A, Stewart CM. Development of functionalized multi-walled carbon-nanotube-based alginate hydrogels for enabling biomimetic technologies. Sci Rep 2016;6:32456.

[110] Kang HW, Lee SJ, Ko IK, Kengla C, Yoo JJ, Atala A. A 3D bioprinting system to produce human-scale tissue constructs with structural integrity. Nat Biotechnol 2016;34:312–9.
[111] Fong EL, Watson BM, Kasper FK, Mikos AG. Building bridges: leveraging interdisciplinary collaborations in the development of biomaterials to meet clinical needs. Adv Mater 2012;24:4995–5013.
[112] Lu L, Zhu X, Valenzuela RG, Currier BL, Yaszemski MJ. Biodegradable polymer scaffolds for cartilage tissue engineering. Clin Orthop Relat Res 2001;S251–70.
[113] Butler DL, Goldstein SA, Guldberg RE, et al. The impact of biomechanics in tissue engineering and regenerative medicine. Tissue Eng Part B Rev 2009;15:477–84.
[114] Silva NA, Cooke MJ, Tam RY, et al. The effects of peptide modified gellan gum and olfactory ensheathing glia cells on neural stem/progenitor cell fate. Biomaterials 2012;33:6345–54.
[115] Vidal G, Blanchi T, Mieszawska AJ, et al. Enhanced cellular adhesion on titanium by silk functionalized with titanium binding and RGD peptides. Acta Biomater 2013;9:4935–43.
[116] Engelhardt EM, Micol LA, Houis S, et al. A collagen-poly(lactic acid-co-varepsilon-caprolactone) hybrid scaffold for bladder tissue regeneration. Biomaterials 2011;32:3969–76.
[117] Serrano MC, Pagani R, Vallet-Regi M, et al. In vitro biocompatibility assessment of poly(epsilon-caprolactone) films using L929 mouse fibroblasts. Biomaterials 2004;25:5603–11.
[118] Sun H, Mei L, Song C, Cui X, Wang P. The in vivo degradation, absorption and excretion of PCL-based implant. Biomaterials 2006;27:1735–40.
[119] Chang CC, Boland ED, Williams SK, Hoying JB. Direct-write bioprinting three-dimensional biohybrid systems for future regenerative therapies. J Biomed Mater Res B Appl Biomater 2011;98:160–70.
[120] Lippens E, Swennen I, Girones J, et al. Cell survival and proliferation after encapsulation in a chemically modified Pluronic(R) F127 hydrogel. J Biomater Appl 2013;27:828–39.
[121] Brivanlou AH, Gage FH, Jaenisch R, Jessell T, Melton D, Rossant J. Stem cells. Setting standards for human embryonic stem cells. Science 2003;300:913–6.
[122] Condic ML, Rao M. Regulatory issues for personalized pluripotent cells. Stem Cells 2008;26:2753–8.
[123] Hochedlinger K, Jaenisch R. Nuclear transplantation, embryonic stem cells, and the potential for cell therapy. N Engl J Med 2003;349:275–86.
[124] Bae H, Puranik AS, Gauvin R, et al. Building vascular networks. Sci Transl Med 2012;4:160ps123.
[125] Lovett M, Lee K, Edwards A, Kaplan DL. Vascularization strategies for tissue engineering. Tissue Eng Part B Rev 2009;15:353–70.
[126] Cilento BG, Freeman MR, Schneck FX, Retik AB, Atala A. Phenotypic and cytogenetic characterization of human bladder urothelia expanded in vitro. J Urol 1994;152:665–70.
[127] Zhang YY, Ludwikowski B, Hurst R, Frey P. Expansion and long-term culture of differentiated normal rat urothelial cells in vitro. In Vitro Cell Dev Biol Anim 2001;37:419–29.
[128] Caplan AI, Correa D. The MSC: an injury drugstore. Cell Stem Cell 2011;9:11–5.
[129] Klebe RJ. Cytoscribing: a method for micropositioning cells and the construction of two- and three-dimensional synthetic tissues. Exp Cell Res 1988;179:362–73.
[130] Xu T, Jin J, Gregory C, Hickman JJ, Boland T. Inkjet printing of viable mammalian cells. Biomaterials 2005;26:93–9.
[131] Cui X, Boland T, D'Lima DD, Lotz MK. Thermal inkjet printing in tissue engineering and regenerative medicine. Recent Pat Drug Deliv Formul 2012;6:149–55.

[132] Cohen DL, Malone E, Lipson H, Bonassar LJ. Direct freeform fabrication of seeded hydrogels in arbitrary geometries. Tissue Eng 2006;12:1325–35.
[133] Iwami K, Noda T, Ishida K, Morishima K, Nakamura M, Umeda N. Bio rapid prototyping by extruding/aspirating/refilling thermoreversible hydrogel. Biofabrication 2010;2: 014108.
[134] Shor L, Guceri S, Chang R, et al. Precision extruding deposition (PED) fabrication of polycaprolactone (PCL) scaffolds for bone tissue engineering. Biofabrication 2009;1: 015003.
[135] Barron JA, Wu P, Ladouceur HD, Ringeisen BR. Biological laser printing: a novel technique for creating heterogeneous 3-dimensional cell patterns. Biomed Microdevices 2004;6:139–47.
[136] Guillemot F, Souquet A, Catros S, et al. High-throughput laser printing of cells and biomaterials for tissue engineering. Acta Biomater 2010;6:2494–500.
[137] Guillotin B, Souquet A, Catros S, et al. Laser assisted bioprinting of engineered tissue with high cell density and microscale organization. Biomaterials 2010;31:7250–6.
[138] Skardal A, Mack D, Kapetanovic E, et al. Bioprinted amniotic fluid-derived stem cells accelerate healing of large skin wounds. Stem Cells Transl Med 2012;1:792–802.
[139] Cui X, Breitenkamp K, Finn MG, Lotz M, D'Lima DD. Direct human cartilage repair using three-dimensional bioprinting technology. Tissue Eng Part A 2012;18:1304–12.
[140] De Coppi P, Bartsch Jr. G, Siddiqui MM, et al. Isolation of amniotic stem cell lines with potential for therapy. Nat Biotechnol 2007;25:100–6.
[141] Duan B, Hockaday LA, Kang KH, Butcher JT. 3D bioprinting of heterogeneous aortic valve conduits with alginate/gelatin hydrogels. J Biomed Mater Res A 2013;101:1255–64.
[142] Norotte C, Marga FS, Niklason LE, Forgacs G. Scaffold-free vascular tissue engineering using bioprinting. Biomaterials 2009;30:5910–7.
[143] Xu F, Celli J, Rizvi I, Moon S, Hasan T, Demirci U. A three-dimensional in vitro ovarian cancer coculture model using a high-throughput cell patterning platform. Biotechnol J 2011;6:204–12.
[144] Michael S, Sorg H, Peck CT, et al. Tissue engineered skin substitutes created by laser-assisted bioprinting form skin-like structures in the dorsal skin fold chamber in mice. PLoS One 2013;8: e57741.
[145] Keriquel V, Guillemot F, Arnault I, et al. In vivo bioprinting for computer- and robotic-assisted medical intervention: preliminary study in mice. Biofabrication 2010;2: 014101.
[146] Okamoto T, Suzuki T, Yamamoto N. Microarray fabrication with covalent attachment of DNA using bubble jet technology. Nat Biotechnol 2000;18:438–41.
[147] Goldmann T, Gonzalez JS. DNA-printing: utilization of a standard inkjet printer for the transfer of nucleic acids to solid supports. J Biochem Biophys Methods 2000;42:105–10.
[148] Xu T, Kincaid H, Atala A, Yoo JJ. High-throughput production of single-cell microparticles using an inkjet printing technology. J Manuf Sci Eng 2008;130: 021017.
[149] Cui X, Dean D, Ruggeri ZM, Boland T. Cell damage evaluation of thermal inkjet printed Chinese hamster ovary cells. Biotechnol Bioeng 2010;106:963–9.
[150] Tekin E, Smith PJ, Schubert US. Inkjet printing as a deposition and patterning tool for polymers and inorganic particles. Soft Matter 2008;4:703–13.
[151] Tasoglu S, Demirci U. Bioprinting for stem cell research. Trends Biotechnol 2013;31:10–9.
[152] Murphy SV, Skardal A, Atala A. Evaluation of hydrogels for bio-printing applications. J Biomed Mater Res A 2013;101:272–84.
[153] Khalil S, Sun W. Biopolymer deposition for freeform fabrication of hydrogel tissue constructs. Mater Sci Eng C 2007;27:469–78.

[154] Fedorovich NE, Swennen I, Girones J, et al. Evaluation of photocrosslinked Lutrol hydrogel for tissue printing applications. Biomacromolecules 2009;10:1689–96.
[155] Chang R, Nam J, Sun W. Effects of dispensing pressure and nozzle diameter on cell survival from solid freeform fabrication-based direct cell writing. Tissue Eng Part A 2008;14:41–8.
[156] Jakab K, Damon B, Neagu A, Kachurin A, Forgacs G. Three-dimensional tissue constructs built by bioprinting. Biorheology 2006;43:509–13.
[157] Visser J, Peters B, Burger TJ, et al. Biofabrication of multi-material anatomically shaped tissue constructs. Biofabrication 2013;5: 035007.
[158] Marga F, Jakab K, Khatiwala C, et al. Toward engineering functional organ modules by additive manufacturing. Biofabrication 2012;4: 022001.
[159] Mironov V, Kasyanov V, Markwald RR. Organ printing: from bioprinter to organ biofabrication line. Curr Opin Biotechnol 2011;22:667–73.
[160] Smith CM, Stone AL, Parkhill RL, et al. Three-dimensional bioassembly tool for generating viable tissue-engineered constructs. Tissue Eng 2004;10:1566–76.
[161] Bohandy J, Kim B, Adrian F. Metal deposition from a supported metal film using an excimer laser. J Appl Phys 1986;60:1538–9.
[162] Barron JA, Ringeisen BR, Kim H, Spargo BJ, Chrisey DB. Application of laser printing to mammalian cells. Thin Solid Films 2004;453:383–7.
[163] Ringeisen BR, Kim H, Barron JA, et al. Laser printing of pluripotent embryonal carcinoma cells. Tissue Eng 2004;10:483–91.
[164] Chrisey DB. MATERIALS PROCESSING: the power of direct writing. Science 2000;289:879–81.
[165] Colina M, Serra P, Fernandez-Pradas JM, Sevilla L, Morenza JL. DNA deposition through laser induced forward transfer. Biosens Bioelectron 2005;20:1638–42.
[166] Hopp B, Smausz T, Kresz N, et al. Survival and proliferative ability of various living cell types after laser-induced forward transfer. Tissue Eng 2005;11:1817–23.
[167] Gruene M, Deiwick A, Koch L, et al. Laser printing of stem cells for biofabrication of scaffold-free autologous grafts. Tissue Eng Part C Methods 2011;17:79–87.
[168] Koch L, Kuhn S, Sorg H, et al. Laser printing of skin cells and human stem cells. Tissue Eng Part C Methods 2010;16:847–54.
[169] Guillotin B, Guillemot F. Cell patterning technologies for organotypic tissue fabrication. Trends Biotechnol 2011;29:183–90.
[170] Bielli A, Scioli MG, Gentile P, et al. Adult adipose-derived stem cells and breast cancer: a controversial relationship. Springerplus 2014;3:345.
[171] Warnke PH, Springer IN, Wiltfang J, et al. Growth and transplantation of a custom vascularised bone graft in a man. Lancet 2004;364:766–70.
[172] Warnke PH, Wiltfang J, Springer I, et al. Man as living bioreactor: fate of an exogenously prepared customized tissue-engineered mandible. Biomaterials 2006;27:3163–7.
[173] Eto H, Suga H, Matsumoto D, et al. Characterization of structure and cellular components of aspirated and excised adipose tissue. Plast Reconstr Surg 2009;124:1087–97.
[174] Christiaens V, Lijnen HR. Angiogenesis and development of adipose tissue. Mol Cell Endocrinol 2010;318:2–9.
[175] Chiu YC, Cheng MH, Uriel S, Brey EM. Materials for engineering vascularized adipose tissue. J Tissue Viability 2011;20:37–48.
[176] Hutley LJ, Herington AC, Shurety W, et al. Human adipose tissue endothelial cells promote preadipocyte proliferation. Am J Physiol Endocrinol Metab 2001;281:E1037–44.
[177] Cao Y. Angiogenesis modulates adipogenesis and obesity. J Clin Invest 2007;117:2362–8.

[178] Rophael JA, Craft RO, Palmer JA, et al. Angiogenic growth factor synergism in a murine tissue engineering model of angiogenesis and adipogenesis. Am J Pathol 2007;171:2048–57.

[179] Tsuji T, Yamaguchi K, Kikuchi R, et al. Promotion of adipogenesis by an EP2 receptor agonist via stimulation of angiogenesis in pulmonary emphysema. Prostaglandins Other Lipid Mediat 2014;112:9–15.

[180] Li J, Qiao X, Yu M, et al. Secretory factors from rat adipose tissue explants promote adipogenesis and angiogenesis. Artif Organs 2014;38:E33–45.

[181] Young DA, Choi YS, Engler AJ, Christman KL. Stimulation of adipogenesis of adult adipose-derived stem cells using substrates that mimic the stiffness of adipose tissue. Biomaterials 2013;34:8581–8.

[182] Ingber DE. Cellular mechanotransduction: putting all the pieces together again. FASEB J 2006;20:811–27.

[183] Juliano RL, Haskill S. Signal transduction from the extracellular matrix. J Cell Biol 1993;120:577–85.

[184] Maniotis AJ, Chen CS, Ingber DE. Demonstration of mechanical connections between integrins, cytoskeletal filaments, and nucleoplasm that stabilize nuclear structure. Proc Natl Acad Sci U S A 1997;94:849–54.

[185] Prusty D, Park BH, Davis KE, Farmer SR. Activation of MEK/ERK signaling promotes adipogenesis by enhancing peroxisome proliferator-activated receptor gamma (PPARgamma) and C/EBPalpha gene expression during the differentiation of 3T3-L1 preadipocytes. J Biol Chem 2002;277:46226–32.

[186] Tanabe Y, Koga M, Saito M, Matsunaga Y, Nakayama K. Inhibition of adipocyte differentiation by mechanical stretching through ERK-mediated downregulation of PPARgamma2. J Cell Sci 2004;117:3605–14.

[187] McBeath R, Pirone DM, Nelson CM, Bhadriraju K, Chen CS. Cell shape, cytoskeletal tension, and RhoA regulate stem cell lineage commitment. Dev Cell 2004;6:483–95.

[188] Shoham N, Gottlieb R, Sharabani-Yosef O, Zaretsky U, Benayahu D, Gefen A. Static mechanical stretching accelerates lipid production in 3T3-L1 adipocytes by activating the MEK signaling pathway. Am J Physiol Cell Physiol 2012;302:C429–41.

[189] Tanabe Y, Saito MT, Nakayama K. Mechanical stretching and signaling pathways in adipogenesis. Stud Mechanobiol Tissue Eng Biomater 2013;16:35–62.

[190] Ruknudin A, Sachs F, Bustamante JO. Stretch-activated ion channels in tissue-cultured chick heart. Am J Physiol 1993;264:H960–72.

[191] Samani A, Bishop J, Luginbuhl C, Plewes DB. Measuring the elastic modulus of ex vivo small tissue samples. Phys Med Biol 2003;48:2183–98.

[192] Comley K, Fleck NA. A micromechanical model for the Young's modulus of adipose tissue. Int J Solids Struct 2010;47:2982–90.

[193] Wiggenhauser PS, Muller DF, Melchels FP, et al. Engineering of vascularized adipose constructs. Cell Tissue Res 2012;347:747–57.

[194] Yu C, Bianco J, Brown C, et al. Porous decellularized adipose tissue foams for soft tissue regeneration. Biomaterials 2013;34:3290–302.

[195] Cheung HK, Han TT, Marecak DM, Watkins JF, Amsden BG, Flynn LE. Composite hydrogel scaffolds incorporating decellularized adipose tissue for soft tissue engineering with adipose-derived stem cells. Biomaterials 2014;35:1914–23.

[196] Bellas E, Lo TJ, Fournier EP, et al. Injectable silk foams for soft tissue regeneration. Adv Healthc Mater 2015;4:452–9.

[197] Kilian KA, Bugarija B, Lahn BT, Mrksich M. Geometric cues for directing the differentiation of mesenchymal stem cells. Proc Natl Acad Sci U S A 2010;107:4872–7.

[198] Stosich MS, Bastian B, Marion NW, Clark PA, Reilly G, Mao JJ. Vascularized adipose tissue grafts from human mesenchymal stem cells with bioactive cues and microchannel conduits. Tissue Eng 2007;13:2881–90.

[199] Kaully T, Kaufman-Francis K, Lesman A, Levenberg S. Vascularization—the conduit to viable engineered tissues. Tissue Eng Part B Rev 2009;15:159–69.

[200] Loh QL, Choong C. Three-dimensional scaffolds for tissue engineering applications: role of porosity and pore size. Tissue Eng Part B Rev 2013;19:485–502.

[201] Abrahamson DR. Recent studies on the structure and pathology of basement membranes. J Pathol 1986;149:257–78.

[202] Patrick Jr. CW. Tissue engineering strategies for adipose tissue repair. Anat Rec 2001;263:361–6.

[203] Kim KJ, Joe YA, Kim MK, et al. Silica nanoparticles increase human adipose tissue-derived stem cell proliferation through ERK1/2 activation. Int J Nanomedicine 2015;10:2261–72.

[204] Choi YS, Vincent LG, Lee AR, et al. The alignment and fusion assembly of adipose-derived stem cells on mechanically patterned matrices. Biomaterials 2012;33:6943–51.

[205] Francis MP, Sachs PC, Madurantakam PA, et al. Electrospinning adipose tissue-derived extracellular matrix for adipose stem cell culture. J Biomed Mater Res A 2012;100:1716–24.

[206] Shadjou N, Hasanzadeh M. Bone tissue engineering using silica-based mesoporous nanobiomaterials: recent progress. Mater Sci Eng C Mater Biol Appl 2015;55:401–9.

[207] Smith CM, Christian JJ, Warren WL, Williams SK. Characterizing environmental factors that impact the viability of tissue-engineered constructs fabricated by a direct-write bioassembly tool. Tissue Eng 2007;13:373–83.

[208] Lee W, Pinckney J, Lee V, et al. Three-dimensional bioprinting of rat embryonic neural cells. Neuroreport 2009;20:798–803.

[209] Jakab K, Norotte C, Marga F, Murphy K, Vunjak-Novakovic G, Forgacs G. Tissue engineering by self-assembly and bio-printing of living cells. Biofabrication 2010;2: 022001.

[210] Chang CC, Krishnan L, Nunes SS, et al. Determinants of microvascular network topologies in implanted neovasculatures. Arterioscler Thromb Vasc Biol 2012;32:5–14.

[211] Wang L, Johnson JA, Zhang Q, Beahm EK. Combining decellularized human adipose tissue extracellular matrix and adipose-derived stem cells for adipose tissue engineering. Acta Biomater 2013;9:8921–31.

[212] Wang M, Yu L. Transplantation of adipose-derived stem cells combined with decellularized cartilage ECM: a novel approach to nasal septum perforation repair. Med Hypotheses 2014;82:781–3.

[213] Liu X, Holzwarth JM, Ma PX. Functionalized synthetic biodegradable polymer scaffolds for tissue engineering. Macromol Biosci 2012;12:911–9.

[214] Hosseinzadeh E, Davarpanah M, Hassanzadeh Nemati N, Tavakoli SA. Fabrication of a hard tissue replacement using natural hydroxyapatite derived from bovine bones by thermal decomposition method. Int J Organ Transplant Med 2014;5:23–31.

[215] Li Y, Meng H, Liu Y, Lee BP. Fibrin gel as an injectable biodegradable scaffold and cell carrier for tissue engineering. ScientificWorldJournal 2015;2015:685690.

[216] Taghiabadi E, Nasri S, Shafieyan S, Jalili Firoozinezhad S, Aghdami N. Fabrication and characterization of spongy denuded amniotic membrane based scaffold for tissue engineering. Cell J 2015;16:476–87.

[217] Mano JF, Silva GA, Azevedo HS, et al. Natural origin biodegradable systems in tissue engineering and regenerative medicine: present status and some moving trends. J R Soc Interface 2007;4:999–1030.

[218] Zhang YS, Gao JH, Lu F, Zhu M, Liao YJ. Cellular compatibility of type collagen I scaffold and human adipose-derived stem cells. Nan Fang Yi Ke Da Xue Xue Bao 2007;27:223–5.

[219] Lequeux C, Oni G, Wong C, et al. Subcutaneous fat tissue engineering using autologous adipose-derived stem cells seeded onto a collagen scaffold. Plast Reconstr Surg 2012;130:1208–17.

[220] Ferraro GA, De Francesco F, Nicoletti G, et al. Human adipose CD34+ CD90+ stem cells and collagen scaffold constructs grafted in vivo fabricate loose connective and adipose tissues. J Cell Biochem 2013;114:1039–49.

[221] Chan EC, Kuo SM, Kong AM, et al. Three dimensional collagen scaffold promotes intrinsic vascularisation for tissue engineering applications. PLoS One 2016;11: e0149799.

[222] Xu FT, Liang ZJ, Li HM, et al. Ginsenoside Rg1 and platelet-rich fibrin enhance human breast adipose-derived stem cell function for soft tissue regeneration. Oncotarget 2016;7:35390–403.

[223] Vashi AV, Abberton KM, Thomas GP, et al. Adipose tissue engineering based on the controlled release of fibroblast growth factor-2 in a collagen matrix. Tissue Eng 2006;12:3035–43.

[224] von Heimburg D, Zachariah S, Low A, Pallua N. Influence of different biodegradable carriers on the in vivo behavior of human adipose precursor cells. Plast Reconstr Surg 2001;108:411–20. discussion 421-412.

[225] Hemmrich K, von Heimburg D, Rendchen R, Di Bartolo C, Milella E, Pallua N. Implantation of preadipocyte-loaded hyaluronic acid-based scaffolds into nude mice to evaluate potential for soft tissue engineering. Biomaterials 2005;26:7025–37.

[226] Wittmann K, Dietl S, Ludwig N, et al. Engineering vascularized adipose tissue using the stromal-vascular fraction and fibrin hydrogels. Tissue Eng Part A 2015;21:1343–53.

[227] Hsueh YS, Chen YS, Tai HC, et al. Laminin-alginate beads as preadipocyte carriers to enhance adipogenesis in vitro and in vivo. Tissue Eng Part A 2017;23(5–6):185–94.

[228] Choi JS, Yang HJ, Kim BS, et al. Human extracellular matrix (ECM) powders for injectable cell delivery and adipose tissue engineering. J Control Release 2009;139:2–7.

[229] Wu I, Nahas Z, Kimmerling KA, Rosson GD, Elisseeff JH. An injectable adipose matrix for soft-tissue reconstruction. Plast Reconstr Surg 2012;129:1247–57.

[230] Debels H, Gerrand YW, Poon CJ, Abberton KM, Morrison WA, Mitchell GM. An adipogenic gel for surgical reconstruction of the subcutaneous fat layer in a rat model. J Tissue Eng Regen Med 2017;11(4):1230–41.

[231] Kawaguchi N, Toriyama K, Nicodemou-Lena E, Inou K, Torii S, Kitagawa Y. De novo adipogenesis in mice at the site of injection of basement membrane and basic fibroblast growth factor. Proc Natl Acad Sci U S A 1998;95:1062–6.

[232] Walton RL, Beahm EK, Wu L. De novo adipose formation in a vascularized engineered construct. Microsurgery 2004;24:378–84.

[233] Fischbach C, Spruss T, Weiser B, et al. Generation of mature fat pads in vitro and in vivo utilizing 3-D long-term culture of 3T3-L1 preadipocytes. Exp Cell Res 2004;300:54–64.

[234] Neubauer M, Hacker M, Bauer-Kreisel P, et al. Adipose tissue engineering based on mesenchymal stem cells and basic fibroblast growth factor in vitro. Tissue Eng 2005;11:1840–51.

[235] Alhadlaq A, Tang M, Mao JJ. Engineered adipose tissue from human mesenchymal stem cells maintains predefined shape and dimension: implications in soft tissue augmentation and reconstruction. Tissue Eng 2005;11:556–66.

[236] Rossi E, Gerges I, Tocchio A, et al. Biologically and mechanically driven design of an RGD-mimetic macroporous foam for adipose tissue engineering applications. Biomaterials 2016;104:65–77.
[237] Flynn L, Prestwich GD, Semple JL, Woodhouse KA. Adipose tissue engineering in vivo with adipose-derived stem cells on naturally derived scaffolds. J Biomed Mater Res A 2009;89:929–41.
[238] Gentleman E, Nauman EA, Livesay GA, Dee KC. Collagen composite biomaterials resist contraction while allowing development of adipocytic soft tissue in vitro. Tissue Eng 2006;12:1639–49.
[239] Rubin JP, Bennett JM, Doctor JS, Tebbets BM, Marra KG. Collagenous microbeads as a scaffold for tissue engineering with adipose-derived stem cells. Plast Reconstr Surg 2007;120:414–24.
[240] Toda S, Uchihashi K, Aoki S, et al. Adipose tissue-organotypic culture system as a promising model for studying adipose tissue biology and regeneration. Organogenesis 2009;5:50–6.
[241] Kim BS, Kim JS, Lee J. Improvements of osteoblast adhesion, proliferation, and differentiation in vitro via fibrin network formation in collagen sponge scaffold. J Biomed Mater Res A 2013;101:2661–6.
[242] Klar AS, Guven S, Biedermann T, et al. Tissue-engineered dermo-epidermal skin grafts prevascularized with adipose-derived cells. Biomaterials 2014;35:5065–78.
[243] Mauney JR, Nguyen T, Gillen K, Kirker-Head C, Gimble JM, Kaplan DL. Engineering adipose-like tissue in vitro and in vivo utilizing human bone marrow and adipose-derived mesenchymal stem cells with silk fibroin 3D scaffolds. Biomaterials 2007;28:5280–90.
[244] Lucero HA, Kagan HM. Lysyl oxidase: an oxidative enzyme and effector of cell function. Cell Mol Life Sci 2006;63:2304–16.
[245] Hong L, Peptan I, Clark P, Mao JJ. Ex vivo adipose tissue engineering by human marrow stromal cell seeded gelatin sponge. Ann Biomed Eng 2005;33:511–7.
[246] Tonello C, Vindigni V, Zavan B, et al. In vitro reconstruction of an endothelialized skin substitute provided with a microcapillary network using biopolymer scaffolds. FASEB J 2005;19:1546–8.
[247] La Gatta A, De Rosa M, Marzaioli I, Busico T, Schiraldi C. A complete hyaluronan hydrodynamic characterization using a size exclusion chromatography-triple detector array system during in vitro enzymatic degradation. Anal Biochem 2010;404:21–9.
[248] Yoon IS, Chung CW, Sung JH, et al. Proliferation and chondrogenic differentiation of human adipose-derived mesenchymal stem cells in porous hyaluronic acid scaffold. J Biosci Bioeng 2011;112:402–8.
[249] Desiderio V, De Francesco F, Schiraldi C, et al. Human Ng2+ adipose stem cells loaded in vivo on a new crosslinked hyaluronic acid-Lys scaffold fabricate a skeletal muscle tissue. J Cell Physiol 2013;228:1762–73.
[250] Mathews S, Mathew SA, Gupta PK, Bhonde R, Totey S. Glycosaminoglycans enhance osteoblast differentiation of bone marrow derived human mesenchymal stem cells. J Tissue Eng Regen Med 2014;8:143–52.
[251] Wang Y, Kim HJ, Vunjak-Novakovic G, Kaplan DL. Stem cell-based tissue engineering with silk biomaterials. Biomaterials 2006;27:6064–82.
[252] Rockwood DN, Preda RC, Yucel T, Wang X, Lovett ML, Kaplan DL. Materials fabrication from Bombyx mori silk fibroin. Nat Protoc 2011;6:1612–31.
[253] Altman GH, Diaz F, Jakuba C, et al. Silk-based biomaterials. Biomaterials 2003;24:401–16.
[254] Bellas E, Marra KG, Kaplan DL. Sustainable three-dimensional tissue model of human adipose tissue. Tissue Eng Part C Methods 2013;19:745–54.

[255] Kang JH, Gimble JM, Kaplan DL. In vitro 3D model for human vascularized adipose tissue. Tissue Eng Part A 2009;15:2227–36.

[256] Hanken H, Gohler F, Smeets R, et al. Attachment, viability and adipodifferentiation of pre-adipose cells on silk scaffolds with and without co-expressed FGF-2 and VEGF. In Vivo 2016;30:567–72.

[257] Abbott RD, Wang RY, Reagan MR, et al. The use of silk as a scaffold for mature, sustainable unilocular adipose 3D tissue engineered systems. Adv Healthc Mater 2016;5:1667–77.

[258] Choi JH, Bellas E, Gimble JM, Vunjak-Novakovic G, Kaplan DL. Lipolytic function of adipocyte/endothelial cocultures. Tissue Eng Part A 2011;17:1437–44.

[259] Currie LJ, Sharpe JR, Martin R. The use of fibrin glue in skin grafts and tissue-engineered skin replacements: a review. Plast Reconstr Surg 2001;108:1713–26.

[260] Feng X, Clark RA, Galanakis D, Tonnesen MG. Fibrin and collagen differentially regulate human dermal microvascular endothelial cell integrins: stabilization of alphav/beta3 mRNA by fibrin1. J Invest Dermatol 1999;113:913–9.

[261] Janmey PA, Winer JP, Weisel JW. Fibrin gels and their clinical and bioengineering applications. J R Soc Interface 2009;6:1–10.

[262] Chung E, Rytlewski JA, Merchant AG, Dhada KS, Lewis EW, Suggs LJ. Fibrin-based 3D matrices induce angiogenic behavior of adipose-derived stem cells. Acta Biomater 2015;17:78–88.

[263] Zimmerlin L, Rubin JP, Pfeifer ME, Moore LR, Donnenberg VS, Donnenberg AD. Human adipose stromal vascular cell delivery in a fibrin spray. Cytotherapy 2013;15:102–8.

[264] Rohringer S, Hofbauer P, Schneider KH, et al. Mechanisms of vasculogenesis in 3D fibrin matrices mediated by the interaction of adipose-derived stem cells and endothelial cells. Angiogenesis 2014;17:921–33.

[265] Hunt NC, Grover LM. Cell encapsulation using biopolymer gels for regenerative medicine. Biotechnol Lett 2010;32:733–42.

[266] Smetana Jr. K. Cell biology of hydrogels. Biomaterials 1993;14:1046–50.

[267] Burdick JA, Vunjak-Novakovic G. Engineered microenvironments for controlled stem cell differentiation. Tissue Eng Part A 2009;15:205–19.

[268] Sasaki T, Fassler R, Hohenester E. Laminin: the crux of basement membrane assembly. J Cell Biol 2004;164:959–63.

[269] Patrick Jr. CW, Wu X. Integrin-mediated preadipocyte adhesion and migration on laminin-1. Ann Biomed Eng 2003;31:505–14.

[270] Noro A, Sillat T, Virtanen I, et al. Laminin production and basement membrane deposition by mesenchymal stem cells upon adipogenic differentiation. J Histochem Cytochem 2013;61:719–30.

[271] Davis GE, Senger DR. Endothelial extracellular matrix: biosynthesis, remodeling, and functions during vascular morphogenesis and neovessel stabilization. Circ Res 2005;97:1093–107.

[272] Gruene M, Pflaum M, Deiwick A, et al. Adipogenic differentiation of laser-printed 3D tissue grafts consisting of human adipose-derived stem cells. Biofabrication 2011;3: 015005.

[273] Yao R, Zhang R, Luan J, Lin F. Alginate and alginate/gelatin microspheres for human adipose-derived stem cell encapsulation and differentiation. Biofabrication 2012;4: 025007.

[274] Flynn LE. The use of decellularized adipose tissue to provide an inductive microenvironment for the adipogenic differentiation of human adipose-derived stem cells. Biomaterials 2010;31:4715–24.

[275] Fu RH, Wang YC, Liu SP, et al. Decellularization and recellularization technologies in tissue engineering. Cell Transplant 2014;23:621–30.
[276] Sano H, Orbay H, Terashi H, Hyakusoku H, Ogawa R. Acellular adipose matrix as a natural scaffold for tissue engineering. J Plast Reconstr Aesthet Surg 2014;67:99–106.
[277] Hruschka V, Saeed A, Slezak P, et al. Evaluation of a thermoresponsive polycaprolactone scaffold for in vitro three-dimensional stem cell differentiation. Tissue Eng Part A 2015;21:310–9.
[278] Omidi E, Fuetterer L, Reza Mousavi S, Armstrong RC, Flynn LE, Samani A. Characterization and assessment of hyperelastic and elastic properties of decellularized human adipose tissues. J Biomech 2014;47:3657–63.
[279] Haddad SM, Omidi E, Flynn LE, Samani A. Comparative biomechanical study of using decellularized human adipose tissues for post-mastectomy and post-lumpectomy breast reconstruction. J Mech Behav Biomed Mater 2016;57:235–45.
[280] Messina A, Bortolotto SK, Cassell OC, Kelly J, Abberton KM, Morrison WA. Generation of a vascularized organoid using skeletal muscle as the inductive source. FASEB J 2005;19:1570–2.
[281] Abberton KM, Bortolotto SK, Woods AA, et al. Myogel, a novel, basement membrane-rich, extracellular matrix derived from skeletal muscle, is highly adipogenic in vivo and in vitro. Cells Tissues Organs 2008;188:347–58.
[282] Kleinman HK, McGarvey ML, Liotta LA, Robey PG, Tryggvason K, Martin GR. Isolation and characterization of type IV procollagen, laminin, and heparan sulfate proteoglycan from the EHS sarcoma. Biochemistry 1982;21:6188–93.
[283] Kleinman HK, Martin GR. Matrigel: basement membrane matrix with biological activity. Semin Cancer Biol 2005;15:378–86.
[284] Findlay MW, Messina A, Thompson EW, Morrison WA. Long-term persistence of tissue-engineered adipose flaps in a murine model to 1 year: an update. Plast Reconstr Surg 2009;124:1077–84.
[285] Baptista LS, Silva KR, Santos MFS, et al. Scalable and reproducible biofabrication of spheroids from human adipose-derived tissue stem cells isolated by mechanical dissociation. In: Tissue engineering and Regenerative Medicine International Society—EU meeting, Genova, Italy; 2014.
[286] Cheng NC, Wang S, Young TH. The influence of spheroid formation of human adipose-derived stem cells on chitosan films on stemness and differentiation capabilities. Biomaterials 2012;33:1748–58.
[287] FitzGerald JF, Kumar AS. Biologic versus synthetic mesh reinforcement: what are the pros and cons? Clin Colon Rectal Surg 2014;27:140–8.
[288] Martello F, Tocchio A, Tamplenizza M, et al. Poly(amido-amine)-based hydrogels with tailored mechanical properties and degradation rates for tissue engineering. Acta Biomater 2014;10:1206–15.
[289] Tocchio A, Martello F, Tamplenizza M, et al. RGD-mimetic poly(amidoamine) hydrogel for the fabrication of complex cell-laden micro constructs. Acta Biomater 2015;18:144–54.
[290] Znaleziona J, Ginterova P, Petr J, et al. Determination and identification of synthetic cannabinoids and their metabolites in different matrices by modern analytical techniques—a review. Anal Chim Acta 2015;874:11–25.
[291] Moroni L, de Wijn JR, van Blitterswijk CA. Integrating novel technologies to fabricate smart scaffolds. J Biomater Sci Polym Ed 2008;19:543–72.
[292] Angelova N, Hunkeler D. Rationalizing the design of polymeric biomaterials. Trends Biotechnol 1999;17:409–21.

[293] Hu Y, Winn SR, Krajbich I, Hollinger JO. Porous polymer scaffolds surface-modified with arginine-glycine-aspartic acid enhance bone cell attachment and differentiation in vitro. J Biomed Mater Res A 2003;64:583–90.
[294] Tanahashi K, Mikos AG. Protein adsorption and smooth muscle cell adhesion on biodegradable agmatine-modified poly(propylene fumarate-co-ethylene glycol) hydrogels. J Biomed Mater Res A 2003;67:448–57.
[295] Kim TG, Park TG. Biomimicking extracellular matrix: cell adhesive RGD peptide modified electrospun poly(D, L-lactic-co-glycolic acid) nanofiber mesh. Tissue Eng 2006;12:221–33.
[296] Blit PH, Shen YH, Ernsting MJ, Woodhouse KA, Santerre JP. Bioactivation of porous polyurethane scaffolds using fluorinated RGD surface modifiers. J Biomed Mater Res A 2010;94:1226–35.
[297] Guarnieri D, De Capua A, Ventre M, et al. Covalently immobilized RGD gradient on PEG hydrogel scaffold influences cell migration parameters. Acta Biomater 2010;6:2532–9.
[298] Tuin SA, Pourdeyhimi B, Loboa EG. Creating tissues from textiles: scalable nonwoven manufacturing techniques for fabrication of tissue engineering scaffolds. Biomed Mater 2016;11: 015017.
[299] Sheikh FA, Ju HW, Moon BM, et al. Hybrid scaffolds based on PLGA and silk for bone tissue engineering. J Tissue Eng Regen Med 2016;10:209–21.
[300] Zangaglia R, Martignoni E, Glorioso M, et al. Macrogol for the treatment of constipation in Parkinson's disease. A randomized placebo-controlled study. Mov Disord 2007;22:1239–44.
[301] Ozcelik B, Blencowe A, Palmer J, et al. Highly porous and mechanically robust polyester poly(ethylene glycol) sponges as implantable scaffolds. Acta Biomater 2014;10:2769–80.
[302] Ruoslahti E, Pierschbacher MD. Arg-Gly-Asp: a versatile cell recognition signal. Cell 1986;44:517–8.
[303] Lin CC, Anseth KS. PEG hydrogels for the controlled release of biomolecules in regenerative medicine. Pharm Res 2009;26:631–43.
[304] Garcia AJ. PEG-maleimide hydrogels for protein and cell delivery in regenerative medicine. Ann Biomed Eng 2014;42:312–22.
[305] Briquez PS, Hubbell JA, Martino MM. Extracellular matrix-inspired growth factor delivery systems for skin wound healing. Adv Wound Care (New Rochelle) 2015;4:479–89.
[306] Patel PN, Gobin AS, West JL, Patrick Jr. CW. Poly(ethylene glycol) hydrogel system supports preadipocyte viability, adhesion, and proliferation. Tissue Eng 2005;11:1498–505.
[307] Clevenger TN, Hinman CR, Ashley Rubin RK, et al. Vitronectin-based, biomimetic encapsulating hydrogel scaffolds support adipogenesis of adipose stem cells. Tissue Eng Part A 2016;22:597–609.
[308] Niedzwiecki L, Teahan J, Harrison RK, Stein RL. Substrate specificity of the human matrix metalloproteinase stromelysin and the development of continuous fluorometric assays. Biochemistry 1992;31:12618–23.
[309] Ferruti P, Bianchi S, Ranucci E, Chiellini F, Piras AM. Novel agmatine-containing poly(amidoamine) hydrogels as scaffolds for tissue engineering. Biomacromolecules 2005;6:2229–35.
[310] Davidenko N, Campbell JJ, Thian ES, Watson CJ, Cameron RE. Collagen-hyaluronic acid scaffolds for adipose tissue engineering. Acta Biomater 2010;6:3957–68.
[311] Lin SD, Huang SH, Lin YN, et al. Engineering adipose tissue from uncultured human adipose stromal vascular fraction on collagen matrix and gelatin sponge scaffolds. Tissue Eng Part A 2011;17:1489–98.

[312] Frydrych M, Roman S, MacNeil S, Chen B. Biomimetic poly(glycerol sebacate)/poly(l-lactic acid) blend scaffolds for adipose tissue engineering. Acta Biomater 2015;18:40–9.
[313] Julier Z, Park AJ, Briquez PS, Martino MM. Promoting tissue regeneration by modulating the immune system. Acta Biomater 2017;53:13–28.
[314] MacDougald OA, Hwang CS, Fan H, Lane MD. Regulated expression of the obese gene product (leptin) in white adipose tissue and 3T3-L1 adipocytes. Proc Natl Acad Sci U S A 1995;92:9034–7.
[315] Pittenger MF, Mackay AM, Beck SC, et al. Multilineage potential of adult human mesenchymal stem cells. Science 1999;284:143–7.
[316] Serlachius M, Andersson LC. Upregulated expression of stanniocalcin-1 during adipogenesis. Exp Cell Res 2004;296:256–64.
[317] Saiki A, Watanabe F, Murano T, Miyashita Y, Shirai K. Hepatocyte growth factor secreted by cultured adipocytes promotes tube formation of vascular endothelial cells in vitro. Int J Obes (Lond) 2006;30:1676–84.
[318] Zebisch K, Voigt V, Wabitsch M, Brandsch M. Protocol for effective differentiation of 3T3-L1 cells to adipocytes. Anal Biochem 2012;425:88–90.
[319] Farmer SR. Transcriptional control of adipocyte formation. Cell Metab 2006;4:263–73.
[320] Frye CA, Patrick CW. Three-dimensional adipose tissue model using low shear bioreactors. In Vitro Cell Dev Biol Anim 2006;42:109–14.
[321] Bouillon R, Carmeliet G, Lieben L, et al. Vitamin D and energy homeostasis: of mice and men. Nat Rev Endocrinol 2014;10:79–87.
[322] Thomson JA, Itskovitz-Eldor J, Shapiro SS, et al. Embryonic stem cell lines derived from human blastocysts. Science 1998;282:1145–7.
[323] Cuaranta-Monroy I, Simandi Z, Nagy L. Differentiation of adipocytes in monolayer from mouse embryonic stem cells. Methods Mol Biol 2016;1341:407–15.
[324] Gucciardo L, Lories R, Ochsenbein-Kolble N, Done E, Zwijsen A, Deprest J. Fetal mesenchymal stem cells: isolation, properties and potential use in perinatology and regenerative medicine. BJOG 2009;116:166–72.
[325] Takahashi K, Okita K, Nakagawa M, Yamanaka S. Induction of pluripotent stem cells from fibroblast cultures. Nat Protoc 2007;2:3081–9.
[326] Aversa F, Tabilio A, Velardi A, et al. Treatment of high-risk acute leukemia with T-cell-depleted stem cells from related donors with one fully mismatched HLA haplotype. N Engl J Med 1998;339:1186–93.
[327] Lindroos B, Suuronen R, Miettinen S. The potential of adipose stem cells in regenerative medicine. Stem Cell Rev 2011;7:269–91.
[328] Apovian CM. The obesity epidemic—understanding the disease and the treatment. N Engl J Med 2016;374:177–9.
[329] Oedayrajsingh-Varma MJ, van Ham SM, Knippenberg M, et al. Adipose tissue-derived mesenchymal stem cell yield and growth characteristics are affected by the tissue-harvesting procedure. Cytotherapy 2006;8:166–77.
[330] Astori G, Vignati F, Bardelli S, et al. "In vitro" and multicolor phenotypic characterization of cell subpopulations identified in fresh human adipose tissue stromal vascular fraction and in the derived mesenchymal stem cells. J Transl Med 2007;5:55.
[331] Zhu X, Shi W, Tai W, Liu F. The comparition of biological characteristics and multilineage differentiation of bone marrow and adipose derived Mesenchymal stem cells. Cell Tissue Res 2012;350:277–87.
[332] Baer PC, Geiger H. Adipose-derived mesenchymal stromal/stem cells: tissue localization, characterization, and heterogeneity. Stem Cells Int 2012;2012:812693.

[333] Illouz YG. Body contouring by lipolysis: a 5-year experience with over 3000 cases. Plast Reconstr Surg 1983;72:591–7.

[334] Ogura F, Wakao S, Kuroda Y, et al. Human adipose tissue possesses a unique population of pluripotent stem cells with nontumorigenic and low telomerase activities: potential implications in regenerative medicine. Stem Cells Dev 2014;23:717–28.

[335] Sterodimas A, de Faria J, Nicaretta B, Pitanguy I. Tissue engineering with adipose-derived stem cells (ADSCs): current and future applications. J Plast Reconstr Aesthet Surg 2010;63:1886–92.

[336] Zuk PA, Zhu M, Mizuno H, et al. Multilineage cells from human adipose tissue: implications for cell-based therapies. Tissue Eng 2001;7:211–28.

[337] Tavazoie M, Van der Veken L, Silva-Vargas V, et al. A specialized vascular niche for adult neural stem cells. Cell Stem Cell 2008;3:279–88.

[338] Halvorsen YD, Bond A, Sen A, et al. Thiazolidinediones and glucocorticoids synergistically induce differentiation of human adipose tissue stromal cells: biochemical, cellular, and molecular analysis. Metabolism 2001;50:407–13.

[339] Halvorsen YD, Franklin D, Bond AL, et al. Extracellular matrix mineralization and osteoblast gene expression by human adipose tissue-derived stromal cells. Tissue Eng 2001;7:729–41.

[340] Tapp H, Hanley Jr. EN, Patt JC, Gruber HE. Adipose-derived stem cells: characterization and current application in orthopaedic tissue repair. Exp Biol Med (Maywood) 2009;234:1–9.

[341] Thesleff T, Lehtimaki K, Niskakangas T, et al. Cranioplasty with adipose-derived stem cells and biomaterial: a novel method for cranial reconstruction. Neurosurgery 2011;68:1535–40.

[342] Pelto J, Bjorninen M, Palli A, et al. Novel polypyrrole-coated polylactide scaffolds enhance adipose stem cell proliferation and early osteogenic differentiation. Tissue Eng Part A 2013;19:882–92.

[343] Sandor GK, Tuovinen VJ, Wolff J, et al. Adipose stem cell tissue-engineered construct used to treat large anterior mandibular defect: a case report and review of the clinical application of good manufacturing practice-level adipose stem cells for bone regeneration. J Oral Maxillofac Surg 2013;71:938–50.

[344] Estes BT, Wu AW, Guilak F. Potent induction of chondrocytic differentiation of human adipose-derived adult stem cells by bone morphogenetic protein 6. Arthritis Rheum 2006;54:1222–32.

[345] Choi YS, Matsuda K, Dusting GJ, Morrison WA, Dilley RJ. Engineering cardiac tissue in vivo from human adipose-derived stem cells. Biomaterials 2010;31:2236–42.

[346] Choi SA, Lee JY, Wang KC, et al. Human adipose tissue-derived mesenchymal stem cells: characteristics and therapeutic potential as cellular vehicles for prodrug gene therapy against brainstem gliomas. Eur J Cancer 2012;48:129–37.

[347] Planat-Benard V, Silvestre JS, Cousin B, et al. Plasticity of human adipose lineage cells toward endothelial cells: physiological and therapeutic perspectives. Circulation 2004;109:656–63.

[348] Fang B, Song Y, Liao L, Zhang Y, Zhao RC. Favorable response to human adipose tissue-derived mesenchymal stem cells in steroid-refractory acute graft-versus-host disease. Transplant Proc 2007;39:3358–62.

[349] Gonzalez-Rey E, Anderson P, Gonzalez MA, Rico L, Buscher D, Delgado M. Human adult stem cells derived from adipose tissue protect against experimental colitis and sepsis. Gut 2009;58:929–39.

[350] Gonzalez-Rey E, Gonzalez MA, Varela N, et al. Human adipose-derived mesenchymal stem cells reduce inflammatory and T cell responses and induce regulatory T cells in vitro in rheumatoid arthritis. Ann Rheum Dis 2010;69:241–8.
[351] Riordan NH, Ichim TE, Min WP, et al. Non-expanded adipose stromal vascular fraction cell therapy for multiple sclerosis. J Transl Med 2009;7:29.
[352] Trivedi HL, Vanikar AV, Thakker U, et al. Human adipose tissue-derived mesenchymal stem cells combined with hematopoietic stem cell transplantation synthesize insulin. Transplant Proc 2008;40:1135–9.
[353] Alvarez PD, Garcia-Arranz M, Georgiev-Hristov T, Garcia-Olmo D. A new bronchoscopic treatment of tracheomediastinal fistula using autologous adipose-derived stem cells. Thorax 2008;63:374–6.
[354] Nie C, Yang D, Morris SF. Local delivery of adipose-derived stem cells via acellular dermal matrix as a scaffold: a new promising strategy to accelerate wound healing. Med Hypotheses 2009;72:679–82.
[355] Nie C, Zhang G, Yang D, et al. Targeted delivery of adipose-derived stem cells via acellular dermal matrix enhances wound repair in diabetic rats. J Tissue Eng Regen Med 2015;9:224–35.
[356] Kosaraju R, Rennert RC, Maan ZN, et al. Adipose-derived stem cell-seeded hydrogels increase endogenous progenitor cell recruitment and neovascularization in wounds. Tissue Eng Part A 2016;22:295–305.
[357] Jurgens WJ, Oedayrajsingh-Varma MJ, Helder MN, et al. Effect of tissue-harvesting site on yield of stem cells derived from adipose tissue: implications for cell-based therapies. Cell Tissue Res 2008;332:415–26.
[358] Iyyanki T, Hubenak J, Liu J, Chang EI, Beahm EK, Zhang Q. Harvesting technique affects adipose-derived stem cell yield. Aesthet Surg J 2015;35:467–76.
[359] Schreml S, Babilas P, Fruth S, et al. Harvesting human adipose tissue-derived adult stem cells: resection versus liposuction. Cytotherapy 2009;11:947–57.
[360] Duscher D, Luan A, Rennert RC, et al. Suction assisted liposuction does not impair the regenerative potential of adipose derived stem cells. J Transl Med 2016;14:126.
[361] Spalding KL, Arner E, Westermark PO, et al. Dynamics of fat cell turnover in humans. Nature 2008;453:783–7.
[362] Kornicka K, Marycz K, Tomaszewski KA, Maredziak M, Smieszek A. The effect of age on osteogenic and adipogenic differentiation potential of human adipose derived stromal stem cells (hASCs) and the impact of stress factors in the course of the differentiation process. Oxid Med Cell Longev 2015;2015:309169.
[363] Weisberg SP, McCann D, Desai M, Rosenbaum M, Leibel RL, Ferrante Jr. AW. Obesity is associated with macrophage accumulation in adipose tissue. J Clin Invest 2003;112:1796–808.
[364] Xu H, Barnes GT, Yang Q, et al. Chronic inflammation in fat plays a crucial role in the development of obesity-related insulin resistance. J Clin Invest 2003;112:1821–30.
[365] Yoshimura K, Shigeura T, Matsumoto D, et al. Characterization of freshly isolated and cultured cells derived from the fatty and fluid portions of liposuction aspirates. J Cell Physiol 2006;208:64–76.
[366] Li H, Zimmerlin L, Marra KG, Donnenberg VS, Donnenberg AD, Rubin JP. Adipogenic potential of adipose stem cell subpopulations. Plast Reconstr Surg 2011;128:663–72.
[367] Markarian CF, Frey GZ, Silveira MD, et al. Isolation of adipose-derived stem cells: a comparison among different methods. Biotechnol Lett 2014;36:693–702.

[368] Tholpady SS, Llull R, Ogle RC, Rubin JP, Futrell JW, Katz AJ. Adipose tissue: stem cells and beyond. Clin Plast Surg 2006;33:55–62. vi.
[369] Quarto N, Longaker MT. FGF-2 inhibits osteogenesis in mouse adipose tissue-derived stromal cells and sustains their proliferative and osteogenic potential state. Tissue Eng 2006;12:1405–18.
[370] Gierloff M, Petersen L, Oberg HH, Quabius ES, Wiltfang J, Acil Y. Adipogenic differentiation potential of rat adipose tissue-derived subpopulations of stromal cells. J Plast Reconstr Aesthet Surg 2014;67:1427–35.
[371] Han TT, Toutounji S, Amsden BG, Flynn LE. Adipose-derived stromal cells mediate in vivo adipogenesis, angiogenesis and inflammation in decellularized adipose tissue bioscaffolds. Biomaterials 2015;72:125–37.
[372] Ghorbani FM, Kaffashi B, Shokrollahi P, Seyedjafari E, Ardeshirylajimi A. PCL/chitosan/Zn-doped nHA electrospun nanocomposite scaffold promotes adipose derived stem cells adhesion and proliferation. Carbohydr Polym 2015;118:133–42.
[373] Mays RW, van't Hof W, Ting AE, Perry R, Deans R. Development of adult pluripotent stem cell therapies for ischemic injury and disease. Expert Opin Biol Ther 2007;7:173–84.
[374] Mimeault M, Hauke R, Batra SK. Stem cells: a revolution in therapeutics-recent advances in stem cell biology and their therapeutic applications in regenerative medicine and cancer therapies. Clin Pharmacol Ther 2007;82:252–64.
[375] Miltenyi S, Muller W, Weichel W, Radbruch A. High gradient magnetic cell separation with MACS. Cytometry 1990;11:231–8.
[376] Valli H, Sukhwani M, Dovey SL, et al. Fluorescence- and magnetic-activated cell sorting strategies to isolate and enrich human spermatogonial stem cells. Fertil Steril 2014;102:566–80 e567.
[377] Indumathi S, Mishra R, Harikrishnan R, Rajkumar JS, Kantawala N, Dhanasekaran M. Lineage depletion of stromal vascular fractions isolated from human adipose tissue: a novel approach towards cell enrichment technology. Cytotechnology 2014;66:219–28.
[378] Bourin P, Bunnell BA, Casteilla L, et al. Stromal cells from the adipose tissue-derived stromal vascular fraction and culture expanded adipose tissue-derived stromal/stem cells: a joint statement of the International Federation for Adipose Therapeutics and Science (IFATS) and the International Society for Cellular Therapy (ISCT). Cytotherapy 2013;15:641–8.
[379] Lauvrud AT, Kelk P, Wiberg M, Kingham PJ. Characterization of human adipose tissue-derived stem cells with enhanced angiogenic and adipogenic properties. J Tissue Eng Regen Med 2016;https://doi.org/10.1002/term.2147.
[380] Chang Q, Lu F. A novel strategy for creating a large amount of engineered fat tissue with an axial vascular pedicle and a prefabricated scaffold. Med Hypotheses 2012;79:267–70.
[381] Khan S, Villalobos MA, Choron RL, et al. Fibroblast growth factor and vascular endothelial growth factor play a critical role in endotheliogenesis from human adipose-derived stem cells. J Vasc Surg 2017;65(5):1483–92.
[382] Awad HA, Halvorsen YD, Gimble JM, Guilak F. Effects of transforming growth factor beta1 and dexamethasone on the growth and chondrogenic differentiation of adipose-derived stromal cells. Tissue Eng 2003;9:1301–12.
[383] Kim WS, Park HS, Sung JH. The pivotal role of PDGF and its receptor isoforms in adipose-derived stem cells. Histol Histopathol 2015;30:793–9.
[384] Behr B, Tang C, Germann G, Longaker MT, Quarto N. Locally applied vascular endothelial growth factor A increases the osteogenic healing capacity of human adipose-derived stem cells by promoting osteogenic and endothelial differentiation. Stem Cells 2011;29:286–96.

[385] Gehmert S, Gehmert S, Hidayat M, et al. Angiogenesis: the role of PDGF-BB on adipose-tissue derived stem cells (ASCs). Clin Hemorheol Microcirc 2011;48:5–13.

[386] Ting AC, Craft RO, Palmer JA, et al. The adipogenic potential of various extracellular matrices under the influence of an angiogenic growth factor combination in a mouse tissue engineering chamber. Acta Biomater 2014;10:1907–18.

[387] Morimoto A, Okamura K, Hamanaka R, et al. Hepatocyte growth factor modulates migration and proliferation of human microvascular endothelial cells in culture. Biochem Biophys Res Commun 1991;179:1042–9.

[388] Davis GE, Stratman AN, Sacharidou A, Koh W. Molecular basis for endothelial lumen formation and tubulogenesis during vasculogenesis and angiogenic sprouting. Int Rev Cell Mol Biol 2011;288:101–65.

[389] Esteva FJ, Sahin AA, Smith TL, et al. Prognostic significance of phosphorylated P38 mitogen-activated protein kinase and HER-2 expression in lymph node-positive breast carcinoma. Cancer 2004;100:499–506.

[390] Wagner EF, Nebreda AR. Signal integration by JNK and p38 MAPK pathways in cancer development. Nat Rev Cancer 2009;9:537–49.

[391] Kim BS, Kang KS, Kang SK. Soluble factors from ASCs effectively direct control of chondrogenic fate. Cell Prolif 2010;43:249–61.

[392] Shigematsu S, Yamauchi K, Nakajima K, Iijima S, Aizawa T, Hashizume K. IGF-1 regulates migration and angiogenesis of human endothelial cells. Endocr J 1999;46(Suppl):S59–62.

[393] Aghdam SY, Eming SA, Willenborg S, et al. Vascular endothelial insulin/IGF-1 signaling controls skin wound vascularization. Biochem Biophys Res Commun 2012;421:197–202.

[394] Shima N, Kimoto M, Yamaguchi M, Yamagami S. Increased proliferation and replicative lifespan of isolated human corneal endothelial cells with L-ascorbic acid 2-phosphate. Invest Ophthalmol Vis Sci 2011;52:8711–7.

[395] Pike DB, Cai S, Pomraning KR, et al. Heparin-regulated release of growth factors in vitro and angiogenic response in vivo to implanted hyaluronan hydrogels containing VEGF and bFGF. Biomaterials 2006;27:5242–51.

[396] Hamed S, Ben-Nun O, Egozi D, et al. Treating fat grafts with human endothelial progenitor cells promotes their vascularization and improves their survival in diabetes mellitus. Plast Reconstr Surg 2012;130:801–11.

[397] Yao R, Du Y, Zhang R, Lin F, Luan J. A biomimetic physiological model for human adipose tissue by adipocytes and endothelial cell cocultures with spatially controlled distribution. Biomed Mater 2013;8: 045005.

[398] Alfieri A, Ong AC, Kammerer RA, et al. Angiopoietin-1 regulates microvascular reactivity and protects the microcirculation during acute endothelial dysfunction: role of eNOS and VE-cadherin. Pharmacol Res 2014;80:43–51.

[399] Haug V, Torio-Padron N, Stark GB, Finkenzeller G, Strassburg S. Comparison between endothelial progenitor cells and human umbilical vein endothelial cells on neovascularization in an adipogenesis mouse model. Microvasc Res 2015;97:159–66.

[400] Koolwijk P, van Erck MG, de Vree WJ, et al. Cooperative effect of TNFalpha, bFGF, and VEGF on the formation of tubular structures of human microvascular endothelial cells in a fibrin matrix. Role of urokinase activity. J Cell Biol 1996;132:1177–88.

[401] Cross MJ, Claesson-Welsh L. FGF and VEGF function in angiogenesis: signalling pathways, biological responses and therapeutic inhibition. Trends Pharmacol Sci 2001;22:201–7.

[402] Silva EA, Mooney DJ. Effects of VEGF temporal and spatial presentation on angiogenesis. Biomaterials 2010;31:1235–41.

[403] Serrero G, Mills D. Physiological role of epidermal growth factor on adipose tissue development in vivo. Proc Natl Acad Sci U S A 1991;88:3912–6.
[404] Vassaux G, Negrel R, Ailhaud G, Gaillard D. Proliferation and differentiation of rat adipose precursor cells in chemically defined medium: differential action of anti-adipogenic agents. J Cell Physiol 1994;161:249–56.
[405] Hauner H, Rohrig K, Petruschke T. Effects of epidermal growth factor (EGF), platelet-derived growth factor (PDGF) and fibroblast growth factor (FGF) on human adipocyte development and function. Eur J Clin Invest 1995;25:90–6.
[406] Haystead TA, Hardie DG. Both insulin and epidermal growth factor stimulate lipogenesis and acetyl-CoA carboxylase activity in isolated adipocytes. Importance of homogenization procedure in avoiding artefacts in acetyl-CoA carboxylase assay. Biochem J 1986;234:279–84.
[407] Moule SK, Edgell NJ, Welsh GI, et al. Multiple signalling pathways involved in the stimulation of fatty acid and glycogen synthesis by insulin in rat epididymal fat cells. Biochem J 1995;311(Pt 2):595–601.
[408] Baba AS, Harper JM, Buttery PJ. Effects of gastric inhibitory polypeptide, somatostatin and epidermal growth factor on lipogenesis in ovine adipose explants. Comp Biochem Physiol B Biochem Mol Biol 2000;127:173–82.
[409] Cigolini M, Smith U. Human adipose tissue in culture. VIII. Studies on the insulin-antagonistic effect of glucocorticoids. Metabolism 1979;28:502–10.
[410] Walton PE, Etherton TD, Evock CM. Antagonism of insulin action in cultured pig adipose tissue by pituitary and recombinant porcine growth hormone: potentiation by hydrocortisone. Endocrinology 1986;118:2577–81.
[411] Huber B, Czaja AM, Kluger PJ. Influence of epidermal growth factor (EGF) and hydrocortisone on the co-culture of mature adipocytes and endothelial cells for vascularized adipose tissue engineering. Cell Biol Int 2016;40:569–78.
[412] Chazaud B. Macrophages: supportive cells for tissue repair and regeneration. Immunobiology 2014;219:172–8.
[413] Daley JM, Brancato SK, Thomay AA, Reichner JS, Albina JE. The phenotype of murine wound macrophages. J Leukoc Biol 2010;87:59–67.
[414] Mokarram N, Merchant A, Mukhatyar V, Patel G, Bellamkonda RV. Effect of modulating macrophage phenotype on peripheral nerve repair. Biomaterials 2012;33:8793–801.
[415] Kharraz Y, Guerra J, Mann CJ, Serrano AL, Munoz-Canoves P. Macrophage plasticity and the role of inflammation in skeletal muscle repair. Mediators Inflamm 2013;2013:491497.
[416] Ben-Mordechai T, Holbova R, Landa-Rouben N, et al. Macrophage subpopulations are essential for infarct repair with and without stem cell therapy. J Am Coll Cardiol 2013;62:1890–901.
[417] Li Z, Xu F, Wang Z, et al. Macrophages undergo M1-to-M2 transition in adipose tissue regeneration in a rat tissue engineering model. Artif Organs 2016;40:E167–78.
[418] Debels H, Galea L, Han XL, et al. Macrophages play a key role in angiogenesis and adipogenesis in a mouse tissue engineering model. Tissue Eng Part A 2013;19:2615–25.
[419] Chandler EM, Seo BR, Califano JP, et al. Implanted adipose progenitor cells as physicochemical regulators of breast cancer. Proc Natl Acad Sci U S A 2012;109:9786–91.
[420] Koellensperger E, Gramley F, Preisner F, Leimer U, Germann G, Dexheimer V. Alterations of gene expression and protein synthesis in co-cultured adipose tissue-derived stem cells and squamous cell-carcinoma cells: consequences for clinical applications. Stem Cell Res Ther 2014;5:65.

[421] Cousin B, Ravet E, Poglio S, et al. Adult stromal cells derived from human adipose tissue provoke pancreatic cancer cell death both in vitro and in vivo. PLoS One 2009;4: e6278.

[422] Zhao W, Ren G, Zhang L, et al. Efficacy of mesenchymal stem cells derived from human adipose tissue in inhibition of hepatocellular carcinoma cells in vitro. Cancer Biother Radiopharm 2012;27:606–13.

[423] Karnoub AE, Dash AB, Vo AP, et al. Mesenchymal stem cells within tumour stroma promote breast cancer metastasis. Nature 2007;449:557–63.

[424] Mizuno H. Adipose-derived stem cells for tissue repair and regeneration: ten years of research and a literature review. J Nippon Med Sch 2009;76:56–66.

[425] Dai R, Wang Z, Samanipour R, Koo KI, Kim K. Adipose-derived stem cells for tissue engineering and regenerative medicine applications. Stem Cells Int 2016;2016:6737345.

[426] Lindroos B, Boucher S, Chase L, et al. Serum-free, xeno-free culture media maintain the proliferation rate and multipotentiality of adipose stem cells in vitro. Cytotherapy 2009;11:958–72.

[427] Atashi F, Jaconi ME, Pittet-Cuenod B, Modarressi A. Autologous platelet-rich plasma: a biological supplement to enhance adipose-derived mesenchymal stem cell expansion. Tissue Eng Part C Methods 2015;21:253–62.

[428] Liao HT, James IB, Marra KG, Rubin JP. The effects of platelet-rich plasma on cell proliferation and adipogenic potential of adipose-derived stem cells. Tissue Eng Part A 2015;21:2714–22.

[429] Ovsianikov A, Deiwick A, Van Vlierberghe S, et al. Laser fabrication of 3D gelatin scaffolds for the generation of bioartificial tissues. Materials 2011;4:288–99.

[430] Williams SK, Touroo JS, Church KH, Hoying JB. Encapsulation of adipose stromal vascular fraction cells in alginate hydrogel spheroids using a direct-write three-dimensional printing system. Biores Open Access 2013;2:448–54.

[431] Huber B, Borchers K, Tovar GE, Kluger PJ. Methacrylated gelatin and mature adipocytes are promising components for adipose tissue engineering. J Biomater Appl 2016;30:699–710.

[432] Zuk PA, Zhu M, Ashjian P, et al. Human adipose tissue is a source of multipotent stem cells. Mol Biol Cell 2002;13:4279–95.

[433] Gomillion CT, Burg KJ. Stem cells and adipose tissue engineering. Biomaterials 2006;27:6052–63.

[434] Findlay MW, Dolderer JH, Trost N, et al. Tissue-engineered breast reconstruction: bridging the gap toward large-volume tissue engineering in humans. Plast Reconstr Surg 2011;128:1206–15.

[435] Dolderer JH, Abberton KM, Thompson EW, et al. Spontaneous large volume adipose tissue generation from a vascularized pedicled fat flap inside a chamber space. Tissue Eng 2007;13:673–81.

[436] Lokmic Z, Stillaert F, Morrison WA, Thompson EW, Mitchell GM. An arteriovenous loop in a protected space generates a permanent, highly vascular, tissue-engineered construct. FASEB J 2007;21:511–22.

[437] Lilja HE, Morrison WA, Han XL, et al. An adipoinductive role of inflammation in adipose tissue engineering: key factors in the early development of engineered soft tissues. Stem Cells Dev 2013;22:1602–13.

[438] Peng Z, Dong Z, Chang Q, et al. Tissue engineering chamber promotes adipose tissue regeneration in adipose tissue engineering models through induced aseptic inflammation. Tissue Eng Part C Methods 2014;20:875–85.

[439] Wan J, Dong Z, Lei C, Lu F. Generating an engineered adipose tissue flap using an external suspension device. Plast Reconstr Surg 2016;138:109–20.

[440] Hofer SO, Knight KM, Cooper-White JJ, et al. Increasing the volume of vascularized tissue formation in engineered constructs: an experimental study in rats. Plast Reconstr Surg 2003;111:1186–92. discussion 1193-1184.

[441] Cao Y, Mitchell G, Messina A, et al. The influence of architecture on degradation and tissue ingrowth into three-dimensional poly(lactic-co-glycolic acid) scaffolds in vitro and in vivo. Biomaterials 2006;27:2854–64.

[442] Dolderer JH, Thompson EW, Slavin J, et al. Long-term stability of adipose tissue generated from a vascularized pedicled fat flap inside a chamber. Plast Reconstr Surg 2011;127:2283–92.

[443] Zhan W, Chang Q, Xiao X, et al. Self-synthesized extracellular matrix contributes to mature adipose tissue regeneration in a tissue engineering chamber. Wound Repair Regen 2015;23:443–52.

[444] Lu Z, Yuan Y, Gao J, Lu F. Adipose tissue extract promotes adipose tissue regeneration in an adipose tissue engineering chamber model. Cell Tissue Res 2016;364:289–98.

[445] Morrison WA, Marre D, Grinsell D, Batty A, Trost N, O'Connor AJ. Creation of a large adipose tissue construct in humans using a tissue-engineering chamber: a step forward in the clinical application of soft tissue engineering. EBioMedicine 2016;6:238–45.

[446] Heit YI, Lancerotto L, Mesteri I, et al. External volume expansion increases subcutaneous thickness, cell proliferation, and vascular remodeling in a murine model. Plast Reconstr Surg 2012;130:541–7.

[447] Liu YS, Lee OK. In search of the pivot point of mechanotransduction: mechanosensing of stem cells. Cell Transplant 2014;23:1–11.

[448] Brorson H. Adipose tissue in lymphedema: the ignorance of adipose tissue in lymphedema. Lymphology 2004;37:175–7.

[449] Wang N, Tytell JD, Ingber DE. Mechanotransduction at a distance: mechanically coupling the extracellular matrix with the nucleus. Nat Rev Mol Cell Biol 2009;10:75–82.

[450] Gretzer C, Emanuelsson L, Liljensten E, Thomsen P. The inflammatory cell influx and cytokines changes during transition from acute inflammation to fibrous repair around implanted materials. J Biomater Sci Polym Ed 2006;17:669–87.

[451] Anderson JM, Rodriguez A, Chang DT. Foreign body reaction to biomaterials. Semin Immunol 2008;20:86–100.

[452] Cronin KJ, Messina A, Knight KR, et al. New murine model of spontaneous autologous tissue engineering, combining an arteriovenous pedicle with matrix materials. Plast Reconstr Surg 2004;113:260–9.

[453] Lokmic Z, Mitchell GM. Engineering the microcirculation. Tissue Eng Part B Rev 2008;14:87–103.

[454] Matsuda K, Falkenberg KJ, Woods AA, Choi YS, Morrison WA, Dilley RJ. Adipose-derived stem cells promote angiogenesis and tissue formation for in vivo tissue engineering. Tissue Eng Part A 2013;19:1327–35.

[455] Chhaya MP, Melchels FP, Holzapfel BM, Baldwin JG, Hutmacher DW. Sustained regeneration of high-volume adipose tissue for breast reconstruction using computer aided design and biomanufacturing. Biomaterials 2015;52:551–60.

[456] Reichert JC, Cipitria A, Epari DR, et al. A tissue engineering solution for segmental defect regeneration in load-bearing long bones. Sci Transl Med 2012;4: 141ra193.

[457] Rohner D, Hutmacher DW, Cheng TK, Oberholzer M, Hammer B. In vivo efficacy of bone-marrow-coated polycaprolactone scaffolds for the reconstruction of orbital defects in the pig. J Biomed Mater Res B Appl Biomater 2003;66:574–80.

[458] Schantz JT, Lim TC, Ning C, et al. Cranioplasty after trephination using a novel biodegradable burr hole cover: technical case report. Neurosurgery 2006;58: ONS-E176; discussion ONS-E176.
[459] Rai B, Oest ME, Dupont KM, Ho KH, Teoh SH, Guldberg RE. Combination of platelet-rich plasma with polycaprolactone-tricalcium phosphate scaffolds for segmental bone defect repair. J Biomed Mater Res A 2007;81:888–99.
[460] Stevens MM, Marini RP, Schaefer D, Aronson J, Langer R, Shastri VP. In vivo engineering of organs: the bone bioreactor. Proc Natl Acad Sci U S A 2005;102:11450–5.

Further Reading

[1] Chae MP, Hunter-Smith DJ, Murphy SV, Atala A, Rozen WM. 3D bioprinting in nipple-areolar complex reconstruction. In: Shiffman MA, editor. Nipple-Areolar complex reconstruction: principles and clinical techniques. Heidelberg, Germany: Springer; 2016.

3D bioprinting nerve

16

S.-h. Hsu, C.-W. Chen
Institute of Polymer Science and Engineering, National Taiwan University, Taipei, Taiwan, ROC

Nerve tissue has complicated structure consisted of neural network and various types of cells. Although there is still limited literature with regard to printing neural tissue, the early results have shown that the techniques are of great potential for the regeneration of impaired nerves as well as the study of neural diseases. These techniques include the scaffolds that facilitate the proliferation and differentiation of neural cells, the construct either with or without embedded cells for the repair of damaged nerves, and the three-dimensional (3D) structure of materials and cells mimicking the tissue in organisms. Nonetheless, still further research is need to be conducted for developing more effective strategies for neural tissue engineering as well as presenting the 3D structure of nerve in organisms more accurately.

16.1 Advantage of 3D vs. 2D neural cell culture by implementing 3D printing technique

16.1.1 2D and 3D cell cultures

Artificial tissues could be produced by using various biofabrication techniques. A predefined pattern for the materials and cells may be obtained. Conventional methods of cell culture are often performed in a 2D environment, consequently limiting the tissue structure to the planar form, which is very different from the complex 3D structure in organisms [1], as illustrated in Fig. 16.1. For the complicated and dynamic 3D neural network structure, mimicking the cell environment by 2D culture may lead to a number of problems. For example, the abnormal contact of neural cells and the formation of networks could overactivate the glial cells that are contaminated [2]. It is generally considered that the coculture of neural cells with astrocytes promotes the growth and development of the neural cells [3]. However, under certain circumstances, astrocytes become more reactive and release stress factor, which is harmful for cells [4]. In particular, this phenomenon often occurs in the 2D cell culture, manifested as the high reactivity and hyperproliferation of astrocytes [5]. Nevertheless, astrocytes were proven to promote synaptogenesis, which was the key process in the nerve regeneration [6]. According to the studies, astrocytes in 3D cell culture were less active, but synaptogenesis is still proceeded as in the case of 2D environments [7]. 3D cell culture in poly(L-lactic acid) nanofibers was reported to enhance the differentiation of mouse neural stem cells (NSCs) and accelerate the growth of neurites [8].

3D Bioprinting for Reconstructive Surgery. https://doi.org/10.1016/B978-0-08-101103-4.00016-8

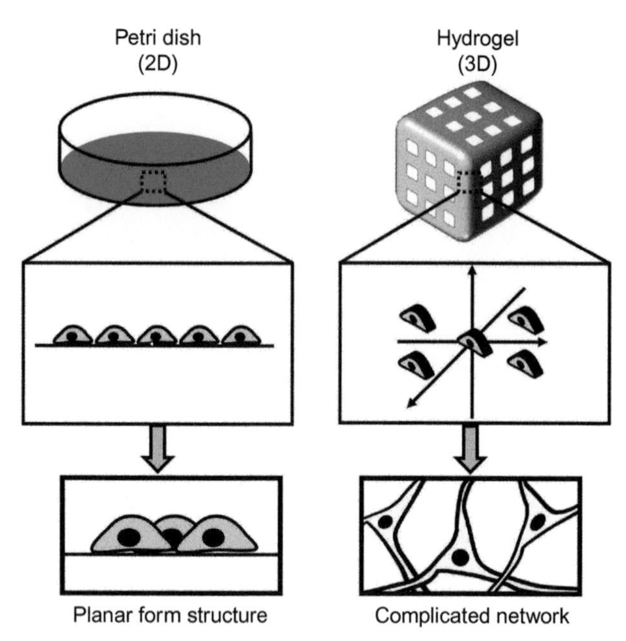

Fig. 16.1 Cell culture methods: 2D vs. 3D.

16.1.2 The implementation of 3D printing in neuronal cell culture

The culture of primary neurons requires specific geometries [9]. The neurons are very sensitive to surface coating [10]. Moreover, supporting cells are often needed to achieve an optimal culturing environment [11]. Currently developed 3D-printing techniques enable the fabrication of customized structures. The structure may be designed not only to support the neural cell culture but also to guide the growth in a defined path, as shown in Fig. 16.2. The technique makes it more convenient to investigate the factors that affect the neural cell growth [12]. The differences between the growth of neural cells in 2D and 3D environments can thus be more thoroughly studied. Commonly used cues include diffusible factors, substrate-bound factors, electrical and magnetic field, and topographical features. Researchers have cultured dorsal root ganglia (DRG) on 2D glass adjacent to the areas of 3D collagen type I gel with laminin (LN) and chondroitin sulfate proteoglycan (CSPG) serving as cell adhesive and cell repulsive cues, respectively [13,14]. The results showed that the neural cells guided by the LN mainly grew on the surface of the 3D structure, whereas those with CSPG cues grew into the 3D gel in random directions. In addition, more significant growth of neurites was observed in the cells with LN cues [15]. It was reported that neurites grew longer in 3D collagen when compared with those cultured in 2D collagen [16]. Moreover, neurons tended to form a more globular structure in 3D agarose [17]. These differences in growth and morphology of cells must be taken into account when it comes to designing nerve guidance conduits. Another related study was done by Li et al.; they employed poly-D-lysine (PDL)-modified polydimethylsiloxane (PDMS) as 2D structure and used Matrigel to construct a 3D environment. The neurites were

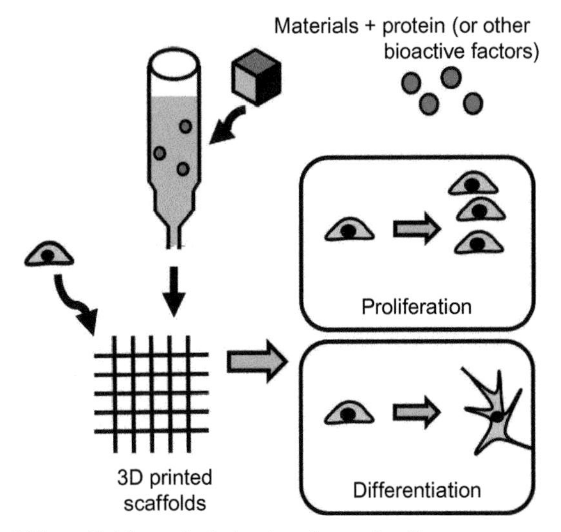

Fig. 16.2 Effect of 3D scaffolds on the behavior of neural cells.

found to grow in the direction of PDL rather than grow into Matrigel, which implied that chemical cues are more important than structural ones in this case [18].

16.2 3D-printing of biomedical materials followed by seeding with cells

16.2.1 3D-printed nerve conduits for peripheral nerve regeneration

Because of the limited regeneration ability of peripheral nerves, additional surgeries are usually required to facilitate the repair process. Conventional methods include direct suturing, autograft, and decellularized allografts. However, these methods have many disadvantages, including additional surgeries, chronic pains, morbidity at the donor sites, and immune responses [19]. Peripheral nerve guidance channels or conduits (NGCs) are advantageous because of the flexibility regarding the choices of materials and the fine tuning of the mechanical or biological properties [20,21], and the fact that only a single surgery was required. NGCs with complex structure and curves could be fabricated by 3D-printing technique, as illustrated in Fig. 16.3, unlike the conventional methods that involve molds, which could mostly be used only to prepare linear cylinders. Cui et al. applied a double-nozzle, low-temperature deposition manufacturing system for preparing double-layered polyurethane (PU)-collagen nerve conduits for peripheral nerve regeneration. The optimal porosity and hydrophilicity could be adjusted by changing the polymer concentration [22]. Recent developments of NGCs have improved from the traditional passive scaffolds that simply provided the protective environment for the nerve regeneration to the ones that could promote neurite outgrowth and axon regeneration [23]. The topographical or chemical cues within the conduits could be modified to affect cell behaviors [24]. For example,

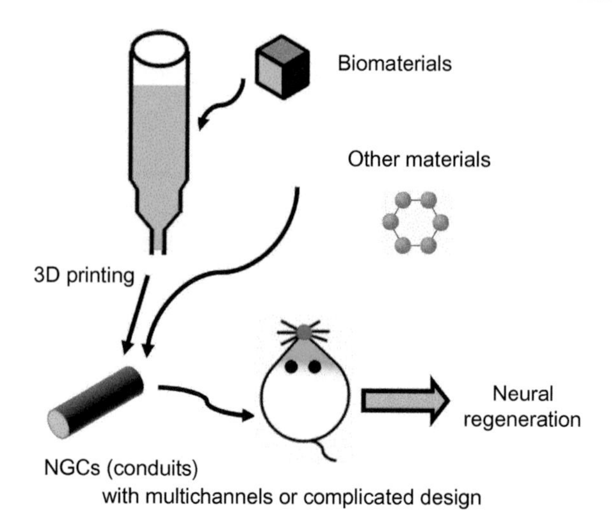

Fig. 16.3 Nerve guidance channels (nerve conduits) fabricated by 3D printing.

cutting off the sciatic nerve bifurcation of mice, and customized NGCs made of silicone were fabricated using the 3D structured light scanning and implanted to promote the nerve regeneration. There were both structural and chemical cues on the interior of the conduit, such as the microgrooves and growth factors, respectively. Both cues were capable of guiding the growth of axons. The growth factors were embedded in the gelatin methacrylate hydrogels with a concentration gradient created by 3D printing. Axons and Schwann cells were found to migrate and grow into two different bifurcating branches by following the specific growth factors [25]. Schwann cells were found to be of great importance in the development of peripheral nerve cells and therefore could facilitate the nerve regeneration [26].

If the width of nerve injury was too large, direct suturing would cause strain, affecting the regeneration process. Therefore, an autograft, allograft, or NGC became the remaining plausible methods. Nevertheless, when the length of the damaged nerves reaches 20 mm, the effect of NGCs on nerve repair would also be limited [18]. To overcome the challenge, more functional materials such as poly(3-hydroxybutyrate), extracellular matrix, growth factors, internal guidance structure, or supporting cells were incorporated [19]. Some commonly used supporting cells are Schwann cells, olfactory ensheathing cells (OECs), and stem cells. Embryonic stem cells (ESCs), NSCs, and mesenchymal stem cells (MSCs) have all been employed. However, considering the source of cells, MSCs have the least ethical issues [26]. On the other hand, the central nervous system has even more limited regeneration ability due to several reasons, including the presence of myelin-associated inhibitors, a generally slower axonal growth rate, and the inhibitory influences of the glial and the extracellular environment [27].

16.2.2 3D-printed scaffolds for neural cells and central nervous system repair

3D printing can be used to produce scaffolds for the central nervous system. Some examples of 3D-printing materials and cells in the central nervous system are described below separately. By alternately, printing collagen and rat embryonic neurons combined with astrocytes layer by layer, by using sodium carbonate as cross-linking agent to make collagen gel; the layered scaffolds could be fabricated. The scaffolds provided the suitable environment for cell culture without harming the cells [28]. 3D printing enabled us to construct predefined special 3D patterns of cells, which were more favorable for the investigation of the interactions between cells or between cells and ECM to understand better the developments of the neural tissues in organisms. ECM proteins such as fibronectin and LN were printed in predefined patterns on the cell-repulsive materials, on which the cells were attached and proliferated [29]. The geometry of the proteins might affect the morphology and functions of cells, such as the growth of neurites. In addition, combining microelectronic materials with the cell patterns helps to detect precisely the signals of cells, and as a result, the variations in their functions could be recorded [30]. The scaffolds that are mainly consisted of graphene, which have excellent electrical, mechanical, and biological properties, were fabricated by extrusion-based direct ink printing under ambient temperature and pressure. Polylactide-co-glycolide (PLG) was incorporated into the ink to improve the biocompatibility, biodegradability, and elastic property of the material. The prepared scaffolds were conductive as well as elastic and supported the human mesenchymal stem cells (hMSCs) to attach and differentiate. The signs of differentiation included the obvious elongation of hMSCs and the neuron-like morphology. Compared with the cells cultured on PLG, those cultured on graphene-PLG scaffolds had better viability and proliferation rate, which might be attributed to the surface roughness, mechanical and electrical properties of the graphene-PLG scaffolds. By changing the ratio of graphene to PLG, the mechanical properties could be adjusted. Specifically, as the PLG content increased, the material became more brittle and less elastic [31]. Graphene is among a few types of materials that have both biocompatibility and conductivity. These materials were found to enhance the signaling between cells, cell differentiation, and migration without causing significant immune responses or fibrous capsule formation [32]. How to distribute growth factors into the surroundings in a controlled release manner has become a challenge because most of the growth factors are water soluble, that is, they diffuse easily and quickly throughout the bodies of organisms. To solve the problem, growth factors were embedded into the biomedical materials, as in the case of vascular endothelial growth factors (VEGF), which were incorporated into fibrin gels before 3D printing. By combing murine NSCs and collagen gels, along with these growth factors using 3D printing under ambient pressure that was favorable for cells, the artificial neural tissues were constructed. The NSCs were attracted by VEGF and showed neurite elongation and cell differentiation [33]. Growth factors were found to be related closely to the development of brains. In addition, when someone had a stroke, multiple types of growth factors would be activated. Therefore, the research on growth factors was a key for understanding the human brain developments as well as assessing the treatments of neurological diseases [34]. Sanjana et al. employed a custom-built inkjet printer to fabricate micropatterns of materials for neural cell culture. Poly(ethylene) glycol and a collagen/poly-d-lysine (PDL) mixture were served as cell-repulsive and cell-adhesive material, respectively. Hippocampal neurons and glia adhered to the materials strongly as patterns for more than 25 days [35]. A well-defined, complex 3D cell structure was constructed by direct printing of NT2 cells and fibrin gel alternately with thrombin as the cross-linker. The pores of the scaffolds allowed nutrients and oxygen to flow through NT2 cells adhered onto the material and showed a uniform distribution. After 12 days of culture, neurites began to grow. The heat shock around the nozzle lasts for <3 μs; thus the electrophysiological properties of the cells were not affected. Moreover, the trypsinized cells used in the study were less prone to damages caused by shear stress [36]. Heat shock has been shown to have a long-term effect on the electrophysiological properties as well as causing neurons to lose their phenotypes, consequently differentiate into other types of cells. Therefore, when it comes to printing neurons, heat shock should be taken into account with care [37]. Zhang et al. fabricated a 3D hydrogel structure consisting of gelatin and hyaluronan for the repair of traumatic brain injury of rats. At 4 weeks after implantation, the structure was well connected with host tissues. Furthermore, the materials completely degraded after 13 weeks by replacing them with cells [38].

16.3 3D printing of biomedical materials with cells (bioprinting)

Bioprinting is defined as a computer-assisted process to pattern and assemble living or nonliving materials in a predefined manner, which can be applied to regenerative medicine or biological studies [39]. When cells and materials are printed simultaneously, the bioink must support cell attachment and maintain the functionalities of cells for the subsequent differentiation, maturation, or proliferation, as shown in Fig. 16.4.

In addition, the rheological and mechanical properties of the ink have to be in the proper ranges so that it is printable, yet it also has some mechanical strength to somewhat maintain the shape right after being printed [1]. Nonetheless, in another new

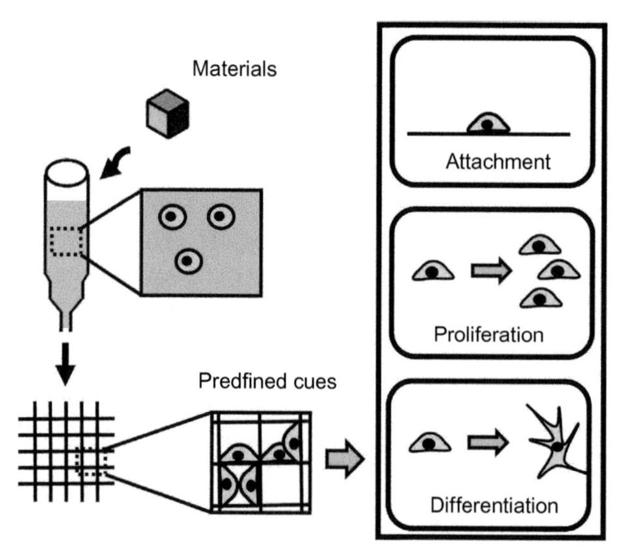

Fig. 16.4 The effect of predefined physicochemical cues produced by 3D printing.

technique called 3D cell bioprinting, instead of bioink, 3D cellular building blocks are the basic units to create the structure. The technique has some advantage because of the avoidance of cell aggregation in the bioink or the clogging at the nozzle [40].

For treating the central nervous system injuries, Hsieh et al. synthesized thermoresponsive water-based biodegradable PU dispersions based on polycaprolactone (PCL) diol and poly(D,L-lactide) diol (PDLLA diol), which formed gels upon increasing temperature to 37°C without any cross-linkers. The mechanical strength could be easily tuned by changing the solid content of the dispersions. In particular, NSCs proliferated and differentiated significantly in PU dispersions with solid contents of 25–30 wt%. Furthermore, the PU hydrogels with a solid content of 25 wt% was injected into zebrafish embryo neural injury and proved to restore the impaired nerves. In addition, the PU dispersion was printed into 3D construct and then implanted into the brain of zebrafish that was traumatically injured and showed a rescue rate of 81% after 6 days [41]. The rheology of the aqueous PU bioink may be further modified by blending with soy protein [42]. In another work on peripheral nerve regeneration, PCL strands were printed first, followed by the deposition of PCL nanofibers onto them. And then, alginate containing PC12 neural cells was printed. After cross-linking agent was added to stabilize the structure, the prepared scaffolds were rolled to form 3D NGCs. The printed PC12 neural cells had a viability of 95% and distributed uniformly within the conduits. Moreover, at the third day of cell culture, the cells started to proliferate. In comparison with the conventional methods involving seeding of cells onto the material after the printing, the distribution of the cells is more uniform. However, a certain period of time was needed for the cells to settle due to the shear stress as well as the hampered diffusion of nutrients. These NGCs could facilitate the peripheral nerve regeneration. Furthermore, arginyl-glycyl-aspartic acid (Arg-Gly-Asp or RGD) peptides or collagen could be incorporated into the NGCs to provide attachment points for cells, which was favorable for cells to proliferate [43]. 3D structure of mammalian cells with high resolution could be directly printed by biological laser printing (BioLP) technique. The solution containing cells was different from the solution being heated by laser, thus the cells were not affected by heat shock and showed a viability of 95%. Rat OECs proved to enhance the growth of neurites in central nervous system and facilitate the nerve regeneration in spinal cord injuries [44]. In a study, OECs were cultured in hydrogels and were found to have high viability as well as to maintain the original phenotype. BioLP required only a small amount of cells to print precise patterns, maximizing the potential of the low-yield harvested cells. The precise locations of cells provided a means to quantify the effect of pressure, nutrient diffusion, and pores on the migration of cells [45]. Conventional 2D cell culture or scaffolds could not represent the complex 3D microstructure of neural networks in human brains. Therefore, some studies have employed bioprinting to construct the brain-like 3D structure composed of layer-by-layer hydrogels containing primary neurons. The bioink consisted of RGD peptide-modified gelatin and primary cortical neurons. Peptides were found to be favorable for cells to attach, proliferate, and promote the formation of neural networks. The results indicated that the cells were viable and showed the signs of differentiation. Moreover, neural networks were established after 7 days of cell culture. Neurons or glial cells are in particular sensitive to chemical buffers. Thus, when ionic solutions were used as cross-linking agents, the effects on the cells must be evaluated [45]. In the above study, calcium chloride or Dulbecco's modified eagle medium (DMEM) was chosen as cross-linkers, and the cell viabilities were not affected [46]. Gly-Arg-Gly-Asp-Ser synthetic peptide (GRGDS)-modified gellam gum (GG)-based hydrogels were also found to support DRG-based neurite outgrowth. In addition, adipose tissue-derived stem cells attached to the modified hydrogels. Therefore, the materials may be used for future applications in peripheral nerve injuries regenerative medicine [47]. Although tissues produced by conventional organotypic culture preserved the original structure in organisms, their long-term viability was unfavorable [48]. Therefore, in a study, human midbrain-derived neuronal progenitor cells (hmNPCs) were cultured in a stirred medium and underwent self-assembly to form a 3D neurosphere structure with a size of 300–400 μm. The morphology of cells gradually changed and differentiated into dopaminergic neurons, astrocytes, and oligodendrocytes, forming ECM-like environment around them. The overall structure became even more similar to that of the brain tissues. Investigations of neurological disorders or the development of brains could be performed on these tissues. The study also demonstrated the importance of the understanding of cell-cell and cell-ECM interactions [49].

16.4 Summary and future perspectives

It is known that cells showed very different behavior in vivo from those observed in conventional 2D cell culture, leading to the concerns about the credibility of the studies on illness, drug tests, and treatment assessment. Moreover, neural networks in organisms are particularly complex and the interactions between cells or cells and the surrounding have become more important. 3D printing allows us to simulate the environment in organism in a more accurate way. By choosing the suitable materials, not only the growth of the neural cells could be enhanced but also the path on which the cells grow may be predefined, which is especially convenient for the study on the factors affecting cell growth and behavior.

3D-printing cell cultures can be classified into two types (Fig. 16.5). One is 3D-printing materials and cells separately, and the other is 3D-printing materials containing cells. In the case of peripheral nerve repair or regeneration, compared with conventional autografts, allografts, or direct suturing, customized NGCs (either with cells or without cells) fabricated by 3D printing have many advantages. These include the adjustment of biological and mechanical properties, and the absence of surgeries performed at the donor sites, chronical pains, and immune responses. Direct 3D printing of cells provided a way to print precise patterns onto the substrate, but the bioink had to be carefully chosen. The bioink must keep the cells active and attached to the ink and be printable itself. However, the printed structure needed to have some mechanical strength to maintain the shape. In addition, the effects of heat shock or shear stress on the cells must be considered since it was reported that they might have influence on the long-term health of cells. When cross-linking agents were employed, those which caused few or no damages on cells were preferred because the neural cells were especially sensitive to the environment.

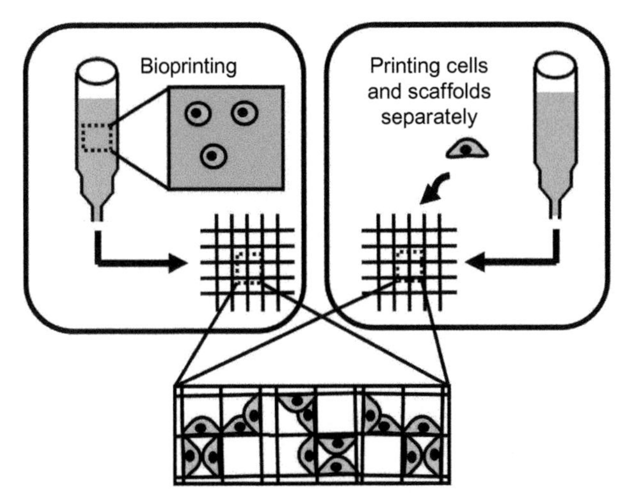

Fig. 16.5 Common ways of 3D printing for neural system cell culture.

Further research or improvements on 3D printing of nerves or the related materials are needed for future development, such as employing more bioactive materials to facilitate the repair of nerves in which the injured gap was too wide to be treated by existing methods, incorporating of ECM, growth factors, and chemical or structural cues to guide cells, adding various types of supporting cells, restoring the central nervous system functions, and better understanding of cell-cell, cell-ECM, and cell-material interactions (Fig. 16.6). With a thorough and detailed knowledge of interactions

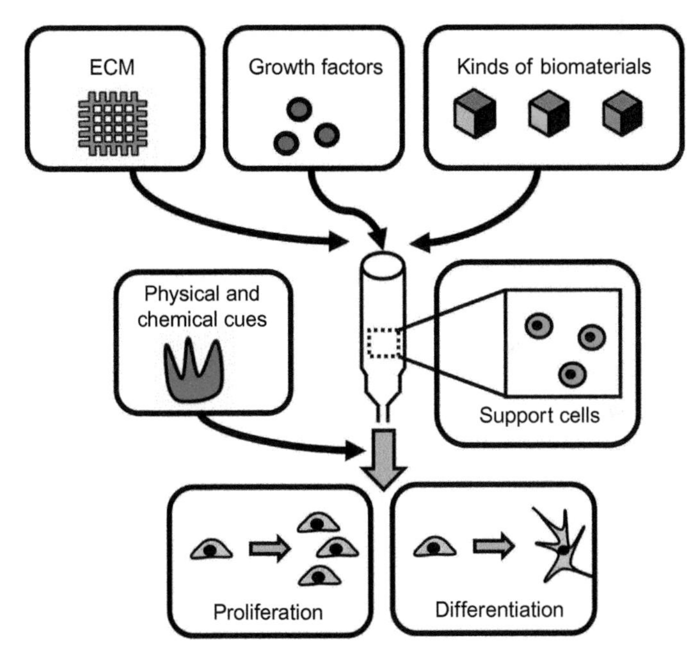

Fig. 16.6 The future perspective of 3D printing for neural regeneration.

between cells and materials, the most effective biomedical devices could be designed and fabricated to enhance the nerve regeneration. In addition, the constructs simulating the tissues in organisms could be used to study diseases and treatments. The 3D-printed neural organoids can be used as drug screening platforms. In the meantime, bioprinted nerve or brain remains to be one of the most exciting and challenging issues in this area that requires interdisciplinary efforts from doctors, biologists, engineers, and other experts for a breakthrough.

References

[1] Wust S, Muller R, Hofmann S. 3D Bioprinting of complex channels—effects of material, orientation, geometry, and cell embedding. J Biomed Mater Res A 2015;103(8):2559.

[2] Puschmann TB, Zanden C, De Pablo Y, Kirchhoff F, Pekna M, Liu J, et al. Bioactive 3D cell culture system minimizes cellular stress and maintains the in vivo-like morphological complexity of astroglial cells. Glia 2013;61:432.

[3] Pfrieger FW. Role of glial cells in the formation and maintenance of synapses. Brain Res Rev 2010;63:39.

[4] Yu P, Wang H, Katagiri Y, Geller HM. An in vitro model of reactive astrogliosis and its effect on neuronal growth. Methods Mol Biol 2012;814:327.

[5] Foo LC, Allen NJ, Bushong EA, Ventura PB, Chung WS, Zhou L, et al. Development of a method for the purification and culture of rodent astrocytes. Neuron 2011;71:799.

[6] Pekna M, Pekny M, Nilsson M. Modulation of neural plasticity as a basis for stroke rehabilitation. Stroke 2012;43:2819.

[7] Puschmann TB, de Pablo Y, Zanden C, Liu J, Pekny M. A novel method for three-dimensional culture of central nervous system neurons. Tissue Eng Part C 2016;20(6):1–8.

[8] Yang F, Murugan R, Wang S, Ramakrishna S. Electrospinning of nano/micro scale poly(L-lactic acid) aligned fibers and their potential in neural tissue engineering. Biomaterials 2005;26:2603.

[9] Wongsarnpigoon A, Grill WM. Computer-based model of epidural motor cortex stimulation: effects of electrode position and geometry on activation of cortical neurons. Clin Neurophysiol 2012;123(1):160–72.

[10] Sun Y, Huang Z, Liu W, Yang K, Sun K, Xing S, et al. Surface coating as a key parameter in engineering neuronal network structures in vitro. Biointerphases 2012;7(29):1–15.

[11] Pixley SK. Purified cultures of keratin-positive olfactory epithelial cells: identification of a subset as neuronal supporting (sustentacular) cells. J Neurosci Res 1992;31(4):693–707.

[12] Wardyn JD, Sanderson C, Swan LE, Stagi M. Low cost production of 3D-printed devices and electrostimulation chambers for the culture of primary neurons. J Neurosci Meth 2015;251:17–23.

[13] Luckenbill-Edds L. Laminin and the mechanism of neuronal outgrowth. Brain Res Brain Res Rev 1997;23(1-2):1–27.

[14] Hoke A, Silver J. Proteoglycans and other repulsive molecules in glial boudaries during development and regeneration of the nervous system. Prog Brain Res 1996;108:149–63.

[15] Kofron CM, Fong VJ, Hoffman-Kim D. Neurite outgrowth at the interface of 2D and 3D growth environments. J Neural Eng 2009;6:1–13.

[16] Li GN, Livi LL, Gourd CM, Deweerd ES, Hoffman-Kim D. Genomic and morphological changes of neuroblastoma cells in response to three-dimensional matrices. Tissue Eng 2007;13(5):1035–47.

[17] Balgude AP, Yu X, Szymanski A, Bellamkonda RV. Agarose gel stiffness determines rate of DRG neurite extension in 3D cultures. Biomaterials 2001;22:1077–84. 13, 1035–47.
[18] Li N, Folch A. Integration of topographical and biochemical cues by axons during growth on microfabricated 3-D substrates. Exp Cell Res 2005;311:307–16.
[19] Pateman CJ, Harding AJ, Glen A, Taylor CS, Christmas CR, Robinson PP, et al. Nerve guides manufactured from photocurable polymers to aid peripheral nerve repair. Biomaterials 2015;49:77–89.
[20] Kim YT, Romero-Ortega MI. Material considerations for peripheral nerve interfacing. MRS Bull 2012;37(6):573–80.
[21] Nectow AR, Marra KG, Kaplan DL. Biomaterials for the development of peripheral nerve guidance conduits. Tissue Eng Part B Rev 2012;18(1):40–50.
[22] Cui T, Yan Y, Zhang R, Liu L, Xu W, Wang X. Rapid prototyping of a double-layer polyurethane–collagen conduit for peripheral nerve regeneration. Tissue Eng Part C 2009;15(1):1–9.
[23] Gu X, Ding F, Williams DF. Neural tissue engineering options for peripheral nerve regeneration. Biomaterials 2014;35:6143–56.
[24] Faroni A, Mobasseri SA, Kingham PJ, Reid AJ. Peripheral nerve regeneration: experimental strategies and future perspectives. Adv Drug Deliv Rev 2015;82–83:160–7.
[25] Johnson BN, Lancaster KZ, Zhen G, He JY, Gupta MK, Kong YL, et al. 3D printed anatomical nerve regeneration pathways. Adv Funct Mater 2015;25:6205–17.
[26] Kalbermatten DF, Erba P, Mahay D, Wiberg M, Pierer G, Terenghi G. Schwann cell strip for peripheral nerve repair. J Hand Surg Eur 2008;33:587–94.
[27] Ferguson TA, Son YJ. Extrinsic and intrinsic determinants of nerve regeneration. J Tissue Eng 2011;2(1):1–10.
[28] Lee W, Pinckney J, Lee V, Lee JH, Fischer K, Polio S, et al. Three-dimensional bioprinting of rat embryonic neural cells. NeuroReport 2009;20(8):798–803.
[29] Chen CS, Mrksich M, Huang S, Whitesides GM, Ingber DE. Micropatterned surfaces for control of cell shape, position, and function. Biotechnol Prog 1998;14:356–63.
[30] Lauer L, Ingebrandt S, Scholl M, Offenhäusser A. Aligned microcontact printing of biomolecules on microelectronic device surfaces. IEEE Trans Biomed Eng 2001;48(7):838–42.
[31] Jakus AE, Secor EB, Rutz AL, Jordan SW, Hersam MC, Shah RN. Three-dimensional printing of high-content graphene scaffolds for electronic and biomedical applications. ACS Nano 2015;9(4):4636–48.
[32] Lee SK, Kim H, Shim BS. Graphene: an emerging material for biological tissue engineering. Carbon Lett 2013;14:63–75.
[33] Lee YB, Polio S, Lee W, Dai G, Menon L, Carroll RS, et al. Bio-printing of collagen and VEGF-releasing fibrin gel scaffolds for neural stem cell culture. Exp Neurol 2010;223:645–52.
[34] Carmeliet P, Storkebaum E. Vascular and neuronal effects of VEGF in the nervous system: implications for neurological disorders. Semin Cell Dev Biol 2002;13:39–53.
[35] Sanjana NE, Fuller SB. A fast flexible ink-jet printing method for patterning dissociated neurons in culture. J Neurosci Methods 2004;136(2):151–63.
[36] Xu T, Gregory CA, Molnar P, Cui XF, Jalota S, Bhaduri SB, et al. Viability and electrophysiology of neural cell structures generated by the inkjet printing method. Biomaterials 2006;27:3580–8.
[37] Dawson-Scully K, Meldrum RR. Heat shock protects synaptic transmission in flight motor circuitry of locusts. Neuroreport 1998;9(11):2589.
[38] Zhang T, Yan Y, Wang X, Xiong Z, Lin F. Three-dimensional gelatin and gelatin/hyaluronan hydrogel structures for traumatic brain injury. J Bioact Compat Polym 2007;22:19–29.

[39] Guillemot F, Guillotin B, Fontaine A, Ali M, Catros S, Kériquel V, et al. Laser-assisted bioprinting to deal with tissue complexity in regenerative medicine. MRS Bull 2011;36(12):1015–9.
[40] Khatiwala C, Law R, Shepherd B, Dorfman S, Csete M. 3D cell bioprinting for regenerative medicine research and therapies. Gene Ther Regul 2012;7(1):1–19.
[41] Hsieh F-Y, Lin H-H, Hsu S-h. 3D bioprinting of neural stem cell-laden thermoresponsive biodegradable polyurethane hydrogel and potential in central nervous system repair. Biomaterials 2015;71:48–57.
[42] Lin H-H, Hsieh F-Y, Tseng C-S, Hsu S-h. Preparation and characterization of a biodegradable polyurethane hydrogel and the hybrid gel with soy protein for 3D cell-laden bioprinting. J Mater Chem B 2016;4:6694–705.
[43] Dreher R, Starly B. Biofabrication of multimaterial three-dimensional constructs embedded with patterned alginate strands encapsulating PC12 neural cell lines. J Nanotechnol Eng Med 2015;6:1–8.
[44] Dombrowski MA, Sasaki M, Lankford KL, Kocsis JD, Radtke C. Myelination and nodal formation of regenerated peripheral nerve fibers following transplantation of acutely prepared olfactory ensheathing cells. Brain Res 2006;1125:1–8.
[45] Othon CM, Wu XJ, Anders JJ, Ringeisen BR. Single-cell printing to form three-dimensional lines of olfactory ensheathing cells. Biomed Mater 2008;3:1–6.
[46] Lozano R, Stevens L, Thompson BC, Gilmore KJ, Gorkin III R, Stewart EM, et al. 3D printing of layered brain-like structures using peptide modified gellan gum substrates. Biomaterials 2015;67:264–73.
[47] Assunção-Silva RC, Oliveira CC, Ziv-Polat O, Gomes ED, Sahar A, Sousa N, et al. Induction of neurite outgrowth in 3D hydrogel-based environments. Biomed Mater 2015;10(5):1–7.
[48] Bodea GO, Blaess S. Organotypic slice cultures of embryonic ventral midbrain: a system to study dopaminergic neuronal development in vitro. J Vis Exp 2012;59:3350.
[49] Simao D, Pinto C, Piersanti S, Weston A, Peddie CJ, Bastos AEP, et al. Modeling human neural functionality in vitro: three-dimensional culture for dopaminergic differentiation. Tissue Eng Part A 2015;21(3–4):654–68.

Further Reading

[1] Pan C, Kumar C, Bohl S, Klingmueller U, Mann M. Comparative proteomic phenotyping of cell lines and primary cells to assess preservation of cell type specific functions. Mol Cell Proteomics 2009;8:443–50.

3D bioprinting skin

17

Z. Li, S. Huang, X. Fu
General Hospital of PLA, Beijing, China

17.1 Introduction

Because of the development of cell culture technique in the 1970s, nearly 6000-fold large area of cultured epithelium could be harvested from a small piece of epidermis in 3 weeks, which was considered as a leap from the passive wound bandage to bioactive skin substitutes [1].Unfortunately, the cultured epithelium could be grafted onto the cleaned wound bed but was prone to failure because of the wound infection or accumulated exudation. Since then, researches on skin substitutes were promoted by large application demand of clinical treatment of burns and skin ulcers instead of autologous split-thickness skin graft and skin allograft [2], resulting in several mature skin products in the latest four decades, especially tissue engineering (TE) skin substitutes.

The integration of dermis equivalents into the TE skin substitutes was a big step for its better compatibility and adhesion to the wound bed. Integra Artificial Skin, the first biosynthetic stratified skin substitute product in the 1980s, primarily serves as a temporary coverage of wound bed preparing for the transplant of autologous split-thickness skin, which drastically reduces the use of auto-skin [3]. Apligraf, the most popular skin substitute product worldwide, is a bilayered living TE skin containing keratinocytes and fibroblasts that are derived from neonatal foreskin [4], which proves to be permanent but dysfunction for lack of hair follicles, sweat glands, sebaceous glands, and so on. TE provides the basic principles on the establishment of stratified, fully functional skin substitute, while the practical and mature methods and techniques, however, are still to be discovered.

Three-dimensional (3D) bioprinting, one of the most eye-catching technologies in the last decade in TE, provides a better application-oriented methodology compared with other TE technologies and an easier technique to deposit multiple types of matrix and cells into well-organized and functional tissue constructs, like skin. Technically, 3D bioprinting can not only mimic the stratified structure and complex composition of natural skin but also contribute to the construction of skin appendages, such as hair follicles, sweat glands, sebaceous glands, and so on, which cannot be achieved by conventional TE technologies but is necessary in clinical application. With the promotion of bioprinters and discovery of new matrix biomaterials and seed cells, 3D bioprinting has been successfully applied in the construction of bone [5], articular cartilage [6], and blood vessel [7], which offers a promising solution for skin substitute.

Recently, the progress of several researches [8–12] on skin substitute indicates that their products of 3D-bioprinted skin constructs vividly mimic the structure of

3D Bioprinting for Reconstructive Surgery. https://doi.org/10.1016/B978-0-08-101103-4.00017-X

natural skin, but they are far away from mature products of skin substitute due to the undesirable physical properties of matrix biomaterials and uncontrolled behaviors of cells, which deserves to be summarized and discussed for the improvement of the future research. Fortunately, new designs [13–16] of TE skin constructs, by 3D bioprinting or not, with the integration of hair follicles or sweat glands spark imaginations as well as boost confidence on functional TE skin products in the future.

Therefore, an overview on current researches of 3D-bioprinted skin constructs is presented in this review, followed by a discussion on future design of TE skin substitutes as well as a summary of the achievements of skin appendages regeneration and prospects of this research field.

17.2 Technical progress in 3D-bioprinted skin constructs

From the very beginning of the researches on 3D-bioprinted skin constructs, a classical procedure or schematic in researches concerning 3D bioprinting that includes cell extraction, expansion in vitro, encapsulation in hydrogel, bioprinting, in vitro culture, and in vivo application is carried out by most available 3D-bioprinted skin researches [17]. Beyond this procedure, these researches are faced with multiple skin-specific problems, such as why adult cells are more popular in a practical way while stem cells provide more proliferate and differentiate activity, how to modify biomaterials to mimic compressive and tensile strength of natural skin tissue, how to physically print epithelial tissue onto connective tissue with biological and chemical interactions, and so on. Much more attention should be paid to the cells, biomaterials, and even printers before an ideal 3D-bioprinted skin construct model is finally accomplished.

17.3 Cell type

As mentioned above, adult cells, especially keratinocytes and fibroblasts, were chosen in most researches to be seeded into 3D-bioprinted constructs rather than stem cells (see Table 17.1). High viability of printed cells maintained in hydrogel, collagen secretion of fibroblasts, and K14 expression of keratinocytes were confirmed in these researches while the drawback was even prominent: little changes were shown in 2 weeks' culture due to the low secretion activity and proliferation rate, which might be the main obstacle for further clinical application.

Extensive resources may be the advantage of adult cells on the other hand, which makes modification of adult cell a solution for its application. Transformation of adult cells into pluripotent stem cells by biological or chemical approaches, named "reprogramming," may work in the future that can also avoid rejection in the application of allograft stem cells, providing that the technique is safe enough for the clinical application [21].

High proliferation rate is the prerequisite of seed cell that, however, is also considered as "inherent difficulty" [21] of adult cells and even stem cells. Nevertheless, high functional activities of seed cells may be the complementary key to the tissue

Table 17.1 **Comprehensive overview of available researches on 3D-bioprinted skin constructs**

Research	Printer	Seed cell		Matrix biomaterial		Process schematic	Post process
		cell	Concentration	material	Concentration		
Lee [8]	An array of four pneumatically driven dispensers (inkjet) and a three-axis Cartesian robotic stage [18]	Keratinocytes hKCs P6 Fibroblasts hFBs, P6	1.0×10^6/mL 1.0×10^6/mL	Type I collagen	2.05 mg/mL (DPBS) pH = 4.5	In vitro or in situ (PDMS wound model) bioprint	In vitro or on PDMS wound model cultured (8d); viability test of dispensed cells (live/dead staining); β-tublin labeled FB (IHC); pankeratin labeled KC (IHC)
Skardal [9]	Pressure-driven, computer controlled (inkjet) bioprinting device [19]	Amniotic fluid-derived stem cells hAFSCs, P16	16.6×10^6/mL	Fibrinogen Type I collagen Thrombin	50 mg/mL (PBS) 2.2 mg/mL (PBS) 20 IU/mL (PBS)	In situ bioprint	Wound closure rate (gross history); wound contraction and re-epithelialization rate (HE); neovascularization and blood vessel maturation (IHC); cells remaining in the regenerating skin (GFP/DAPI)
Michael [10,11]	Laser-assisted bioprinting (LaBP) [20]	Keratinocytes (HaCaT) Fibroblasts (NIH3T3)	Per sublayer 1.5×10^6 cells Per sublayer 1.5×10^6 cells/	Per sublayer: Type I collagen (3.0 mg/mL) 37 uL + PBS × 5 uL + 1 N × NaOH 0.85 ± 0.5 uL (pH = 7.1 ± 0.3)		In vitro bioprint	Submerged (1 night) + in vivo skin fold chamber (11d) or submerged (1 night) + in vitro ALI culture (11d); wound history (Masson's trichrome); e-cadherin, cytokeratin 14, Ki-67 and collagen IV (IF)
		Stabilization matrix: matriderm 2.3 × 2.3 cm					
Lee [12]	Pneumatic pressure driven (inkjet) bio-printing platform with 8 cell-dispensing channels [18]	Keratinocytes (HaCaT) Fibroblasts (HFF-1)	$0.5–5 \times 10^6$/mL $0.5–5 \times 10^6$/mL	Type I collagen	3.0 mg/mL (PBS)	In vitro bioprint	Submerged culture (8d) + ALI culture (14d); printed skin ALI culture d7 and d14 (HE/DAPI); N-cadherin junctions in ALI culture d14 (IF)

construction in vitro and regeneration in vivo, which makes stem cells the best choice, as conducted in Skardal's research [9]. Amniotic fluid-derived stem cells (AFSCs) are broadly multipotent and have been shown to differentiate into cells of all three lineages. Although few of the AFSCs remained in the wound after 14 days, their high secretion activity of cytokines and growth factors make great contribution to the wound closure and angiogenesis. Epidermal stem cells (ESCs), one of committed progenitor cells in epithelial tissue, is another promising seed cell in 3D-bioprinting skin construct for its directional differentiation into adult cells in epidermis, hair follicles, and sweat glands. It has been reported by Huang et al. [15] that 3D-bioprinted constructs seeded with ESCs makes remarkable contribution to the heal of burned skin and the regeneration of sweat glands, which would be discussed below.

17.4 Biomaterials

Biomaterial is one of the main challenges in 3D-bioprinting field. An ideal biomaterial that can be used in 3D-bioprinted skin constructs should provide desired compressive and tensile strength as well as printability and biocompatibility. Synthetic polymers, such as poly(ε-caprolactone) (PCL) and poly(lactic-*co*-glycolic acid) (PLGA), provide scaffold with high compressive strength to the construction of TE hard tissue such as bone and teeth, while its prolonged remain in open wound may cause infection, which makes them unsuitable for the 3D-bioprinted skin constructs. Acellular tissue matrices and naturally derived polymers are more biocompatible materials which promote cell recognition and adhesion. Acellular tissue matrices have been used to cover temporarily the severe burn, which showed positive therapeutic effect and reduce the scar formation. The role of acellular tissue matrices in TE skin is still to be discovered for few instance of application is available [22].

Naturally derived materials arc thc most popular biomaterials for their better performance in bioprinted constructs. Hypothetically, type I collagen, one of main components in extracellular matrix (ECM) of dermis secreted by fibroblast, may be the first choice for research. However, its application in these researches indicates that type I collagen performs poor mechanical strength without chemical cross-linking by glutaraldehyde, and its dilution is not dispensable until the pH is 4.5 that may affect the cell viability and narrow its application. Alginate, another popular biomaterial in this field, performs better in mechanical strength when cross-linked by calcium chloride as well as for the thermoregulated printability. The research conducted by Lee et al. [23] introduced the RGD (Arg-Gly-Asp) (cell adhesion ligand)-modified alginate promoted the cell recognition and adhesion that overcome the absence of cell activating sites on pure alginate, which indicated that modification of naturally derived biomaterials can be efficient to promote material property and cell function, while its best approach is still to be verified and discussed.

The addition of bioactive cytokines and growth factors may be another solution to overcome the drawbacks of undesirable biomaterial microenvironment of seed cells in bioprinted constructs. Huang et al. [15] demonstrated that dermal homogenate originated from 5-day-old mouse plantar dermis (PD) that contains most bioactive constituents in dermis significantly promoted cell inductivity of hydrogel that consisted of alginate and gelatin. Lee et al. [23] provided a beneficial approach that seed cells cultured on collagen for a short period before they are mixed into the bioink, which not only enriched ECM, bioactive cytokines, and growth factors secreted by seed cells but also promoted the cell spreading that contributes to cell functional activity.

17.5 Bioprinter and bioprinting process

Inkjet bioprinter is more popular in 3D-bioprinted skin researches for its low cost, high resolution, and high cell viability [24]. The specific advantage of inkjet bioprinter for skin bioprinting is that high-resolution and densified droplet "array" forms 2D flat surfaces with flexible position of cells, which may be a better model of natural skin tissue. However, available biomaterials restricts to collagen in these researches because of the request of liquefaction, which limits the further application of inkjet bioprinter in this field. Laser-assisted bioprinter is faced with the same limitations of confined biomaterials to be used besides the high cost. Cross-linking of liquid biomaterials is a solution of poor mechanical strength, such as glutaraldehyde, pH regulation, and UV, whereas this procedure involves modification of natural biomaterials and affects the cell viability and functional activity irreversibly.

Microextrusion bioprinter is more suitable for bioprinting of 3D scaffolds in most solid organs, blood vessels, and hard tissues by yielding continuous beads of materials rather than droplets or skin. But the corresponding biomaterials with relative high mechanical strength may provide a better solution, beyond which the space among scaffolds makes room for the regeneration of skin appendages. The potential applied advantages of microextrusion bioprinter make it a promising printer in the construction of bioprinted skin substitutes.

Bilayered skin-like construct of epidermis and dermis with multiple cells, which is more popular in these researches, require multiple channels or printing nozzles, which makes the printing process more complicated. However, the homogeneously bioprinted constructs seeded with AFSCs [9] make good performance in the treatment of wound, which arise a question about the design of skin constructs.

17.6 Design of skin constructs

Most of the researches on 3D-bioprinted skin constructs favor the bilayered skin-like structure (Fig. 17.1A and E) that consists of epidermis-like surface layer with high density of keratinocytes and dermis-like bottom with low density of fibroblasts, and this classic structure has been well established. However, the prolonged culture in vitro (Fig. 17.1D) after bioprinting witnessed obvious cell spreading, junction, and secretary activity as well as little progression of tissue formation while the keratinocytes of the constructs cultured in vivo partly grew on top of the normal epidermis (Fig. 17.1F). In contrast to the rapid wound closure assisted by the constructs seeded with AFSCs, the undesired results of these researches may be a sign of a potential revolution in design of skin constructs.

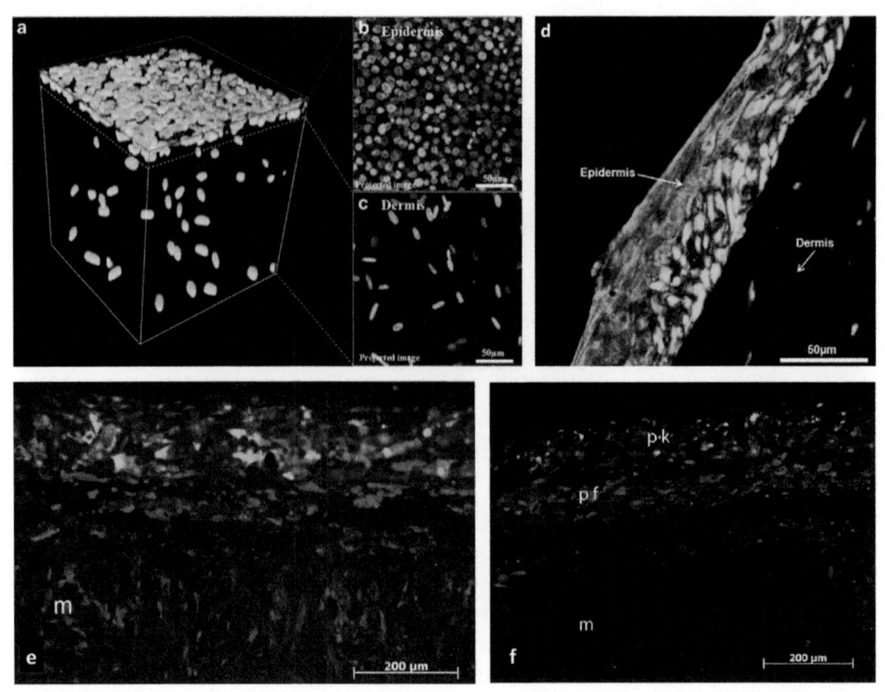

Fig. 17.1 Immunofluorescence of bioprinted skin constructs: (A–C) [12] submerged culture for 7 days, live cell nuclei are stained green in B and C; (D) [12] stained for *N*-cadherin (green) tight junctions at day 14 of ALI culture; (E) [10] LaBP formed skin-like structure; and (F) [10] skin-like structure in vivo culture for 11 days. *pk*, printed keratinocytes; *pf*, printed fibroblasts; *m*, Matriderm; *ne*, normal epidermis; and *n*, native mouse skin.

Moreover, in Skardal's research [9], the wound closurc provcs to bc AFSCs assisted rather than regeneration, which makes the mechanism of TE-assisted regeneration more confused. Recently, the research of a new area—the spatiotemporal control of dynamic cellular interactions—may provide a reasonable explanation: the behavior of stem cells, when they work collectively, can be much more sophisticated than one [25]. That is to say, tissue regeneration may require self-organization of stem cell population in specific spatial structures and environments. The detailed mechanism of this explanation is to be discovered in a long research period, whereas the concept is of great value nowadays. The construction of an assistant constructs and microenvironment for the promotion of functional activity of cells may be the main principle of next generation of 3D bioprinting that applied in regenerative medicine.

17.7 Regeneration of skin appendages

Completely functional organs have been considered as the final goal of regenerative medicine. Therefore, it is far from satisfaction that burn and skin ulcers are healed with a multilayered skin without appendages, which are lack of function of thermoregulation and excretion, and finally it results in secondary autologous skin graft transplant. Regeneration of skin appendages is impossible without the development of stem cell and TE technology. Recently, several researches [13–16] with completely different designs and methods on this field make great contribution to the final goal of completely functional skin.

A research conducted by Wu et al. [13] (Fig. 17.2A and B) introduced human reconstructed skin (hRSK) that consists of intact epidermis and dermis with mature hair follicles and sebaceous glands, primarily uses tissue culture cells that are expanded from fetal epidermis and dermis, with or without adult skin cells. It was remarkable that the delivery of human skin cells into wound was accomplished by PET membrane (polyethylene terephthalate, 3.0 μm pore size) rather than by any other ECM components. Another research conducted by Takagi et al. [14] (Fig. 17.2C and D) achieved functional skin with hair follicles and sebaceous glands based on induced pluripotent stem cells (iPSCs) and clustering-dependent embryoid body (CDB) transplantation method, which have been proven to be safe without tumorigenesis. These two researches give us new views on the methods concerning regeneration of organs that

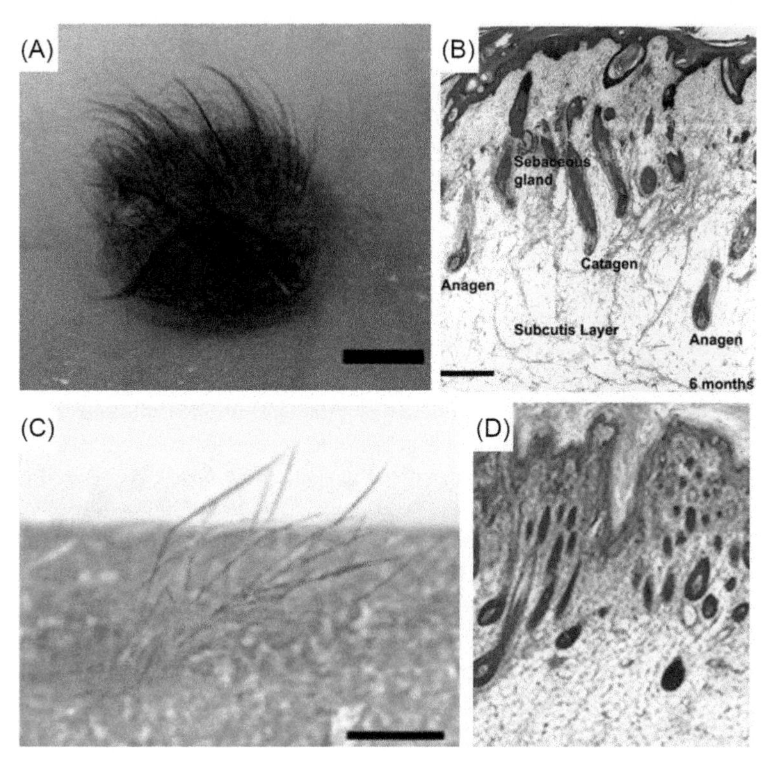

Fig. 17.2 Skin regeneration with hair and its tissue sections: (A and B) [13] hRSK with numerous hair shafts at 3 months after grafting dissociated human cells, a:hRSK covers about 2 cm^2 (2 months after cell grafting); (C and D) [14] macromorphological observations of engraftment by CDB into the dorsoventral skin of nude mice showing the eruption and growth of iPSC-derived hair follicles (Scale bars: 1 mm).

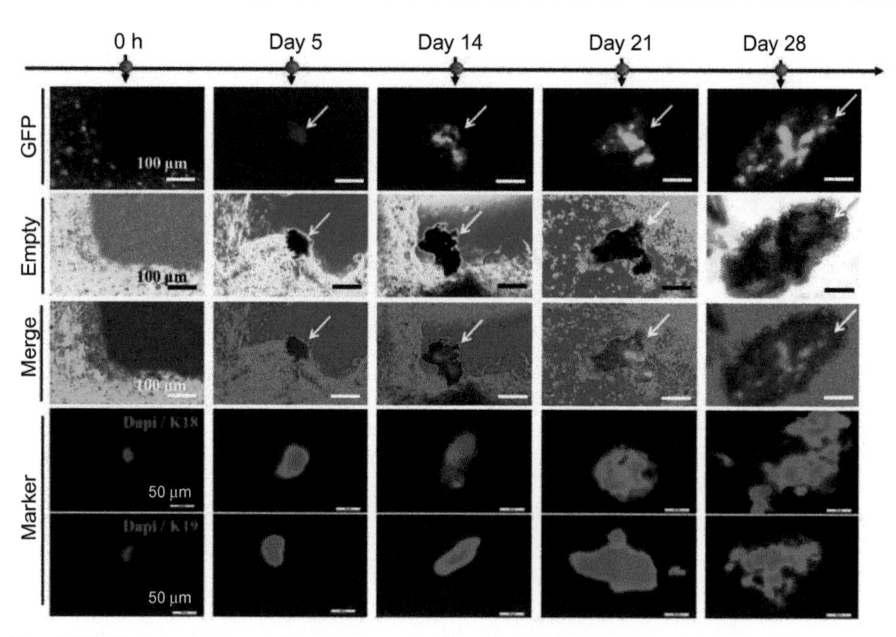

Fig. 17.3 The morphology of sweat gland tissue after day 28 of culture detected in the group of 300 μm and PD+construct. The 300-μm-printed constructs conferred a significant increase in sweat gland cell aggregates from weeks 1–2, and the specific sweat gland-bud emerged at week 3 [16].

may be valuable for the future researches based on TE, especially for researches on 3D bioprinting (Fig. 17.3).

Meanwhile, the absence of sweat glands in those stem cell-assisted and CDB-derived full-thickness skin tissue suggests that the regeneration of sweat glands is more difficult than that of hair follicles and sebaceous glands that may be conducted by different mechanism. Fortunately, the research conducted by Huang et al. [15] introduced enhanced specific differentiation of epidermal progenitors to regenerate sweat glands through a design of 3D-bioprinted ECM that provides the spatial inductive cues. Based on these 3D-bioprinted ECM scaffolds with the seed cells of epidermal progenitors, Liu et al. [16] achieved the regeneration of sweat gland-like tissue that was discovered to express the sweat glands-specific keratins and just right in the space among the scaffolds after prolonged culture in vitro. Therefore, 3D bioprinting-mediated spatial directional differentiation in vitro may be the pathway to the regeneration of sweat glands, the mechanobiological mechanism of which is still to be further discovered.

Hair follicles, sebaceous glands, and sweat glands are main appendages in natural skin tissue, the regeneration of which is far from completely functional regeneration of skin. The formation of vessels in a skin equivalent is equally important, which might be achieved by the direct interaction of endothelial cells with fibroblasts and their secreted ECM proteins and growth factors in vitro. In addition, the integration of melanocytes, nerve tissues, and even immune cells in the TE skin constructs is still to be achieved.

17.8 Prospects of researches on skin substitutes

The development of additive manufacturing technology [26] and materials science promote their application in regenerative medicine, such as 3D bioprinting. The last decade has witnessed the construction of basic principles and methods of 3D bioprinting in its application of in vitro reconstruction of organs, based on which several model of 3D-bioprinted skin has been constructed. However, it remains a challenge to create a precise and complex new tissue consisting of several cell types that are arranged in a specific 3D pattern. Furthermore, the different tissue functions strongly depend on its specific structure and on the cells that are influenced by their distinct microenvironment. Due to the limitations of seed cells, biomaterials, printers, and even design mentioned above, the 3D-bioprinted skin constructs is far from perfection even in the next decades, which indicates that application-oriented skin models rather than completely regeneration of skin tissue may be a more practical goal of researches in this field [27]. Recently, 3D-bioprinted skin has been considered as a promising model of melanoma in vitro for drug development test, whereas several mature 3D-bioprinted skin models have been used in the toxicity test of cosmetics and chemistry [28]. These researches can not only meet the application requirements but also make potential contribution to the regeneration of completely functional skin tissue in the future.

References

[1] Green H, Kehinde O, Thomas J. Growth of cultured human epidermal cells into multiple epithelia suitable for grafting. Proc Natl Acad Sci U S A 1979;76(11):5665–8.

[2] Supp DM, Boyce ST. Engineered skin substitutes: practices and potentials. Clin Dermatol 2005;23(4):403–12.

[3] Bello YM, Falabella AF, Eaqlstein WH. Tissue-engineered skin: current status in wound healing. Am J Clin Dermatol 2001;2(5):305–13.

[4] Heimbach D, Luterman A, Burke JF, et al. Artificial dermis for major burns: a multicenter randomized clinical trial. Ann Surg 1988;208(3):313–20.

[5] Kang HW, Lee SJ, Ko IK, et al. A 3D bioprinting system to produce human-scale tissue constructs with structural integrity. Nat Biotechnol 2016;34(3):312–9.

[6] Cui X, Breitenkamp K, Finn MG, et al. Direct human cartilage repair using three-dimensional bioprinting technology. Tissue Eng Part A 2012;18(11-12):1304–12.

[7] Jia W, Gungor-Ozkerim PS, Zhang YS, et al. Direct 3D bioprinting of perfusable vascular constructs using a blend bioink. Biomaterials 2016;106:58–68.

[8] Lee W, Debasitis JC, Lee VK, et al. Multi-layered culture of human skin fibroblasts and keratinocytes through three-dimensional freeform fabrication. Biomaterials 2009;30(8):1587–95.

[9] Skardal A, Mack D, Kapetanovic E, et al. Bioprinted amniotic fluid-derived stem cells accelerate healing of large skin wounds. Stem Cells Transl Med 2012;1(11):792–802.

[10] Michael S, Sorg H, Peck C-T, et al. Tissue engineered skin substitutes created by laser-assisted bioprinting form skin-like structures in the dorsal skin fold chamber in mice. PLoS One 2013;8(3) e57741.

[11] Koch L, Deiwick A, Schlie S, et al. Skin tissue generation by laser cell printing. Biotechnol Bioeng 2012;109(7):1855–63.

[12] Lee V, Singh G, Trasatti JP, et al. Design and fabrication of human skin by three-dimensional bioprinting. Tissue Eng Part C Meth 2014;20(6):473–84.
[13] Wu XW, Scott Jr. L, Washenik K, et al. Full-thickness skin with mature hair follicles generated from tissue culture expanded human cells. Tissue Eng Part A 2014;20(23-24):3314–21.
[14] Takaqi R, Ishimaru J, Sugawara A, et al. Bioengineering a 3D integumentary organ system from iPS cells using an in vivo transplantation model. Sci Adv 2016;2(4) e1500887.
[15] Huang S, Yao B, Xie JF, et al. 3D bioprinted extracellular matrix mimics facilitate directed differentiation of epithelial progenitors for sweat gland regeneration. Acta Biomater 2016;32:170–7.
[16] Liu N, Huang S, Yao B, et al. 3D bioprinting matrices with controlled pore structure and release function guide in vitro self-organization of sweat gland. Sci Rep 2016;6:34410.
[17] Murphy SV, Atala A. 3D bioprinting of tissues and organs. Nat Biotechnol 2014;32(8):773–85.
[18] Lee w, Lee V, Polio S, et al. On-demand three-dimensional freeform fabrication of multi-layered hydrogel scaffold with fluidic channels. Biotechnol Bioeng 2010;105(6):1178–86.
[19] Yoo JJ, Atala A, Binder KW, et al. Delivery system. U.S. patent application 20110172611. July 14, 2011.
[20] Unger C, Gruene M, Koch L, et al. Time-resolved imaging of hydrogel printing via laser-induced forward transfer. Appl Phys A Mater Sci Process 2011;103:271–7.
[21] Atala A. Regenerative medicine strategies. J Pediatr Surg 2012;47:17–28.
[22] Fang T, Lineaweaver WC, Sailes FC, et al. Clinical application of cultured epithelial autografts on acellular dermal matrices in the treatment of extended burn injuries. Ann Plast Surg 2014;73(5):509–15.
[23] Lee HJ, Kim YB, Ahn SH, et al. A new approach for fabricating collagen/ECM-based bioinks using preosteoblasts and human adipose stem cells. Adv Healthc Mater 2015;4(9):1359–68.
[24] Mandrycky C, Wang Z, Kim K, et al. 3D bioprinting for engineering complex tissues. Biotechnol Adv 2016;34(4):422–34.
[25] Sasai Y. Next-generation regenerative medicine: organogenesis from stem cells in 3D culture. Cell Stem Cell 2013;12(5):520–30.
[26] Zadpoor AA, Malda J. Additive manufacturing of biomaterials, tissues and organs. Ann Biomed Eng 2017;45:1–11.
[27] Ng WL, Wang S, Yeong WY, et al. Skin bioprinting: impending reality or fantasy? Trends Biotechnol 2016;34(9):689–99.
[28] Vultur A, Schanstra T, Herlyn M. The promise of 3D skin and melanoma cell bioprinting. Melanoma Res 2016;26(2):205–6.

3D bioprinting blood vessels

18

N. Hu, Y.S. Zhang
Harvard Medical School, Cambridge, MA, United States

18.1 Introduction

Blood vessels play a significant role in the circulatory system [1]. The main function of blood vessels is transporting blood from the heart to the rest of the tissues and organs throughout the body and then bringing it back to the heart. Blood vessels include three main types: arteries, veins, and capillaries. Arteries transport blood from the heart to other parts of the human body, and veins carry blood from other parts of the body back to the heart; capillaries, which are the smallest blood vessels, are the connective parts between arteries and veins, serving a function of enabling substance exchange between the blood and the surrounding tissues [2]. There are also other types of blood vessels such as elastic arteries, distributing arteries, arterioles, venules, large collecting vessels (e.g., subclavian vein, jugular vein, renal vein, and iliac vein), and venae cavae, which can be still defined however, in general as arteries (for those transporting blood away from the heart) and veins (for those carrying blood toward the heart) [2].

The structures of blood vessels are crucial to their physiological functions [3,4]. As shown in Fig. 18.1, arteries and veins both have three layers, and the middle layer of arteries is much thicker than that of veins. The intima consists of squamous endothelial cells (ECs), which are intertwined with a polysaccharide intercellular matrix to form the lumen for blood transportation, whereas the veins contain valve structure to avoid backflow. Internal elastic lamina surrounds the exterior of the intima, which is composed by subendothelial tissues and elastic banding. The media is the middle layer in the vessels, where the elastic fibers, connective tissues, polysaccharides, and vascular smooth muscle cells (SMCs) are mainly located. The SMCs control the contraction and expansion of vessels, enabling the modulation of the hemodynamics. The adventitia contains nutrient capillaries and nerves to support the vessels. There is also an external elastic lamina between the media and adventitia. Capillaries are the connections between arteries and veins, composed of EC and connective tissues, which provide alternative pathways for transporting blood from arteries to veins.

Compared with the heart, blood vessels seldom actively pump blood to tissues. However, certain blood vessels can adjust inner diameters by the slight peristalsis from the contraction of the smooth muscle layer, which is controlled by the autonomic nervous system [5,6]. Vasoconstriction and vasodilation can also be regulated by vasoconstrictors (e.g., prostaglandins, vasopressin, angiotensin, and epinephrine) and vasodilator (e.g., nitric oxide). Oxygen bound to hemoglobin of red blood cells is one of the main nutrients transported by the blood vessel. Oxygen saturation of hemoglobin is typically in the range 95%–100% in all the arteries except the pulmonary

3D Bioprinting for Reconstructive Surgery. https://doi.org/10.1016/B978-0-08-101103-4.00018-1

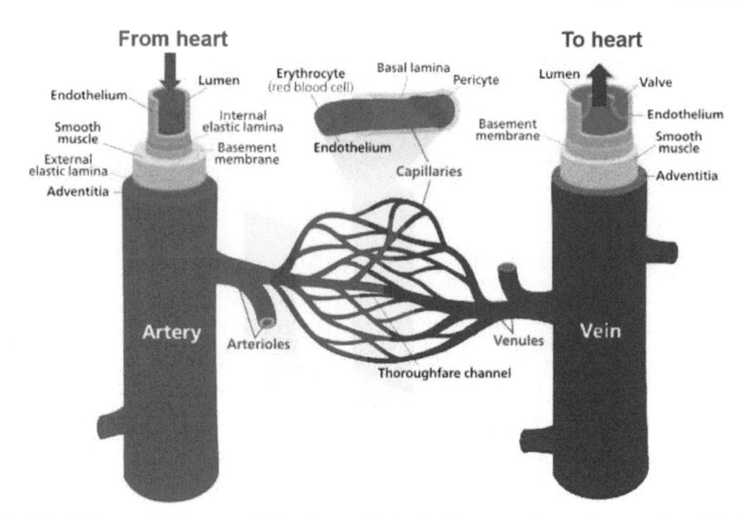

Fig. 18.1 Physiology structures of blood vessels including arteries, veins, and capillaries.

artery, whereas oxygen saturation of blood is about 75% in all the arteries except the pulmonary artery [2].

Blood vessels are significantly related to some common diseases. Angiogenesis in a tumor accelerates the progress of cancer by supporting the nutrition of the cancer cells [7]. The lipid lumps formed in the blood vessel wall can lead to atherosclerosis [8], which is the most common cardiovascular disease causing many deaths. Moreover, permeability of blood vessels increases during inflammation [9], and common hemorrhage induced by damage or in spontaneous cases is usually due to mechanical injury of the vascular endothelium. Besides, blockage of the blood vessels caused by internal and external factors (e.g., atherosclerotic plaque, embolized blood clot, or foreign body) probably leads to ischemia of tissue, sometimes even necrosis [10].

Tissue engineering has provided a promising strategy to repair and replace portions of tissues (e.g., bone, cartilage, blood vessels, and skin), using a combinatory approach including biology, medicine, materials, chemistry, and engineering [11–13]. Among the different tissue types, blood vessels are one of the most important yet challenging tissues to engineer, which are mainly fabricated by seeding proper cell types such as ECs and SMCs on suitable biomaterial substrates [14–16]. However, engineered blood vessels using conventional strategies based on scaffolds are usually produced using relatively sophisticated procedures, and cannot be easily applied to vessels with complex architectures. The recent advances in the three-dimensional (3D) bioprinting technology on the contrary, have provided unprecedented flexibility in engineering blood vessels with high resolution, strong fidelity, and high complexity [17]. In this chapter, vascular tissue engineering based on 3D extrusion bioprinting is discussed. We first introduce the different strategies for fabricating blood vessels, including sacrificial bioprinting, embedded bioprinting, hollow tube bioprinting, and microtissue bioprinting. Finally, we summarize with brief conclusions and future perspectives.

18.2 3D bioprinting strategies

Many strategies based on 3D bioprinting have been innovated over the decade for the fabrication of engineered blood vessels, where most commonly used include for example, sacrificial bioprinting, embedded bioprinting, hollow tube bioprinting, and microtissue bioprinting, among others.

Sacrificial bioprinting is probably the most established method to produce engineered blood vessels (Fig. 18.2A) [18–20]. In this approach, the sacrificial materials (e.g., agarose [21], carbohydrate glass [20], Pluronic F127 [19,22,23], and gelatin [24,25]) are usually used to fabricate the template channels within hydrogel matrices, which can be selectively removed or dissolved for subsequent seeding of vascular cells. The sacrificial bioprinting strategy can achieve the rapid fabrication of perfusable blood vessels in an efficient manner, potentially enabling the generation of complex vascular patterns. However, sacrificial bioprinting is limited in its ability to engineer spatially distributed blood vessel constructs due to the lack of support in the 3D volume, and therefore cannot satisfy some needs where the creation of sophisticated 3D vasculature is required.

To address the limitation associated with sacrificial bioprinting, embedded bioprinting has been proposed as a novel strategy to produce freeform 3D vascular constructs by taking advantage of the antigravity feature of the supporting hydrogels, which significantly improves the complexity of the blood vessels that can be bioprinted [26–28]. In a typical embedded bioprinting strategy (Fig. 18.2B), the shear-thinning bioink can be directly extruded into a self-healing supporting hydrogel matrix to build the arbitrarily shaped vessel-like structures.

Hollow tube bioprinting represents an additional strategy that can directly deposit hollow vascular structures in a single step, typically through the use of a bioprinting system in conjunction with the wet-spinning technique with a multilayer printhead (Fig. 18.2C) [29,30]. These bioprinted hollow tubes can transport fluids within the hollow interiors, which have similar characteristics with blood vessels. These hollow blood vessel-like structures can be fabricated by first bioprinting the tubes followed by culturing vascular cells or by directly bioprinting vascular cell-laden bioinks into the hollow structures.

Besides, other bioprinting strategies have also been developed to fabricate engineered blood vessels such as the use of spheroids and microtissues as the bioinks for printing, which take the spheroid or microtissues as the basic building units to enable

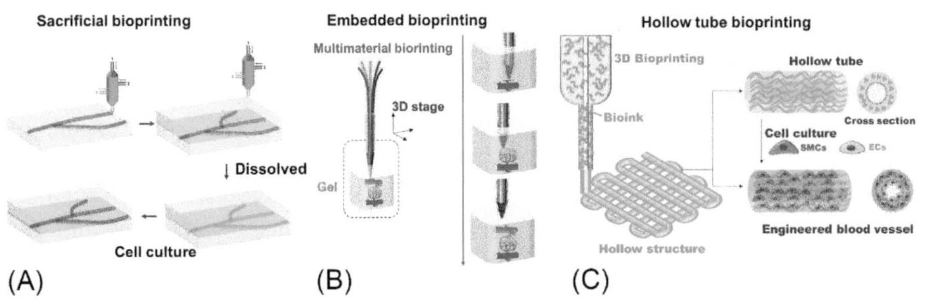

Fig. 18.2 Main types of 3D bioprinting strategies for engineering blood vessels: (A) sacrificial bioprinting, (B) embedded bioprinting, and (C) hollow tube bioprinting.

engineering of blood vessels at a relatively large scale than other strategies mentioned above [31]. All these bioprinting strategies will be separately discussed in details in the following sections.

18.3 Sacrificial bioprinting for vascular engineering

The common strategy of sacrificial bioprinting is based on the employment of sacrificial material to fabricate a sacrificial template, and a hydrogel matrix will be subsequently cast or bioprinted surrounding the template. Finally, this sacrificial template can be removed by mechanical pulling or dissolution. As such, the well-defined microchannels will be formed for vascular engineering. As aforementioned, a variety of biomaterials can be used as the sacrificial templates such as agarose [21], gelatin [24,25], Pluronic F127 [19,22,23], and carbohydrate glass [20], among others [32,33].

The Khademhosseini group reported a sacrificial bioprinting strategy based on agarose microfibers to achieve vascular structures via mechanical removal [21]. As shown in Fig. 18.3A, agarose was used as a sacrificial material, and the agarose microfibers were bioprinted in predefined vessel-like patterns. The gelatin methacryloyl (GelMA) hydrogel was cast over the bioprinted agarose microfibrous templates. The bioprinted agarose microfibers were subsequently removed by manual pulling or using mild vacuum from the photo-cross-linked GelMA hydrogels. The perfusable microchannels were formed without an additional dissolution step, and GelMA hydrogel containing intrinsic cell-binding sites ensured strong attachment of human umbilical vein endothelial cells (HUVECs). Taking advantage of this sacrificial strategy, various vessel-like structures could be rapidly generated (Fig. 18.3B).

To improve further the biocompatibility of bioprinted vascular structure, Dai and colleagues developed a collagen-based vascular construct [24], which adopted a hydrogel of endothelial cell-laden gelatin as the sacrificial material for bioprinting at a lower temperature (Fig. 18.3C). Following a subsequent culture at 37°C in an incubator, the endothelial cells would attach to the surface of the microchannels as the gelatin liquefies and dissolves out over time, forming the engineered vessels presenting vascular morphology and barrier functions (Fig. 18.3D).

Lewis and coworkers developed a cell-laden, vascularized tissues with a thickness of more than 1 cm, including HUVECs, human mesenchymal stem cells (hMSCs), and human neonatal dermal fibroblasts (HNDFs) embedded in extracellular matrix materials, where high cell viability could be maintained for more than 6 weeks in a perfusable chip (Fig. 18.3E) [19]. These thick vascularized tissues were fabricated by sacrificial bioprinting using a composite fugitive bioink. The bioink containing Pluronic and thrombin were used to bioprint the template in a vascular pattern, where the cell-laden hydrogel containing gelatin and fibrinogen were used to surround the template to form the thick tissue. The fugitive bioink was then liquefied and removed to generate a large vascular network, which could be further endothelialized and perfused to achieve long-term survival of the thick tissue using a perfusion system (Fig. 18.3F–M).

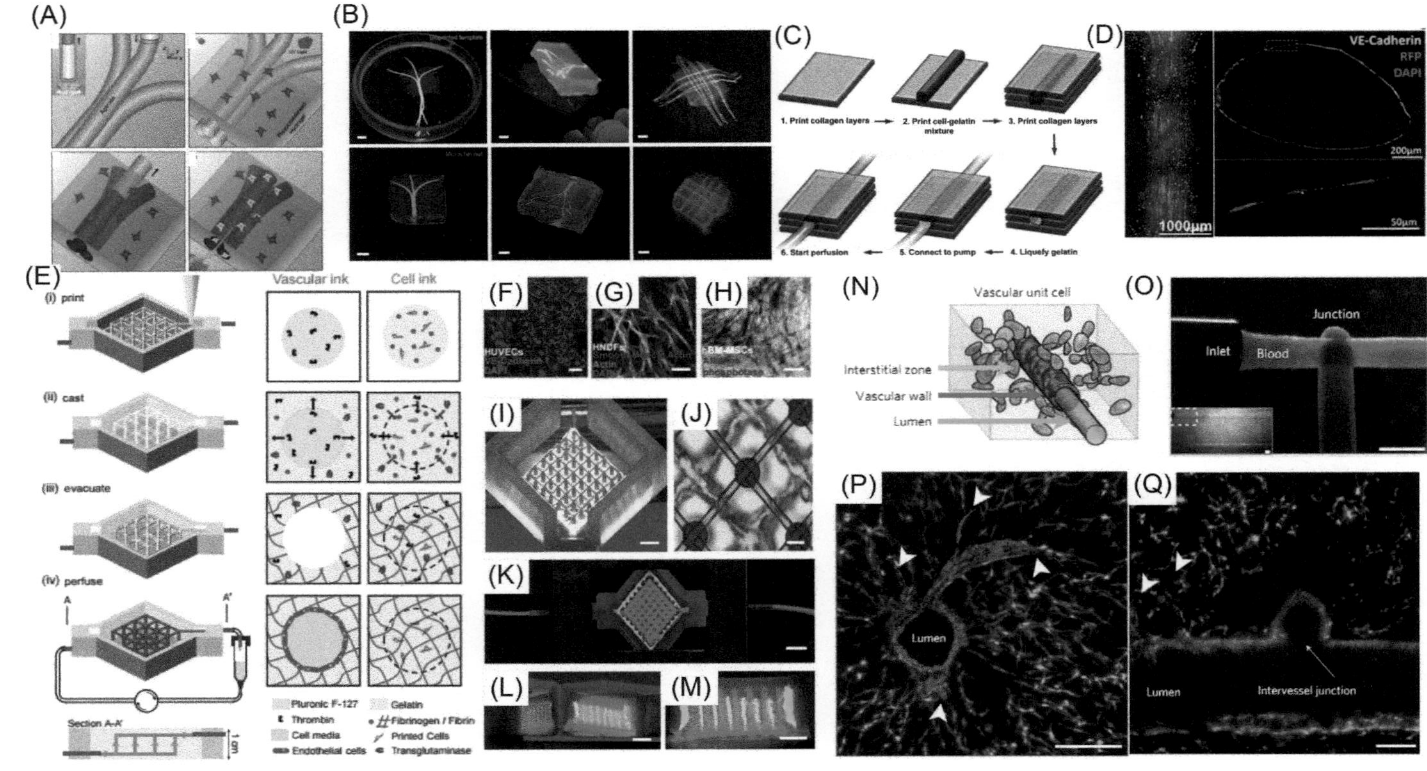

Fig. 18.3 See figure legend in next page

Similarly, the Chen Group reported the use of a network of rigid 3D filaments as the sacrificial template, printed with carbohydrate glass [20]. The formed vascular structure could be lined with endothelial cells, which included the lumen, the endothelialized vascular wall, the matrix, and cell-based interstitial zone (Fig. 18.3N). The vascular structure supported the pulsatile blood flow at high pressure (Fig. 18.3O). The endothelial cells formed patterned vasculature and intervascular junctions (Fig. 18.3P and Q).

These sacrificial bioprinting strategies made the construction of the vascular network, the endothelialization, and the extravascular tissue independent, potentially compatibile with various types of cells and extracellular matrices for engineering various types of tissues. However, the sacrificial materials have been usually associated with different degrees of cytotoxicity originating from template dissolution. For example, the bioprinted sacrificial carbohydrate glass templates required coating with poly(lactide-*co*-glycolide) (PLGA) to prevent osmotic damage to cells enclosed

Fig. 18.3 continued Sacrificial bioprinting for vascular engineering. (A) Schematic of agarose template fiber-based bioprinting. The agarose template is bioprinted by a bioprinter with a piston fitted inside a glass capillary aspirating at predefined locations after gelation, which is subsequently cast over by a hydrogel. Fully perfusable microchannels were formed after the agarose template is removed from the photo-cross-linked gel. (B) Photographs of the bioprinted templates (green) enclosed in GelMA hydrogels and the respective microchannels perfused with a fluorescent microbead suspension (pink) (Scale bars: 3 mm). (C) Construction processes of the vascular channel using cell-gelatin mixture as a sacrificial material. (D) Fluorescence images of the flow of green beads and bioprinted endothelial cells on Day 1 under dynamic flow culture. After 5 days of dynamic culture, it could be found from a cross section that endothelial cells formed a monolayer along the channel surface. (E) Thick vascularized structure is bioprinted in a perfusion chip by fugitive bioink and cell-laden bioink. The extracellular matrix (ECM) material is then cast over the bioprinted vascular structure. The sacrificial bioink is then liquefied and evacuated. The vascular structure is subsequently endothelialized and perfused. (F–H) HUVECs, HNDFs, and hMSCs growing in the bioprinted matrix (Scale bar: 50 μm). (I) Photographs of sacrificial structure (red) on chip (Scale bar: 2 mm). (J) Top view image of sacrificial structure (red) and cell inks (green) (Scale bar: 50 μm). (K–M) Image of a 3D bioprinted tissue with vascular unit cell in a perfusion chamber (K) and their cross sections (L and M) (Scale bars: 5 mm). (N) Schematic of vascular unit cell including the lumen, the endothelialized vascular wall, and the matrix and cell-based interstitial zone. (O) Vascular channels and intervessel junctions support pulsatile flow of human blood. (P) Endothelial cells formed patterned vasculature at the arrow position. (Q) Intervessel junctions near the larger vessels.
(A–B) Reproduced with permission from Bertassoni LE, Cecconi M, Manoharan V, Nikkhah M, Hjortnaes J, Cristino AL, et al. Hydrogel bioprinted microchannel networks for vascularization of tissue engineering constructs. Lab Chip 2014;14(13):2202–2211. Copyright 2014 Royal Society of Chemistry; (C–D) Reproduced with permission from Lee VK, Kim DY, Ngo H, Lee Y, Seo L, Yoo S-S, et al. Creating perfused functional vascular channels using 3D bio-printing technology. Biomaterials 2014;35(28):8092–8102. Copyright 2014 Elsevier; (E–M) Reproduced with permission from Kolesky DB, Homan KA, Skylar-Scott MA, Lewis JA. Three-dimensional bioprinting of thick vascularized tissues. Proct Natl Acad Sci U S A 2016;113(12):3179–3184. Copyright 2016 National Academy of Sciences; (N–Q) Reproduced with permission from Miller JS, Stevens KR, Yang MT, Baker BM, Nguyen D-HT, Cohen DM, et al. Rapid casting of patterned vascular networks for perfusable engineered three-dimensional tissues. Nat Mater 2012;11(9):768–774. Copyright 2012 Nature Publishing Group.

inside the hydrogel during the dissolution process [20]. However, highly concentrated Pluronic has shown cytotoxic effects for cells embedded within the surrounding matrices [22]. Therefore, it is of paramount importance to further optimize the formulations of the sacrificial bioinks to improve biocompatibility and expand the applicability of sacrificial bioprinting in vascular engineering.

18.4 Embedded bioprinting for vascular engineering

Conventional extrusion bioprinting can deposit structures in a layer-by-layer manner, which, however, is difficult to achieve freeform structures due to lack of support in the volumetric space, precluding the fabrication of complex blood vessels. To address such a challenge, embedded bioprinting recently emerged to provide a new strategy to bioprint freeform patterns in 3D space at high resolution.

Burdick and colleagues initially developed the embedded bioprinting that allowed direct extrusion of shear-thinning hydrogel bioinks into self-healing support hydrogels [26]. They used the supramolecular hydrogels based on conjugation of adamantane (Ad) and β-cyclodextrin (CD), respectively, to hyaluronic acid (HA), which showed good biocompatibility [34] (Fig. 18.4A). Specifically, Ad-HA and CD-HA could rapidly form guest-host complexes immediately after mixing (Fig. 18.4B). It is therefore possible that this guest-host hydrogel system be directly writable due to their noncovalent and reversible bonds that could be disrupted by a physical stimulus (e.g., shear stress) and reform rapidly after the removal of the stimulus [35]. These properties qualified their use for embedded bioprinting, where the support matrix could accommodate the bioprinted material to maintain the structure through self-healing [36,37]. Consequently, 3D filaments could be deposited within the support hydrogel, and their diameters and patterns might also be easily controlled by the movement and size of needles (Fig. 18.4C–F).

The Feinberg Group further reported a similar strategy by applying the soft protein and polysaccharide hydrogels to bioprint the complex 3D biological structures in freeform, termed *freeform reversible embedding of suspended hydrogels* (FRESH) [27]. FRESH-enabled bioprinting of complex structures in a thermoreversible support bath using computer-aided design (CAD) models derived from optical computed tomography (OCT) and magnetic resonance imaging (MRI) data at a resolution of ~200 μm (Fig. 18.4G and H). Alternatively, microparticles-based soft granular hydrogels also appear to be efficient support materials for embedded bioprinting of arbitrary structures due to their shear-thinning behaviors [38–40]. Granular gels made from soft microparticles can smoothly transit between the solid and fluid states, making them ideal for holding complex and high-resolution bioprinted structures. Accordingly, the Angelini Group bioprinted complex 3D objects consisting of branching vessel-like networks using hydrogels, colloids, silicones, and living cells [28]. As shown in Fig. 18.4K, as a fine injection tip extruded the bioink in a predefined spatial path, the support granular gel could rapidly solidify after temporarily fluidizing locally, leading to stabilization of the bioprinted pattern at the injection site. Bioprinted freeform structures could be cross-linked and removed from the granular gel. This approach provides a versatile way to engineer diverse vascular structures with varying diameters and branches (Fig. 18.4L).

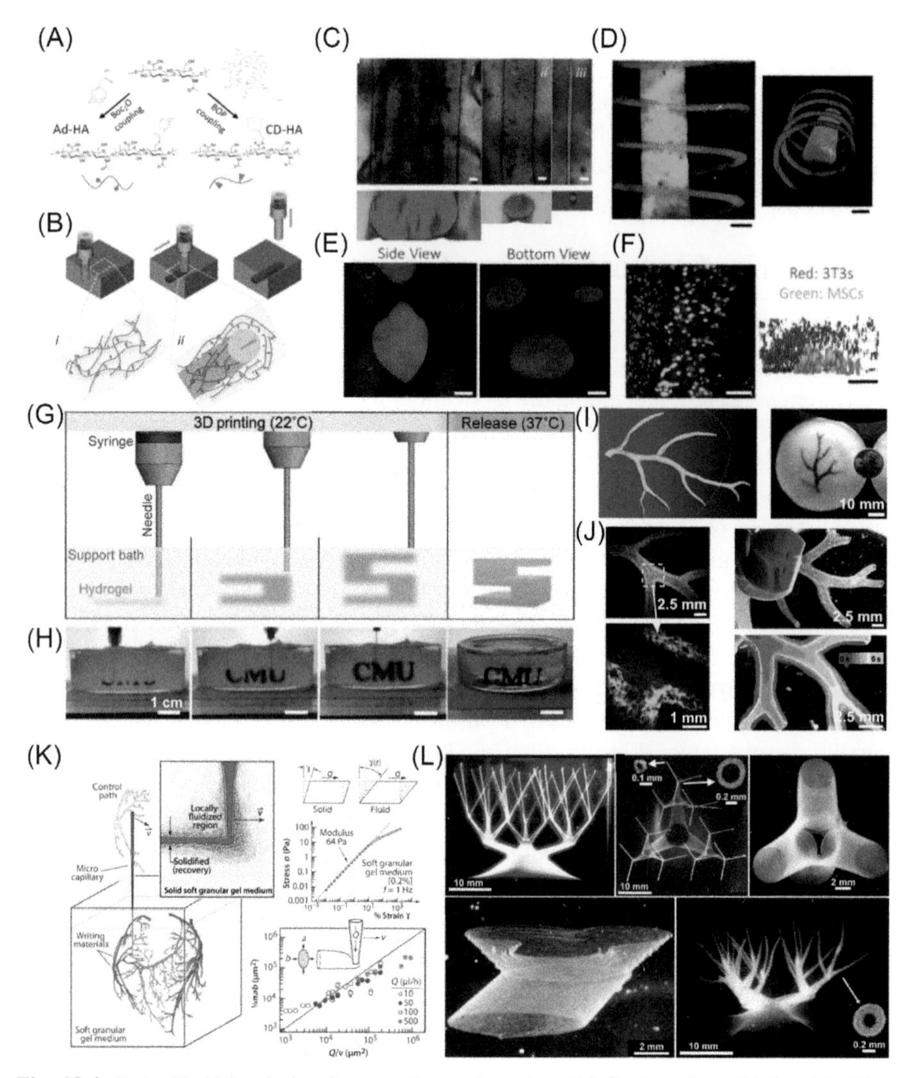

Fig. 18.4 Embedded bioprinting for vascular engineering. (A) Conjugation of Ad and β-CD to HA. (B) Schematics of embedded bioprinting by injecting a supramolecular ink (red) into a supramolecular support gel (green) from the original network (i) to partial cross-link pattern (ii). (C) Rhodamine-labeled filaments extruded into a fluorescein-labeled support gel by needles with different diameters of 20 G (i), 27 G (ii), and 34 G (iii). (Scale bars: 100 μm). (D) A filament of a fluorescein-labeled ink with a continuous spiral pattern (Scale bars: 200 μm). (E) Discrete pockets of a rhodamine-labeled ink printed into an unlabeled support gel (Scale bars: 200 μm). (F) hMSCs (green) bioprinted within a bioink into a support gel containing NIH/3T3 fibroblasts (red) (Scale bars: 200 μm). (G) Schematic of the FRESH processes, showing the bioprinted and cross-linked hydrogel (green) in the gelatin support bath (yellow). The 3D structure is fabricated layer by layer and the support gelatin melted to release a bioprinted structure by heating at 37°C. (H) Alginate "CMU" were built by FRESH bioprinted in the font of Times New Roman. The letters are still not deforming when the gelatin support

18.5 Hollow tube bioprinting for vascular engineering

Although the sacrificial and embedded bioprinting strategies enable the generation of vascular structures, they generally involve complicated multistep procedures. To this end, hollow tube bioprinting provides an alternative to build vessel-like channel, better imitating the anatomy and physiology of blood vessels [29,41,42,30].

Based on alginate, Ozbolat and colleagues developed the technique to bioprint tubular microchannels mimicking the blood vessels, which could be constantly perfused to provide nutrients for encapsulated cells [29]. For this hollow tube bioprinting, a coaxial nozzle was designed, where alginate bioink was extruded from the sheath flow whereas the $CaCl_2$ solution was delivered from the interior (Fig. 18.5A–E). Vessel-like hollow microfibers were formed on cross-linking of the sheath alginate bioink by Ca^{2+}. The cell viability could be assured by optimization of the bioprinting conditions such as the geometry of the coaxial nozzle as well as the concentrations of the bioinks and the cross-linking agent. Due to the shear stress, the encapsulated cells would experience certain damage during the bioprinting process (Fig. 18.5F); however, the cells could gradually recover during subsequent culture (Fig. 18.5G), where their functionality was also validated by evaluating the tissue-specific biomarker expression and extracellular matrix production.

The bioink formulation is another critical parameter in hollow tube bioprinting especially when improved functions of the bioprinted cells are required. Hydrogel materials such as gelatin, collagen, hyaluronic acid, alginate, and poly(ethylene glycol) (PEG),

---

Fig. 18.4 continued melts the change in optical, fluidic properties. (I) A section of human right coronary arterial tree from 3D MRI is translated into machine code for FRESH bioprinting, and was bioprinted with alginate (black) in the gelatin support bath. (J) The bioprinted arterial trees showing the vessel wall, well-formed lumen, and multiple bifurcations under fluorescent alginate (green). The engineered vessel was validated to be well sealed by perfusion. (K) Granular gel as an alternative embedded bioprinting support materials. A capillary tip extruded out material into the granular gel to form a complex pattern. The granular gel locally fluidizes and then rapidly solidifies when the tip moves across it, so the complex structures can be built without solidifying or cross-linking to support themselves. Stress-strain measurements presented a shear modulus of 64 Pa and a yield stress of 9 Pa for 0.2% (w/v) Carbopol gel. The cross-sectional area of embedded bioprinting showed nearly ideal behavior $Q = \frac{1}{4}\pi abv$ over a wide range of tip speeds, v, and flow rates, Q. (L) Diverse branched vascular networks based on the granular gel support materials. A continuous hollow vessel network with several orders of magnitude in diameter and aspect ratio.

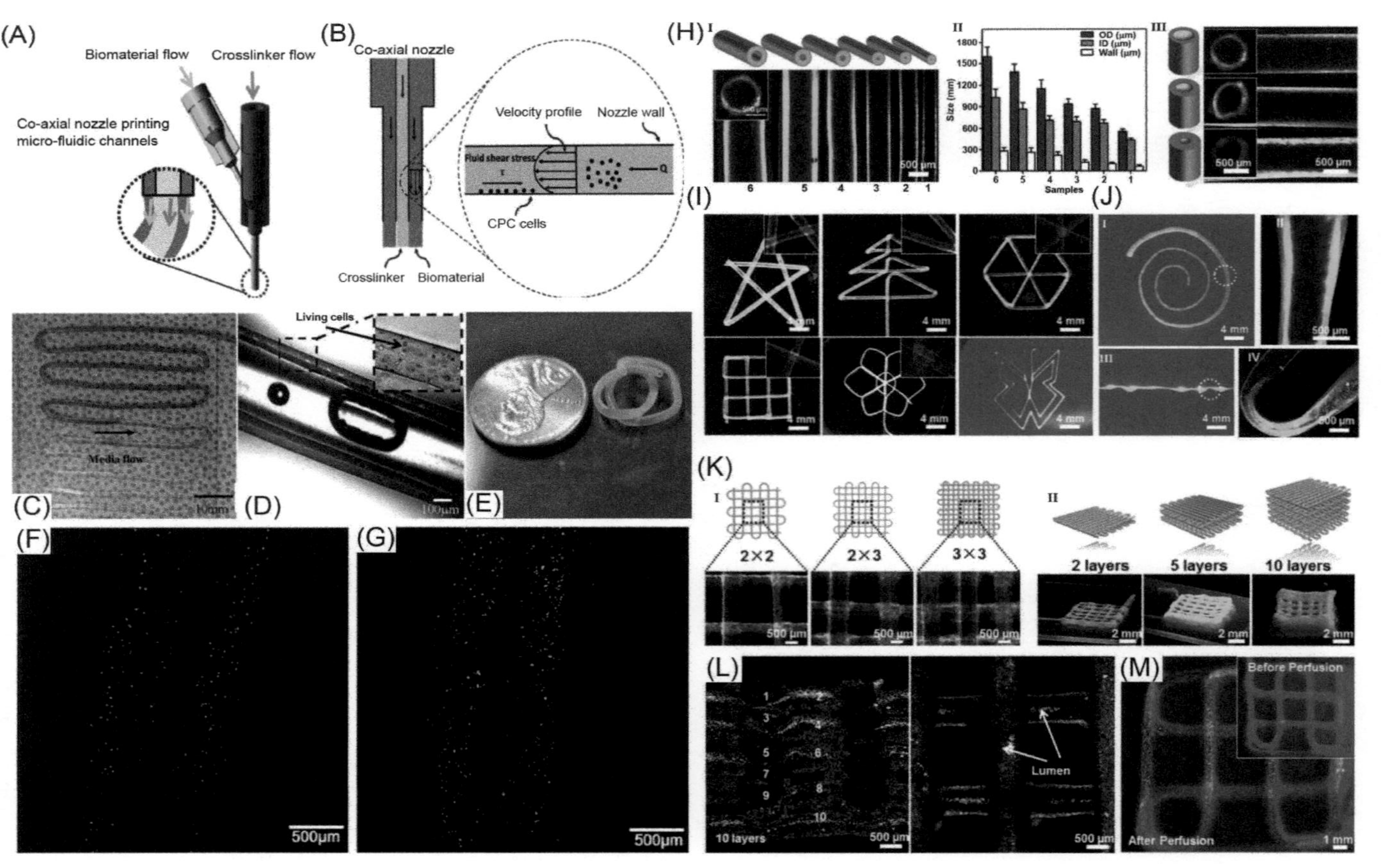

Fig. 18.5 Hollow tube bioprinting for vascular engineering: (A) schematic of coaxial nozzle for hollow tube bioprinting, (B) shear stress generated in the coaxial nozzle system, (C) cell-laden tubular channels were bioprinted in the form of zigzag with medium perfusion,

have been used as bioinks based on their biocompatibility and sufficient viscosity for bioprinting [34,43,44]. Nevertheless, none of them alone offers suitable properties for direct bioprinting of perfusable hollow tubes. The Khademhosseini Group recently reported a hollow tube bioprinting strategy by engineering a blend bioink containing alginate, GelMA, and 4-arm PEG-tetra-acrylate (PEGTA) [30], which not only possessed suitable rheological properties for use as the bioink, but also had sufficient mechanical strength and biocompatibility to support the functionality of the bioprinted hollow tubes. The alginate component in the blend bioink could be cross-linked by the delivered Ca^{2+} in both the core and the sheath to maintain the initial shape of extruded hollow tubes. Following bioprinting, the morphologies of the hollow tubes were chemically fixed by photo-cross-linking of the GelMA and PEGTA components. This blend bioink-based hollow tube bioprinting could easily achieve control over the size and pattern of the generated hollow tubes (Fig. 18.5H–M). The PEGTA molecules in the bioink led to the formation of a loose network with large pores to support both the strong mechanics of the blood vessels and the proliferation/stretching of the embedded vascular cells, whereas the biocompatible nature of GelMA further ensured functionality of the cells.

18.6 Scaffold-free multicellular spheroid bioprinting for vascular engineering

For vascular engineering, the choice of matrix materials plays a significant role. The immunogenicity, degradation rate, toxicity of degradation products, host inflammatory responses, and mechanical properties should all be considered. These potential problems associated with biomaterials may affect the applications of engineered vascular constructs, and may directly interfere with their biological functions in the long term,

Fig. 18.5 continued (D) images of vessel-like hollow structure and cell encapsulation in the channel wall, (E) cell-laden hollow tube showing promising mechanical and structural integrity after a 1-week culture, (F) confocal image for live/dead staining of at 12-h postprinting, massive cell death (red fluorescent) was observed all over the printed structure, (G) after a 72-h incubation, a few dead cells were scattered among increased live cells, (H) bioprinted hollow tube with different outer diameters (I) and quantification data (II), with same outer diameters and different inner diameters (III), (I) different bioprinted perfusable tubes with various shapes, (J) hollow tube with a gradually increasing size and periodically varying sizes, (K) images of the bioprinted hollow tube-based grids with different aspect ratios and numbers of layers, (L) confocal images showing a uniform 3D structure with 10 layers of bioprinted hollow tubes, which were perfused with red fluorescent microbeads, and (M) fluorescence images before (inset) and after the injection with red fluorescent microbeads into the lumen of the continuous bioprinted hollow tube.
(A–G) Reproduced with permission from Yu Y, Zhang Y, Martin JA, Ozbolat IT. Evaluation of cell viability and functionality in vessel-like bioprintable cell-laden tubular channels. J Biomech Eng 2013;135(9):091011. Copyright 2013 American Society of Mechanical Engineers; (H–M) Reproduced with permission from Jia W, Gungor-Ozkerim PS, Zhang YS, Yue K, Zhu K, Liu W, et al. Direct 3D bioprinting of perfusable vascular constructs using a blend bioink. Biomaterials 2016;106:58–68. Copyright 2016 Elsevier.

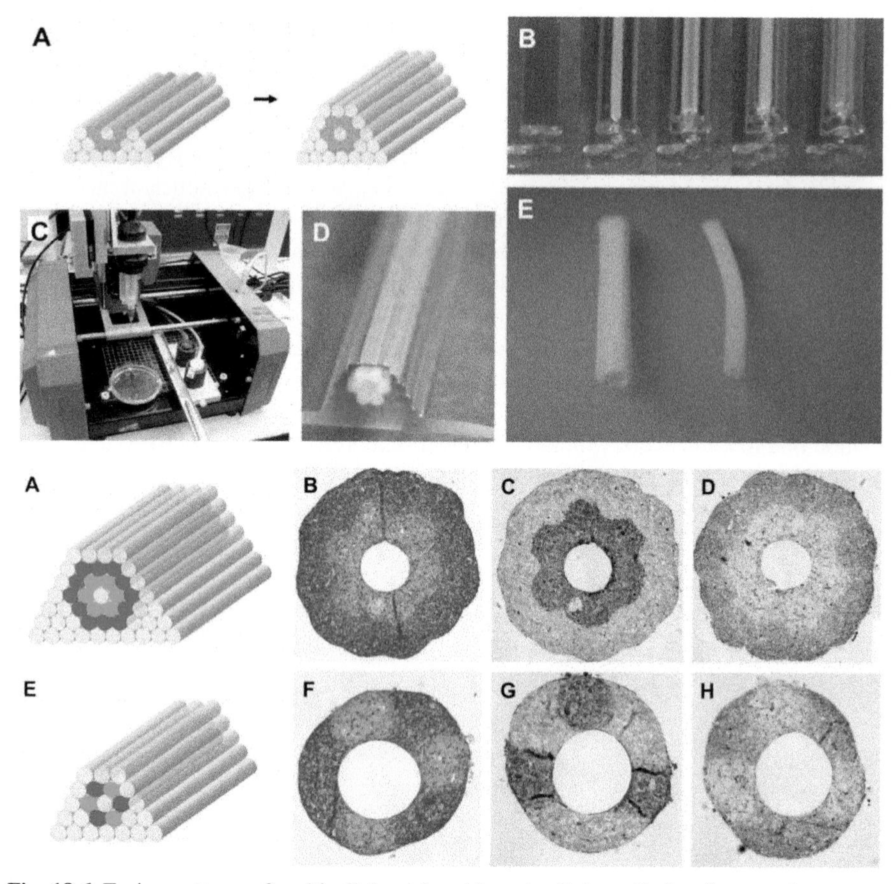

Fig. 18.6 Fusion patterns of multicellular spheroids and cellular cylinders for vascular engineering. (A) Schematics of deposition of the smallest diameter tube by agarose rods (pink) and multicellular spheroids (orange) with same diameters. (B) Different complex tubular structures. (C) Branched structure. (D) HSF spheroids bioprinted per template in (A). (E) Fusion pattern labeled red and green CHO spheroids in a tube after 7 days. (F) Branched structure based on HSF spheroids with 1.2 and 0.9 mm of diameters. (G) The fused branched structure after 6 days. (H, I) Design template by deposition of agarose cylinders (blue) and multicellular SMC cylinders. (J) Bioprinter with two printheads. (K) Bioprinted construct. (L) Engineered pig SMC tubes after 3 days of postprinted fusion (left: 2.5 mm OD; right: 1.5 mm OD). (M, Q) A double-layered vascular wall. HUVSMC (green) and HSF (red) multicellular cylinders were assembled per specific patterns. (N, R) Hematoxylin and eosin, (O, S) smooth muscle α-actin (brown), and (P, T) caspase-3 (brown) staining showing the results of the respective structures in (M) and (Q) after 3 days of fusion.
(A–T) Reproduced with permission from Norotte C, Marga FS, Niklason LE, Forgacs G. Scaffold-free vascular tissue engineering using bioprinting. Biomaterials 2009;30(30):5910–5917. Copyright 2009 Elsevier.

especially when clinical translation is the immediate goal [45]. Therefore, scaffold-free bioprinting has emerged as another strategy to fabricate vascular structures.

In one example, Forgacs and colleagues introduced a rapid prototyping technique based on 3D-automated deposition of multicellular spheroids or cylinders by a bioprinter (Fig. 18.6A–C, H, M, Q) [31]. The 3D tissue structures were formed through the postbioprinting fusion of the multicellular spheroids or cylinders (Fig. 18.6E, G, N–P, R–T), which is similar to the self-assembly phenomenon observed during early morphogenesis [46]. Delivery of multicellular spheroids by bioprinting was rapid and accurate, and assured maximal cell density leading to minimal cell damage. A fully biological self-assembly approach was thus developed by a rapid scaffold-free prototyping bioprinting for fabrication of blood vessels of various sizes (OD: 0.9–2.5 mm) (Fig. 18.6D–G). Multiple types of cells such as SMCs, human skin fibroblasts (HSF), Chinese Hamster Ovary (CHO) cells, and human umbilical vein smooth muscle cells (HUVSMCs) were integrated into discrete units. The multicellular spheroids or cylinders had controllable diameters (300–500 μm), and agarose rods were used as a template in this bioprinting strategy. This unique scaffold-free bioprinting strategy enabled the engineering of vascular structures without the use of any biomaterials and can be potentially extended to fabricate blood vessels at clinically relevant scales.

18.7 Conclusion

In this chapter, we focused on the discussions of various 3D bioprinting strategies for vascular engineering, including sacrificial bioprinting, embedded bioprinting, hollow tube bioprinting, and scaffold-free multicellular spheroid-based bioprinting. Key examples of each bioprinting strategy were illustrated. It is anticipated that further improvements of the bioink formulations would play a crucial role in assuring the stability of the bioprinted vascular structures as well as the maintenance of good cell viability and functionality. A combination of different strategies might also generate additional advantages for vascular bioprinting, such as deposition of multiscale patterns. We believe that 3D extrusion bioprinting will become a promising and powerful approach in engineering sophisticated and functional blood vessels for widespread applications in both in vivo transplantation and in vitro drug screening.

Acknowledgments

Y.S.Z. acknowledges the National Cancer Institute of the National Institutes of Health Pathway to Independence Award (K99CA201603) and the LUSH Prize.

References

[1] Welsh J. Heart, circulation and blood cells. In: Physiology of mollusca, vol. 2; 2013. p. 125–74.

[2] Carola R, Harley JP, Noback CR. Human anatomy and physiology. Columbus, OH: McGraw-Hill College; 1992.

[3] Lafleur MA, Handsley MM, Dylan R. Blood vessel structure. Expert Rev Mol Med 2003;5.
[4] Bendayan M, Sandborn E, Rasio E, Inoue S, Michel RP, Hogg JC, et al. Blood vessel structure. In: Microcirculation. Medford, MA: Springer; 1976. p. 149–62.
[5] Vanhoutte PM, Rubanyi GM, Miller VM, Houston DS. Modulation of vascular smooth muscle contraction by the endothelium. Annu Rev Physiol 1986;48(1):307–20.
[6] Piascik MT, Guarino RD, Smith MS, Soltis EE, Saussy D, Perez DM. The specific contribution of the novel alpha-1D adrenoceptor to the contraction of vascular smooth muscle. J Pharmacol Exp Ther 1995;275(3):1583–9.
[7] Welti J, Loges S, Dimmeler S, Carmeliet P. Recent molecular discoveries in angiogenesis and antiangiogenic therapies in cancer. J Clin Invest 2013;123(8):3190–200.
[8] Bentzon JF, Otsuka F, Virmani R, Falk E. Mechanisms of plaque formation and rupture. Circ Res 2014;114(12):1852–66.
[9] Wardlaw JM, Smith C, Dichgans M. Mechanisms of sporadic cerebral small vessel disease: insights from neuroimaging. Lancet Neurol 2013;12(5):483–97.
[10] Gidaspow D. Multiphase flow and fluidization: continuum and kinetic theory descriptions. Cambridge, MA: Academic Press; 1994.
[11] Langer R, Vacanti JP. Tissue engineering. Science 1993;260(5110):920–6.
[12] Atala A, Kasper FK, Mikos AG. Engineering complex tissues. Sci Transl Med 2012;4(160). 160rv12-rv12.
[13] Zhang YS, Xia Y. Multiple facets for extracellular matrix mimicking in regenerative medicine. Nanomedicine 2015;10(5):689–92.
[14] Weinberg CB, Bell E. A blood vessel model constructed from collagen and cultured vascular cells. Science 1986;231(4736):397–400.
[15] Tranquillo RT, Ross J, Reyes M. Engineered blood vessels. Google Patents; 2012.
[16] L'heureux N, Pâquet S, Labbé R, Germain L, Auger FA. A completely biological tissue-engineered human blood vessel. FASEB J 1998;12(1):47–56.
[17] Datta P, Ayan B, Ozbolat IT. Bioprinting for vascular and vascularized tissue biofabrication. Acta Biomater 2017;51:1–20.
[18] Lee W, Lee V, Polio S, Keegan P, Lee J-H, Fischer K, et al. On-demand three-dimensional freeform fabrication of multi-layered hydrogel scaffold with fluidic channels. Biotechnol Bioeng 2010;105(6):1178.
[19] Kolesky DB, Homan KA, Skylar-Scott MA, Lewis JA. Three-dimensional bioprinting of thick vascularized tissues. Proc Natl Acad Sci U S A 2016;113(12):3179–84.
[20] Miller JS, Stevens KR, Yang MT, Baker BM, Nguyen D-HT, Cohen DM, et al. Rapid casting of patterned vascular networks for perfusable engineered three-dimensional tissues. Nat Mater 2012;11(9):768–74.
[21] Bertassoni LE, Cecconi M, Manoharan V, Nikkhah M, Hjortnaes J, Cristino AL, et al. Hydrogel bioprinted microchannel networks for vascularization of tissue engineering constructs. Lab Chip 2014;14(13):2202–11.
[22] Kolesky DB, Truby RL, Gladman AS, Busbee TA, Homan KA, Lewis JA. 3D bioprinting of vascularized, heterogeneous cell-laden tissue constructs. Adv Mater 2014;26(19):3124–30.
[23] Zhang YS, Davoudi F, Walch P, Manbachi A, Luo X, Dell'Erba V, et al. Bioprinted thrombosis-on-a-chip. Lab Chip 2016;16:4097–105.
[24] Lee VK, Kim DY, Ngo H, Lee Y, Seo L, Yoo S-S, et al. Creating perfused functional vascular channels using 3D bio-printing technology. Biomaterials 2014;35(28):8092–102.
[25] Lee VK, Dai G, Zou H, Yoo S-S, editors. Generation of 3-D glioblastoma-vascular niche using 3-D bioprinting. Biomedical engineering conference (NEBEC), 41st annual northeast; IEEE; 2015.

[26] Highley CB, Rodell CB, Burdick JA. Direct 3D printing of shear-thinning hydrogels into self-healing hydrogels. Adv Mater 2015;27(34):5075–9.
[27] Bhattacharjee T, Zehnder SM, Rowe KG, Jain S, Nixon RM, Sawyer WG, et al. Writing in the granular gel medium. Sci Adv 2015;1(8). e1500655.
[28] Hinton TJ, Jallerat Q, Palchesko RN, Park JH, Grodzicki MS, Shue H-J, et al. Three-dimensional printing of complex biological structures by freeform reversible embedding of suspended hydrogels. Sci Adv 2015;1(9). e1500758.
[29] Yu Y, Zhang Y, Martin JA, Ozbolat IT. Evaluation of cell viability and functionality in vessel-like bioprintable cell-laden tubular channels. J Biomech Eng 2013;135(9). 091011.
[30] Jia W, Gungor-Ozkerim PS, Zhang YS, Yue K, Zhu K, Liu W, et al. Direct 3D bioprinting of perfusable vascular constructs using a blend bioink. Biomaterials 2016;106:58–68.
[31] Norotte C, Marga FS, Niklason LE, Forgacs G. Scaffold-free vascular tissue engineering using bioprinting. Biomaterials 2009;30(30):5910–7.
[32] Bellan LM, Kniazeva T, Kim ES, Epshteyn AA, Cropek DM, Langer R, et al. Fabrication of a hybrid microfluidic system incorporating both lithographically patterned microchannels and a 3D fiber-formed microfluidic network. Adv Healthc Mater 2012;1(2):164–7.
[33] Bellan LM, Pearsall M, Cropek DM, Langer R. A 3D interconnected microchannel network formed in gelatin by sacrificial shellac microfibers. Adv Mater 2012;24(38):5187–91.
[34] Burdick JA, Prestwich GD. Hyaluronic acid hydrogels for biomedical applications. Adv Mater 2011;23(12):H41–56.
[35] Appel EA, del Barrio J, Loh XJ, Scherman OA. Supramolecular polymeric hydrogels. Chem Soc Rev 2012;41(18):6195–214.
[36] Guvendiren M, Lu HD, Burdick JA. Shear-thinning hydrogels for biomedical applications. Soft Matter 2012;8(2):260–72.
[37] Seiffert S, Sprakel J. Physical chemistry of supramolecular polymer networks. Chem Soc Rev 2012;41(2):909–30.
[38] Mattsson J, Wyss HM, Fernandez-Nieves A, Miyazaki K, Hu Z, Reichman DR, et al. Soft colloids make strong glasses. Nature 2009;462(7269):83–6.
[39] Saunders BR, Vincent B. Microgel particles as model colloids: theory, properties and applications. Adv Colloid Interface Sci 1999;80(1):1–25.
[40] Dimitriou CJ, Ewoldt RH, McKinley GH. Describing and prescribing the constitutive response of yield stress fluids using large amplitude oscillatory shear stress (LAOStress). J Rheol 2013;57(1):27–70.
[41] Zhang Y, Yu Y, Ozbolat IT. Direct bioprinting of vessel-like tubular microfluidic channels. J Nanotechnol Eng Med 2013;4(2). 020902.
[42] Zhang Y, Yu Y, Akkouch A, Dababneh A, Dolati F, Ozbolat IT. In vitro study of directly bioprinted perfusable vasculature conduits. Biomater Sci 2015;3(1):134–43.
[43] Slaughter BV, Khurshid SS, Fisher OZ, Khademhosseini A, Peppas NA. Hydrogels in regenerative medicine. Adv Mater 2009;21(32-33):3307–29.
[44] Khademhosseini A, Langer R. Microengineered hydrogels for tissue engineering. Biomaterials 2007;28(34):5087–92.
[45] Williams DF. On the mechanisms of biocompatibility. Biomaterials 2008;29(20):2941–53.
[46] Marga F, Neagu A, Kosztin I, Forgacs G. Developmental biology and tissue engineering. Birth Defects Res C Embryo Today 2007;81(4):320–8.

3D bioprinting composite tissue

19

Z. Wang, K. Sakthivel, X. Jin, K. Kim
University of British Columbia, Kelowna, BC, Canada

19.1 Introduction

Organ shortage crisis has become a severe problem in North America in recent 10 years [1]. According to the statistics by the U.S. Department of Health & Human Service [2] in 2015, 121,418 people needed a lifesaving organ transplant, while 78,020 (more than 60%) people were on the waiting list. In the same year, the total number of organ donors was only 15,064, which could not meet even 20% of the demands in the waiting list. The deficit of organs, as shown by the statistics, seems unlikely to be met with the organ donors. In this perspective, tissue-engineered artificial organs have been emerged as one of the most promising solutions to alleviate organ shortage problems [3]. At the current stage, tissue engineers have used biofabrication techniques to fabricate artificial organs. Biofabrication is defined as the production of complex living and nonliving biological products from raw materials such as living cells, molecules, extracellular matrices, and biomaterials [4]. Currently, biofabrication can be realized through various techniques, including but not limited to microfluidics [5], bioprinting [6,7], and bioreactors [8]. Among all those techniques, a biocompatible additive manufacturing, called "bioprinting", has attracted most of the research interests because of the ease of fabrication process, high resolution, and good compatibility with 3D printing [6]. The bioprinting technique can control various parameters accurately to fabricate artificial organs. Those parameters include cell type, cell distribution, and tissue structure, allowing composition for mimicking in vivo native tissues [9]. Because of its high compatibility, the bioprinting technique has also been introduced to many research areas in biomedical engineering, including quantitative biology [10], stem cell research [11,12], organ-on-a-chip [13], and drug delivery [14]. However, most of the bioprinting studies were carried out with the aim to generate functional tissues. Till now, bioprinting has successfully fabricated numerous types of tissues, including bone, cartilage, fat, nerve, skin, muscle, tendons, and vessels [6,7]. However, a variety of tissues in our body are not independent from each other. One example is the circulation system. Blood vessels are distributed everywhere in the human body and they interact with the local tissues for providing oxygen and nutrition and exchanging biosignals. Without the support of blood capillaries for providing efficient nutrients, biosignal factors, and waste transportation, most cells within the tissue-like construct cannot live [15]. Therefore, to fabricate transplantable organs in vitro, integrating tissues from a variety of cells is required. In addition to the type of cells, the various internal and external architectures of tissues must be taken into consideration [16]. For

3D Bioprinting for Reconstructive Surgery. https://doi.org/10.1016/B978-0-08-101103-4.00019-3

example, to fabricate a vascular network, the desired lumen diameter, branching pattern, and vessel stiffness need to be optimized [16]. In short, the dramatic increment in anatomical complexity and cell types makes the fabrication of composite tissues extremely complicated. In this chapter, we discuss the strategies, methodologies, and reported studies regarding the 3D composite tissue fabrication.

19.2 Characteristics of composite tissues

Before discussing the 3D composite tissue fabrication, it is better to discuss the skin tissue as a good example to show the complexity of the human body. Fig. 19.1 shows a typical section view of a human skin. As can be seen, the human skin is composed of two major layers—epidermis and dermis. The epidermis is located at the outermost part of the skin and protects the inner body parts of human beings from outside environment. It consists of mostly keratinocytes, which are responsible for the strength, resistance, and stretchability of the skin surface [17]. Within keratinocytes, filaments of keratin are embedded in a gelatin-like matrix. The substance rich in proteins and lipids fill the narrow spaces between cells. The keratinocytes are attached to each other to form a layer of cells and then build up the epidermis layer-by-layer. Under the epidermis, there is a second layer, which is a thick layer of dense connective tissues called the dermis. The dermis represents most of the thickness of the human skin. The complex combination of collagen and elastin fibers in the dermis helps the human body in avoiding any mechanical injury [18]. In addition to providing mechanical support, the dermis, which consists of various components, also plays an important role in maintaining skin functions. Blood capillaries deliver nutrition to the cells in the skin. Various types of glands help the surface of the skin in maintaining humidity. Nerve endings in the skin make it one of the most important sensory portals of the human body. Also, hair grows from the hair follicles in the dermis. Therefore, the dermis contains a variety of different types of cells, including, but not limited to, endothelial cells in veins, muscle cells near hair follicles, neurocytes in the nerve endings, and fibroblasts to form complex fiber combination.

19.3 Strategies for tissue fabrication

19.3.1 General inquiries

From the example of the skin, many in vivo tissues contain multiple types of cells and various extracellular matrix (ECM) proteins (i.e., collagen, fibronectin, laminin, hyaluronic acid, and so on) [19]. Well-organized cells and ECM proteins have different functions in the tissues [20]. Moreover, in most of the tissues, there are numerous vessel, gland, and nerve endings that are involved in the survival and information exchange of the tissues [21]. To realize the composite tissues in vitro, we have to consider a variety of factors mentioned previously as illustrated in Fig. 19.2. The multiple types of cells and biomaterials with the accurate control of the fabrication process

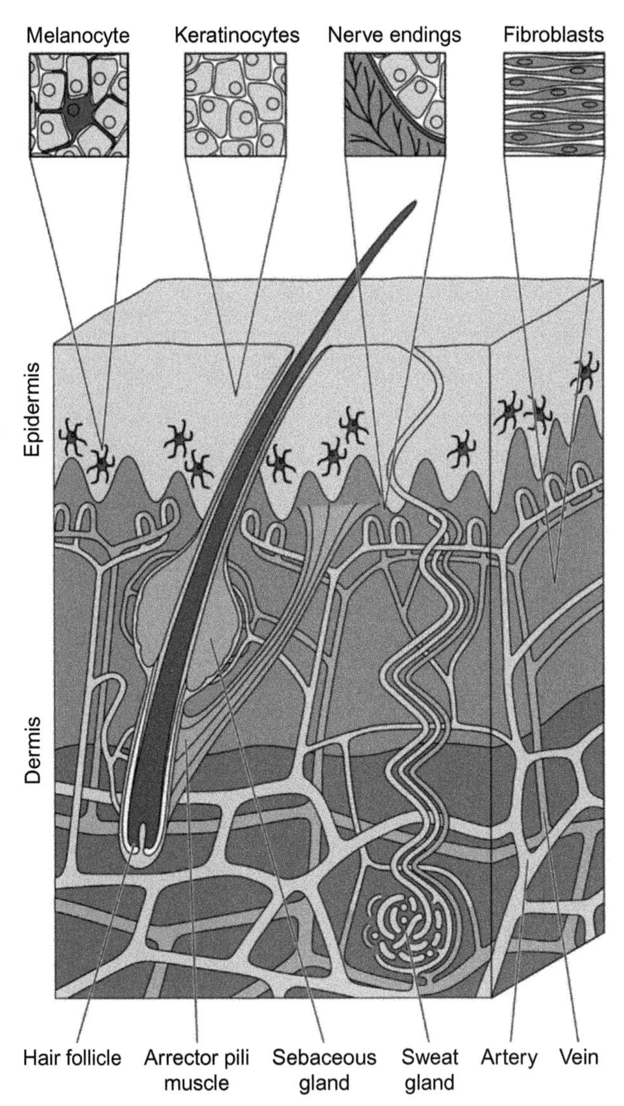

Fig. 19.1 A typical section through human skin, showing the epidermis and dermis layers. Adapted from Jablonski NG. Skin: a natural history. Berkeley, CA: University of California Press; 2006.

are required to build cellular microenvironments (i.e., cell-cell and cell-ECM interactions), as well as anatomically realistic tissue structures. In addition, the integration of vessels and nerve endings is critical to achieve long-term cell proliferation and form functional tissues.

Currently, bioprinting, a method in biofabrication techniques, is still in its early stage. It faces many hurdles in fabricating perfect composite tissues to resemble in vivo tissues. Despite of those hurdles, researchers have recently started to work on the

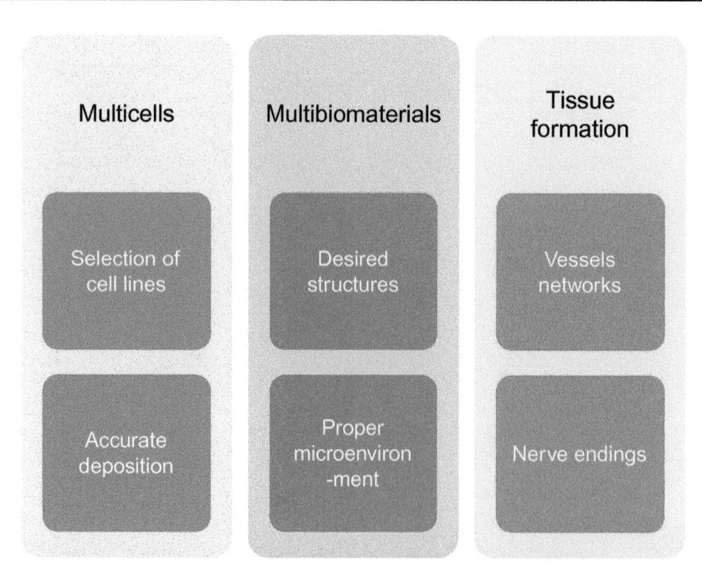

Fig. 19.2 Key factors to consider for the fabrication of highly biomimetic composite tissues.

generation of complex and heterogeneous tissues with various types of cell, biomaterials, and fabrication methods. The successful fabrication of heterogeneous tissues may be a breakthrough in biofabrication techniques and close to the fabrication of functional tissues in vitro. The rest of this chapter introduces and reviews the reported strategies and successful examples of the fabrication of complex heterogeneous tissues.

19.3.2 Multinozzle-based bioprinting

The first step to fabricate heterogeneous composite tissues is to have the capability of printing multiple types of cells and biomaterials in one bioprinting process. To meet such a requirement, the bioprinting system must be capable of printing multibioinks (which contain different types of cells and different biomaterials). Among all the bioprinting techniques, extrusion and inkjet bioprinting can support to print multibioinks easily by adding more printing heads and nozzles. However, the inkjet bioprinting is not frequently used to generate heterogeneous tissues because it has limited capability to print complex 3D structures [6]. Therefore, extrusion bioprinting system with multinozzles has become a major solution to print heterogeneous tissues.

19.3.2.1 Bioprinting system and process

Fig. 19.3A shows the photography of a four nozzle-based extrusion bioprinting system [22]. Each nozzle was integrated with a *Z*-direction linear stage to control the vertical position and fixed on a gantry linear stage system for *X* and *Y* axes positioning. During printing, the computer first determines the nozzles used and then calculates the distance that the stages need to travel. After the *X*-*Y* stage moves to the desired position, the selected nozzle moves down to the printing area. The *X*-*Y* stage moves again, while

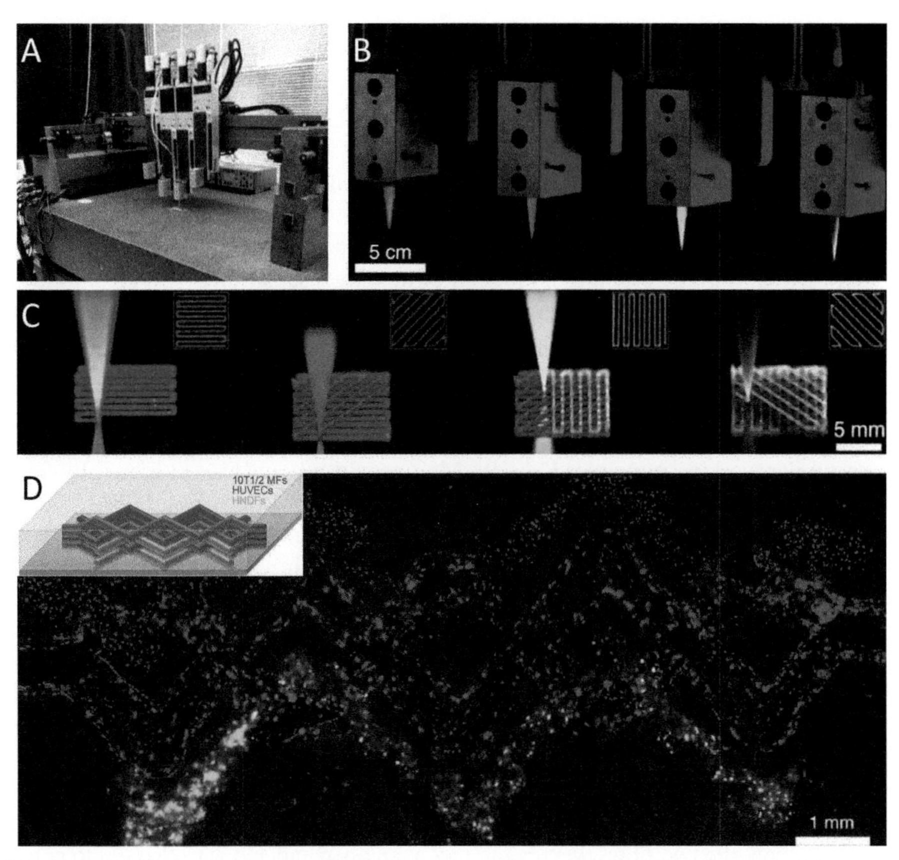

Fig. 19.3 Multinozzles-based biofabrication of vascularized heterogeneous tissues. (A) A photograph of a typical multinozzles-based bioprinting system. (B) Fluorescence images of four printing nozzles containing different bioinks. Scale bar = 5 cm. (C) Fluorescence image of four-layer lattice printed by sequentially depositing four bioinks. Scale bar = 5 mm. (D) Composite fluorescence image of the 3D-printed heterogeneous tissue constructed by three different cells: 10T1/2 fibroblasts (*blue*), HNDFs (*green*), and HUVECs (*red*). Scale bar = 1 mm.
Adapted from Kolesky DB, Truby RL, Gladman AS, Busbee TA, Homan KA, Lewis JA. 3D bioprinting of vascularized, heterogeneous cell-laden tissue constructs. Adv Mater 2014; 26: 3124–3130. doi:10.1002/adma.201305506.

the dispenser of the nozzle starts to extrude the bioinks. After patterning with the bioinks, the Z-direction stage moves up. This process keeps repeating until the entire 3D pattern is completed. Fig. 19.3B shows fluorescence images of four printing nozzles that contain different hydrogels, which can be easily distinguished under a fluorescence camera. The authors visualized layer-by-layer multinozzle extrusion printing process in Fig. 19.3C. One nozzle was used to print a layer of square patterns. It can be clearly seen how different bioinks were printed to form a complex 3D structure with the accurate control of the shape of each bioink.

19.3.2.2 Applications

The system described in Section 19.3.2.1 was further used to fabricate 3D vascularized heterogeneous cell-laden tissues (Fig. 19.3D). The heterogeneity of printed tissues was realized through a complex tissue structure and multiple cell types. Fibroblasts (10T1/2 MFs), and human umbilical vein endothelial cells (HUVEC) were printed, respectively, to form the 3D vascularized tissues. The HUVEC mimicked the vein of in vivo tissues and promoted the viability of nearby fibroblasts. This phenomenon was matched well with the promoting function of in vivo vascular niches to local tissues [23]. Hence, successful fabrication of the 3D vascularized heterogeneous tissues not only provided a new concept of using the bioprinting technique to generate complex tissues, but also verified that such heterogeneous tissues were able to function corresponding to in vivo tissues.

In addition to the aforementioned system, there are many multinozzle systems that have similar working principles [24–29]. Multinozzle systems have successfully fabricated numerous tissues for tissue engineering applications, including high-stiffness dual cell-laden structures mimicking osteochondral tissues [25], clinically compliant biomaterial-based complex cartilaginous structures [27], in vitro mandible bone, calvarial bone, ear cartilage, and skeletal muscle tissues [28]. Xu et al. [30] employed a multireservoir inkjet printing system to fabricate simple heterogeneous pie-shape tissues. Because of the infeasibility of inkjet printing techniques for 3D complex structures, the fabricated tissue structures were not as complicated as most extrusion-based systems. Recently, Colosi et al. [31] introduced a novel microfluidic-based printing nozzle system that was capable of printing two different bioinks simultaneously. This nozzle system was based on a *Y*-junction of two laminar flows, which could allow the flows with similar viscosities passing through a channel without mixing. At the end of the *Y*-junction, the two flows were being met at an open end and extruded out to complete the printing process. Such a *Y*-junction can be easily modified to accommodate more bioinks at a time by increasing the number of flows to fabricate more complex cell-laden microtubes or microtube-based heterogeneous tissues.

19.3.3 Combinatorial bioprinting system

Although the multinozzle bioprinting system successfully generated a variety of heterogeneous tissues, it still has many shortcomings due to the extrusion-based working mechanism [6]. These shortcomings include relatively low cell viability resulted from the large external forces applied during extrusion, the limitation to achieve very fine resolution (i.e., $<100\,\mu m$), and slow fabrication speed taking around several hours. In the meantime, there are several newly emerged high resolution, rapid, and highly biocompatible bioprinting techniques, such as stereolithography [32–35] and laser-based bioprinting [36,37].

19.3.3.1 Stereolithography- and laser-based bioprinting system

The stereolithography-based bioprinting uses an array of light beams with controllable intensities to selectively crosslink a patterning area of photo-sensitive biomaterials.

Therefore, stereolithography can print the entire layer of biomaterials at a time, which accelerates the bioprinting process significantly. To our knowledge, most of the printing process can be done within 1 h [33]. The rapid printing process does not affect the resolution of printed patterns. The stereolithography-based bioprinting systems generally have a very fine resolution down to 25–50 μm, two fold smaller than the extrusion-based systems. The laser-based bioprinting uses a strong pulse or continuous laser to crosslink biomaterials within a few seconds and maximally preserve the cell viability (>90% as reported in Ref. [37]). The drawback of those new methods is the limitation to print multibioinks simultaneously. However, in many composite tissues, most of the areas are filled with specific types of cells (e.g., keratinocytes on the epidermis layer of skin). It would be great to combine the stereolithography or laser-based printing system with the extrusion-based printing system to print composite tissues. The stereolithography and laser-based system have the advantages of high speed, high resolution, and high cell viability, while the extrusion-based system has the advantages of accommodating multibioinks to fabricate heterogeneous tissues.

19.3.3.2 Combinatorial bioprinting system

Shanjani et al. [38] recently integrated the visible light-based stereolithography system with the extrusion system to build a combinatorial bioprinting system for fabricating heterogeneous tissues. Fig. 19.4A shows the hybrid bioprinting system. The stereolithography-based bioprinting system is placed on top of the hybrid system while the nozzles and stages for the extrusion system are set in the middle. With the hybrid system, the vascularized bone grafts were fabricated using endothelial cells and stem cells. The bone grafts can be divided into three sections: a hydrogel conduit for vascularization, a collagen hydrogel for encapsulating stem cells, and cell-free poly-(ε-capro-lactone) (PCL) filaments to provide stiffer microenvironments to induce the osteogenesis of stem cells (Fig. 19.4B). During the fabrication, the cell-free PCL filaments were printed by the extrusion-based system, while the stereolithography-based system was responsible for printing the vascular conduit by photocrosslinking the HUVEC-laden poly(ethylene glycol) diacrylate (PEGDA) hydrogel. In the end, pores in the PCL filaments were filled with the collagen encapsulating C3H10T1/2 mouse mesenchymal stem cells. The extrusion and stereolithography hybrid system printed the aforementioned structures in a commutative way, as illustrated in Fig. 19.4C. The hybrid system first started to print PCL filaments on the build platform (Fig. 19.4C-1). Once the extrusion printing process was completed, the platform was moved down and submerged into the PEGDA prepolymer solution to prepare the stereolithography printing process (Fig. 19.4C-2). After the stereolithography process was done, the build platform was lifted up and the extrusion process started again to print a new layer of PCL filaments. This workflow was repeated until the entire heterogeneous structure was completed. Fig. 19.4D shows a fluorescence image of live/dead assayed bone grafts for 5 days after printing. It can be seen that the printed composite tissues maintained the designed shape and most of the cells were alive. The high cell viability after 5 days and the integrity of the printed structures verified the feasibility of the hybrid bioprinting system.

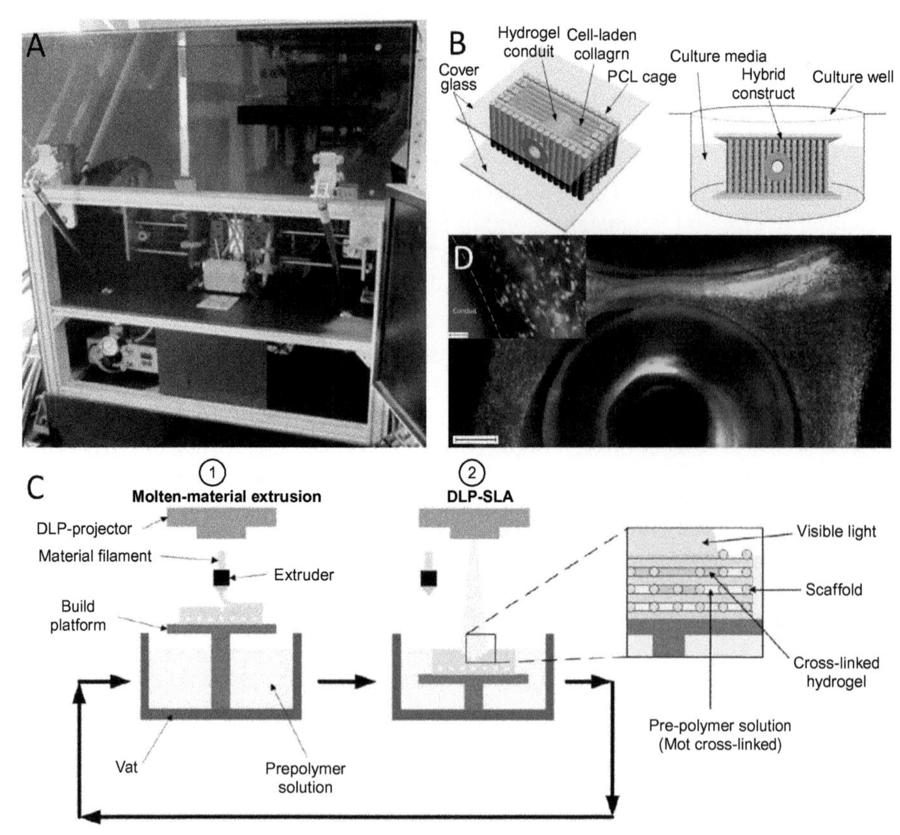

Fig. 19.4 Hybrid-bioprinting-based biofabrication of vascularized bone grafts. (A) Photograph of hybrid bioprinting system consisting of stereolithography and extrusion. (B) Schematic diagram of conduit-collagen-scaffold hybrid model for bone grafts regeneration. (C) Schematic diagram of the printing process of the hybrid bioprinting system. (D) Fluorescence image of live-dead assayed cells encapsulated in the heterogeneous tissues at day 5. Scale bar = 50 μm for the zoom-in image. Scale bar = 1 mm for the rest of image.
Adapted from Shanjani Y, Pan CC, Elomaa L, Yang Y. A novel bioprinting method and system for forming hybrid tissue engineering constructs. Biofabrication 2015; 7: 45008. doi:10.1088/1758-5090/7/4/045008.

19.3.4 Microfluidic-based biofabrication

In Sections 19.3.2 and 19.3.3, the heterogeneous tissues were fabricated by bioprinting techniques. Although the bioprinting techniques have been considered as the most powerful and versatile tool for biofabrication, the microfluidic-based biofabrication methods also have great advantages in generating a specific type of tissue in a high throughput and highly controllable manner [39], especially for line-based tissues, such as nerve endings and vessels. In this section, the strategy of using microfluidic techniques to fabricate heterogeneous tissues is discussed. Microfluidic devices have been widely used for manipulating cell-laden hydrogels to form various building blocks for tissue engineering [40] and quantitative biology [41,42]. The microfluidic channels are extremely suitable for the rapid fabrication of cell-laden microdroplets (also referred as microbeads) [43] and microtubes (also referred as microfibers) [44]. Both of them play important roles in fabricating tissue scaffolds [40].

19.3.4.1 Microdroplets

Microdroplets are well-accepted toolkits for many engineering applications including chemical reactions, delivery agents, molecule synthesis, and photonics [45]. The microdroplets with controllable sizes in diameter (e.g., 50–200 μm) can be generated in a high-throughput way by using microfluidic T-shape, flow-focusing, and co-axial devices [46]. During the droplet generation process, two immiscible solutions, such as a hydrogel-prepolymer solution as an aqueous phase and an oil phase, are injected into the device using syringe pumps and met at the crossjunction of the microchannel. Due to the large interfacial force created by the oil phase, the hydrogel prepolymer solution is dispersed to form microdroplets [47]. The droplets are collected and then polymerized through a thermal, photo, or chemical crosslinking process [48]. Fabricated microdroplets are widely used in lower level biology, such as the quantitative analysis of cell response to microenvironments [49]. For tissue engineering, the microdroplets are mainly considered as a "point"-based building blocks (Fig. 19.5A) [39]. In addition, the microdroplets can be molded to build a complex 3D structure or to form "plane"-based cell sheets for further molding, stacking, or rolling (Fig. 19.5A). For example, the microdroplets have been used directly as building blocks for tissue fabrication [43]. In this study, cell-laden microdroplets were used to fill a customized mold made from polydimethylsiloxane (PDMS). After several days of culturing, all the droplets merged to form microtissues with uniform cell distribution. The shape of the tissue was fully dependent on the geometry of the PDMS mold. The microdroplet-based molding method is an excellent way to rapidly fabricate the relatively large size of tissue structures with uniform microenvironments. However, such macroscale tissues may suffer from the lack of blood vessels for delivering oxygen and nutrition to the cells in the structures, which may cause significant cell death as time goes on for long-term cell culturing [43]. This problem can be solved by introducing "artificial vessels" during molding. One possible way to build the artificial vessels is that we first fabricate HUVEC-laden microvessels by molding droplets into a filament shape and then mix the microvessels with the microdroplets encapsulating other cells for molding bulk tissues [39].

19.3.4.2 Microtubes

Tubular tissues, such as blood vessels or neurons can be easily fabricated using the microfluidic co-axial channel devices [44,50], as shown in Fig. 19.5B and C. The structure of the junction is similar to that used for generating droplets. However, for generating microtubes, the oil phase was replaced with a chemical crosslinker (e.g., $CaCl_2$ solution for crosslinking alginate hydrogel) as shown in Fig. 19.5B and C. Once the crosslinker meets the hydrogel prepolymer solution, the hydrogel prepolymer solution is polymerized to form a hydrogel network. Due to the stability of laminar flows

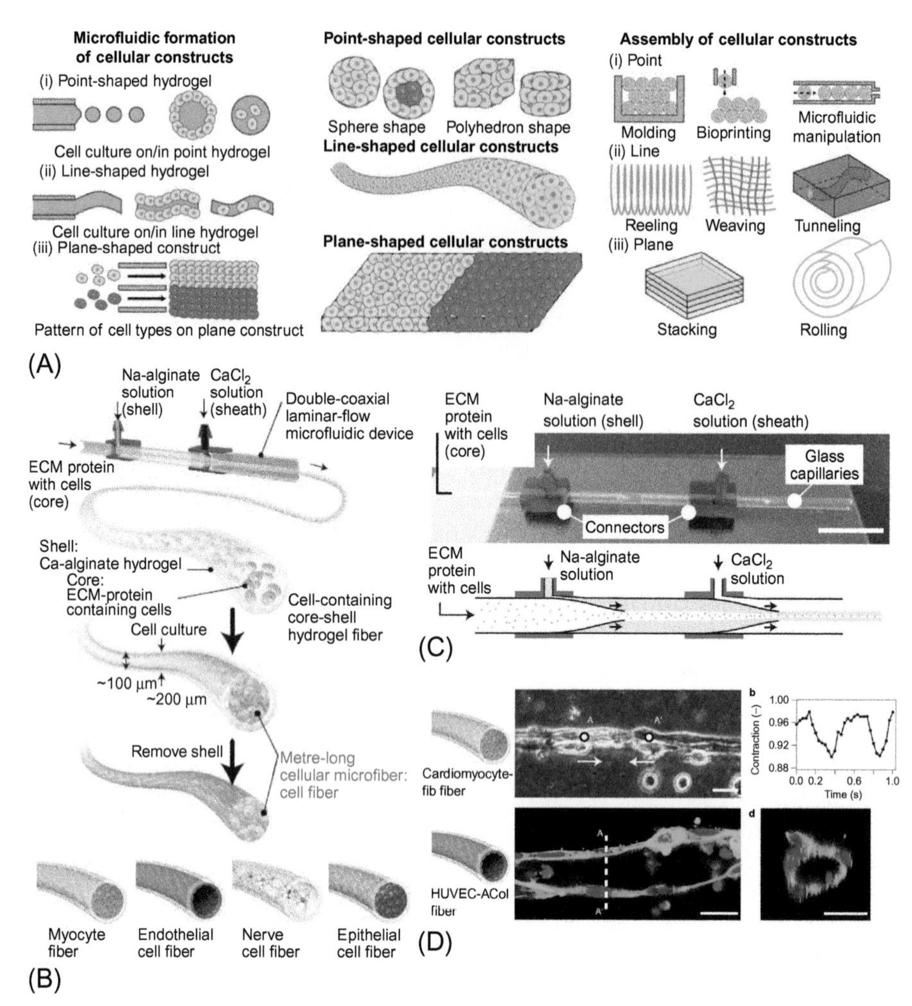

Fig. 19.5 Microfluidic-based biofabrication. (A) Concepts of microfluidic-based biofabrication. The point (microdroplets), and line (microfiber) building blocks can be used to fabricate plane or more complex 3D cell-laden structures through several assembly methods (Adapted from Morimoto Y, Hsiao AY, Takeuchi S. Point-, line-, and plane-shaped cellular constructs for 3D tissue assembly. Adv Drug Deliv Rev 2015; 95: 29–39. doi:10.1016/j.addr.2015.09.003). (B) Schematic and (C) photograph of a double-coaxial microfluidic device to fabricate cell-laden microfibers. The alginate hydrogel was crosslinked by the $CaCl_2$ solution and could be removed by an enzymatic reaction. Microfibers were capable of encapsulating various types of cells. Scale bar = 100 μm. (D) Image of cardiomyocyte microfiber with constant beating frequency. Scale bar = 20 μm and image of HUVEC microfibers cultured for 3 days. The HUVECs form a monolayer covering the surface of fiber. Scale bar = 20 μm.

Adapted from Morimoto Y, Hsiao AY, Takeuchi S. Point-, line-, and plane-shaped cellular constructs for 3D tissue assembly. Adv Drug Deliv Rev 2015; 95: 29–39. doi:10.1016/j.addr.2015.09.003; Onoe H, Okitsu T, Itou A, Kato-Negishi M, Gojo R, Kiriya D, et al. Metre-long cell-laden microfibres exhibit tissue morphologies and functions. Nat Mater 2013; 12: 584–590. doi:10.1038/nmat3606.

in the microfluidic channels, the generation process of microtubes in the co-axial junction is fairly stable, and the length of the microtubes can be a meter-long maximally [44]. It is also possible to achieve a core-shell heterogeneous microtube by combining several co-axial junctions subsequently. The materials and cells used for the shells can be varied. Therefore, it is possible to mimic the anatomical layer-by-layer structure of vessels by using multishell microfibers. Onoe et al. [44] have used this idea to fabricate various types of cell-laden microtubes, including a myocyte fiber, endothelial fiber, nerve cell fiber, and epithelial cell fiber (Fig. 19.5C). They also observed the beating of cardiomyocyte fibers after three days of culturing and the formation of uniform cell monolayers of HUVECs at day 4 (Fig. 19.5D). Comparing with the bioprinted vessels, microfluidic-based microtubes have several advantages, such as high throughput, high controllability, and better uniformity. More interestingly, the microtubes can be used to build complex structures through several engineering methods, such as tunneling [51], reeling, and weaving [39] (Fig. 19.5A). Taken together, the microfluidic-based biofabrication methods are advantageous to generate microdroplets or fiber-based cell-laden hydrogels for fabricating the composite tissues as a bottom-up approach. Considering the ease of generating multilayer structures to mimic in vivo vessels and nerves, the vascularization and functionalization of the bioprinted heterogeneous tissues may be benefited from integrating with the microfluidic-based microfibers.

19.4 Future works

The fabrication strategies discussed in Section 19.3 have led to the successful fabrication of several composite tissues. However, there is still a long way to go to fabricate ready-to-use transplantable tissues in vitro. The limitations may come from the difficulties of realizing complex in vivo tissue structures [29] and understanding the interaction between native tissues and artificial tissues after transplantation [28].

19.4.1 Design and fabrication of anatomically realistic models

Fabricating transplantable, patient-specific tissues to repair the damaged tissues or organs is the common goal of the bioprinting techniques in clinical applications. To achieve this goal, researchers need to derive optimized tissue structures with the help of noninvasive diagnosing techniques (e.g., medical imaging techniques) and fabricate such structures accurately by using an advanced design/manufacturing process.

19.4.1.1 Modeling tissue structures from medical images

For practical applications, the accuracy of bioprinting is highly related to the quality of medical imaging techniques, which visualize the shape and type of damaged tissues in vivo. The most commonly used medical imaging techniques are computed tomography (CT) and magnetic resonance imaging (MRI), which will be the best candidates to provide the details of in vivo tissues with high resolution and availability. The CT is based on the variable absorption of X-rays by different tissues [52]. However, it is unable to detect the signals from the tissues that have dense background tissues [53] (e.g., distinguishing between small breast gland tissues and dense breast fats is difficult). The MRI detects the small fraction of nuclei of tissues resulted from a strong magnetic field and provides better imaging contrast than CT. However, it has a limited resolution [54]. The conversion from CT or MRI to corresponding anatomically realistic tissue model can be performed manually through open-access software [55] or automatically through threshold-based image segmentation, interpolation, and reconstruction [56–58] (Fig. 19.6A). After the conversion, medical images become the numerical 3D modeling

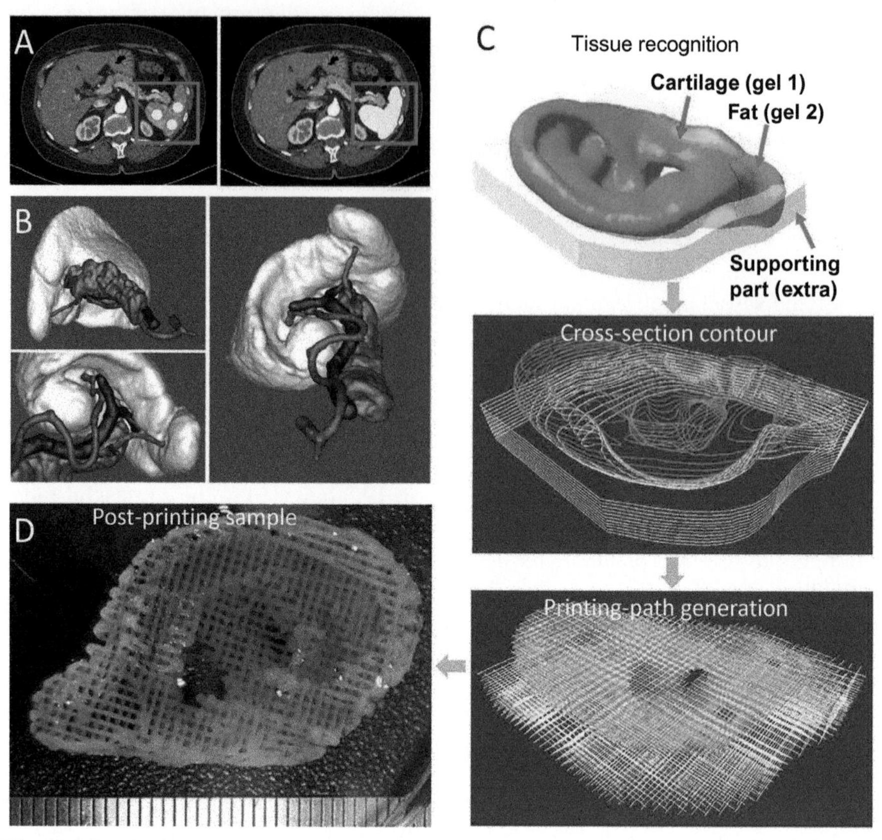

Fig. 19.6 Design and fabrication of anatomically realistic tissue models. (A) and (B) Modeling 3D tissue structure from medical images (Adapted from Pietrabissa A, Marconi S, Peri A, Pugliese L, Cavazzi E, Vinci A, et al. From CT scanning to 3-D printing technology for the preoperative planning in laparoscopic splenectomy. Surg Endosc 2015. doi:10.1007/s00464-015-4185-y). (A) MRI images undergoing segmentation during image processing. Red squares indicate the growth of a specific tissue. (B) 3D tissue model with various types and anatomically realistic shapes of tissues. (C) Bio-CAD/CAM process to covert the 3D tissue model to the printing path for bioprinting, including tissue recognition, cross-section contouring, and printing-path generation. (D) Image of a printed sample using the printing path generated in (C). Adapted from Jung JW, Lee J-S, Cho D-W. Computer-aided multiple-head 3D printing system for printing of heterogeneous organ/tissue constructs. Sci Rep 2016; 6: 21685. doi:10.1038/srep21685.

files, such as STL (stereolithography) or STEP (Standard for the Exchange of Product model data) for manufacturing [59]. Currently, the resolution of MRI and the low contrast of CT could be a problem for visualizing the details of the tissue structures. CT/MRI correlation and fusion algorithm [60] may benefit the conversion process and generate highly accurate in vivo tissue model for bioprinting. Though the fusion algorithm works well, the currently commercialized CT or MRI techniques are difficult to obtain clear images at the cellular level. Cellular microenvironments are critical for cell growth and differentiation. The poor quality of information about the cellular microenvironments may affect the feasibility of printed tissues after transplantation. Recent advances in micro-CT [61,62] and cellular and molecular MRI [63,64] may offer tissue engineers high quality images with both tissue level integrity and cellular level resolution.

19.4.1.2 Bio-CAD/CAM

Subsequently, the modeling data is processed by the computer aided design/computer aided manufacturing (CAD/CAM) program for structure analysis and printing path calculation to fabricate proper tissue structures. Jung et al. [65] recently demonstrated a comprehensive bio-CAD/CAM algorithm for the multinozzle extrusion-based bioprinting system. In the system, the modeling data underwent tissue recognition, cross-section contour construction, and printing-path generation successively (Fig. 19.6B) [65]. The tissue recognition process assigned different bioinks to corresponding tissues and ensured the printability of models. It is noted that the derived tissue models may have a poor printability due to the irregular edges or the loss of supporting structures. To improve the printability, the algorithm during the tissue recognition was required to define an extra part (the yellow part in Fig. 19.6B) to support the desired structure and convert the irregular bottom to the printable flat surface while maintaining the desired structure of tissues. The cross-section contouring process was used to calculate the closed surfaces of every bioink at different layers. The contouring details were used to generate the printing path of each bioink layer-by-layer. During the printing-path generation, the algorithm generates the paths of supporting material and bioinks to fill up the closed surfaces. In the end, this process was coded in the format of numerical control programing language (G-code) and sent to operate the stages and dispensers for printing. Currently, there are no well-established Bio-CAD/CAM platforms for other bioprinting techniques except the extrusion-based technique, which makes the printing process of realistic tissue models labor-intensive. Even the reported platform for the extrusion-based printing system still requires many human interventions to operate, which slow down the fabrication process. Developing a robust and automated Bio-CAD/CAM platform is highly demanded to overcome current and future challenges in clinical applications of the bioprinting systems.

19.4.2 Tissue maturation and fusion

In addition to the design of accurate bioprinting structures for the clinical usage, at the current stage, it is more important to understand the interactions between native and artificial tissues. There are several questions needed to be answered as follows:

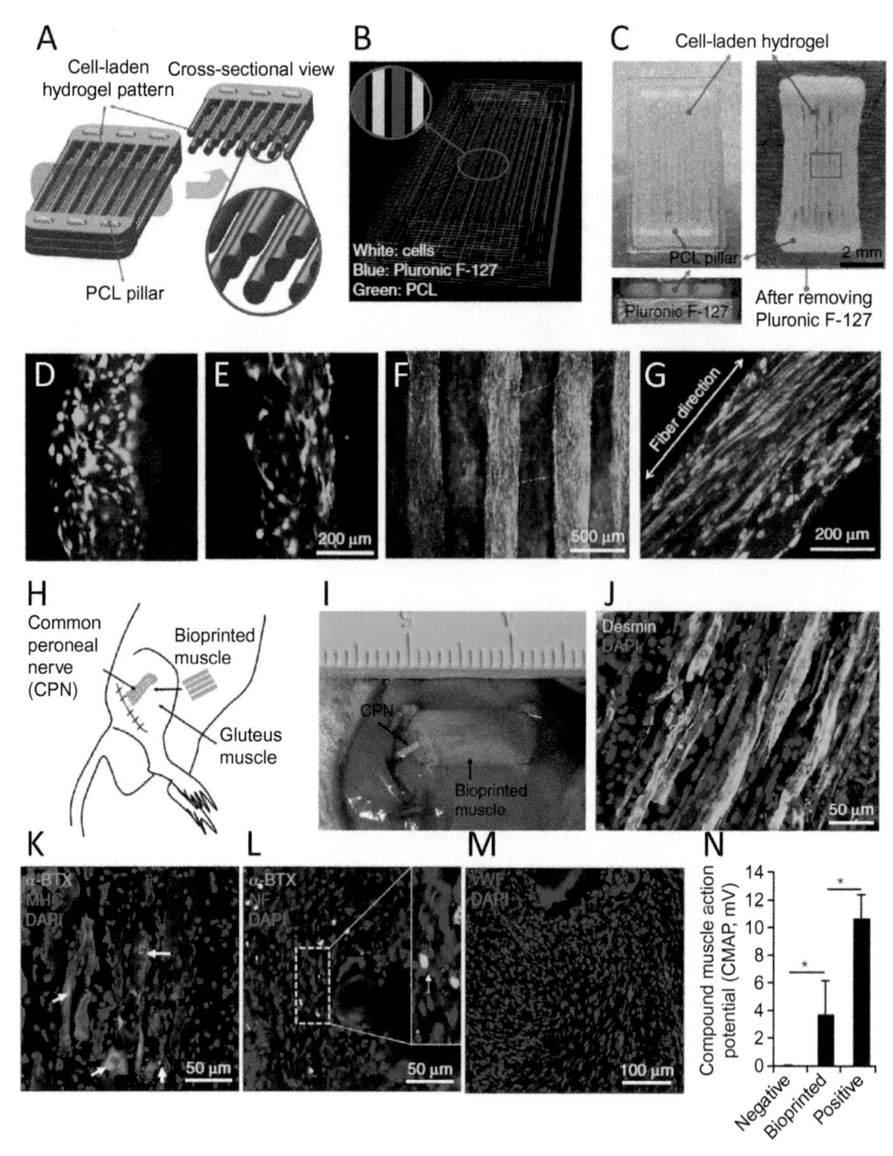

Fig. 19.7 Bioprinted skeletal muscle tissues: examining in vitro fabrication and in vivo maturation: (A) and (B) schematic diagrams of the design of bioprinted muscle tissues; (C) photographs of bioprinted muscle before and after removing the sacrificial hydrogel; (D) myocytes were not aligned without the presence of PCL and sacrificial hydrogel. Scale bar = 100 μm; (E) myocytes were aligned well in the presence of PCL and sacrificial hydrogel. Scale bar = 100 μm; (F) fluorescence image of live-dead assayed bioprinted muscles at day 7. Scale bar = 500 μm; (G) fluorescence image of stained myosin heavy chain at day 7, showing the well-aligned and dense myocytes. Scale bar = 200 μm; (H) schematic diagram of the implantation of the bioprinted muscles to rats. The bioprinted muscles were placed above the gluteus muscle and connected to CPN; (I) partly fused and matured bioprinted muscles two

Can the printed tissues maintain their structures in vivo? How much degree will the artificial tissues be fused with local native tissues? Are there any long-term responses from native tissues? How much degree can the artificial tissues function in the similar way of native tissues? Kang et al. [28] have provided preliminary data to answer the above questions systematically. This study used a multinozzle extrusion-based system to fabricate several tissues that were based on defects found in the medical imaging data. Those tissues include mandible bone, calvarial bone, ear cartilage, and skeletal muscle tissues. The study also examined the function of the printed tissues both in vitro and in vivo.

The fabricated skeletal muscle tissues are discussed further as an example. In the muscle tissues in vivo, myoblasts are well-aligned in one direction to form multinucleated myotubes [66]. With this anatomical structure, mouse myoblasts were bioprinted with Pluronic F-127 hydrogel in a fiber-like bundle structure supported by PCL pillars at both ends (Fig. 19.7A and B). The Pluronic F-127 was employed as a sacrificial material to align myoblast fibers. The sacrificial material was removed after 3 days (Fig. 19.7C). The viability and alignment of myoblasts were examined at day 3 through immunostaining. It is noted that without the help of the PCL pillars and Pluronic F-127, the alignment of myoblasts was inappropriate (Fig. 19.7D). On the contrary, the cell alignment was much improved with the presence of the PCL pillars and Pluronic F-127 (Fig. 19.7E). The aligned myoblast bundle was further cultured and examined to check the cell viability and density at day 7 (Fig. 19.7F and G). The cell viability and density were very high at day 7, demonstrating the high feasibility of the biomimetic skeleton muscle tissues in vitro. To investigate the in vivo maturation of the printed skeleton muscle tissues, the printed tissues at day 7 were implanted into the gluteus muscle of nude rats. The dissected distal end of the proximal stump from the common peroneal nerve (CPN) was embedded in the implanted tissues to promote the integration and maturation (Fig. 19.7H and I). The bioprinted and implanted tissues were harvested in vivo for two weeks and then immunostained to examine the degree of maturation and fusion. Well-organized muscle fiber structures (Fig. 19.7J) and the integration of various tissues, which include the bioprinted muscle fibers (MHC$^+$ and α-BTX$^+$, Fig. 19.7K), neurofilament (NF$^+$, Fig. 19.7I), and vessels (vWF, Fig. 19.7M), were observed. The presence and integration of vessels and neurons with the bioprinted muscles indicated that the transplanted tissues underwent the maturation and fusion process with the native tissues. The functionality of the transplanted muscles after 4 weeks of implantation was also examined by electromyography, which

Fig. 19.7 Continued weeks after implantation;. (J) and (K) fluorescent images showing (J) the high density, (K) well-aligned myocytes, (L) the fusion between neurons and myocytes, and (M) the appearance of endothelial cells. Scale bar = 50 μm for (J)–(L). Scale bar = 100 μm for (M); (N) functional assessment of the bioprinted muscles constructed after 4 weeks of implantation ($*p<0.05$). Positive control was the normal gastrocnemius muscle, while negative control was the gluteus muscle after dissected from CPN.
Adapted from Kang H-W, Lee SJ, Ko IK, Kengla C, Yoo JJ, Atala A. A 3D bioprinting system to produce human-scale tissue constructs with structural integrity. Nat Biotechnol 2016. doi:10.1038/nbt.3413.

evaluates the electrical and neurological activation of muscle tissues (Fig. 19.7N). Both native muscle tissues (positive) and bioprinted muscles were able to respond to the stimulation of nerves while the subcutaneous tissues (negative) had no response. Although the action potential of the bioprinted muscles was significantly lower than the native tissues, it was still a breakthrough outcome to have immature but functional muscles through the bioprinting process after only 4 weeks of implantation. With these results, the in vivo functionalization and maturation of the bioprinted composite tissues may not be achieved without the help of neurons and vessels, which provide the printed tissues with various stimulations, such as electrical, mechanical, and biochemical cues. Although the in vivo implantation is the best way to study tissue maturation, the high cost, low throughput, and the low controllability of the in vivo implantation may limit our study and understanding. Hence, integrating the bioprinted tissues with microengineered platforms, such organ-on-a-chip [67] and high-through cellular (or tissue level) microarrays [68,69] may be the alternative way of the fast, inexpensive characterization process for the tissue maturation in vitro.

19.5 Conclusion

We systematically introduced the biofabrication methods of 3D composite tissues in this chapter. Due to the complexity and heterogeneity of the native tissues, we discussed the key factors, which consider the composite tissue fabrication. Several methods to fabricate heterogeneous composite tissues were reviewed in detail, including the multinozzle-based bioprinting, hybrid bioprinting, and microfluidic-based generation of microfibers. In addition to the bioprinting process, this chapter covered recent studies in the in vivo maturation of the bioprinted tissues. In the end, we discussed the perspectives of the methods to study tissue maturation in a high-throughput way and the sources to derive anatomically-realistic models for the bioprinting process. In summary, currently, there are no one-for-all solutions for fabricating the composite tissues and the tissue maturation after bioprinting and implantation is limited. A few recent studies showed promising results of fabricating the composite tissues for transplantation. However, there is still a long way to go to fabricate clinically applicable composite tissues in vitro.

Acknowledgment

This work was supported by the Natural Sciences and Engineering Research Council of Canada (NSERC) Discovery Grant (RGPIN-2014-04010) and the Canada Foundation for Innovation.

References

[1] Abouna GM. Organ shortage crisis: problems and possible solutions. Transplant Proc 2008;40:34–8. https://doi.org/10.1016/j.transproceed.2007.11.067.
[2] U.S.D. of H.& H. Services, United Network for Organ Sharing: OPTN report, 2015.

[3] Vacanti JP, Langer R. Tissue engineering: the design and fabrication of living replacement devices for surgical reconstruction and transplantation. Lancet 1999;354(Suppl.):SI32–4. https://doi.org/10.1016/S0140-6736(99)90247-7.
[4] Mironov V, Trusk T, Kasyanov V, Little S, Swaja R, Markwald R. Biofabrication: a 21st century manufacturing paradigm. Biofabrication 2009;1:22001. https://doi.org/10.1088/1758-5082/1/2/022001.
[5] Andersson H, van den Berg A. Microfabrication and microfluidics for tissue engineering: state of the art and future opportunities. Lab Chip 2004;4:98–103. https://doi.org/10.1039/b314469k.
[6] Mandrycky C, Wang Z, Kim K, Kim DH. 3D bioprinting for engineering complex tissues. Biotechnol Adv 2016;34:422–34. https://doi.org/10.1016/j.biotechadv.2015.12.011.
[7] Murphy SV, Atala A. 3D bioprinting of tissues and organs. Nat Biotechnol 2014;32:773–85. https://doi.org/10.1038/nbt.2958.
[8] Martin I, Wendt D, Heberer M. The role of bioreactors in tissue engineering. Trends Biotechnol 2004;22:80–6. https://doi.org/10.1016/j.tibtech.2003.12.001.
[9] Jana S, Lerman A. Bioprinting a cardiac valve. Biotechnol Adv 2015;33:1503–21. https://doi.org/10.1016/j.biotechadv.2015.07.006.
[10] Ma Y, Ji Y, Huang G, Ling K, Zhang X, Xu F. Bioprinting 3D cell-laden hydrogel microarray for screening human periodontal ligament stem cell response to extracellular matrix. Biofabrication 2015;7:44105. https://doi.org/10.1088/1758-5090/7/4/044105.
[11] Tasoglu S, Demirci U. Bioprinting for stem cell research. Trends Biotechnol 2013;31:10–9. https://doi.org/10.1016/j.tibtech.2012.10.005.
[12] Dai R, Wang Z, Samanipour R, Koo K, Kim K. Adipose-derived stem cells for tissue engineering and regenerative medicine applications. Stem Cells Int 2016.https://doi.org/10.1155/2016/6737345.
[13] Wang Z, Samanipour R, Koo K, Kim K. Organ-on-a-chip platforms for drug delivery and cell characterization: a review. Sens Mater 2015;27:487–506.
[14] Stanton MM, Samitier J, Sánchez S. Bioprinting of 3D hydrogels. Lab Chip 2015;15:3111–5. https://doi.org/10.1039/C5LC90069G.
[15] Radisic M, Yang L, Boublik J, Cohen RJ, Langer R, Freed LE, et al. Medium perfusion enables engineering of compact and contractile cardiac tissue. Am J Physiol Heart Circ Physiol 2004;2139:507–16.
[16] Visser J, Peters B, Burger TJ, Boomstra J, Dhert WJA, Melchels FPW, et al. Biofabrication of multi-material anatomically shaped tissue constructs. Biofabrication 2013;5:35007. https://doi.org/10.1088/1758-5082/5/3/035007.
[17] Jablonski NG. Skin: a natural history. Berkeley, CA: University of California Press; 2006.
[18] Kusuma S, Vuthoori RK, Piliang M, Zins JE. Skin anatomy and physiology. In: Plastic and Reconstructive Surgery. Springer; 2010. p. 161–71.
[19] Frantz C, Stewart KM, Weaver VM, Frantz C, Stewart KM, Weaver VM. The extracellular matrix at a glance. J Cell Sci 2010;2010:4195–200. https://doi.org/10.1242/jcs.023820.
[20] Midwood KS, Williams LV, Schwarzbauer JE. Tissue repair and the dynamics of the extracellular matrix. Int J Biochem Cell Biol 2004;36:1031–7. https://doi.org/10.1016/j.biocel.2003.12.003.
[21] Carmeliet P. Blood vessels and nerves: common signals, pathways and diseases. Nat Rev Genet 2003;4:710–20. https://doi.org/10.1038/nrg1158.
[22] Kolesky DB, Truby RL, Gladman AS, Busbee TA, Homan KA, Lewis JA. 3D bioprinting of vascularized, heterogeneous cell-laden tissue constructs. Adv Mater 2014;26:3124–30. https://doi.org/10.1002/adma.201305506.

[23] Butler JM, Kobayashi H, Rafii S. Instructive role of the vascular niche in promoting tumour growth and tissue repair by angiocrine factors. Nat Rev Cancer 2010;10:138–46. https://doi.org/10.1038/nrc2791.
[24] Khalil S, Nam J, Sun W. Multi-nozzle deposition for construction of 3D biopolymer tissue scaffolds. Rapid Prototyp J 2005;11:9–17. https://doi.org/10.1108/13552540510573347.
[25] Shim J-H, Lee J-S, Kim JY, Cho D-W. Bioprinting of a mechanically enhanced three-dimensional dual cell-laden construct for osteochondral tissue engineering using a multi-head tissue/organ building system. J Micromech Microeng 2012;22:85014. https://doi.org/10.1088/0960-1317/22/8/085014.
[26] Wüst S, Godla ME, Müller R, Hofmann S. Tunable hydrogel composite with two-step processing in combination with innovative hardware upgrade for cell-based three-dimensional bioprinting. Acta Biomater 2014;10:630–40. https://doi.org/10.1016/j.actbio.2013.10.016.
[27] Kesti M, Eberhardt C, Pagliccia G, Kenkel D, Grande D, Boss A, et al. Bioprinting complex cartilaginous structures with clinically compliant biomaterials. Adv Funct Mater 2015;25:7406–17. https://doi.org/10.1002/adfm.201503423.
[28] Kang H-W, Lee SJ, Ko IK, Kengla C, Yoo JJ, Atala A. A 3D bioprinting system to produce human-scale tissue constructs with structural integrity. Nat Biotechnol 2016.https://doi.org/10.1038/nbt.3413.
[29] Jung JW, Lee J-S, Cho D-W. Computer-aided multiple-head 3D printing system for printing of heterogeneous organ/tissue constructs. Sci Rep 2016;6:21685. https://doi.org/10.1038/srep21685.
[30] Xu T, Zhao W, Zhu J-M, Albanna MZ, Yoo JJ, Atala A. Complex heterogeneous tissue constructs containing multiple cell types prepared by inkjet printing technology. Biomaterials 2013;34:130–9. https://doi.org/10.1016/j.biomaterials.2012.09.035.
[31] Colosi C, Shin SR, Manoharan V, Massa S, Costantini M, Barbetta A, et al. Microfluidic bioprinting of heterogeneous 3D tissue constructs using low-viscosity bioink. Adv Mater 2015.https://doi.org/10.1002/adma.201503310.
[32] Gauvin R, Chen Y-C, Lee JW, Soman P, Zorlutuna P, Nichol JW, et al. Microfabrication of complex porous tissue engineering scaffolds using 3D projection stereolithography. Biomaterials 2012;33:3824–34. http://www.sciencedirect.com/science/article/pii/S0142961212000944 [Accessed 26 December 2015].
[33] Wang Z, Abdulla R, Parker B, Samanipour R, Ghosh S, Kim K. A simple and high-resolution stereolithography-based 3D bioprinting system using visible light crosslinkable bioinks. Biofabrication 2015;7:1–29. https://doi.org/10.1088/1758-5090/7/4/045009.
[34] Gou M, Qu X, Zhu W, Xiang M, Yang J, Zhang K, et al. Bio-inspired detoxification using 3D-printed hydrogel nanocomposites. Nat Commun 2014;5:3774. https://doi.org/10.1038/ncomms4774.
[35] Zorlutuna P, Jeong JH, Kong H, Bashir R. Stereolithography-based hydrogel microenvironments to examine cellular interactions. Adv Funct Mater 2011;21:3642–51. https://doi.org/10.1002/adfm.201101023.
[36] Hribar KC, Meggs K, Liu J, Zhu W, Qu X, Chen S. Three-dimensional direct cell patterning in collagen hydrogels with near-infrared femtosecond laser. Sci Rep 2015;5:17203. https://doi.org/10.1038/srep17203.
[37] Wang Z, Jin X, Dai R, Holzman JF, Kim K. An ultrafast hydrogel photocrosslinking method for direct laser bioprinting. RSC Adv 2016;6:21099–104. https://doi.org/10.1039/C5RA24910D.
[38] Shanjani Y, Pan CC, Elomaa L, Yang Y. A novel bioprinting method and system for forming hybrid tissue engineering constructs. Biofabrication 2015;7:45008. https://doi.org/10.1088/1758-5090/7/4/045008.

[39] Morimoto Y, Hsiao AY, Takeuchi S. Point-, line-, and plane-shaped cellular constructs for 3D tissue assembly. Adv Drug Deliv Rev 2015;95:29–39. https://doi.org/10.1016/j.addr.2015.09.003.

[40] Chung BG, Lee K-H, Khademhosseini A, Lee S-H. Microfluidic fabrication of microengineered hydrogels and their application in tissue engineering. Lab Chip 2012;12:45–59. https://doi.org/10.1039/c1lc20859d.

[41] Huebner A, Sharma S, Srisa-Art M, Hollfelder F, Edel JB, deMello AJ. Microdroplets: a sea of applications? Lab Chip 2008;8:1244. https://doi.org/10.1039/b806405a.

[42] Theberge AB, Courtois F, Schaerli Y, Fischlechner M, Abell C, Hollfelder F, et al. Microdroplets in microfluidics: an evolving platform for discoveries in chemistry and biology. Angew Chem Int Ed 2010;49:5846–68. https://doi.org/10.1002/anie.200906653.

[43] Matsunaga YT, Morimoto Y, Takeuchi S. Molding cell beads for rapid construction of macroscopic 3D tissue architecture. Adv Mater 2011;23:90–4. https://doi.org/10.1002/adma.201004375.

[44] Onoe H, Okitsu T, Itou A, Kato-Negishi M, Gojo R, Kiriya D, et al. Metre-long cell-laden microfibres exhibit tissue morphologies and functions. Nat Mater 2013;12:584–90. https://doi.org/10.1038/nmat3606.

[45] Teh S-Y, Lin R, Hung L-H, Lee AP. Droplet microfluidics. Lab Chip 2008;8:198–220. https://doi.org/10.1039/b715524g.

[46] Christopher GF, Anna SL. Microfluidic methods for generating continuous droplet streams. J Phys D Appl Phys 2007;40:R319–36. https://doi.org/10.1088/0022-3727/40/19/R01.

[47] Wang Z, Samanipour R, Gamaleldin M, Sakthivel K, Kim K. An automated system for high-throughput generation and optimization of microdroplets. Biomicrofluidics 2016;10:54110. https://doi.org/10.1063/1.4963666.

[48] Samanipour R, Wang Z, Ahmadi A, Kim K. Computational and experimental study of microfluidic flow-focusing generation of hydrogel droplets. J Appl Polym Sci 2016;133:43701. https://doi.org/10.1002/app.43701.

[49] Tumarkin E, Tzadu L, Csaszar E, Seo M, Zhang H, Lee A, et al. High-throughput combinatorial cell co-culture using microfluidics. Integr Biol 2011;3:653–62. https://doi.org/10.1039/c1ib00002k.

[50] Onoe H, Takeuchi S. Cell-laden microfibers for bottom-up tissue engineering. Drug Discov Today 2015;20:236–46. https://doi.org/10.1016/j.drudis.2014.10.018.

[51] Bertassoni LE, Cecconi M, Manoharan V, Nikkhah M, Hjortnaes J, Cristino AL, et al. Hydrogel bioprinted microchannel networks for vascularization of tissue engineering constructs. Lab Chip 2014;14:2202–11. https://doi.org/10.1039/c4lc00030g.

[52] Xia J, Ip HH, Samman N, Wang D, Kot CS, Yeung RW, et al. Computer-assisted three-dimensional surgical planning and simulation: 3D virtual osteotomy. Int J Oral Maxillofac Surg 2000;29:11–7. https://doi.org/10.1016/S0901-5027(00)80116-2.

[53] Chen B, Ning R. Cone-beam volume CT breast imaging: feasibility study. Med Phys 2002;29:755–70. https://doi.org/10.1118/1.1461843.

[54] Pykett IL, Newhouse JH, Buonanno FS, Brady TJ, Goldman MR, Kistler JP, et al. Principles of nuclear magnetic resonance imaging. Radiology 1982;143:157–68. https://doi.org/10.1148/radiology.143.1.7038763.

[55] Pietrabissa A, Marconi S, Peri A, Pugliese L, Cavazzi E, Vinci A, et al. From CT scanning to 3-D printing technology for the preoperative planning in laparoscopic splenectomy. Surg Endosc 2015.https://doi.org/10.1007/s00464-015-4185-y.

[56] Zastrow E, Davis SK, Lazebnik M, Kelcz F, Van Veen BD, Hagness SC. Development of anatomically realistic numerical breast phantoms with accurate dielectric properties

for modeling microwave interactions with the human breast. IEEE Trans Biomed Eng 2008;55:2792–800. https://doi.org/10.1109/TBME.2008.2002130.

[57] Wang Z, Xiao X, Song H, Wang L, Li Q. Development of anatomically realistic numerical breast phantoms based on T1- and T2-weighted MRIs for microwave breast cancer detection. IEEE Antennas Wirel Propag Lett 2014;13:1757–60. https://doi.org/10.1109/LAWP.2014.2353852.

[58] Song H, Xiao X, Wang Z, Kikkawa T. UWB microwave breast cancer detection with MRI-derived 3-D realistic numerical breast model. In: IEEE Antennas Propag. Soc. AP-S Int. Symp. 2015 October; 2015. p. 544–5. https://doi.org/10.1109/APS.2015.7304658.

[59] Nam J, Starly B, Darling A, Sun W. Computer aided tissue engineering for modeling and design of novel tissue scaffolds. Comput Aided Des Applic 2004;1:633–40. https://doi.org/10.1016/j.cad.2005.02.002.

[60] Laine AF. In the spotlight: biomedical imaging. IEEE Rev Biomed Eng 2008;1:4–7. https://doi.org/10.1109/RBME.2008.2008221.

[61] Badea CT, Johnston SM, Qi Y, Ford NL, Wheatley AR, Holdsworth DW, et al. In vivo small-animal imaging using micro-CT and digital subtraction angiography. Phys Med Biol 2008;53:R319–50. https://doi.org/10.1088/0031-9155/53/19/R01.

[62] Schambach SJ, Bag S, Schilling L, Groden C, Brockmann MA. Application of micro-CT in small animal imaging. Methods 2010;50:2–13. https://doi.org/10.1016/j.ymeth.2009.08.007.

[63] Shapiro EM, Skrtic S, Koretsky AP. Sizing it up: cellular MRI using micron-sized iron oxide particles. Magn Reson Med 2005;338:329–38. https://doi.org/10.1002/mrm.20342.

[64] Serres S, Soto MS, Hamilton A, Martina A, Carbonell WS, Robson MD, et al. Molecular MRI enables early and sensitive detection of brain metastases. PNAS 2012;109:6674–9. https://doi.org/10.1073/pnas.1117412109.

[65] Jung JW, Lee J-S, Cho D-W. Computer-aided multiple-head 3D printing system for printing of heterogeneous organ/tissue constructs. Sci Rep 2016;6:21685. https://doi.org/10.1038/srep21685.

[66] Das M, Wilson K, Molnar P, Hickman JJ. Differentiation of skeletal muscle and integration of myotubes with silicon microstructures using serum-free medium and a synthetic silane substrate. Nat Protoc 2007;2:1795–801. https://doi.org/10.1038/nprot.2007.229.

[67] Wang Z, Samanipour R, Kim K. Organ-on-a-chip platforms for drug screening and tissue engineering. In: Biomedical engineering: frontier research and converging technologies. 2016. p. 293–325. https://doi.org/10.1007/978-3-319-21813-7.

[68] Wang Z, Calpe B, Zerdani J, Lee Y, Oh J, Bae H, et al. High-throughput investigation of endothelial-to-mesenchymal transformation (EndMT) with combinatorial cellular microarrays. Biotechnol Bioeng 2015;113:1403–12. https://doi.org/10.1002/bit.25905.

[69] Fernandes TG, Diogo MM, Clark DS, Dordick JS, Cabral JM. High-throughput cellular microarray platforms: applications in drug discovery, toxicology and stem cell research. Trends Biotechnol 2009;27:342–9. https://doi.org/10.1016/j.tibtech.2009.02.009. Pii: S0167-7799(09)00070-5.

The commercial 3D bioprinting industry

E. Combellack[*,†], *Z.M. Jessop*[*,†], *I.S. Whitaker*[*,†]
[*]Reconstructive Surgery and Regenerative Medicine Research Group, Swansea University Medical School, Swansea, United Kingdom, [†]The Welsh Centre for Burns and Plastic Surgery, Morrison Hospital, Swansea, United Kingdom

20.1 Introduction

Modern three-dimensional (3D) bioprinting promises the creation of bespoke tissue-engineered constructs that would herald the end of donor site morbidity, creation of solid complex organs "made to order," and facilitate a revolution in both biological sciences and medical research. Biotechnology companies along with major laboratories worldwide, in response to this potential, have invested decades and millions of dollars in the development of printing technology and its applications across a number of disciplines. The relatively simple core principle of stereolithography developed in the 1980s has sparked a materials and technological revolution that has seen the development of more precise printers, intelligent bioinks, and scaffold materials capable of supporting de novo tissue growth. The original aims of producing solid 3D polymer models have been superseded, and today 3D bioprinting is used in a range of scientific and commercial fields such as in drug delivery systems, cosmetic testing, and in the printing of customized 3D scaffolds to support tissue growth [1–3]. While much of the activity worldwide is research based, there is already a significant commercial element, which incorporates these new technologies into modern practice. The current uses of 3D printing include the creation of devices and implants which have been available for a number of years and allow the production of personalized implants to augment operative technique and bespoke 3D prosthesis for reconstruction [4–6].

The added biological component to 3D printing moves this relatively established technology into a different sphere of potential manufacturing and revenue streams. There are a number of key areas within which bioprinting may convey a paradigm shift and alter the way in which companies, health-care providers, and consumers interact. Broadly speaking, some of these key areas involve tissue engineering of solid organ transplants, cosmetic and consumer testing, and drug discovery. Along with food and animal products, biosensors and implants, bioprinting has the potential to transform the way in which personalized manufacturing might shape and impact our lives over the next 30 years. In this chapter, we give a brief overview of the current global market, key areas of significant investment and growth, and discuss some of the companies leading the way in 3D bioprinting. The bioprinting landscape and key translational steps needed to deliver this technology to market will be discussed

3D Bioprinting for Reconstructive Surgery. https://doi.org/10.1016/B978-0-08-101103-4.00029-6

further and touching on future challenges, we discuss how these barriers might be overcome to ensure that the full potential of this novel technology is delivered.

20.2 The market

Each of the bioprinting companies on the market, whether developing their own platform technology or adapting the existing hardware, uses one of the four main print-head technologies: extrusion, laser-induced forward transfer (LIFT), inkjet, and microvalve [7–14].

Between 2014 and 2015, the 3D bioprinting market saw the entry of a number of new companies, start-ups, spinouts, and subsidiaries all vying for a place in this developing field. Investment and initial valuations have been driven by future promises and early demonstration of potential application of their bioprinting technology [15,16]. As such it is difficult to estimate the value of these companies who all promise target delivery of printed tissue first with astronomical revenue potential.

20.2.1 3D printing: The evolution of the base technology

To put 3D bioprinting into context, it is important to understand the development of the 3D printing and additive manufacturing (AM) industry over the last 30 years. In 2013, it was estimated that this market was worth around $1.3 billion with forecasts that would rise to between $4 and 6 billion by 2018 [16,17]. A number of major companies have announced a significant investment into 3D printing and AM in the last 2 years. Stryker announced in January 2016 that it would spend nearly $400 million building an AM facility for the production of implants, with Siemens in Sweden, and UK-based Metalysis investing €21.4 and £20 million, respectively, in printing technology and infrastructure [16]. Comparably 3D bioprinting is still in its infancy; however, market and tech forecast of leading companies suggest that ultimately the market will be measured in the tens of billions of dollars with an initial value of around $1.8 billion achieved by 2027 [15].

With the introduction of a new technology there is variable uptake and a need for significant and sustained investment in addition to a fundamental consumer need or perceived gap, which must exist for a product to be fully integrated into the marketplace. Each year Gartner produces its technology maturation curve or the "hype cycle" for new and emerging technologies as a way of informing the market and separating fleeting technologies from those, which are commercially viable with transformational potential. Cycling through the initial interest and influx of investment during the primary phase of innovation trigger, technologies must weather the peak of inflated expectation through the trough of disillusionment before emerging into the plateau of productivity.

3D bioprinting has relied on the success of 3D printing to garner interest and investment. Hype for the base technology, which was developed in the 1980s really began in 2014 when the first industrial-grade 3D printers entered the marketplace and demonstrated that it was possible to print in high-definition resin or metal [18]. The consumer market responded with smaller "desktop" printers shortly after, which

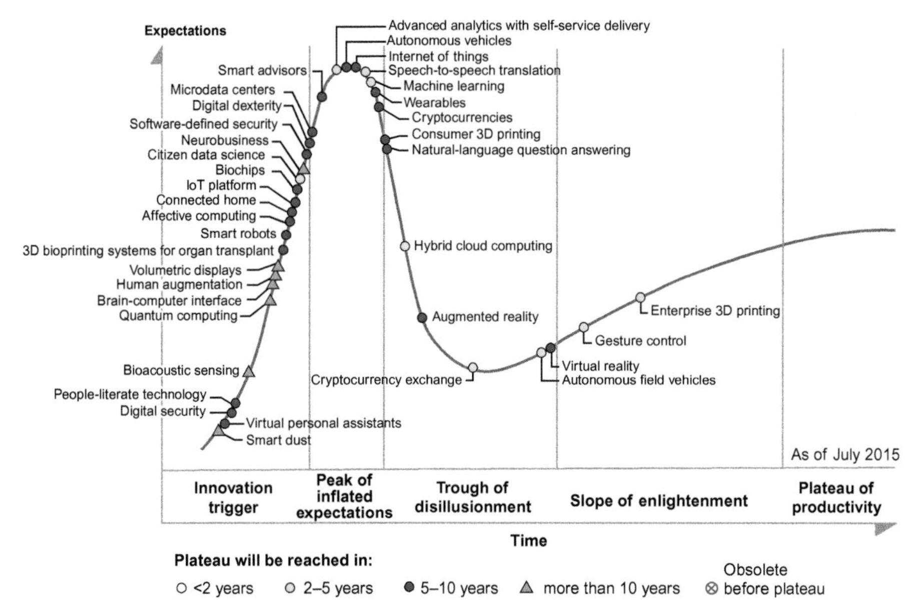

Fig. 20.1 2015 Hype cycle for emerging technologies demonstrating the position of 3D printing having peaked the previous year, now heading toward the trough of disillusionment [19].

was made possible by the expiration of a number of key patents detailing the original 3D printing process. The consumer demand for the technology increased exponentially as did the applications in medical device and implants resulting in a spike in unit production [18]. The lack of surrounding infrastructure however, complexity of programming, and poor user interface meant that the consumers were left disillusioned and the initial investment wavered meaning that only a year later it was on the downward slope [19] (Fig. 20.1).

In the 24 months that followed, however, a number of companies including P&G, HP, and GE have all invested heavily in 3D printing and bioprinting as expectations have been reset and the application of this technology has focused on the industrial and commercial applications within the marketplace. 3D printing is expanding commercially with scale no longer being a barrier, most recently demonstrated by the San Francisco-based start-up Apis Cor, who printed an entire house in 24 h [20]. Similarly, bioprinting has now become a key investment priority for Big Pharma, medical devices, and cosmetic companies, which has seen the technology diverge from traditional printing, and coupled with biological advancements in cell source and scaffold material creates opportunities to meet potentially significant gaps in the market.

20.2.2 Bioprinting: The gap in the market

Initially focusing on the commercial and industrial market it becomes possible to identify strategic gaps, which create a potential opportunity for bioprinting companies. The cost of drug development is becoming prohibitively expensive and stricter guidance around animal testing means that the pharmaceutical and cosmetic companies are looking for more cost-effective alternatives without compromising results. Procter and Gamble (P&G) has already banned animal testing on all but the 20% of its products which require it by law [21] mandating an alternative approach to product testing. In health care, waiting lists for patients on long-term dialysis or approaching end-stage liver failure present an economic health burden felt by every health system in the world. Tissue engineering promises the creation of solid organ transplants that would save thousands each year and lift the chronic health burden. In addition, tissue engineering holds the potential to restore form and function to patients born with congenital deformity or who acquire it later in life secondary to trauma or malignancy. Bioprinting companies have recognized the opportunity to supply the health-care sector with personalized tissue-engineered constructs and the investment is reflected in the development of advanced printer systems and the creation of novel biomaterials [1,22–24].

20.2.2.1 Tissue engineering

Tissue engineering is a promising technology but not as mature as the cellular therapy sector. Replacement of damaged or missing tissue in pursuit of restoration of form and function is the holy grail of tissue engineering. An interdisciplinary field, it relies on an optimized cell source, scaffold, and printing technology to create de novo tissue fit-for-purpose. A decade on from the Vacanti mouse, we saw the first report of a tissue-engineered organ (bladder) being implanted into a patient [25]. Since then, the technology has advanced significantly and several organs and tissues have been engineered and implanted into patients; however, the materials engineering and cell biology require further work [26]. Flat structures such as the skin, which are the most efficient to engineer, have seen the most progress and examples of technology being offered to patients for wound healing include Integra (collagen scaffolds) as well as cellular matrices such as Apligraft. Bioprinting is a platform technology that has the potential to expand the tissue engineering field by allowing production of more complex tissues for reconstructive surgery using the bottom-up approach rather than decellularizing the existing structures. Bioprinting firms have diversified to include a range of cell source, scaffold material, and printing techniques in order to address some of the key hurdles and translational steps, but the same fundamental scientific questions exist to allow production of native-like tissues.

20.2.2.2 Organ transplant

With the development of intelligent bioinks, novel scaffold materials, and diverse cell sources, 3D bioprinting has the potential to answer a number of clinical and research-based questions. Significant commercial potential lies within the field of tissue engineering and the creation of bioprinted 3D solid organs. The need for organ donation globally presents a set of unique challenges that mean an ever-increasing population of patients are waiting for a donor match. In 2014 more than 119,000 transplants were performed worldwide, an increase of almost 2% from the year previously and desperately short of total global need accounting for <10% of all transplant patients waiting [27]. The current system is a resource-intensive service, which has been under increasing scrutiny in the environment of austerity and economic constraints [28].

The promise to eliminate the waiting time, need for immunosuppression, and a donor-matched organ has attracted significant financial, research, and technological investment with the potential to change lives significantly [29–32]. In current tech reports which are forecasting into the late 2020s, it is theorized that this technology may be achievable within our lifetimes. However, it is more likely that simple structural tissues such as the cartilage and bone used in reconstructive surgeries will be some of the first transplanted due to the inherent complexity in architecture and function of solid organs which is more difficult to reproduce. A 30-year horizon view estimates that it is conceivable that the health-care burden created by organ transplant will be substantially reduced with issues revolving around a route to market, up-scaling production to meet complex global demand, and regulation of manufacturing and implantation quality control [15,33].

20.2.2.3 Drug development

Drug discovery is a highly expensive and resource-heavy process that often results in failure to reach market and clear final regulatory hurdles due to lack of sufficient preclinical data, which accurately demonstrates the testing methodologies [33]. With the estimated cost of a new drug to market at around $2.6 billion, reliance on animal testing and 2D human cell assays is often a poor substitute for the human model and results in a significant cost to the pharmaceutical company [34]. In 2014, the bioprinting company Organovo's first product to market was a 3D-bioprinted liver assay which survived for 40 days and allowed more accurate toxicity testing and metabolic studies of pharmaceuticals [35]. With more companies now offering 3D-bioprinted tissue models including liver, kidney, lung, and thyroid, drug companies are better positioned to complete early phase testing before applying to proceed into first-phase human trials.

20.2.2.4 Cosmetic testing

In 2013, the European Union issued a ban covering both the testing of cosmetic products and ingredients on animals, and bringing to market any product which had been tested on animals [36]. While the ban on cosmetic component testing had been in effect since 2009, the ban on testing finished cosmetics came into effect the following year in September 2014. This raised a number of issues with cosmetic companies manufacturing and selling their products within the EU. To address this, companies such as L'Oreal have turned to 3D bioprinting as part of the solution. In 2013, in advance of the ban enforcement, it had set up a partnership with Organovo to create skin models for cosmetic testing [21,33]. Organovo was able to print live cells at a cellular concentration of 100% creating the skin models required to gain essential information during the process of product development. Proctor & Gamble has also set up a $44 million fund as part of a wider 5-year research plan to invite researchers (primarily in Singapore) to submit proposals and bring bioprinting expertise into P&G. As the 3D models printed become more robust, the scope of testing will increase, significantly reducing the need for external testing, while meeting the legal and ethical obligations required [2,37,38].

20.3 Key companies

With a market expected to be worth around $1.8 billion by 2027, it is understandable that each year more companies are entering the market and established 3D printing companies are expanding to offer bioprinters and hardware to support this emerging field. Broadly speaking, companies are diverging into one of the two categories: focusing on the printers and support for the physical machinery with and without bioink supply, and those focusing on the production of 3D-bioprinted products such as skin models, assays, and tissue for cosmetic and drug testing. Companies with more than a decade of 3D printing experience such as EnvisionTEC have diversified and are now manufacturing bioprinters for biofabrication and implants which cost just under $200,000 which have been available since the early 2000s. They have an established supply chain and are simply adapting existing technology to "keep up" with the changing landscape while maintaining their core business,continuing to supply 3D printers to a number of commercial and industrial sectors.

Start-ups focusing solely on 3D bioprinting such as BioBots which was founded by two friends at the University of Pennsylvania in 2013 have already developed two versions of their desktop 3D bioprinter, the latest of which costs around $5000. In 2015, they raised $1.5 million in funding and with more than 100 units of their BioBot 1 printer sold, they have made an estimated $1 million in revenue to date [39]. The concept is clear; they are offering an accessible entry-level product, which allows smaller research institutions to buy the technology into their organizations without any significant research investment. By providing on-going tech-support BioBots are able to access a wealth of free user data as it refines its technology ready for the next model to be released. In addition, they produce a range of support and sacrificial biopinks to use specifically in their machines; conceivably, you only need to add the cellular component and you are ready to go.

There are a number of spin offs such as OxSyBio from Oxford University whose aim is to develop 3D printing techniques to produce a functional range of tissues for the repair and replacement of organs and skin. Although they raised £1 million in 2015 from the IP Group PLC to develop their proprietary droplet printing technology they are still in the research phase and as of yet do not have a product to take to market. Another company, CELLINK, initially focused on developing bioinks for extrusion bioprinting, is now expanding into producing bioprinters aimed at life science companies and researchers. Despite the substantial investment to date, many of the more strategically placed companies are yet to post significant profits, rather generating revenues from research grants and kick-starter or open platform fundraising.

Organovo, the only publically traded bioprinting company, generated huge interest when it first went public and at one point trading at an estimated value of $1 billion despite having generated minimal profit. There is a fine balance however, and after their CEO and cofounder Keith Murphy left in April 2017 their share price bore the brunt, falling nearly 8% in the hours following the announcement. Their new CEO Taylor Crouch has previous experience with eStudySite, which in part conducted clinical trials, will be essential as Organovo aims to bring more of its bioprinted products to market and boost its revenue streams. It is important however to bear in mind that without extensive track records or significant profits, all bioprinting companies, public or privately owned, rely on the hype and promise of what is to come. The key will be sustaining the interest and investment while the technology catches up with the imagination and aspirations of the commercial markets.

20.4 Translational steps

The refinement of current technologies will be an essential part of the progression not only of the technology but also of the logistical process, which will support the product to market. Regardless of the estimated value of these innovative tech start-ups, all they need is a sustainable way of generating revenue in the medium to long term by reliably bringing a product to market. No product can get to market without a robust end-to-end supply chain and as the technology develops, so too the regulation and infrastructure that surrounds this new technology. Regulatory boards that monitor and manage implantable devices will be quick to halt any process, which does not meet the standards designed to protect patient safety and guarantee traceability. Large-scale production to meet regional, national, or international demand must also be considered, with raw materials, processes, and storage not generating a total item cost that would put it beyond the reach of most health-care systems.

20.5 Future development

Incremental advancements in 3D printing technology are getting ever smaller as the techniques are refined and the market place becomes more crowded. At some point producing a smaller printer will no longer be enough to gain market share and reliability along with a range of scaffolds and more diverse cell sources will become the economic differentiators. Market analysts are already talking about the next evolution within bioprinting, hinting that the addition of an extra dimension will alter the way in which bioprinting is ultimately utilized [40]. 4D printing hints at advanced technology that allows printed constructs or implants to respond to their environment promising wearable or implantable technology that adapts to the underling host system, external signal, or disease process [41–44]. The 3D bioprinting field offers unprecedented opportunity to change the paradigm in health care, but requires significant investment as well as the conviction to push the research through to clinical translation.

References

[1] Chia HN, Wu BM. Recent advances in 3D printing of biomaterials. J Biol Eng 2015;9(1):1728.

[2] Vijayavenkataraman S, Lu WF, Fuh JYH. 3D bioprinting of skin: a state-of-the-art review on modelling, materials, and processes. Biofabrication 2016;8(3):032001.

[3] Hsieh FY, Lin HH, Hsu S. 3D bioprinting of neural stem cell-laden thermoresponsive biodegradable polyurethane hydrogel and potential in central nervous system repair. Biomaterials 2015;71:48–57.

[4] Ventola CL. Medical applications for 3D printing: current and projected uses. P T 2014;39:704–11.

[5] Prakash BB. 3D printing and its applications. Int J Sci Res 2016;5(3):1532–5.

[6] Nale SB, Kalbande AG. A review on 3D printing technology. Int J Innov Emerg Res Eng 2015;2(9):33–6.

[7] Gudapati H, Dey M, Ozbolat I. A comprehensive review on droplet-based bioprinting: past, present and future. Biomaterials 2016;102:20–42.

[8] Ng WL, Yeong WY, Naing MW. Microvalve bioprinting of cellular droplets with high resolution and consistency. Proceedings of the 2nd international conference on progress in additive manufacturing (Pro-AM 2016); 2016. p. 397–402.

[9] Catros S, Fricain JC, Guillotin B, Pippenger B. Laser-assisted bioprinting for creating on-demand patterns of human osteoprogenitor cells and nano-hydroxyapatite. Biofabrication 2011;3:025001. IOPscience.

[10] Xu T, Jin J, Gregory C, Hickman JJ, Boland T. Inkjet printing of viable mammalian cells. Biomaterials 2005;26(1):93–9.

[11] Arai K, Iwanaga S, Toda H, Genci C, Nishiyama Y, Nakamura M. Three-dimensional inkjet biofabrication based on designed images. Biofabrication 2011;3(3):034113.

[12] Roth EA, Xu T, Das M, Gregory C, Hickman JJ, Boland T. Inkjet printing for high-throughput cell patterning. Biomaterials 2004;25(17):3707–15.

[13] Michael S, Sorg H, Peck C-T, Koch L, Deiwick A, Chichkov B, et al. Tissue engineered skin substitutes created by laser-assisted bioprinting form skin-like structures in the dorsal skin fold chamber in mice. PLoS One 2013;8(3):e57741.

[14] Lee JM, Yeong WY. Design and printing strategies in 3D bioprinting of cell-hydrogels: a review. Adv Healthc Mater 2016;5(22):2856–65.

[15] Tsao N. 3D Bioprinting 2017-2027: technologies, markets, forecasts. www.idtechex.com; 2017. Available from: http://www.idtechex.com/research/reports/3d-bioprinting-2017-2027-technologies-markets-forecasts-000537.asp [cited 16 September 2016].

[16] Associates W. Additive manufacturing: the state of the industry. Wohlers Report; 2016 April p. 1–6.

[17] Gonzalez EE, la Cerda de R. Quo Vadis; 2015. p. 6. Available from: https://www.pwc.com/mx/es/industrias/archivo/2015-04-quovadis.pdf [cited 24 July 2017].

[18] Hilkene C. 3D printing in 2017: moving past disillusionment and into production applications. www.idtechex.com; 2017. Available from: https://www.ame.org/target/articles/2017/3d-printing-2017-moving-past-disillusionment-and-production-applications [cited 25 July 2017].

[19] Gartner. Gartner's 2015 Hype cycle for emerging technologies identifies the computing innovations that organizations should monitor; 2015. Available from: http://www.gartner.com/newsroom/id/3114217 [cited 27 July 2017].

[20] Apis Cor. The first on-site house has been printed in Russia. www.apis-cor.com; 2017. Available from: http://apis-cor.com/en/about/news/first-house [cited 21 July 2017].

[21] Gordon R. P&G, L'Oreal push 3D bioprinting from research to commercial products. www.idtechex.com. Available from: http://www.idtechex.com/research/articles/p-g-loreal-push-3d-bioprinting-from-research-to-commercial-products-00007888.asp?donotredirect=true [cited 27 July 2017].

[22] Chen F-M, Liu X. Advancing biomaterials of human origin for tissue engineering. Prog Polym Sci 2016;53:86–168.

[23] Zhang YS, Arneri A, Bersini S, Shin SR, Zhu K, Goli-Malekabadi Z, et al. Bioprinting 3D microfibrous scaffolds for engineering endothelialized myocardium and heart-on-a-chip. Biomaterials 2016;110:45–59.

[24] Lee J-S, Hong JM, Jung JW, Shim J-H, Oh J-H, Cho D-W. 3D printing of composite tissue with complex shape applied to ear regeneration. Biofabrication 2014;6(2):024103.
[25] Atala A, Bauer SB, Soker S, Yoo JJ, Retik AB. Tissue-engineered autologous bladders for patients needing cystoplasty. Lancet 2006;367(9518):1241–6.
[26] Vacanti CA, Cima LG, Ratkowski D, Upton J, Vacanti JP. Tissue engineered growth of new cartilage in the shape of a human ear using synthetic polymers seeded with chondrocytes. MRS Proc 2011;252:475.
[27] Reports 2014—GODT. 2016. Available from: http://www.transplant-observatory.org/data-reports-2014/.
[28] Axelrod DA. Economic and financial outcomes in transplantation. Curr Opin Organ Transplant 2013;18(2):222–8.
[29] Mironov V, Boland T, Trusk T, Forgacs G. Organ printing: computer-aided jet-based 3D tissue engineering. Trends Biotechnol 2003;21(4):157–61.
[30] Marx V. Tissue engineering: organs from the lab. Nature 2015;522(7556):373–7.
[31] Marga F, Jakab K, Khatiwala C, Shepherd B, Dorfman S, Hubbard B, et al. Toward engineering functional organ modules by additive manufacturing. Biofabrication 2012;4(2):022001.
[32] Murphy SV, Atala A. 3D bioprinting of tissues and organs. Nat Biotechnol 2014;32(8):773–85. Nature Publishing Group.
[33] IDTechX. 3D Bioprinting 2014-2024: Applications, Markets, Players. www.idtechex.com; 2015. p. 1–104. Available from: http://www.idtechex.com/research/articles/3d-bioprinting-2014-2024-applications-and-markets-00006479.asp?donotredirect=true [cited 26 July 2017].
[34] Mullard A. New drugs cost US$2.6 billion to develop. Nat Rev Drug Discov 2014;13(12):877.
[35] Murphy K. Organovo's bioprinting platform: enabling 3D, architecturally correct, fully human tissue for drug discovery and transplantation. Organovo. Available from: http://organovo.com/cell-gene-meeting-mesa/ [cited 27 July 2017].
[36] European Commision. Ban on animal testing. European Commision; 2013. Available from: https://ec.europa.eu/growth/sectors/cosmetics/animal-testing_en [cited 20 July 2017].
[37] Velasquillo C, Galue EA, Rodriquez L, Ibarra C. Skin 3D bioprinting. J Cosmet Dermatol Sci Appl 2013;3:85–9.
[38] Cubo N, Garcia M, del Cañizo JF, Velasco D, Jorcano JL. 3D bioprinting of functional human skin: production and in vivo analysis. Biofabrication 2016;9(1):1–12. IOP Publishing.
[39] Dill K. Inside the company that wants to put a $10,000 3D bioprinter on every lab bench. Forbes; 2016. Available from: https://www.forbes.com/sites/kathryndill/2016/11/02/inside-the-company-that-wants-to-put-a-10000-3d-bioprinter-on-every-lab-bench/#3e961f4edf30 [cited 21 July 2017].
[40] Panetta K. 3 Trends appear in the gartner hype cycle for emerging technologies; 2016. Available from: http://www.gartner.com/smarterwithgartner/3-trends-appear-in-the-gartner-hype-cycle-for-emerging-technologies-2016/ [cited 22 July 2017].
[41] Li Y-C, Zhang YS, Akpek A, Shin SR, Khademhosseini A. 4D bioprinting: the next-generation technology for biofabrication enabled by stimuli-responsive materials. Biofabrication 2016;9(1):012001.
[42] Gao B, Yang Q, Zhao X, Jin G, Ma Y, Xu F. 4D bioprinting for biomedical applications. Trends Biotechnol 2016;34(9):746–56.
[43] An J, Chua CK, Mironov V. A perspective on 4D bioprinting. Int J Bioprint 2016;2:3–5.
[44] Leist SK, Zhou J. Current status of 4D printing technology and the potential of light-reactive smart materials as 4D printable materials. Virtual Phys Prototyp 2016;11(4):249–62. 1st ed., Taylor & Francis.

Index

Note: Page numbers followed by *f* indicate figures, and *t* indicate tables.